# Stochastic models, estimation, and control

VOLUME 3

This is Volume 141-3 in
MATHEMATICS IN SCIENCE AND ENGINEERING
A Series of Monographs and Textbooks
Edited by RICHARD BELLMAN, *University of Southern California*

The complete listing of books in this series is available from the Publisher upon request.

# Stochastic models, estimation, and control

VOLUME 3

*PETER S. MAYBECK*

DEPARTMENT OF ELECTRICAL ENGINEERING
AIR FORCE INSTITUTE OF TECHNOLOGY
WRIGHT-PATTERSON AIR FORCE BASE
OHIO

1982

ACADEMIC PRESS
A Subsidiary of Harcourt Brace Jovanovich, Publishers
New York London
Paris San Diego San Francisco São Paulo Sydney Tokyo Toronto

ACADEMIC PRESS, INC.
111 Fifth Avenue, New York, New York 10003

*United Kingdom Edition published by*
ACADEMIC PRESS, INC. (LONDON) LTD.
24/28 Oval Road, London NW1 7DX

Library of Congress Cataloging in Publication Data

Maybeck, Peter S.
Stochastic models, estimation, and control.

(Mathematics in science and engineering)
Includes bibliographies and indexes.
1. System analysis. 2. Control theory. 3. Estimation theory. I. Title. II. Series.
QA402.M37 519.2 78-8836
ISBN 0-12-480703-8 (v.3) AACR2

PRINTED IN THE UNITED STATES OF AMERICA

82 83 84 85 9 8 7 6 5 4 3 2 1

To my Parents

# Contents

## VOLUME 3

# Preface

As was true of Volumes 1 and 2, the purpose of this book is twofold. First, it attempts to develop a thorough understanding of the fundamental concepts incorporated in stochastic processes, estimation, and control. Second, and of equal importance, it provides experience and insights into applying the theory to realistic practical problems. Basically, it investigates the theory and derives from it the tools required to reach the ultimate objective of systematically generating effective designs for estimators and stochastic controllers for operational implementation.

Perhaps most importantly, the entire text follows the basic principles of Volumes 1 and 2 and concentrates on presenting material in the most lucid, best motivated, and most easily grasped manner. It is oriented toward an engineer or an engineering student, and it is intended both to be a textbook from which a reader can *learn* about estimation and stochastic control and to provide a good reference source for those who are deeply immersed in these areas. As a result, considerable effort is expended to provide graphical representations, physical interpretations and justifications, geometrical insights, and practical implications of important concepts, as well as precise and mathematically rigorous development of ideas. With an eye to practicality and eventual implementation of algorithms in a digital computer, emphasis is maintained on the case of continuous-time dynamic systems with sampled-data measurements available; nevertheless, corresponding results for discrete-time dynamics or for continuous-time measurements are also presented. These algorithms are developed in detail to the point where the various design trade-offs and performance evaluations involved in achieving an efficient, practical configuration can be understood. Many examples and problems are used throughout the text to aid comprehension of important concepts. Furthermore, there is an extensive set of references in each chapter to allow pursuit of ideas in the open literature once an understanding of both theoretical concepts and practical implementation issues has been established through the text.

This volume builds upon the foundations set in Volumes 1 and 2. Chapter 13 introduces the basic concepts of stochastic control and dynamic programming as the fundamental means of synthesizing optimal stochastic control laws. Subsequently, Chapter 14 concentrates attention on the important LQG synthesis of controllers, based upon linear system models, quadratic cost criteria for defining optimality, and Gaussian noise models. This chapter and much of Chapter 13 can be understood solely on the basis of modeling and estimation concepts from Volume 1. It covers the important topics of stability and robustness, and synthesis and realistic performance analysis of digital (and analog) controllers, including many practically useful controller forms above and beyond the basic LQG regulator. Finally, Chapter 15 develops practical nonlinear controllers, exploiting not only the linear control insights from the preceding two chapters and Volume 1, but also the nonlinear stochastic system modeling and both adaptive and nonlinear filtering of Volume 2.

Thus, these three volumes form a self-contained and integrated source for studying stochastic models, estimation, and control. In fact, they are an outgrowth of a three-quarter sequence of graduate courses taught at the Air Force Institute of Technology; and thus the text and problems have received thorough class testing. Students had previously taken a basic course in applied probability theory, and many had also taken a first control theory course, linear algebra, and linear system theory, but the required aspects of these disciplines have also been developed in Volume 1. The reader is assumed to have been exposed to advanced calculus, differential equations, and some vector and matrix analysis on an engineering level. Any more advanced mathematical concepts are developed within the text itself, requiring only a willingness on the part of the reader to deal with new means of conceiving a problem and its solution. Although the mathematics becomes relatively sophisticated at times, efforts are made to motivate the need for, and to stress the underlying basis of, this sophistication.

The author wishes to express his gratitude to the many students who have contributed significantly to the writing of this book through their feedback to me—in the form of suggestions, questions, encouragement, and their own personal growth. I regard it as one of God's many blessings that I have had the privilege to interact with these individuals and to contribute to their growth. The stimulation of technical discussions and association with Professors Michael Athans, John Deyst, Nils Sandell, Wallace Vander Velde, William Widnall, and Alan Willsky of the Massachusetts Institute of Technology, Professor David Kleinman of the University of Connecticut, and Professors Jurgen Gobien, James Negro, J. B. Peterson, and Stanley Robinson of the Air Force Institute of Technology has also had a profound effect on this work. I deeply appreciate the continual support provided by Dr. Robert Fontana, Chairman of the Department of Electrical Engineering at AFIT, and the painstaking care with which many of my associates have reviewed the manuscript. Finally, I wish to thank my wife, Beverly, and my children, Kristen and Keryn, without whose constant love and support this effort could not have been fruitful.

# Notation

*Vectors, Matrices*

*Scalars* are denoted by upper or lower case letters in italic type.

*Vectors* are denoted by lower case letters in boldface type, as the vector $\mathbf{x}$ made up of components $x_i$.

*Matrices* are denoted by upper case letters in boldface type, as the matrix $\mathbf{A}$ made up of elements $A_{ij}$ ($i$th row, $j$th column).

*Random Vectors (Stochastic Processes), Realizations (Samples), and Dummy Variables*

*Random vectors* are set in boldface sans serif type, as $\mathsf{x}(\cdot)$ or frequently just as $\mathsf{x}$ made up of scalar components $\mathsf{x}_i$; $\mathsf{x}(\cdot)$ is a mapping from the sample space $\Omega$ into real Euclidean $n$-space $R^n$: for each $\omega_k \in \Omega$, $\mathsf{x}(\omega_k) \in R^n$.

*Realizations* of the random vector are set in boldface roman type, as $\mathbf{x}$: $\mathsf{x}(\omega_k) = \mathbf{x}$.

*Dummy variables* (for arguments of density or distribution functions, integrations, etc.) are denoted by the equivalent Greek letter, such as $\xi$ being associated with $\mathsf{x}$: e.g., the density function $f_{\mathsf{x}}(\xi)$. The correspondences are $(\mathsf{x}, \xi)$, $(\mathsf{y}, \rho)$, $(\mathsf{z}, \zeta)$, $(\mathsf{Z}, \mathscr{Z})$.

*Stochastic processes* are set in boldface sans serif type, just as random vectors are. The $n$-vector stochastic process $\mathsf{x}(\cdot,\cdot)$ is a mapping from the product space $T \times \Omega$ into $R^n$, where $T$ is some time set of interest: for each $t_j \in T$ and $\omega_k \in \Omega$, $\mathsf{x}(t_j, \omega_k) \in R^n$. Moreover, for each $t_j \in T$, $\mathsf{x}(t_j, \cdot)$ is a random vector, and for each $\omega_k \in \Omega$, $\mathsf{x}(\cdot, \omega_k)$ can be thought of as a particular time function and is called a *sample* out of the process. In analogy with random vector realizations, such samples are set in boldface roman type: $\mathsf{x}(\cdot, \omega_k) = \mathbf{x}(\cdot)$ and $\mathsf{x}(t_j, \omega_k) = \mathbf{x}(t_j)$. Often the second argument of a stochastic process is suppressed: $\mathsf{x}(t, \cdot)$ is often written as $\mathsf{x}(t)$, and this stochastic process evaluated at time $t$ is to be distinguished from a process sample $\mathbf{x}(t)$ at that same time.

*Subscripts*

| | | | |
|---|---|---|---|
| a: | augmented | I: | ideal |
| c: | continuous-time; or controller | m: | model (command generator model) |
| d: | discrete-time; or desired | n: | nominal; or noise disturbance |
| e: | error | r: | reference variable |
| f: | final time; or filter (shaping filter) | ss: | steady state |
| | | t: | truth model |
| | | 0: | initial time |

*Superscripts*

| | | | |
|---|---|---|---|
| $^{\mathrm{T}}$: | transpose (matrix) | $^{\mathrm{R}}$: | right inverse |
| $^{*}$: | complex conjugate transpose; or optimal | $^{\#}$: | pseudoinverse |
| $^{-1}$: | inverse (matrix) | $\hat{}$ : | estimate |
| $^{\mathrm{L}}$: | left inverse | $\bar{}$ : | Fourier transform; or steady state solution |

*Matrix and Vector Relationships*

$\mathbf{A} > \mathbf{0}$: $\mathbf{A}$ is positive definite.
$\mathbf{A} \geq \mathbf{0}$: $\mathbf{A}$ is positive semidefinite.
$\mathbf{x} \leq \mathbf{a}$: componentwise, $x_1 \leq a_1, x_2 \leq a_2, \ldots,$ and $x_n \leq a_n$.

*Commonly Used Abbreviations and Symbols*

| | |
|---|---|
| $E\{\cdot\}$ | expectation |
| $E\{\cdot\mid\cdot\}$ | conditional expectation |
| exp($\cdot$) | exponential |
| lim. | limit |
| l.i.m. | limit in mean (square) |
| ln($\cdot$) | natural log |
| m.s. | mean square |
| max. | maximum |
| min. | minimum |
| $R^n$ | Euclidean $n$-space |
| sgn($\cdot$) | signum (sign of) |
| tr($\cdot$) | trace |
| w.p.1 | with probability of one |
| $\lvert\cdot\rvert$ | determinant of |
| $\lVert\cdot\rVert$ | norm of |
| $\sqrt{\cdot}$ | matrix square root of (see Volume 1) |
| $\in$ | element of |
| $\subset$ | subset of |
| $\{\cdot\}$ | set of; such as $\{\mathbf{x} \in \mathbf{X} : \mathbf{x} \leq \mathbf{a}\}$, i.e., the set of $\mathbf{x} \in \mathbf{X}$ such that $x_i \leq a_i$ for all $i$ |

*List of symbols and pages where they are defined or first used*

| | |
|---|---|
| $\mathbf{A}_{ij}$ | 153 |
| $\mathbf{a}$ | 247 |
| $\mathscr{A}$ | 105 |
| $\mathbf{B}$ | 6; 88 |
| $\mathbf{B}_{cy}$ | 167 |
| $\mathbf{B}_{cz}$ | 167 |
| $\mathbf{B}_{d}$ | 6; 76 |
| $\bar{\mathbf{B}}$ | 74; 76 |
| $\mathscr{B}$ | 105 |
| $\mathbf{C}$ | 69; 231 |
| $\mathbf{c}$ | 223 |
| $\mathscr{C}$ | 13; 26 |
| $\mathscr{C}^*$ | 13; 15; 47; 48 |
| $\mathbf{D}_y$ | 81; 89; 231 |
| $\mathbf{D}_z$ | 81 |
| $\mathbf{d}$ | 131; 132 |
| $\mathbf{dx}$ | 5 |
| $\mathbf{d\beta}$ | 5 |
| $\mathbf{d\beta}_m$ | 5 |
| $\mathbf{E}$ | 137 |
| $\mathbf{E}_x$ | 151 |
| $\mathbf{E}_y$ | 151 |
| $E\{\cdot\}$ | 9 |
| $E\{\cdot\|\cdot\}$ | 13 |
| $\mathbf{e}$ | 50; 100; 115 |
| $\exp\{\cdot\}$ | 24 |
| $\mathbf{F}$ | 6; 88 |
| $\mathbf{f}$ | 5 |
| $f_{\mathbf{x}}$ | 31 |
| $f_{\mathbf{x}}(\xi, t \| \rho, t')$ | 15 |
| $f_{\mathbf{x},\mathbf{y}}$ | 54 |
| $f_{\mathbf{x}\|\mathbf{y}}$ | 15; 24 |
| $\mathbf{G}$ | 6 |
| $\mathbf{G}[\mathbf{x}(t), t]$ | 5 |
| $\mathbf{G}_L$ | 104 |
| $\mathbf{G}_c$ | 96 |
| $\mathbf{G}_c^*$ | 16; 69; 73; 186 |
| $\bar{\mathbf{G}}_c^*$ | 71; 78; 95; 242 |
| $\mathbf{G}_{cx}$ | 167 |
| $\mathbf{G}_{cy}$ | 167 |
| $\mathbf{G}_{cz}$ | 167 |
| $\mathbf{G}_d$ | 6 |
| $\mathbf{G}_d[\mathbf{x}(t_i), t_i]$ | 6 |
| $g$ | 16; 188 |
| $\mathscr{G}$ | 148 |
| $\mathscr{G}_L$ | 105 |
| $\mathbf{H}$ | 6; 184 |
| $\mathbf{h}$ | 5 |
| $\mathsf{I}$ | 45 |
| $\mathbf{I}$ | 76 |
| $\mathbf{I}_i$ | 45 |
| $I$ | 105 |
| $J$ | 10 |
| $J_c$ | 74 |
| $J_r$ | 75 |
| $\bar{J}_r$ | 76 |
| $\mathbf{K}$ | 16; 19; 185 |
| $\bar{\mathbf{K}}$ | 100 |
| $\tilde{\mathbf{K}}$ | 101 |
| $\mathbf{K}_c$ | 15; 70; 73; 186 |
| $\bar{\mathbf{K}}_c$ | 71; 95; 188 |
| $\mathbf{K}_{ss}$ | 79 |
| $\mathbf{K}_x$ | 144; 145 |
| $\mathbf{K}_y$ | 144 |
| $\mathbf{K}_\xi$ | 144; 145 |
| $\mathscr{K}_i$ | 148 |
| $\mathscr{K}_p$ | 148 |
| $\mathsf{L}_r$ | 75 |
| $L$ | 26; 40 |
| $\bar{L}$ | 46; 47 |
| $\bar{L}'$ | 48 |
| $L_f$ | 21 |
| $\bar{L}_f$ | 46; 47 |
| $\bar{L}_f'$ | 48 |
| $\mathbf{M}$ | 90 |
| $\mathbf{M}_D$ | 92 |
| $\mathbf{M}_{DTI}$ | 93 |
| $\mathbf{M}_{TI}$ | 93 |

| | |
|---|---|
| $\mathbf{Y}_{\mathrm{f}}$ | 69 |
| $\mathbf{y}_{\mathrm{a}}$ | 94; 170 |
| $\mathbf{y}_{\mathrm{c}}$ | 2; 69; 81; 89; 223 |
| $\mathbf{y}_{\mathrm{d}}$ | 2 |
| $\hat{\mathbf{y}}_{\mathrm{c}}(t_i^-)$ | 133 |
| $\hat{\mathbf{y}}_{\mathrm{c}}(t_i^+)$ | 133 |
| $\mathbf{Z}$ | 12; 45 |
| $\mathbf{Z}_i$ | 12; 45 |
| $\mathbf{z}(t)$ | 184 |
| $\mathbf{z}(t_i)$ | 5; 6; 81 |
| $\mathbf{z}_i$ | 12; 18 |
| $\mathbf{z}_0$ | 126; 238 |
| $z$ | 97 |
| $\boldsymbol{\beta}$ | 5 |
| $\boldsymbol{\beta}_{\mathrm{m}}$ | 5; 184 |
| $\Delta\mathbf{G}$ | 104 |
| $\Delta\mathbf{u}$ | 133 |
| $\Delta\mathscr{G}$ | 105 |
| $\boldsymbol{\delta}\mathbf{x}$ | 88; 159 |
| $\boldsymbol{\delta}\mathbf{u}$ | 88; 159 |
| $\boldsymbol{\delta}\mathbf{z}$ | 126 |
| $\boldsymbol{\delta}\mathbf{q}$ | 136 |
| $\boldsymbol{\delta}\mathbf{y}_{\mathrm{c}}$ | 124; 159 |
| $\boldsymbol{\varepsilon}$ | 148 |
| $\lambda$ | 109 |
| $\lambda_{\mathrm{max}}$ | 109 |
| $\boldsymbol{\xi}$ | 147; 164 |
| $\boldsymbol{\Pi}$ | 123 |
| $\boldsymbol{\Sigma}$ | 57 |
| $\sigma$ | 171 |
| $\sigma^2$ | 171 |
| $\sigma_{\mathrm{max}}$ | 106 |
| $\sigma^*_{\mathrm{max}}$ | 107; 111 |
| $\sigma_{\mathrm{min}}$ | 106 |
| $\boldsymbol{\Phi}$ | 6; 76 |
| $\boldsymbol{\Phi}_{\mathrm{c}}$ | 167 |
| $\boldsymbol{\phi}$ | 6; 40 |
| $\phi_{\mathbf{x}}$ | 64 |
| $\boldsymbol{\Psi}_{xx}$ | 171 |
| $\bar{\Psi}_{xx}(\omega)$ | 182 |
| $\boldsymbol{\Psi}_{y\mathrm{a}}$ | 171; 174 |
| $\Omega$ | xiii; 15 |
| $\omega$ | 12; 107 |
| $\omega_{\mathrm{s}}$ | 107 |

CHAPTER **13**

# Dynamic programming and stochastic control

## 13.1 INTRODUCTION

Up to this point, we have considered *estimation* of uncertain quantities on the basis of both sampled-data (or continuous) measurements from a dynamic system and mathematical models describing that system's characteristics. Now we wish to consider exerting appropriate *control* over such a system, so as to make it behave in a manner that we desire. This chapter will formulate the *optimal stochastic control* problem, provide the theoretical foundation for its solution, and discuss the potential structure for such solutions. The two subsequent chapters will then investigate the practical design of stochastic controllers for problems adequately described by the "LQG" assumptions (*L*inear system model, *Q*uadratic cost criterion for optimality, and *G*aussian noise inputs; see Chapter 14) and for problems requiring nonlinear models (see Chapter 15).

Within this chapter, Sections 13.2 and 13.3 are meant to be a basic overview of the optimal stochastic control problem, providing insights into important concepts before they are developed in detail later. *Stochastic dynamic programming* is the fundamental tool for solving this problem, and this algorithm is developed in the remainder of the chapter. It will turn out to be inherently a backward recursion in time, and the backward Kolmogorov equation useful for reverse-time propagations is presented in Section 13.4. Dynamic programming itself is then used to generate the optimal stochastic control function, first in Section 13.5 assuming that perfect knowledge of the entire state is available from the system at each sample time, and then in Section 13.6 under the typically more realistic assumption that only incomplete, noise-corrupted measurements are available.

## 13.2 BASIC PROBLEM FORMULATION

Fundamentally, control problems of interest can be described by the block diagram in Fig. 13.1. There is some *dynamic system* of interest, whose behavior is to be affected by $r$ applied *control* inputs $\mathbf{u}(t)$ in such a way that $p$ specified *controlled variables* $\mathbf{y}_c(t)$ exhibit desirable characteristics. These characteristics are prescribed, in part, as the controlled variables $\mathbf{y}_c(t)$ matching a desired $p$-dimensional *reference signal* $\mathbf{y}_d(t)$ as closely, quickly, and stably as possible, either over time intervals (as in tracking a dynamic $\mathbf{y}_d(t)$ or maintaining a piecewise constant setpoint $\mathbf{y}_d$) or at specific time instants (as at some given terminal time in a problem). Stability of the controlled system is an essential requirement, after which additional performance measures can be considered. However, the system responds not only to the control inputs, but also *dynamics disturbances* $\mathbf{n}(t)$ from the environment, over which we typically cannot exert any control. These usually cause the system to behave in an other-than-desired fashion. In order to observe certain aspects of the actual system behavior, sensor devices are constructed to output *measurements* $\mathbf{z}(t)$, which unfortunately may not correspond directly to the controlled variables, and which generally are not perfect due to *measurement corruptions* $\mathbf{n}_m(t)$. These typically incomplete and imperfect measurements are then provided as inputs to a controller, to assist in its generation of appropriate controls for the dynamic system.

EXAMPLE 13.1 For instance, the dynamic system might be the internal environment of a building, with controlled variables of temperature and humidity at various locations to be maintained at desired values through the control inputs to a furnace, air conditioner, and flow control dampers in individual ducts. Environmental disturbances would include heat transfer with external and internal sources and sinks, such as atmospheric conditions and human beings. Direct measurements of temperature and humidity at all locations may not be available, and additional measurements such as duct flow rates might be generated, and all such indications are subject to sensor dynamics, biases, imprecision, readout quantization, and other errors.

The case of matching the controlled variables to desired values at a single time might be illustrated by an orbital rendezvous between two space vehicles. On the other hand, continuously

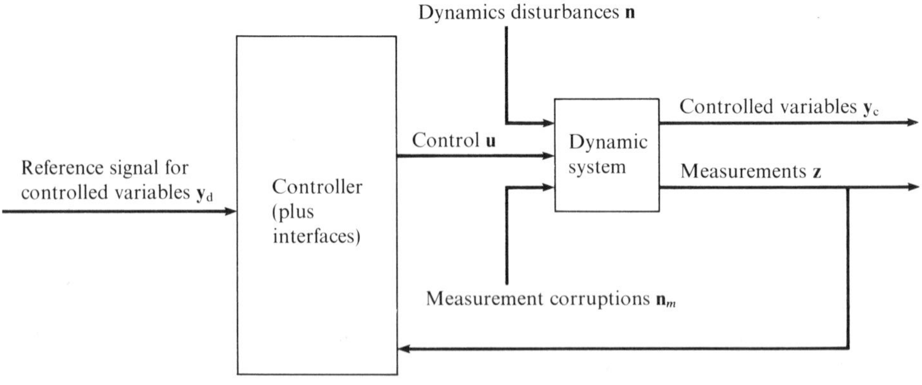


FIG. 13.1 Controlled system configuration.

matching $\mathbf{y}_c(t)$ to a dynamic $\mathbf{y}_d(t)$ is displayed in the problem of tracking airborne targets with communication antennas, cameras, or weapon systems. ■

Let us consider the "dynamic system" block in Fig. 13.1 in more detail. The first task in generating a controller for such a system is to develop a mathematical model that adequately represents the important aspects of the actual system behavior. Figure 13.2 presents a decomposition of the "dynamic system" block assuming that an adequate model can be generated in the form of a stochastic *state* model, with associated controlled variable and measurement output relations expressed in terms of that state. As shown, the controller outputs, $\mathbf{u}$, command actuators to impact the continuous-time state dynamics, which are also driven by disturbances, $\mathbf{n}$. The sensors generate measurements that are functions of that state, corruptions $\mathbf{n}_m$, and possibly of the controls as well. Similarly, the controlled variables are some function of the system state; the dashed line around "controlled variable output function" denotes the fact that it does not necessarily correspond to any physical characteristic or hardware, but simply a functional relationship.

A number of different types of measurements might be available. In the trivial case, nothing is measured, so that the controller in Fig. 13.1 would be an *open-loop* controller. However, in most cases of interest, there are enough disturbances, parameter variations, modeling inadequacies, and other uncertainties associated with the dynamic system that it is highly desirable to feed back observed values of some quantities relating to *actual* system response. One might conceive of a system structure in which *all states* could be measured *perfectly*; although this is not typically possible in practice, such a structure will be of use to gain insights into the properties of more physically realistic *feedback* control systems. Most typically, there are fewer measuring devices than states, and each of these sensors produces a signal as a nonlinear or linear function of the states, corruptions, and in some cases, controls. Physical sensors are often analog devices, so that continuous-time measurements from continuous-time systems are physically realistic. However, in most of the applications of interest to us, the controller itself will be implemented as software in a digital computer, so we will concentrate on *continuous-time state* descriptions with *sampled-data measurements*, compatible as inputs to the computer. We will also consider continuous-time measurements, and discrete-time measurements from systems described naturally only in discrete time, but the sampled-data case will be emphasized. On the other hand, it will be of importance to consider the controlled variables not only at the sampling instants, but throughout the entire interval of time of interest: reasonable behavior at these discrete times is not sufficient to preclude highly oscillatory, highly undesirable performance between sample times.

The noises $\mathbf{n}$ and $\mathbf{n}_m$ in Fig. 13.2 correspond to disturbances, corruptions, and sources of error that can be physically observed. Since physical continuous-time noises cannot truly be white, shaping filter models driven by fictitious

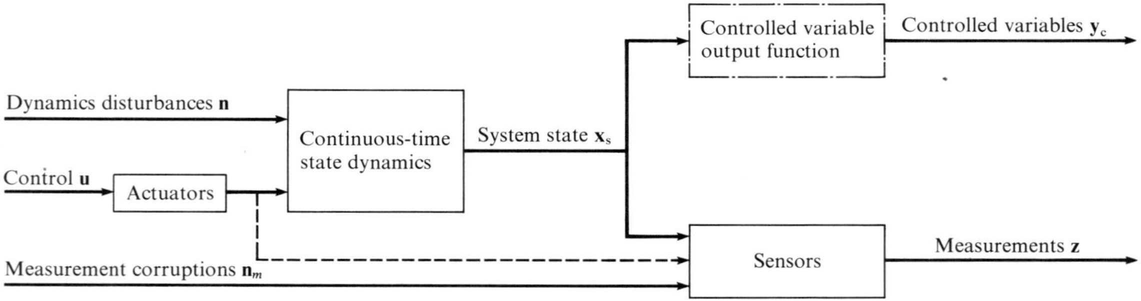


FIG. 13.2 Dynamic system block diagram.

white noise might be required to generate good representations of some components of these noises (see Chapter 11 of Volume 2 and Chapter 4 of Volume 1). White noises might suffice for very wideband physical continuous-time noises, and measurement corruptions are often modelable as uncorrelated and/or independent from one sample time to the next. Augmenting the states of any required shaping filters to the original system state $\mathbf{x}_s$ yields an augmented state $\mathbf{x}$ with which to describe the characteristics of the overall system. As described in Chapter 11, we shall assume that an adequate portrayal can be provided by an $n$-vector process $\mathbf{x}(\cdot,\cdot)$ as a solution to the Itô equation

$$\mathbf{dx}(t) = \mathbf{f}[\mathbf{x}(t), \mathbf{u}(t), t]\,dt + \mathbf{G}[\mathbf{x}(t), t]\,\mathbf{d\beta}(t) \tag{13-1}$$

where $\boldsymbol{\beta}(\cdot,\cdot)$ is $s$-dimensional Brownian motion of diffusion $\mathbf{Q}(t)$, as described in Section 11.5 [26, 45]. If the control were a deterministic open-loop $\mathbf{u}(t)$ or a function of perfectly known current state $\mathbf{u}[\mathbf{x}(t), t]$, then the solution to (13-1) is in fact Markov, as discussed in Section 12.2. However, we wish to address the more practical case in which the entire state is not known without error, but in which sampled-data measurements of the form of the $m$-vector process

$$\mathbf{z}(t_i) = \mathbf{h}[\mathbf{x}(t_i), t_i] + \mathbf{v}(t_i) \tag{13-2}$$

are available at specified sample times, with $\mathbf{v}(\cdot,\cdot)$ zero-mean white Gaussian discrete-time noise usually, but not necessarily, independent of $\boldsymbol{\beta}$, and of covariance $\mathbf{R}(t_i)$ for all $t_i$. Or, in the case of continuous-time measurements, the description would be

$$\mathbf{dy}(t) = \mathbf{h}[\mathbf{x}(t), t]\,dt + \mathbf{d\beta}_{\mathrm{m}}(t) \tag{13-3a}$$

with $\boldsymbol{\beta}_{\mathrm{m}}(\cdot,\cdot)$ $m$-dimensional Brownian motion (again, often assumed independent of $\boldsymbol{\beta}$) of diffusion $\mathbf{R}_{\mathrm{c}}(t)$ for all $t$, or heuristically,

$$\mathbf{z}(t) \triangleq \mathbf{dy}(t)/dt = \mathbf{h}[\mathbf{x}(t), t] + \mathbf{v}_{\mathrm{c}}(t) \tag{13-3b}$$

with $\mathbf{v}_{\mathrm{c}}(\cdot,\cdot)$ zero-mean white Gaussian noise of strength $\mathbf{R}_{\mathrm{c}}(t)$ for all $t$. In these practical cases, we might choose to let $\mathbf{u}$ be a function of the entire measurement history up to the current time, as for instance a function of the conditional mean of $\mathbf{x}(t)$ conditioned on that measurement history. Then, provided $\mathbf{f}$ is admissible (see Section 11.5), the solution to (13-1) is still well defined as an Itô process, but it need not be Markov. Note that $\mathbf{h}$ might also be a function of the measurement history, raising similar concerns. Such aspects will require special care in subsequent developments.

Note that (13-1) admits multiplicative noises driving the dynamics, whereas (13-2) and (13-3) prescribe only additive measurement noise. Although this is not overly restrictive for most problems, an extension could be made by admitting a premultiplier of the measurement noise analogous to $\mathbf{G}$ in (13-1).

For the purposes of control algorithm development, it may be convenient to view the solution to (13-1) between sample times as an "equivalent discrete-time system model" (see Chapter 4 of Volume 1),

$$\mathbf{x}(t_{i+1}) = \boldsymbol{\phi}[\mathbf{x}(t_i), \mathbf{u}(t_i), t_i] + \mathbf{G}_{\mathrm{d}}[\mathbf{x}(t_i), t_i]\mathbf{w}_{\mathrm{d}}(t_i) \tag{13-4}$$

where $\mathbf{w}_{\mathrm{d}}(\cdot,\cdot)$ is zero-mean white Gaussian discrete-time noise with covariance $\mathbf{Q}_{\mathrm{d}}(t_i)$ for all $t_i$, and $\mathbf{u}$ is assumed to be held constant from one sample time to the next. In practice, this model form may have to be approximated with numerical integration. Such a form will in fact be exploited subsequently.

A problem of special interest is one in which the adequate model is of *linear* or *linearized perturbation* form, with state dynamics

$$\mathbf{dx}(t) = [\mathbf{F}(t)\mathbf{x}(t) + \mathbf{B}(t)\mathbf{u}(t)]\,dt + \mathbf{G}(t)\,\mathbf{d}\boldsymbol{\beta}(t) \tag{13-5}$$

and either sampled-data measurements of the form

$$\mathbf{z}(t_i) = \mathbf{H}(t_i)\mathbf{x}(t_i) + \mathbf{v}(t_i) \tag{13-6}$$

or continuous-time measurements as

$$\mathbf{dy}(t) = \mathbf{H}(t)\mathbf{x}(t)\,dt + \mathbf{d}\boldsymbol{\beta}_{\mathrm{m}}(t) \tag{13-7a}$$

or heuristically

$$\mathbf{z}(t) \triangleq \mathbf{dy}(t)/dt = \mathbf{H}(t)\mathbf{x}(t) + \mathbf{v}_{\mathrm{c}}(t) \tag{13-7b}$$

In this particular case, the equivalent discrete-time model can be evaluated without approximation as (see Section 4.9 of Volume 1)

$$\mathbf{x}(t_{i+1}) = \boldsymbol{\Phi}(t_{i+1}, t_i)\mathbf{x}(t_i) + \mathbf{B}_{\mathrm{d}}(t_i)\mathbf{u}(t_i) + \mathbf{G}_{\mathrm{d}}(t_i)\mathbf{w}_{\mathrm{d}}(t_i) \tag{13-8}$$

where $\boldsymbol{\Phi}(t, t_i)$ is the state transition matrix associated with $\mathbf{F}(t)$, $\mathbf{B}_{\mathrm{d}}(t_i)$ equals $[\int_{t_i}^{t_{i+1}} \boldsymbol{\Phi}(t_{i+1}, \tau)\mathbf{B}(\tau)\,d\tau]$, and $[\mathbf{G}_{\mathrm{d}}(t_i)\mathbf{w}_{\mathrm{d}}(t_i)]$ is a zero-mean white Gaussian discrete-time noise of covariance

$$\left[\int_{t_i}^{t_{i+1}} \boldsymbol{\Phi}(t_{i+1}, \tau)\mathbf{G}(\tau)\mathbf{Q}(\tau)\mathbf{G}^{\mathrm{T}}(\tau)\boldsymbol{\Phi}^{\mathrm{T}}(t_{i+1}, \tau)\,d\tau\right]$$

Under assumptions of a quadratic cost to be minimized, as discussed later, the *optimal stochastic controller* for a system described by such a model turns out to be readily synthesized and implemented, and of considerable practical importance.

Recall the controller in Fig. 13.1. Its structure will be based upon state and measurement *models* of the form (13-1)–(13-8). Even the most complex and inclusive models will not be perfect, and purposely simplified and reduced-order models are usually employed to yield a computationally practical control algorithm. For this reason, the measurement feedback from the *actual* system as depicted in the figure is often essential to adequate performance. Figure 13.3 presents a typical sampled-data controller structure to be used in conjunction

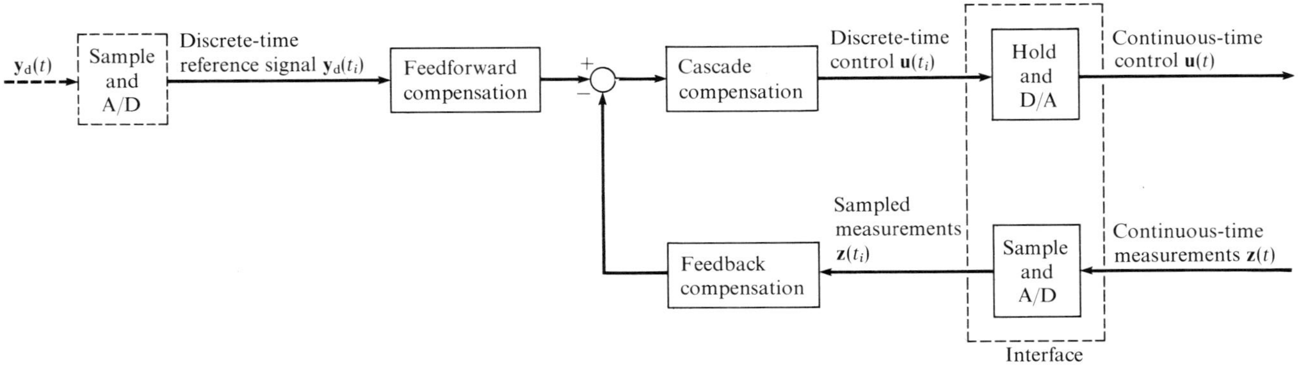


FIG. 13.3 Typical sampled-data controller, with interfaces to continuous-time environment.

with the overall system of Fig. 13.1. The sampled-data measurements from the real system are shown as being generated by passing analog sensor outputs through a sampler and analog-to-digital conversion (A/D) in the "interface" between continuous-time and discrete-time environments. These measurements are processed by a *feedback compensation* algorithm, which may involve simple multiplications by appropriate gains or may entail dynamic compensation and filtering operations. To the output of this compensation block is added the result of similar processing by the *feedforward compensation* of the discrete-time reference signal for the controlled variables, which themselves might be the output of a sampler and analog-to-digital conversion of an analog signal. Operating on the sum of signals by a *cascade compensation* algorithm yields the *discrete-time control signal* $\mathbf{u}(t_i)$ at sample time $t_i$. In fact, in a linear controller algorithm, the cascade compensation can be viewed as the operations common to feedforward and feedback compensations, put in cascade instead for efficiency. Finally, to provide a control signal compatible with a continuous-time system, $\mathbf{u}(t_i)$ is passed through a "hold" and digital-to-analog conversion (D/A). Typically, one uses a "zero order hold," a device that holds the value of $\mathbf{u}(t_i)$ constant over the ensuing sample period $[t_i, t_{i+1})$; i.e., $\mathbf{u}(t)$ is piecewise constant. There are other possibilities, such as a first order hold, which uses the current and preceding value to establish the slope of a piecewise linear $\mathbf{u}(t)$ emanating from the current $\mathbf{u}(t_i)$ value over $[t_i, t_{i+1})$. Different interfacing hold functions have different characteristics, but we will assume the practically significant zero order hold to be used for all developments henceforth.

Our objective now is to generate *optimal stochastic controllers*, optimal in the sense that they minimize some appropriate scalar cost function, or "performance index," associated with the control problem [13]. For instance, in a stochastic tracking problem, one might like to generate the control function of time and measurement history out of the admissible class of functions (perhaps the class of piecewise-constant control functions that do not exceed certain tracker gimbal stops or other constraints) that minimizes integral mean square tracking error (mean square value of the difference between controlled $\mathbf{y}_c$ and commanded $\mathbf{y}_d$, integrated over the entire time interval of interest). This might be modified by a desire not to expend too much energy, or to generate a closed-loop system whose stability [23, 27, 29] and performance characteristics are relatively insensitive to variations in uncertain system-defining parameters, or to restrict attention to dynamic controllers of order no greater than some specified value $n_c$, or to yield solutions in the form of time-invariant dynamic controllers and multiplications by constant gains only. The point is, the optimality of the resulting design is *only with respect to the criterion chosen* to evaluate performance. What is mathematically optimal may not be at all useful in practice. Moreover, there is considerable "art" to choosing a criterion that yields (1) a tractable optimization problem solution and (2) a controller solution that indeed provides all of the desired characteristics in the closed-loop system configuration. To avoid misconstruing the result, perhaps we should call this "stochastic controller design via mathematical optimization" rather than "optimal stochastic control," but we will adopt the accepted latter terminology.

The optimization problem definition itself can differ considerably from one application to another. A problem might have a fixed terminal time of interest, or a set of termination conditions to be reached in a finite, but a priori uncertain, interval of time; these are known as "finite horizon" problems. In comparison, "infinite horizon" problems are envisioned as having no natural terminal time; this is a useful mathematical construct for generating constant-gain steady state controllers for problems described by time-invariant system and stationary noise models. Another aspect of problem definition that can vary significantly is the manner in which constraints are handled. Establishing an admissible class of control functions as those which do not violate certain inequality (or equality) constraints on control and/or state variables may be an attractive way to formulate a problem, but a significantly more tractable problem is produced by penalizing oneself heavily for such violations by appropriate definition of the cost function.

In this chapter, we investigate the basic mathematical tool for formulating and solving these various optimal stochastic control problems, known as *dynamic programming*. Subsequent chapters emphasize the *synthesis* and *performance* of *practical controllers* as depicted by Figs. 13.1–13.3, using this theoretical approach.

## 13.3 INTRODUCTION TO CONCEPTS: OVERVIEW OF SIMPLE LQG PROBLEM

To provide additional physical motivation and insights, we consider the simplest form of "LQG" controller problem: design of the optimal stochastic controller for a problem described in terms of *l*inear system models, *q*uadratic cost criterion, and *G*aussian noise models. The special synthesis capability associated with the solution to this problem makes it particularly attractive to control designers. Throughout this section, both the uniqueness of this special case and the fundamental, more generalized concepts and thought processes will be emphasized.

Let a system be adequately described by the $n$-dimensional stochastic state difference equation

$$\mathbf{x}(t_{i+1}) = \mathbf{\Phi}(t_{i+1}, t_i)\mathbf{x}(t_i) + \mathbf{B}_d(t_i)\mathbf{u}(t_i) + \mathbf{G}_d(t_i)\mathbf{w}_d(t_i) \tag{13-8}$$

where the $r$-dimensional $\mathbf{u}(t_i)$ is the control input to be applied and $\mathbf{w}_d(\cdot,\cdot)$ is $s$-dimensional zero-mean white Gaussian discrete-time noise with

$$E\{\mathbf{w}_d(t_i)\mathbf{w}_d^{\mathrm{T}}(t_j)\} = \mathbf{Q}_d(t_i)\,\delta_{ij} \tag{13-9}$$

and assumed to be independent of the initial state condition. This initial $\mathbf{x}(t_0)$ is modeled as a Gaussian random vector with mean $\bar{\mathbf{x}}_0$ and covariance $\mathbf{P}_0$. Assume this is an equivalent discrete-time model for some underlying continuous-time physical system, so that, for instance, $\mathbf{\Phi}(t_{i+1}, t_i)$ is assured to be invertible for all $t_i$. Available from the system are $m$-dimensional sampled-data measurements of the form

$$\mathbf{z}(t_i) = \mathbf{H}(t_i)\mathbf{x}(t_i) + \mathbf{v}(t_i) \tag{13-6}$$

where $\mathbf{v}(\cdot,\cdot)$ is $m$-dimensional zero-mean white Gaussian discrete-time noise with

$$E\{\mathbf{v}(t_i)\mathbf{v}^{\mathrm{T}}(t_j)\} = \mathbf{R}(t_i)\,\delta_{ij} \tag{13-10}$$

and assumed independent of both $\mathbf{x}(t_0,\cdot)$ and $\mathbf{w}_{\mathrm{d}}(\cdot,\cdot)$.

The objective is to determine the optimal control function $\mathbf{u}^*$, optimal in the sense that it minimizes a quadratic cost function

$$J = E\left\{\sum_{i=0}^{N} \frac{1}{2}\left[\mathbf{x}^{\mathrm{T}}(t_i)\mathbf{X}(t_i)\mathbf{x}(t_i) + \mathbf{u}^{\mathrm{T}}(t_i)\mathbf{U}(t_i)\mathbf{u}(t_i)\right] + \frac{1}{2}\mathbf{x}^{\mathrm{T}}(t_{N+1})\mathbf{X}_{\mathrm{f}}\mathbf{x}(t_{N+1})\right\} \tag{13-11}$$

where $t_0$ is the initial time and $t_{N+1}$ is the known final time of interest: the last time a control is computed and applied as a constant over the succeeding sample interval is at time $t_N$, $N$ sample periods from time $t_0$. First consider the final term in (13-11): it assigns a quadratic penalty to the magnitude of terminal state deviations from zero. If the cost weighting matrix associated with the final state $\mathbf{X}_{\mathrm{f}}$ is diagonal, then these diagonal terms are chosen to reflect the relative importance of maintaining each component of $\mathbf{x}(t_{N+1})$ near zero; the more important the state minimization is, the larger the associated $\mathbf{X}_{\mathrm{f}}$ term.

EXAMPLE 13.2 Let the state $\mathbf{x}(\cdot,\cdot)$ be a two-dimensional process, and let

$$\mathbf{X}_{\mathrm{f}} = \begin{bmatrix} 100 & 0 \\ 0 & 1 \end{bmatrix}$$

Then the final state cost in (13-11) would be

$$E\{\tfrac{1}{2}[100\mathrm{x}_1{}^2(t_{N+1}) + \mathrm{x}_2{}^2(t_{N+1})]\}$$

Note that rms deviations in $\mathrm{x}_2(t_{N+1})$ would have to be ten times as large as $\mathrm{x}_1(t_{N+1})$ rms deviations to accrue the same cost penalty. Since we are generating a cost-minimizing controller, the result will be a controller that exerts more effort in keeping $\mathrm{x}_1(t_{N+1})$ small than $\mathrm{x}_2(t_{N+1})$: it can afford to let $\mathrm{x}_2(t_{N+1})$ grow substantially larger for the same cost. ■

EXAMPLE 13.3 If one considers an air-to-air missile intercept problem, it is crucial to minimize the rms miss distance at the terminal time, but the missile's velocity at that time need not be so tightly controlled about some preselected nominal velocity. Therefore, if this problem formulation were used to generate the missile guidance law, a diagonal $\mathbf{X}_{\mathrm{f}}$ might be chosen with very large entries corresponding to position error states and small values corresponding to velocity error states.

On the other hand, for an orbital rendezvous, matching position and velocity states at the terminal time are both important. So, $\mathbf{X}_{\mathrm{f}}$ would have large entries corresponding to both position and velocity states.

One means of initially choosing $\mathbf{X}_f$ diagonal entries is to determine a maximum allowable value for each state $x_k(t_{N+1})$, and then setting [9]

$$[\mathbf{X}_f]_{kk} = 1/[\text{max allowable } x_k(t_{N+1})]^2$$

By so doing, if all final states were to reach their maximum allowable values simultaneously, each term in the final quadratic cost expression would contribute equally to the overall cost. This would indicate that the cost-minimizing controller should exert equal effort over each of the state components, as is appropriate.

Such a choice of weighting matrix $\mathbf{X}_f$ is only an initial value. In analogy with filter tuning, one must observe the performance capabilities of a controller based upon this choice of design parameters, and alter them iteratively until desired characteristics are achieved. This will be discussed in detail in the next chapter. ■

Now consider the summation term in (13-11). It allows a quadratic penalty to be assigned to the magnitude of state deviations from zero and to the amount of control energy expended over each of the $(N + 1)$ sample periods in the problem of interest. Note that if $u_k$ is a control variable, $\frac{1}{2}u_k^2$ would be the associated generalized control energy. As for the terminal cost, the entries in the weighting matrices $\mathbf{X}(t_i)$ and $\mathbf{U}(t_i)$ reflect the relative importance of maintaining individual state and control component deviations at small values over each sample period in the problem. If these weighting matrices are diagonal, then the larger the diagonal entry, the closer the corresponding variable will be kept to zero.

EXAMPLE 13.4 Consider an autopilot design for terrain avoidance. A typical scenario might call for an aircraft to fly at a cruise altitude, descend to very low altitude for a period of time, and then climb out to cruise altitude again. While at very low altitude, it is critical to maintain very tight control over the vertical position state, so over this interval of time, the corresponding diagonal entry in $\mathbf{X}(t_i)$ would be very large. One would be willing to expend large amounts of control energy to accomplish this objective, so the appropriate entries in $\mathbf{U}(t_i)$ might be small during this time. However, while at cruise altitude, there is no need to maintain such tight vertical position control, so that $\mathbf{X}(t_i)$ diagonal term could be decreased significantly, allowing proportionately greater control over other states. Moreover, one would probably be less inclined to allow as much overall control activity as when the aircraft is very low, so the $\mathbf{U}(t_i)$ entries might be increased at this time too.

The same initial parameter value selection procedure as in Example 13.3 could be used for each sample period on states and controls. However, a "tuning" procedure on $\mathbf{X}(t_i)$ and $\mathbf{U}(t_i)$ for all $t_i$ and on $\mathbf{X}_f$ is usually required, in analogy to tuning of $\mathbf{Q}_d(t_i)$ and $\mathbf{R}(t_i)$ for all $t_i$ and of $\mathbf{P}_0$ in a Kalman filter. This "duality" relationship will be discussed to a greater degree in Chapter 14. ■

The weighting matrices $\mathbf{X}_f$ and $\mathbf{X}(t_i)$ for all $t_i$ are assumed to be $n$-by-$n$, real symmetric (without loss of generality in a quadratic form, since any real matrix can be decomposed into the sum of a symmetric and a skew-symmetric matrix, the latter contributing nothing to a quadratic), and positive semidefinite. Since all eigenvalues are then nonnegative, this precludes lower costs being assigned to larger deviations, but allows zero cost to be assigned to certain state variables of no significance. By comparison, $\mathbf{U}(t_i)$ is assumed $r$-by-$r$, real, symmetric, and positive definite for all $t_i$: this prevents negative or zero cost being associated with control energy expenditure, thereby physically precluding controller solutions that command an infinite amount of control energy to be used at any time. Note that there is no quadratic penalty associated with control at terminal time $t_{N+1}$: this is logical since the problem is finished at this time, and there is no reason to conceive of generating a control to hold constant over a sample period past time $t_{N+1}$.

In the context of the previous section, this cost function expresses the desire to consider the controlled variables $\mathbf{y}_c(t_i)$ to be the entire state $n$-vector $\mathbf{x}(t_i)$ and their desired values $\mathbf{y}_d(t_i)$ to be zero for all time, if $\mathbf{X}(t_i)$ is in fact positive definite for all $t_i$. If $\mathbf{X}(t_i)$ were of rank $p < n$ for all $t_i$, then $\mathbf{y}_c(t_i)$ would be of dimension $p$, corresponding to the linear combinations of states with which are associated positive eigenvalues. The case of trying to drive $\mathbf{x}(t_i)$ to *nonzero* desired values can be handled by defining perturbation states from desired values, but this will be developed in greater detail in the next chapter.

The cost function proposed in (13-11) is the *expected value* of a quadratic function of states and controls throughout the interval of interest and of states at the terminal time. This properly reflects the stochastic nature of the problem: given all the possible sample paths we might receive from the defining stochastic processes, we want to choose the control function that minimizes the *ensemble average* of costs corresponding to each sample. Moreover, for this specific problem, the admissible class of discrete-time controls is unrestricted: no inequality constraints are imposed, and true physical constraints are handled by "proper" choice of weighting matrices in the cost definition.

One might ask why a *quadratic* function of states and controls is specifically chosen for a cost to be associated with a problem described by *linear* system and *Gaussian* noise models. Although other types of cost functions are useful, quadratics are a good physical description of many control objectives, such as minimizing mean square errors while not expending inordinate amounts of control energy. Moreover, if one views most linear models as being derived as linear perturbation models, by developing a controller that specifically attempts to maintain small values of quadratics in state and control deviations, one is inherently enhancing the adequacy of the linear perturbation model itself. However, perhaps the most important reason is that this combination of modeling assumptions yields a *tractable* problem whose solution is in the form of a *readily synthesized, efficiently implemented, feedback* control law. These traits unfortunately are not shared by problems defined under other assumptions.

Let us pursue the control algorithm development further. Consider being placed in the middle of the control problem just defined, at some sample time $t_j$. We have already taken the measurements

$$\mathbf{z}(t_1,\omega_k) = \mathbf{z}_1, \qquad \mathbf{z}(t_2,\omega_k) = \mathbf{z}_2, \qquad \ldots, \qquad \mathbf{z}(t_j,\omega_k) = \mathbf{z}_j$$

or, more compactly, the measurement history $\mathbf{Z}(t_j,\omega_k) = \mathbf{Z}_j$. Furthermore, the control inputs $\mathbf{u}(t_0)$, $\mathbf{u}(t_1)$, . . . , $\mathbf{u}(t_{j-1})$ have already been applied in the past and are immutable. Now we wish to determine the optimal control to apply at time $t_j$ and hold constant until the next sample instant, but we need to reexamine our cost criterion and cause it to embody precisely what we wish to minimize at this point in time. First of all, since we are at $t_j$ and can do nothing to alter what happened in the past, it is logical to minimize the expected *cost to complete the process* from $t_j$ forward to $t_{N+1}$, instead of from $t_0$ forward as in (13-11). If we were to include the earlier terms in the cost definition, then the solution to the optimization problem would generally yield the unrealizable decision to change the controls applied before time $t_j$. Second, the cost in (13-11) involves an unconditional expectation, as would be appropriate at time $t_0$, but now we would want to exploit the additional knowledge about the system gained by taking the history of measurements $\mathbf{Z}_j$. This can be accomplished by evaluating the *conditional expectation* of the cost to complete the process, conditioned on $\mathbf{Z}(t_j, \omega_k) = \mathbf{Z}_j$. Thus, a meaningful cost functional to use in determining the optimal control to apply at time $t_j$ would be

$$\mathscr{C}[\mathbf{Z}_j, t_j] \triangleq E\left\{\sum_{i=j}^{N} \frac{1}{2}[\mathbf{x}^{\mathrm{T}}(t_i)\mathbf{X}(t_i)\mathbf{x}(t_i) + \mathbf{u}^{\mathrm{T}}(t_i)\mathbf{U}(t_i)\mathbf{u}(t_i)] + \frac{1}{2}\mathbf{x}^{\mathrm{T}}(t_{N+1})\mathbf{X}_{\mathrm{f}}\mathbf{x}(t_{N+1}) \,\middle|\, \mathbf{Z}(t_j, \omega_k) = \mathbf{Z}_j\right\} \tag{13-12}$$

If we could find the control to minimize this expected cost-to-go, i.e., if we could solve that optimization problem, then we would have the $r$-dimensional vector of inputs $\mathbf{u}^*[\mathbf{Z}_j, t_j]$ that would yield a minimum value of $\mathscr{C}[\mathbf{Z}_j, t_j]$, denoted as $\mathscr{C}^*[\mathbf{Z}_j, t_j]$. If we could accomplish this for the range of possible values of $\mathbf{Z}_j$, we could specify the *function* $\mathbf{u}^*[\cdot, t_j]$ which, when given any particular history of measurements $\mathbf{Z}_j$, could be evaluated to yield the optimum $r$-vector of controls to apply to the system. Doing this for all times $t_j$ yields the optimal control function $\mathbf{u}^*[\cdot,\cdot]$ as a mapping from $R^{mj} \times T$ into $R^r$, i.e., a function that accepts an argument $\mathbf{Z}_j$ from $mj$-dimensional Euclidean space and an argument $t_j$ from a time set $T$, to yield an $r$-dimensional control vector. It is this function that we seek as a solution to the stochastic optimal control problem.

The question is, how does one perform the optimization? Because of the stochastic nature of the problem, such useful deterministic methods as the calculus of variations and Pontryagin's maximum principle cannot be employed. These approaches are based on the premise that there exists a *single* optimal state and control trajectory in time. Such an assumption cannot be made here: the dynamic driving noise term $\mathbf{G}_{\mathrm{d}}(t_i)\mathbf{w}_{\mathrm{d}}(t_i)$ in (13-8) prevents our knowing ahead of time where we will arrive at time $t_{i+1}$ if we apply a specified $\mathbf{u}(t_i)$, *even* if we knew the current state $\mathbf{x}(t_i)$ *exactly* (which generally we *do not*; the simpler problem in which $\mathbf{x}(t_i)$ is assumed known at $t_i$ instead of $\mathbf{Z}_i$ will be considered separately to gain insight into the solution for the more realistic formulation).

However, *dynamic programming* and the closely related *optimality principle* *can* be employed [1, 2, 4–8, 14, 16–18, 21, 24, 25, 27, 28, 31, 33, 37–39, 42, 43].

One statement of this principle is "Whatever any initial states and decision [or control law] are, all remaining decisions must constitute an optimal policy with regard to the state which results from the first decision" [4].

EXAMPLE 13.5 To appreciate the fundamental concept of the "optimality principle" and dynamic programming, consider a scalar-state, scalar-control problem lasting two sample periods, assuming that perfect knowledge of the current state $x(t_i)$ is available. Thus the system might be described by

$$\mathsf{x}(t_{i+1}) = \Phi \mathsf{x}(t_i) + B_d u(t_i) + G_d \mathsf{w}_d(t_i)$$

and control is to be applied at times $t_0$ and $t_1$ so as to minimize the expected cost to complete the process to final time $t_2$, conditioned on perfect knowledge of the state value at $t_0$, and at $t_1$ once the system arrives there. Assume that you are able to solve for the cost-minimizing control at time $t_1$, $u^*[x(t_1), t_1]$ for all possible $\mathsf{x}(t_1)$ values, i.e., you know the optimal control function $u^*[\cdot, t_1]$ to minimize the expected cost from $t_1$ forward.

Now consider solving for the optimal function at the initial time, $u^*[\cdot, t_0]$. To do so, we *must* assume that when we get to time $t_1$, at whatever realization $\mathsf{x}(t_1, \omega_k) = x(t_1)$ results from the system dynamics and applied control at $t_0$, we will use the cost-minimizing control $u^*[x(t_1), t_1]$, and not some other nonoptimal control $u[x(t_1), t_1]$, from time $t_1$ to time $t_2$. Otherwise, we cannot hope to minimize the total cost from $t_0$ through $t_1$ to $t_2$.

When there are more than two sample periods, or stages, in a problem, we must assume that *all* controls applied downstream will be controls that minimize the expected cost from that time forward. In this manner, we are able to solve for the control functions $u^*[\cdot, t_N]$, $u^*[\cdot, t_{N-1}], \ldots,$ $u^*[\cdot, t_0]$ sequentially. ■

Dynamic programming applied to the current problem yields the result that the optimal control function is obtained from the solution [20] to

$$\mathscr{C}^*[\mathscr{Z}_i, t_i] = \min_{\mathbf{u}(t_i)} \left[ E\left\{\tfrac{1}{2}[\mathbf{x}^{\mathrm{T}}(t_i)\mathbf{X}(t_i)\mathbf{x}(t_i) + \mathbf{u}^{\mathrm{T}}(t_i)\mathbf{U}(t_i)\mathbf{u}(t_i) \middle| \mathbf{Z}(t_i) = \mathscr{Z}_i\right\} \right.$$
$$\left. + E\{\mathscr{C}^*[\mathbf{Z}(t_{i+1}), t_{i+1}] | \mathbf{Z}(t_i) = \mathscr{Z}_i\} \right] \tag{13-13}$$

as solved backwards for all $t_i$ from the terminal condition

$$\mathscr{C}^*[\mathscr{Z}_{N+1}, t_{N+1}] = E\{\tfrac{1}{2}\mathbf{x}^{\mathrm{T}}(t_{N+1})\mathbf{X}_f \mathbf{x}(t_{N+1}) | \mathbf{Z}(t_{N+1}) = \mathscr{Z}_{N+1}\} \tag{13-14}$$

The notation in (13-13) means that $\mathscr{C}^*[\mathscr{Z}_i, t_i]$ is found by minimizing the bracketed right hand side term by choice of control function $\mathbf{u}(t_i)$ to apply at time $t_i$. The first expectation in that bracketed term is the expected cost associated with traversing the next single sample period, and the second expectation is the expected cost to complete the process from $t_{i+1}$ to the final time, *assuming* that *optimal* controls will be used from $t_{i+1}$ forward (note the asterisk on $\mathscr{C}^*$ in that expectation). Thus, $\mathscr{C}^*[\mathbf{Z}(t_{i+1}), t_{i+1}]$ must be known before $\mathscr{C}^*[\mathbf{Z}(t_i), t_i]$ can be obtained, so that (13-13) is *naturally a backward running recursion relation.* When the minimization in (13-13) is performed, the control function that yields that minimum cost is the desired optimal control $\mathbf{u}^*[\cdot, t_i]$, which can be evaluated as $\mathbf{u}^*[\mathbf{Z}_i, t_i]$ for any given realization $\mathbf{Z}(t_i, \omega_k) = \mathbf{Z}_i$. Note that in (13-13), any control function of $\mathbf{Z}(t_i)$ is properly a random vector,

$$\mathbf{u}(t_i, \cdot) = \mathbf{u}[\mathbf{Z}(t_i, \cdot), t_i] \tag{13-15}$$

mapping the sample space $\Omega$ into $R^r$, but conditioning the quadratic of this *function* of $\mathbf{Z}(t_i, \cdot)$ on a *known realization* $\mathbf{Z}(t_i, \omega_k)$ yields the quadratic of the realization $\mathbf{u}(t_i) = \mathbf{u}(t_i, \omega_k)$ itself.

Consider the terminal cost or the cost contribution associated with any single stage. Inherent in Eq. (13-13) is the evaluation of the conditional expectation of each of these cost contributions, sequentially conditioned on measurements up through earlier and earlier times, so that evaluation of $f_{\mathbf{x}(t_j)|\mathbf{Z}(t_k)}(\boldsymbol{\xi} | \mathcal{Z}_k)$ as $t_k$ moves backwards in time from $t_j$ is of fundamental concern. Intimately related to this is the evaluation of the transition probability density

$$f_{\mathbf{x}}(\boldsymbol{\xi}, t | \boldsymbol{\rho}, t') \triangleq f_{\mathbf{x}(t)|\mathbf{x}(t')}(\boldsymbol{\xi} | \boldsymbol{\rho}) \tag{13-16}$$

as $t'$ moves backwards from $t$. The partial differential equation that characterizes this functional behavior is the *backward Kolmogorov equation* [1, 18, 26, 33], and it will be developed later in this chapter.

In a general optimal stochastic control problem, one often cannot progress much further than setting up a dynamic programming algorithm similar to (13-13) and (13-14) and describing behavior of some pertinent conditional densities by means of the backward Kolmogorov equation. Solving the minimization as shown in (13-13) and solving the Kolmogorov partial differential equation are formidable, often intractable, tasks. Even if they could be solved, the result is in the form of a control law $\mathbf{u}^*[\cdot,\cdot]$ that requires explicit knowledge of the entire measurement history $\mathbf{Z}_i$ and of the current time $t_i$: a very burdensome growing memory requirement for implementation.

However, under the LQG assumptions, a vast simplification occurs in the form of *sufficient statistics* [11, 12, 40, 41]. To evaluate the indicated expectations in (13-13), one must establish the conditional density $f_{\mathbf{x}(t_i)|\mathbf{Z}(t_i)}(\boldsymbol{\xi} | \mathbf{Z}_i)$. For the LQG controller problem, this density is Gaussian and completely specified by its mean $\hat{\mathbf{x}}(t_i^+)$ and covariance $\mathbf{P}(t_i^+)$: if these are available, we need *not* remember all of the measurement values $\mathbf{Z}_i$ explicitly. Thus, we can equivalently solve the optimization problem as a function of $\hat{\mathbf{x}}(t_i^+)$ and $\mathbf{P}(t_i^+)$, the latter of which does *not* depend on $\mathbf{Z}_i$, as provided by a Kalman filter. The great advantage is that we do not have to remember an ever-increasing number of values in order to generate the optimal control.

Performing the optimization of (13-13) and (13-14) results in the optimal cost-to-go being the quadratic

$$\mathscr{C}^*[\mathbf{Z}_i, t_i] = \mathscr{C}^*[\hat{\mathbf{x}}(t_i^+), t_i] = \tfrac{1}{2}[\hat{\mathbf{x}}^{\mathrm{T}}(t_i^+)\mathbf{K}_{\mathrm{c}}(t_i)\hat{\mathbf{x}}(t_i^+) + g(t_i)] \tag{13-17}$$

where $\mathbf{K}_{\mathrm{c}}(t_i)$ is the $n$-by-$n$ symmetric matrix satisfying the *backward Riccati difference equation*

$$\begin{aligned}\mathbf{K}_{\mathrm{c}}(t_i) = \mathbf{X}(t_i) &+ \boldsymbol{\Phi}^{\mathrm{T}}(t_{i+1}, t_i)\mathbf{K}_{\mathrm{c}}(t_{i+1})\boldsymbol{\Phi}(t_{i+1}, t_i)\\ &- [\boldsymbol{\Phi}^{\mathrm{T}}(t_{i+1}, t_i)\mathbf{K}_{\mathrm{c}}(t_{i+1})\mathbf{B}_{\mathrm{d}}(t_i)][\mathbf{U}(t_i) + \mathbf{B}_{\mathrm{d}}^{\mathrm{T}}(t_i)\mathbf{K}_{\mathrm{c}}(t_{i+1})\mathbf{B}_{\mathrm{d}}(t_i)]^{-1}\\ &\times [\mathbf{B}_{\mathrm{d}}^{\mathrm{T}}(t_i)\mathbf{K}_{\mathrm{c}}(t_{i+1})\boldsymbol{\Phi}(t_{i+1}, t_i)]\end{aligned} \tag{13-18}$$

solved backwards from the terminal condition

$$\mathbf{K}_c(t_{N+1}) = \mathbf{X}_f \tag{13-19}$$

and where the scalar $g(t_i)$ satisfies the backward recursion

$$g(t_i) = g(t_{i+1}) + \operatorname{tr}\{\mathbf{X}(t_i)\mathbf{P}(t_i^+) + \mathbf{K}(t_{i+1})\mathbf{H}(t_{i+1})\mathbf{P}(t_{i+1}^-)\mathbf{K}_c(t_{i+1})\} \tag{13-20}$$

with the terminal condition

$$g(t_{N+1}) = \operatorname{tr}\{\mathbf{X}_f\mathbf{P}(t_{N+1}^+)\} \tag{13-21}$$

In (13-19) and (13-20), tr denotes trace and **P** and **K** are the state error covariance and gain, respectively, from the Kalman filter associated with this problem: note that the controller Riccati matrix $\mathbf{K}_c$ is completely *distinct* from the Kalman filter gain **K**. Moreover, the optimal control that achieves this minimal cost-to-go is given by

$$\mathbf{u}^*[\hat{\mathbf{x}}(t_i^+), t_i] = -\mathbf{G}_c^*(t_i)\hat{\mathbf{x}}(t_i^+) \tag{13-22}$$

where $\hat{\mathbf{x}}(t_i^+)$ is the state estimate output of the Kalman filter and the optimal feedback controller gain $\mathbf{G}_c^*(t_i)$ is given by

$$\mathbf{G}_c^*(t_i) = [\mathbf{U}(t_i) + \mathbf{B}_d^T(t_i)\mathbf{K}_c(t_{i+1})\mathbf{B}_d(t_i)]^{-1}[\mathbf{B}_d^T(t_i)\mathbf{K}_c(t_{i+1})\mathbf{\Phi}(t_{i+1}, t_i)] \tag{13-23}$$

which is recognized as part of the last term in (13-18).

Consider the structure of the Riccati difference equation (13-18) and terminal condition (13-19). The matrices $\mathbf{X}(t_i)$ and $\mathbf{U}(t_i)$ for all $t_i$ and the matrix $\mathbf{X}_f$ are descriptors of the quadratic cost (13-11), and the matrices $\mathbf{\Phi}(t_{i+1}, t_i)$ and $\mathbf{B}_d(t_i)$ for all $t_i$ define the deterministic part of the dynamic system state model (13-8). Only these matrices are required to evaluate $\mathbf{K}_c(t_i)$ and $\mathbf{G}_c^*(t_i)$ backward in time. *All of the stochastic nature of the problem*—the statistical information about the effects of the sources of uncertainty modeled by the Gaussian $\mathbf{x}(t_0)$, dynamic driving noise $\mathbf{w}_d(\cdot,\cdot)$, and measurement noise $\mathbf{v}(\cdot,\cdot)$—*is totally confined to the* $g(t_i)$ term in the expected cost-to-go (13-17). Furthermore, it turns out that the result of setting $g(t_i) \equiv 0$ for all $t_i$ is just the cost solution to the corresponding *deterministic optimal control problem assuming perfect state knowledge at all time*: the same problem as posed here, but with all of the sources of uncertainty removed and $\mathbf{H}(t_i)$ being set to the identity matrix.

Especially note that the optimal control given in (13-22) is *not* affected in any way by $g(t_i)$. The expected cost to complete the process from time $t_i$, $\mathscr{C}^*[\hat{\mathbf{x}}(t_i^+), t_i]$ *is* affected by $g(t_i)$—the expected value of the cost increases as one progresses from the deterministic optimal control problem with perfect state knowledge, to the stochastic optimal control problem with perfect state knowledge, to the current stochastic optimal control problem with only knowledge of incomplete and imperfect measurements, because of the additional uncertainty in each succeeding case. Nevertheless, the optimal control function itself has the *same identical form* for the three cases. For the first two cases that assume perfect state knowledge, the optimal control law is a state feedback given by

$$\mathbf{u}^*[\mathbf{x}(t_i), t_i] = -\mathbf{G}_c^*(t_i)\mathbf{x}(t_i) \tag{13-24}$$

If perfect knowledge is not available, but only measurements of the form (13-6), then the state $\mathbf{x}(t_i)$ is replaced by the conditional mean of $\mathbf{x}(t_i)$, conditioned on $\mathbf{Z}(t_i, \omega_k) = \mathbf{Z}_i$, i.e., the estimate $\hat{\mathbf{x}}(t_i^+)$ from a Kalman filter that has processed the measurement history $\mathbf{Z}_i$, as seen in Eq. (13-22). The *same* feedback gains $\mathbf{G}_c^*(t_i)$ are valid for all three cases, and these gains are independent of the stochastic nature of any problem. Thus, the optimal control *function* $\mathbf{u}^*[(\cdot), t_i]$ for all $t_i$ in these various cases can be expressed as

$$\mathbf{u}^*[(\cdot), t_i] = -\mathbf{G}_c^*(t_i)(\cdot) \tag{13-25}$$

and thus $\mathbf{u}^*[\mathbf{x}(t_i), t_i]$ and $\mathbf{u}^*[\hat{\mathbf{x}}(t_i^+), t_i]$ are recognized as *identical linear* functions of the *current state* $\mathbf{x}(t_i)$ or the *current estimate* $\hat{\mathbf{x}}(t_i^+)$. We have achieved a *linear memoryless feedback control law* that *does not* vary from case to case.

That the optimal stochastic controller in this case is equivalent to the associated optimal deterministic controller function, but with the state replaced by the conditional mean of the state given the observed measurements, is often described by saying the LQG optimal controller has the *certainty equivalence* property. This is a special case of the *separation* property, in which the optimal stochastic controller is in the form of some deterministic function of the conditional mean state estimate, though not necessarily the same function of state as found in the associated deterministic optimal control problem [11, 22, 29, 34, 40, 43, 44]. In general, without the LQG assumptions, such a separation and/or certainty equivalence does *not* occur in the optimal stochastic controller, which usually cannot be generated in a convenient form at all (other than descriptively as the solution to the dynamic programming optimization) without imposing approximating assumptions.

Thus, *the optimal stochastic controller for a problem described by linear system models driven by white Gaussian noise, subject to a quadratic cost criterion, consists of an optimal linear Kalman filter cascaded with the optimal feedback gain matrix of the corresponding deterministic optimal control problem.* The importance of this is the *synthesis* capability it yields. Under the LQG assumptions, the design of the optimal stochastic controller can be *completely separated* into the design of the appropriate Kalman filter and the design of an optimal deterministic controller associated with the original problem. The feedback control gain matrix is *independent* of all uncertainty, so a controller can be designed assuming that $\mathbf{x}(t_i)$ is known perfectly for all time (including $t_0$) and that $\mathbf{w}_d(t_i) \equiv \mathbf{0}$ and $\mathbf{v}(t_i) \equiv \mathbf{0}$ for all time. Similarly, the filter is *independent* of the matrices that define the controller performance measure, i.e., cost criterion, so the filter can be designed ignoring the fact that a control problem is under

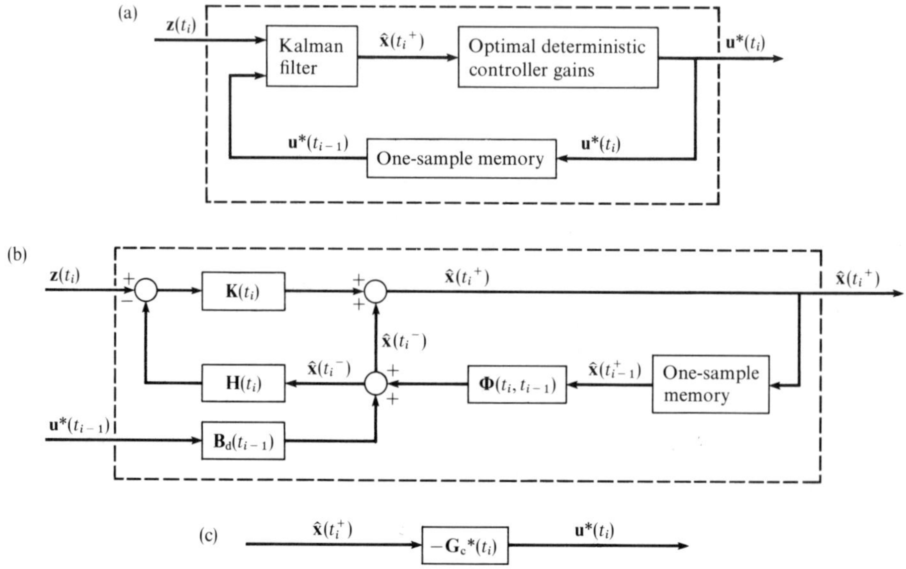


FIG. 13.4 Structure of optimal stochastic controller for the LQG problem. (a) Overall structure of LQG controller. (b) Kalman filter block in (a). (c) Optimal deterministic controller gains block in (a).

consideration. The separation of the design procedure into these two independent components is a substantial simplification from the general case in which the overall stochastic optimal controller must be generated at once, which is generally an unsolvable design problem. The existence of such a *systematic* design synthesis capability, as enhanced by readily available software tools, makes the LQG problem formulation attractive. Efforts are often made, as through linearized perturbation equations, to fit the LQG modeling assumptions in order to exploit this capability.

Figure 13.4 diagrams the structure of the optimal stochastic controller for the LQG problem. As seen in part (a), it is an algorithm that accepts sampled-data measurements $\mathbf{z}(t_i, \omega_k) = \mathbf{z}_i$ as inputs, and outputs the optimal control $\mathbf{u}^*(t_i)$ to apply as constant values to the system over the next sample period, from $t_i$ to $t_{i+1}$. As shown in part (b), the Kalman filter is an algorithm that accepts knowledge of previously applied control $\mathbf{u}^*(t_{i-1})$ and current measurement $\mathbf{z}_i$ and outputs the optimal state estimate $\hat{\mathbf{x}}(t_i^+)$. Finally, as shown in part (c), the optimal deterministic controller gains block simply entails premultiplication of $\hat{\mathbf{x}}(t_i^+)$ by $[-\mathbf{G}_c^*(t_i)]$.

Algorithmically, one can think of three consecutive steps: (1) state estimate propagation from one sample time to the next,

$$\hat{\mathbf{x}}(t_i^-) = \mathbf{\Phi}(t_i, t_{i-1})\hat{\mathbf{x}}(t_{i-1}^+) + \mathbf{B}_d(t_{i-1})\mathbf{u}^*(t_{i-1}) \tag{13-26}$$

(2) measurement update of the state estimate,

$$\hat{\mathbf{x}}(t_i^+) = \hat{\mathbf{x}}(t_i^-) + \mathbf{K}(t_i)[\mathbf{z}_i - \mathbf{H}(t_i)\hat{\mathbf{x}}(t_i^-)] \tag{13-27}$$

and (3) control generation

$$\mathbf{u}^*(t_i) = -\mathbf{G}_c^*(t_i)\hat{\mathbf{x}}(t_i^+) \tag{13-28}$$

In the Kalman filter update, the gain $\mathbf{K}(t_i)$ can be precomputed along with the state error covariance via the forward recursion

$$\mathbf{P}(t_i^-) = \mathbf{\Phi}(t_i, t_{i-1})\mathbf{P}(t_{i-1}^+)\mathbf{\Phi}^{\mathrm{T}}(t_i, t_{i-1}) + \mathbf{G}_{\mathrm{d}}(t_{i-1})\mathbf{Q}_{\mathrm{d}}(t_{i-1})\mathbf{G}_{\mathrm{d}}^{\mathrm{T}}(t_{i-1}) \tag{13-29}$$

$$\mathbf{K}(t_i) = \mathbf{P}(t_i^-)\mathbf{H}^{\mathrm{T}}(t_i)[\mathbf{H}(t_i)\mathbf{P}(t_i^-)\mathbf{H}^{\mathrm{T}}(t_i) + \mathbf{R}(t_i)]^{-1} \tag{13-30}$$

$$\mathbf{P}(t_i^+) = \mathbf{P}(t_i^-) - \mathbf{K}(t_i)\mathbf{H}(t_i)\mathbf{P}(t_i^-) \tag{13-31}$$

starting from the initial condition $\mathbf{P}_0$ at $t_0$, or some factored covariance form equivalent to this (as discussed in Chapter 7 of Volume 1). Similarly, $\mathbf{G}_c^*(t_i)$ can be precomputed and stored (perhaps via curve-fitting or other approximation) using (13-18), (13-19), and (13-23). In actual implementation, the computations involved in the LQG controller are typically rearranged and expressed in a different but equivalent manner, so as to minimize the destabilizing delay time between inputting $\mathbf{z}_i$ and outputting $\mathbf{u}^*(t_i)$, with "setup" computations accomplished in the remainder of the sample period before the next measurement arrives. Further discussion of computationally efficient forms is deferred to later chapters.

To relate this LQG controller to the general structure of Fig. 13.3 of the previous section, Fig. 13.4 can be recast in the form given in Fig. 13.5. As shown, the LQG controller composes the feedback compensation, and the cascade compensation is simply an identity mapping, i.e., direct feedthrough. The output of an as-yet-unspecified feedforward linear compensation is an $r$-dimensional $\mathbf{0}$, so that $\mathbf{u}^*(t_i)$ is generated at the summing junction as $[\mathbf{0} - \mathbf{G}_c^*(t_i)\hat{\mathbf{x}}(t_i^+)]$. This form of controller is also known as a *regulator*, since it

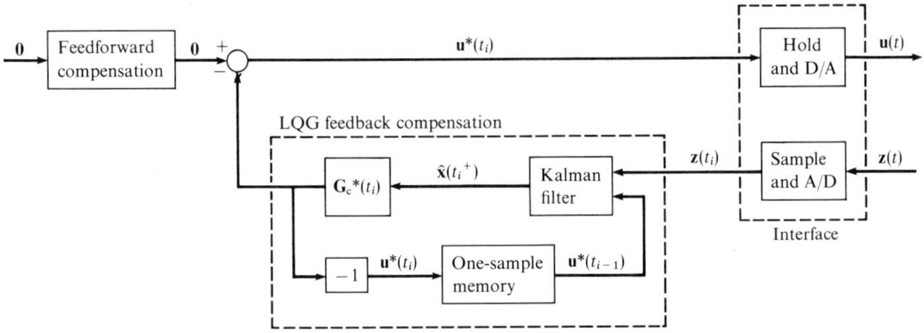


FIG. 13.5 Sampled-data LQG state regulator.

attempts to regulate the state to zero. Nonzero setpoint and dynamic variable tracking controllers, involving nontrivial feedforward compensation outputs, will be discussed in detail in subsequent chapters.

## 13.4 THE BACKWARD KOLMOGOROV EQUATION

In nonlinear sampled-data filtering theory, the basic means of propagating the transition (conditional) probability density (13-16) for the Markov solution $\mathbf{x}(\cdot,\cdot)$ to an Itô stochastic differential equation forward in time was the forward Kolmogorov equation (11-96). This expression for $\partial f_{\mathbf{x}}(\xi, t \mid \boldsymbol{\rho}, t')/\partial t$ allowed the evaluation of $f_{\mathbf{x}}(\xi, t \mid \boldsymbol{\rho}, t')$ for $t'$ a fixed point in time and $t$ growing larger. Similarly, a partial of $f_{\mathbf{x}}(\xi, t \mid \boldsymbol{\rho}, t')$ with respect to the time $t'$ can be developed. As motivated by the discussion of Eq. (13-16), this backward propagation of the density, for fixed time $t$ of current interest as the time $t'$ used for conditioning moves backwards from $t$, occurs naturally in the dynamic programming solution to the optimal stochastic controller problem. The desired result, known as the *backward Kolmogorov equation* for the system described by the nonlinear Itô stochastic differential equation

$$\mathbf{dx}(t) = \mathbf{f}[\mathbf{x}(t), t]\,dt + \mathbf{G}[\mathbf{x}(t), t]\,\mathbf{d\beta}(t) \tag{13-32}$$

is given by [1, 18, 26, 33]

$$\begin{aligned}\frac{\partial f_{\mathbf{x}}(\xi, t \mid \boldsymbol{\rho}, t')}{\partial t'} = &-\sum_{i=1}^{n} f_i[\boldsymbol{\rho}, t'] \frac{\partial}{\partial \rho_i} f_{\mathbf{x}}(\xi, t \mid \boldsymbol{\rho}, t') \\ &-\frac{1}{2}\sum_{i=1}^{n}\sum_{j=1}^{n} \{\mathbf{G}[\boldsymbol{\rho}, t']\mathbf{Q}(t')\mathbf{G}^{\mathrm{T}}[\boldsymbol{\rho}, t']\}_{ij} \frac{\partial^2}{\partial \rho_i \partial \rho_j} f_{\mathbf{x}}(\xi, t \mid \boldsymbol{\rho}, t')\end{aligned} \tag{13-33}$$

This can be proven in a manner similar to that used for the forward Kolmogorov equation. Note the structural difference between (13-33) and (11-96): the partials on the right hand side of (13-33) do not involve $\mathbf{f}$ and $\mathbf{G}$. Because of this structure, (13-33) can also be written conveniently as

$$\begin{aligned}\frac{\partial f_{\mathbf{x}}(\xi, t \mid \boldsymbol{\rho}, t')}{\partial t'} = &-\frac{\partial f_{\mathbf{x}}(\xi, t \mid \boldsymbol{\rho}, t')}{\partial \boldsymbol{\rho}} \mathbf{f}[\boldsymbol{\rho}, t'] \\ &-\frac{1}{2}\operatorname{tr}\left[\{\mathbf{G}[\boldsymbol{\rho}, t']\mathbf{Q}(t')\mathbf{G}^{\mathrm{T}}[\boldsymbol{\rho}, t']\} \frac{\partial^2 f_{\mathbf{x}}(\xi, t \mid \boldsymbol{\rho}, t')}{\partial \boldsymbol{\rho}^2}\right]\end{aligned} \tag{13-34}$$

where, by convention, $\partial f_{\mathbf{x}}/\partial \boldsymbol{\rho}$ is a *row* vector,

$$\partial f_{\mathbf{x}}/\partial \boldsymbol{\rho} \triangleq [(\partial f_{\mathbf{x}}/\partial \rho_1)(\partial f_{\mathbf{x}}/\partial \rho_2) \cdots (\partial f_{\mathbf{x}}/\partial \rho_n)]$$

and $\partial^2 f_{\mathbf{x}}/\partial \boldsymbol{\rho}^2$ is an $n$-by-$n$ symmetric matrix whose $ij$ component is $(\partial^2 f_{\mathbf{x}}/\partial \rho_i \partial \rho_j)$.

To understand the application of the backward Kolmogorov equation, suppose that we have a system described by (13-32) and that we are interested

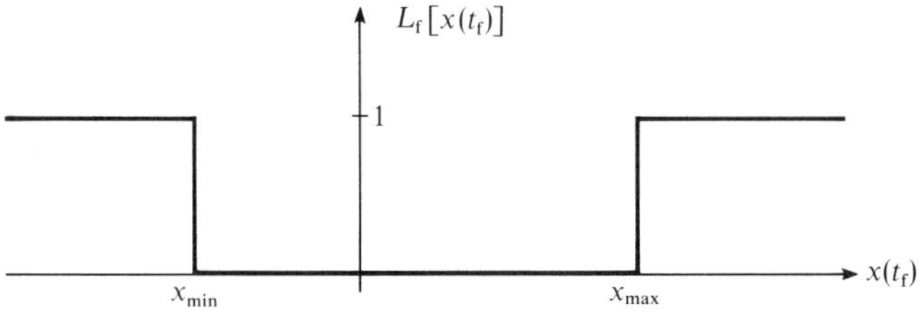


FIG. 13.6 One potential choice of penalty function $L_f$.

in the state of the system at some selected final time $t_f$, such as for the air-to-air missile intercept problem mentioned previously. In particular, we will weigh the final state by a scalar penalty or loss function, $L_f[\mathbf{x}(t_f)]$, to express our desires about possible values of $\mathbf{x}(t_f)$. If we wanted a scalar terminal state to be inside some region bounded by $x_{min}$ and $x_{max}$, we might choose a penalty function as depicted in Fig. 13.6. Or, if we wanted the components of a vector $\mathbf{x}(t_f)$ to be as close to zero as possible, we might choose the quadratic penalty function or final-time loss function

$$L_f[\mathbf{x}(t_f)] = \mathbf{x}(t_f)^T\mathbf{x}(t_f) \tag{13-35}$$

To be assured that $L_f[\mathbf{x}(t_f, \cdot)]$ is in fact a scalar *random variable*, we will assume $L_f$ to be a Baire function of its argument, i.e., a continuous function or a limit of continuous functions (although the function in Fig. 13.6 is not continuous, it *can* be generated as the limit of functions with the four corners rounded off, in the limit as the radii of curvature go to zero). Then it makes sense to define a cost criterion as the expected value of that scalar random variable,

$$\mathscr{C}[\mathbf{x}_0, t_0] = E\{L_f[\mathbf{x}(t_f)] \,|\, \mathbf{x}(t_0) = \mathbf{x}_0\} \tag{13-36}$$

In other words, the cost associated with being at some particular state $\mathbf{x}_0$ at time $t_0$ is the conditional expectation of the value of the penalty $L_f[\mathbf{x}(t_f)]$, conditioned on the fact that $\mathbf{x}(t_0)$ has taken on the realized value of $\mathbf{x}_0$. We can also consider $\mathbf{x}(t_0)$ as random, to generate

$$\begin{aligned}\mathscr{C}[\mathbf{x}(t_0), t_0] &= E\{L_f[\mathbf{x}(t_f)] \,|\, \mathbf{x}(t_0) = \mathbf{x}(t_0)\}\\ &= \int_{-\infty}^{\infty} \cdots \int_{-\infty}^{\infty} L_f[\boldsymbol{\xi}] f_{\mathbf{x}}(\boldsymbol{\xi}, t_f \,|\, \mathbf{x}(t_0), t_0)\, d\xi_1 \cdots d\xi_n \end{aligned}\tag{13-37}$$

and possibly the unconditionally expected cost, averaged over all possible initial values,

$$E\{\mathscr{C}[\mathbf{x}(t_0), t_0]\} = E_{\mathbf{x}(t_0)}\{E\{L_f[\mathbf{x}(t_f)] \,|\, \mathbf{x}(t_0) = \mathbf{x}(t_0)\}\} \tag{13-38}$$

Now we can generalize the concept of the cost involved in being at a particular state at initial time, $t_0$, to the cost of being at some state at any time $t$

in the interval from $t_0$ to $t_f$:

$$\begin{aligned}\mathscr{C}[\mathbf{x}(t), t] &= E\{L_f[\mathbf{x}(t_f)] \,|\, \mathbf{x}(t) = \mathbf{x}(t)\} \\ &= \int_{-\infty}^{\infty} \cdots \int_{-\infty}^{\infty} L_f[\boldsymbol{\xi}] f_{\mathbf{x}}(\boldsymbol{\xi}, t_f \,|\, \mathbf{x}(t), t)\, d\xi_1 \cdots d\xi_n \end{aligned} \tag{13-39}$$

and similarly for $\mathscr{C}[\mathbf{x}(t), t]$. We note that, in the limit as $t \to t_f$, $f_{\mathbf{x}}(\boldsymbol{\xi}, t_f \,|\, \mathbf{x}(t), t)$ becomes a Dirac delta function, so that the cost satisfies the terminal condition

$$\mathscr{C}[\mathbf{x}(t_f), t_f] = E\{L_f[\mathbf{x}(t_f)] \,|\, \mathbf{x}(t_f) = \mathbf{x}(t_f)\} = L_f[\mathbf{x}(t_f)] \tag{13-40}$$

and similarly for $\mathscr{C}[\mathbf{x}(t_f), t_f]$. The validity of (13-40) is also established fundamentally by the fact that $L_f$ is a *function* of $\mathbf{x}(t_f)$.

In view of (13-36)–(13-40), consider multiplying the backward Kolmogorov equation term-by-term by $L_f[\boldsymbol{\xi}]$ and integrating over $\boldsymbol{\xi}$ (treating partials as limits of defining quotients, to be precise). This yields

$$\begin{aligned}\frac{\partial \mathscr{C}[\boldsymbol{\rho}, t']}{\partial t'} &= -\frac{\partial \mathscr{C}[\boldsymbol{\rho}, t']}{\partial \boldsymbol{\rho}} \mathbf{f}(\boldsymbol{\rho}, t') \\ &\quad -\frac{1}{2} \operatorname{tr}\left[ \{\mathbf{G}[\boldsymbol{\rho}, t']\mathbf{Q}(t')\mathbf{G}^{\mathrm{T}}[\boldsymbol{\rho}, t']\} \frac{\partial^2 \mathscr{C}[\boldsymbol{\rho}, t']}{\partial \boldsymbol{\rho}^2} \right]\end{aligned} \tag{13-41}$$

which is then an equation to propagate the cost function backward in time, from the terminal condition

$$\lim_{t' \to t_f} \mathscr{C}[\boldsymbol{\rho}, t'] = L_f[\boldsymbol{\rho}] \tag{13-42}$$

Having chosen an appropriate $L_f$ function, we can use Eqs. (13-41) and (13-42) to evaluate the function $\mathscr{C}[\boldsymbol{\rho}, t']$ for all time $t'$ of interest, as depicted in Fig. 13.7 for a scalar state problem.

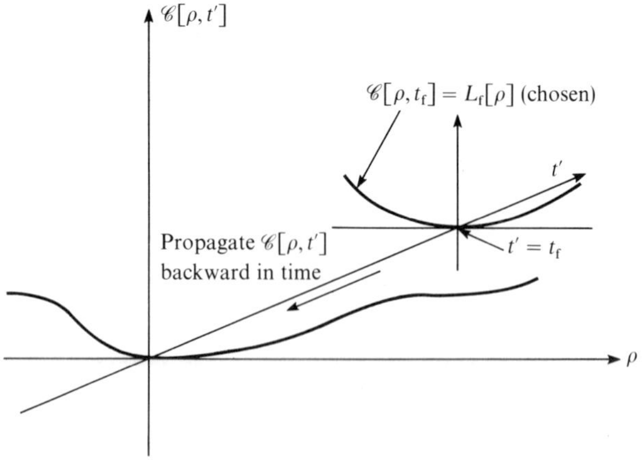


FIG. 13.7 Propagating $\mathscr{C}[\rho, t']$ backward in time from $\mathscr{C}[\rho, t_f]$.

EXAMPLE 13.6 Consider a system described by a linear scalar state model

$$\dot{\mathsf{x}}(t) = F(t)\mathsf{x}(t) + G(t)\mathsf{w}(t)$$

where $\mathsf{w}(\cdot,\cdot)$ is zero-mean white Gaussian noise of strength $Q(t): E\{\mathsf{w}(t)\mathsf{w}(t+\tau)\} = Q(t)\,\delta(\tau)$. This can also be written more rigorously as

$$d\mathsf{x}(t) = F(t)\mathsf{x}(t)\,dt + G(t)\,d\beta(t)$$

where $\beta(\cdot,\cdot)$ is a Brownian motion of diffusion $Q(t)$, or

$$d\mathsf{x}(t) = F(t)\mathsf{x}(t)\,dt + G(t)\sqrt{Q(t)}\,d\beta'(t)$$

with unit-diffusion Brownian motion $\beta'(\cdot,\cdot)$. Let the terminal penalty be quadratic,

$$L_f[\mathsf{x}(t_f)] = \tfrac{1}{2}X_f\mathsf{x}(t_f)^2$$

For this problem, the backward equation for $\mathscr{C}$, derived from the backward Kolmogorov equation, is (13-41),

$$\frac{\partial\mathscr{C}[\rho,t']}{\partial t'} = -\frac{\partial\mathscr{C}[\rho,t']}{\partial\rho}F(t')\rho - \frac{1}{2}G^2(t')Q(t')\frac{\partial^2\mathscr{C}[\rho,t']}{\partial\rho^2}$$

and its terminal condition is (13-42),

$$\mathscr{C}[\rho,t_f] = L_f[\rho] = \tfrac{1}{2}X_f\rho^2$$

In this special case, we can solve this partial differential equation. With considerable hindsight, we assume a solution of the form

$$\mathscr{C}[\rho,t'] = \tfrac{1}{2}[K_c(t')\rho^2 + g(t')]$$

If this assumed form is correct, then

$$\frac{\partial\mathscr{C}[\rho,t']}{\partial t'} = \frac{1}{2}\left[\frac{\partial K_c(t')}{\partial t'}\rho^2 + \frac{\partial g(t')}{\partial t'}\right]$$

$$\frac{\partial\mathscr{C}[\rho,t']}{\partial\rho} = K_c(t')\rho, \qquad \frac{\partial^2\mathscr{C}[\rho,t']}{\partial\rho^2} = K_c(t')$$

Substituting these back into the backward equation for $\mathscr{C}[\rho,t']$ yields

$$\frac{1}{2}\frac{\partial K_c(t')}{\partial t'}\rho^2 + \frac{\partial g(t')}{\partial t'} = -K_c(t')F(t')\rho - \frac{1}{2}G^2(t')Q(t')K_c(t')$$

with the terminal conditions

$$K_c(t_f) = X_f \qquad g(t_f) = 0$$

since $\mathscr{C}[\rho,t_f] = \frac{1}{2}X_f\rho^2$. Now, since this must be true for all $\rho$, we can separately equate the coefficients of $\rho^2$ and $\rho^0$, to obtain

$$\frac{dK_c(t')}{dt'} = -2F(t')K_c(t'); \qquad K_c(t_f) = X_f$$

$$\frac{dg(t')}{dt'} = -G^2(t')Q(t')K_c(t'); \qquad g(t_f) = 0$$

These simple equations can be used to propagate $K_c(t')$ and $g(t')$ backward in time. Once these are evaluated for all $t' \in [t_0, t_f]$, $\mathscr{C}[x(t),t]$ can be evaluated for any time in that interval.

Note that the state variance $P(t)$ satisfies the forward equation $dP/dt = 2FP + G^2Q$ for this problem, whereas $K_c$ satisfies the backward equation $dK_c/dt' = -2FK_c$. These are *adjoint* differential equations and will be discussed further in later sections. Note also that if the homogeneous $dP/dt = 2FP$ yields decreasing $P(t)$, i.e., if the system is stable, $dK_c/dt' = -2FK_c$ assures us of a cost that grows as we go backward in time. ■

## 13.5 OPTIMAL STOCHASTIC CONTROL WITH PERFECT KNOWLEDGE OF THE STATE

This section considers the dynamic programming solution for optimal control of systems described by stochastic models, assuming that perfect knowledge of the entire state is available at discrete sample times [1, 3, 7–9, 12, 17, 18, 24, 25, 28, 29, 31, 37]. For many practical applications, this is an unrealistic simplifying assumption, but the solution to this simpler problem will provide beneficial insights into the more realistic problem in which only incomplete, noise-corrupted measurements are available. This problem will be discussed in the next section.

Because of the practical consideration of eventually implementing the controller with a digital computer, the sampled-data case is developed. First consider the case in which the system of interest can be adequately modeled by the discrete-time equation

$$\mathbf{x}(t_{i+1}) = \boldsymbol{\phi}_{\mathrm{h}}[\mathbf{x}(t_i), t_i] + \mathbf{G}_{\mathrm{d}}[\mathbf{x}(t_i), t_i]\mathbf{w}_{\mathrm{d}}(t_i) \tag{13-43}$$

where $\mathbf{x}(\cdot,\cdot)$ is an $n$-dimensional state process, $\boldsymbol{\phi}_{\mathrm{h}}$ is the $n$-dimensional discrete-time homogeneous system dynamics vector, $\mathbf{G}_{\mathrm{d}}$ is the $n$-by-$s$ dynamics noise input matrix, and $\mathbf{w}_{\mathrm{d}}(\cdot,\cdot)$ is $s$-dimensional zero-mean white Gaussian discrete-time noise of covariance $\mathbf{Q}_{\mathrm{d}}(t_i)$ for all $t_i$:

$$E\{\mathbf{w}_{\mathrm{d}}(t_i)\mathbf{w}_{\mathrm{d}}^{\mathrm{T}}(t_j)\} = \begin{cases} \mathbf{Q}_{\mathrm{d}}(t_i) & i = j \\ \mathbf{0} & i \neq j \end{cases} \tag{13-44}$$

Such a model might be obtained as a discretized model for a continuous-time system described by an Itô stochastic differential equation, such as (13-32). Note that there are no requirements of continuity or the like on the $\boldsymbol{\phi}_{\mathrm{h}}$ and $\mathbf{G}_{\mathrm{d}}$ functions in (13-43). Consider the conditional probability density for $\mathbf{x}(t_{i+1})$, given that $\mathbf{x}(t_i, \omega_k) = \boldsymbol{\rho}$:

$$f_{\mathbf{x}(t_{i+1})|\mathbf{x}(t_i)}(\boldsymbol{\xi}\,|\,\boldsymbol{\rho}) = f_{\mathbf{x}}(\boldsymbol{\xi}, t_{i+1}\,|\,\boldsymbol{\rho}, t_i)$$

$$= \frac{1}{(2\pi)^{n/2}\left|\mathbf{G}_{\mathrm{d}}(\boldsymbol{\rho}, t_i)\mathbf{Q}_{\mathrm{d}}(t_i)\mathbf{G}_{\mathrm{d}}^{\mathrm{T}}(\boldsymbol{\rho}, t_i)\right|^{1/2}} \exp\{\cdot\} \tag{13-45a}$$

$$\{\cdot\} = \{-\tfrac{1}{2}[\boldsymbol{\xi} - \boldsymbol{\phi}_{\mathrm{h}}(\boldsymbol{\rho}, t_i)]^{\mathrm{T}}[\mathbf{G}_{\mathrm{d}}(\boldsymbol{\rho}, t_i)\mathbf{Q}_{\mathrm{d}}(t_i)\mathbf{G}_{\mathrm{d}}^{\mathrm{T}}(\boldsymbol{\rho}, t_i)]^{-1}[\boldsymbol{\xi} - \boldsymbol{\phi}_{\mathrm{h}}(\boldsymbol{\rho}, t_i)]\} \tag{13-45b}$$

This is a Gaussian density, and thus (13-43) is *not* the most general solution form to an Itô stochastic differential equation from one sample time to the next, but a special case. The existence of the inverse of $[\mathbf{G}_d(\boldsymbol{\rho}, t_i)\mathbf{Q}_d(t_i)\mathbf{G}_d^T(\boldsymbol{\rho}, t_i)]$ in (13-45) depends upon the system model being completely controllable from the points of entry of $\mathbf{w}_d(\cdot,\cdot)$, i.e., upon the dynamics noise affecting all states of the system in the interval between $t_i$ and $t_{i+1}$: this requires the dimension of the noise vector $\mathbf{w}_d(\cdot,\cdot)$ in the discrete-time model to be at least $n$ ($s \geq n$). In problems where this condition is not met, a reformulation in terms of characteristic functions can be developed, but since we view (13-43) as a *discretized* model, it is not overly restrictive to assume that the inverse exists. Thus, we have a Gaussian density centered at $\boldsymbol{\phi}_h(\boldsymbol{\rho}, t_i)$ and having covariance $[\mathbf{G}_d(\boldsymbol{\rho}, t_i)\mathbf{Q}_d(t_i)\mathbf{G}_d^T(\boldsymbol{\rho}, t_i)]$. Moreover, since $\mathbf{w}_d(\cdot,\cdot)$ is white, and thus not dependent upon the past, we can write

$$f_{\mathbf{x}(t_{i+1})|\mathbf{x}(t_i), \mathbf{x}(t_{i-1}), \ldots, \mathbf{x}(t_0)} = f_{\mathbf{x}(t_{i+1})|\mathbf{x}(t_i)} \tag{13-46}$$

and thus the *discrete-time* $\mathbf{x}(\cdot,\cdot)$ process is *Markov*.

Now consider addition of a feedback control as an $r$-dimensional function of the current state and time, $\mathbf{u}[\mathbf{x}(t_i), t_i]$, to be used to control the system between time $t_i$ and $t_{i+1}$, i.e., a sampled-data control held constant over that interval. Thus, we alter the system dynamics description to

$$\mathbf{x}(t_{i+1}) = \boldsymbol{\phi}[\mathbf{x}(t_i), \mathbf{u}\{\mathbf{x}(t_i), t_i\}, t_i] + \mathbf{G}_d[\mathbf{x}(t_i), t_i]\mathbf{w}_d(t_i) \tag{13-47}$$

Once again we obtain a *Markov discrete-time* $\mathbf{x}(\cdot,\cdot)$ process, because we could just define

$$\boldsymbol{\phi}_h[\mathbf{x}(t_i), t_i] = \boldsymbol{\phi}[\mathbf{x}(t_i), \mathbf{u}\{\mathbf{x}(t_i), t_i\}, t_i] \tag{13-48}$$

We can now write the transition probability density for $\mathbf{x}(t_{i+1})$ as

$$f_{\mathbf{x}}(\boldsymbol{\xi}, t_{i+1} | \boldsymbol{\rho}, t_i) = \frac{1}{(2\pi)^{n/2}|\mathbf{G}_d(\boldsymbol{\rho}, t_i)\mathbf{Q}_d(t_i)\mathbf{G}_d^T(\boldsymbol{\rho}, t_i)|^{1/2}} \exp\{\cdot\} \tag{13-49a}$$

$$\{\cdot\} = \{-\tfrac{1}{2}(\boldsymbol{\xi} - \boldsymbol{\phi}[\boldsymbol{\rho}, \mathbf{u}(\boldsymbol{\rho}, t_i), t_i])^T[\mathbf{G}_d(\boldsymbol{\rho}, t_i)\mathbf{Q}_d(t_i)\mathbf{G}_d^T(\boldsymbol{\rho}, t_i)]^{-1}(\boldsymbol{\xi} - \boldsymbol{\phi}[\boldsymbol{\rho}, \mathbf{u}(\boldsymbol{\rho}, t_i), t_i])\} \tag{13-49b}$$

From this expression, we can see that the only effect of the control is to be able to move the conditional *mean*, while the conditional covariance is unchanged.

Now assume that (13-43) and (13-47) are replaced by the more general solution to Itô stochastic differential equations (13-32) and (13-1), respectively, with $\mathbf{u}(t)$ defined by

$$\mathbf{u}(t) \triangleq \mathbf{u}[\mathbf{x}(t_i), t_i] \qquad \text{for all } t \in [t_i, t_{i+1}) \tag{13-50}$$

In practice, online numerical integration would yield approximate solutions. The transition probability densities would no longer be Gaussian in general, nor would the effect of the control generally be confined to only the first moment, but the *discrete-time* $\mathbf{x}(\cdot,\cdot)$ process corresponding to the continuous $\mathbf{x}(\cdot,\cdot)$ considered only at the sample times would *still be Markov*, and, for instance, $f_{\mathbf{x}(t)|\mathbf{x}(t_i)}(\xi\,|\,\boldsymbol{\rho})$ would still satisfy the forward Kolmogorov equation for all $t \in [t_i, t_{i+1})$. However, that does not say that the continuous $\mathbf{x}(\cdot,\cdot)$ is Markov: for $t_i < t' < t \leq t_{i+1}$, $f_{\mathbf{x}(t)|\mathbf{x}(t'),\,\mathbf{x}(t_i)}(\xi\,|\,\xi', \boldsymbol{\rho})$ does not equal $f_{\mathbf{x}(t)|\mathbf{x}(t')}(\xi\,|\,\xi')$ since knowledge of $\boldsymbol{\rho}$ is required to evaluate $\mathbf{u}(\boldsymbol{\rho}, t_i)$.

To compose a meaningful discrete-time control problem, define the scalar cost function:

$$\mathscr{C}[\mathbf{x}(t_i), t_i] = E\left\{\sum_{j=i}^{N} L[\mathbf{x}(t_j), \mathbf{u}(t_j), t_j] + L_{\mathrm{f}}[\mathbf{x}(t_{N+1})]\,\middle|\,\mathbf{x}(t_i) = \mathbf{x}(t_i)\right\} \tag{13-51}$$

where $L$ is an explicit scalar (Baire) loss function of the state, control, and time; $L_{\mathrm{f}}$ is an explicit scalar (Baire) final loss function of the terminal state; $t_{N+1}$ is the final time; and $t_N$ is the last sample time that the control is applied and held constant over a sample period, as shown in Fig. 13.8. By assuming $L$ and $L_{\mathrm{f}}$ to be Baire functions, we are assured of (13-51) being the conditional expectation of a scalar *random variable*. Moreover, $\mathscr{C}[\mathbf{x}(t_i, \cdot), t_i]$ would then also be a scalar random variable. Physically, $\mathscr{C}[\mathbf{x}(t_i), t_i]$ is the expected cost-to-go to the terminal time $t_{N+1}$, conditioned on the knowledge that we are at state $\mathbf{x}(t_i, \omega_k) = \mathbf{x}(t_i)$ at time $t_i$, where the individual $L$ terms provide the "running cost" from one sample time to the next and $L_{\mathrm{f}}$ provides the cost at the final time. Note that $\mathbf{u}(t_j)$ in (13-51) is the control to be applied at time $t_j$, and this is still a *particular function of the current state and time*,

$$\mathbf{u}(t_j) \triangleq \mathbf{u}[\mathbf{x}(t_j), t_j] \tag{13-52}$$

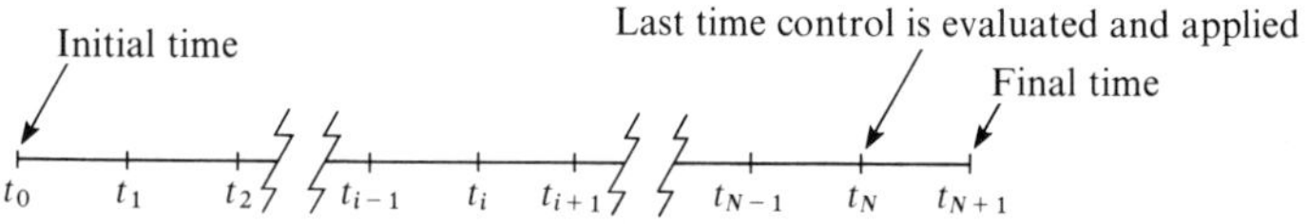


FIG. 13.8 Times of interest in control problem.

We are interested in minimizing the cost function (13-51) by choice of an optimal control function $\mathbf{u}^*[\cdot,\cdot]$ out of some admissible class of functions $\mathbf{u}[\cdot,\cdot]$, denoted as $\mathscr{U}$. Then $\mathscr{U}$ is a *class of functions* $\mathbf{u}[\cdot,\cdot]$ of $(R^n \times T)$ into $R^r$ defined to be admissible: $\mathscr{U}$ might be the set of all such control functions confined to a hypersphere defined by some maximum allowable norm, or any other sort of constrained set, or there may be no constraint at all. Conceptually, it will be useful to think of *fixing the functional form* $\mathbf{u}[\cdot,\cdot]$ and solving for the cost due to using *any* such admissible control function; solving for the cost due to each and every $\mathbf{u}[\cdot,\cdot] \in \mathscr{U}$ and then choosing the *one function* $\mathbf{u}^*[\cdot,\cdot]$ that yields the lowest cost comprises the optimization problem. In this optimization, we shall be considering how the overall cost propagates backward in time, as described by the backward Kolmogorov equation of the previous section.

First consider the cost function at the terminal time $t_{N+1}$:

$$\begin{aligned}\mathscr{C}[\mathbf{x}(t_{N+1}), t_{N+1}] &= E\{L_f[\mathbf{x}(t_{N+1})] \,|\, \mathbf{x}(t_{N+1}) = \mathbf{x}(t_{N+1}, \cdot)\} \\ &= L_f[\mathbf{x}(t_{N+1})] \end{aligned} \tag{13-53}$$

as in (13-40), since $L_f$ is a function of $\mathbf{x}(t_{N+1})$ itself. We can arbitrarily denote this as $\mathscr{C}^*[\mathbf{x}(t_{N+1}), t_{N+1}]$, where the superscript $*$ denotes optimal, or minimum achievable cost with an admissible control, since it is *independent* of what control might be used to attain the final state $\mathbf{x}(t_{N+1})$. Although this is a trivial and meaningless notational manipulation at this point in time, it foretells a structure that will become more meaningful as we work backward in time.

Now take a step backward to time $t_N$, to formulate the optimal control for the last stage, from $t_N$ to $t_{N+1}$. Consider the cost associated with being at $\mathbf{x}(t_N)$ at time $t_N$:

$$\begin{aligned}\mathscr{C}[\mathbf{x}(t_N), t_N] &\triangleq E\{L[\mathbf{x}(t_N), \mathbf{u}(t_N), t_N] + L_f[\mathbf{x}(t_{N+1})] \,|\, \mathbf{x}(t_N) = \mathbf{x}(t_N)\} \\ &= L[\mathbf{x}(t_N), \mathbf{u}(t_N), t_N] + E\{L_f[\mathbf{x}(t_{N+1})] \,|\, \mathbf{x}(t_N) = \mathbf{x}(t_N)\} \end{aligned} \tag{13-54}$$

since $L[\mathbf{x}(t_N), \mathbf{u}[\mathbf{x}(t_N), t_N], t_N]$ is completely determined by knowledge of $\mathbf{x}(t_N, \omega_k) = \mathbf{x}(t_N)$ for *any* chosen fixed $\mathbf{u} \in \mathscr{U}$ that is a *function* of $\mathbf{x}(t_N)$ and $t_N$. By the same reasoning,

$$\mathscr{C}[\mathbf{x}(t_N), t_N] = L[\mathbf{x}(t_N), \mathbf{u}(t_N), t_N] + E\{L_f[\mathbf{x}(t_{N+1})] \,|\, \mathbf{x}(t_N) = \mathbf{x}(t_N, \cdot)\} \tag{13-55}$$

The cost (13-54) can be written in terms of the state transition probability density $f_{\mathbf{x}}(\boldsymbol{\xi}, t_{N+1} \,|\, \mathbf{x}(t_N), t_N)$ as

$$\begin{aligned}\mathscr{C}[\mathbf{x}(t_N), t_N] &= L[\mathbf{x}(t_N), \mathbf{u}(t_N), t_N] \\ &\quad + \int_{-\infty}^{\infty} \cdots \int_{-\infty}^{\infty} L_f(\boldsymbol{\xi}) f_{\mathbf{x}}(\boldsymbol{\xi}, t_{N+1} \,|\, \mathbf{x}(t_N), t_N) \, d\xi_1 \cdots d\xi_n \end{aligned} \tag{13-56}$$

By (13-53), $L_f(\boldsymbol{\xi}) = \mathscr{C}[\boldsymbol{\xi}, t_{N+1}]$, and from the discussion following (13-53), we can also set this equal to $\mathscr{C}^*[\boldsymbol{\xi}, t_{N+1}]$, trivially, to yield

$$\begin{aligned}\mathscr{C}[\mathbf{x}(t_N), t_N] &= L[\mathbf{x}(t_N), \mathbf{u}(t_N), t_N] \\ &\quad + \int_{-\infty}^{\infty} \cdots \int_{-\infty}^{\infty} \mathscr{C}^*[\boldsymbol{\xi}, t_{N+1}] f_{\mathbf{x}}(\boldsymbol{\xi}, t_{N+1} \,|\, \mathbf{x}(t_N), t_N) \, d\xi_1 \cdots d\xi_n \end{aligned} \tag{13-57}$$

The value of this cost $\mathscr{C}[\mathbf{x}(t_N), t_N]$ is completely determined once we specify the particular control function $\mathbf{u}[\cdot, t_N]$ to be used in the transition from $t_N$ to $t_{N+1}$: $\mathbf{u}[\mathbf{x}(t_N), t_N]$ affects $L$ explicitly in (13-57) and also implicitly affects the density $f_{\mathbf{x}}$ in the integral term. We want to control optimally, and the minimum attainable cost is

$$\mathscr{C}^*[\mathbf{x}(t_N), t_N] = \min_{\mathbf{u}\in\mathscr{U}} \Big\{ L[\mathbf{x}(t_N), \mathbf{u}(t_N), t_N] + \int_{-\infty}^{\infty} \cdots \int_{-\infty}^{\infty} \mathscr{C}^*[\boldsymbol{\xi}, t_{N+1}] f_{\mathbf{x}}(\boldsymbol{\xi}, t_{N+1} | \mathbf{x}(t_N), t_N)\, d\xi_1 \cdots d\xi_n \Big\} \tag{13-58}$$

which is to say that this is the minimum expected cost to complete the process from the state $\mathbf{x}(t_N)$ at time $t_N$, out of the entire range of costs generated by the entire set of admissible control functions $\mathbf{u}$. If the minimum exists, then an optimal one-stage feedback control law function $\mathbf{u}^*[\cdot, t_N]$ is obtained: note again that before the minimization is performed, $\mathbf{u}$ is considered to be some *fixed* function of $\mathbf{x}(t_N)$ and $t_N$ and is treated as a dummy variable, and the minimization then yields $\mathbf{u}^*$ as a function of $\mathbf{x}(t_N)$ and $t_N$ only. If the minimum does not exist [2, 7, 20], one can still seek a greatest lower bound (infimum) on the bracketed term in (13-58) instead of a minimum, but an optimal control could not be found to attain that bounding value—one could then seek an "$\varepsilon$-optimal" control that yields a cost $\mathscr{C}[\mathbf{x}(t_N), t_N]$ within an acceptable $\varepsilon$ of that lower bound. We will assume that the minimum exists in our developments. Thus, we have calculated the optimal *function* $\mathbf{u}^*[\cdot, t_N]$ such that, when we get to time $t_N$ in real time and obtain the *perfectly known* value of $\mathbf{x}(t_N)$, we can evaluate the control $\mathbf{u}^*[\mathbf{x}(t_N), t_N]$ to apply to the system. We need such a function of state at time $t_N$: even if we had applied an "optimal" control at time $t_{N-1}$, we generally would not arrive at the $\mathbf{x}(t_N)$ to which we were attempting to drive because of the dynamic driving noise, and we must then have a control function $\mathbf{u}^*[\cdot, t_N]$ to evaluate the optimal control to apply from *wherever* we arrived at time $t_N$. Having thus obtained the functions $\mathscr{C}^*[\cdot, t_N]$ and $\mathbf{u}^*[\cdot, t_N]$, one can now readily conceive of the scalar random variable $\mathscr{C}^*[\mathbf{x}(t_N, \cdot), t_N]$ and the random vector $\mathbf{u}^*[\mathbf{x}(t_N, \cdot), t_N]$ for subsequent use in the dynamic programming algorithm.

Although the replacement of $\mathscr{C}$ by $\mathscr{C}^*$ in the right hand side of (13-58) accomplishes nothing substantive here, this equation provides insight into the structure of the general dynamic programming step. Imagine taking another step backward in time to time $t_{N-1}$, and consider $\mathscr{C}[\mathbf{x}(t_{N-1}), t_{N-1}]$. We will use the *Markov* property of discrete-time $\mathbf{x}(\cdot, \cdot)$ to show that the same form of $\mathscr{C}^*$ equation as (13-58) can be written down for time $t_{N-1}$ as well, and induction will then lead to its validity for *any* time $t_i$.

From the first step backward, we obtained (13-58), where $\mathbf{u}$ affects both $L$ and $f$. This minimization provided a mapping of $\mathbf{x}(t_N) \in R^n$ to $\mathbf{u}(t_N) \in R^r$: an optimal feedback control law, along with the optimal cost associated with using that particular law. Now step back to time $t_{N-1}$. By definition,

$$\begin{aligned}\mathscr{C}[\mathbf{x}(t_{N-1}), t_{N-1}] = E\{&L[\mathbf{x}(t_{N-1}), \mathbf{u}\{\mathbf{x}(t_{N-1}), t_{N-1}\}, t_{N-1}] \\ &+ L[\mathbf{x}(t_N), \mathbf{u}\{\mathbf{x}(t_N), t_N\}, t_N] \\ &+ L_f[\mathbf{x}(t_{N+1})] | \mathbf{x}(t_{N-1}) = \mathbf{x}(t_{N-1})\}\end{aligned} \tag{13-59}$$

Now assume some fixed function $\mathbf{u}[\cdot, t_{N-1}]$ so that we can completely specify $f_{\mathbf{x}}(\xi, t_N | \boldsymbol{\eta}, t_{N-1})$. Then $\mathscr{C}[\mathbf{x}(t_{N-1}), t_{N-1}]$ can be written as

$$\begin{aligned}\mathscr{C}[\mathbf{x}(t_{N-1}), t_{N-1}] = &L[\mathbf{x}(t_{N-1}), \mathbf{u}\{\mathbf{x}(t_{N-1}), t_{N-1}\}, t_{N-1}] \\ &+ \int_{-\infty}^{\infty} L[\xi, \mathbf{u}\{\xi, t_N\}, t_N] f_{\mathbf{x}}(\xi, t_N | \mathbf{x}(t_{N-1}), t_{N-1})\, d\xi \\ &+ \int_{-\infty}^{\infty} L_f[\rho, t_{N+1}] f_{\mathbf{x}}(\rho, t_{N+1} | \mathbf{x}(t_{N-1}), t_{N-1})\, d\rho\end{aligned} \tag{13-60}$$

where the first term simplification is as explained previously. Because discrete-time $\mathbf{x}(\cdot, \cdot)$ is Markov, the conditional density in the last term of (13-60) can be evaluated using the Chapman–Kolmogorov equation (see Chapter 11, Volume 2),

$$f_{\mathbf{x}}(\rho, t_{N+1} | \mathbf{x}(t_{N-1}), t_{N-1}) = \int_{-\infty}^{\infty} f_{\mathbf{x}}(\rho, t_{N+1} | \xi, t_N) f_{\mathbf{x}}(\xi, t_N | \mathbf{x}(t_{N-1}), t_{N-1})\, d\xi \tag{13-61}$$

Substituting this into (13-60), and assuming that we can change the order of integration, we obtain

$$\begin{aligned}\mathscr{C}[\mathbf{x}(t_{N-1}), t_{N-1}] = L[\mathbf{x}(t_{N-1}), \mathbf{u}\{\mathbf{x}(t_{N-1}), t_{N-1}\}, t_{N-1}] + \int_{-\infty}^{\infty} \Big\{ &L[\xi, \mathbf{u}\{\xi, t_N\}, t_N] \\ + \int_{-\infty}^{\infty} \mathscr{C}[\rho, t_{N+1}] f_{\mathbf{x}}(\rho, t_{N+1} | \xi, t_N)\, d\rho \Big\}& \\ \times f_{\mathbf{x}}(\xi, t_N | \mathbf{x}(t_{N-1}), t_{N-1})\, d\xi&\end{aligned} \tag{13-62}$$

But the term in the $\{\ \}$ braces is just $\mathscr{C}[\xi, t_N]$, so

$$\begin{aligned}\mathscr{C}[\mathbf{x}(t_{N-1}), t_{N-1}] = &L[\mathbf{x}(t_{N-1}), \mathbf{u}\{\mathbf{x}(t_{N-1}), t_{N-1}\}, t_{N-1}] \\ &+ \int_{-\infty}^{\infty} \mathscr{C}[\xi, t_N] f_{\mathbf{x}}(\xi, t_N | \mathbf{x}(t_{N-1}), t_{N-1})\, d\xi\end{aligned} \tag{13-63}$$

Now consider solving for the minimum cost $\mathscr{C}^*[\mathbf{x}(t_{N-1}), t_{N-1}]$ and the control that achieves that cost. For *any* choice of admissible function $\mathbf{u}[\cdot, t_{N-1}]$, both the leading term and the conditional density $f_{\mathbf{x}}(\xi, t_N | \mathbf{x}(t_{N-1}), t_{N-1})$ are *fixed*. Since $f_{\mathbf{x}}$ is nonnegative, we want to choose the *optimal* value of $\mathscr{C}[\xi, t_N]$ in order to minimize the integral, and thus be able to minimize $\mathscr{C}[\mathbf{x}(t_{N-1}), t_{N-1}]$. Thus, *independent of the choice of* $\mathbf{u}[\cdot, t_{N-1}]$, the value of $\mathscr{C}[\xi, t_N]$ should always be $\mathscr{C}^*[\xi, t_N]$ in order to minimize the overall cost. This is a statement of the *optimality principle*: "An optimal policy has the property that whatever any initial states and decision (or control law) are, all remaining decisions must constitute an optimal policy with regard to the state which results from the first decision" [4].

In particular, to determine the optimal control function $\mathbf{u}^*$, expressed as a function of $\mathbf{x}(t_{N-1})$ and $t_{N-1}$, to be applied at time $t_{N-1}$ as $\mathbf{u}^*\{\mathbf{x}(t_{N-1}), t_{N-1}\}$, we must solve for

$$\mathscr{C}^*[\mathbf{x}(t_{N-1}), t_{N-1}] = \min_{\mathbf{u} \in \mathscr{U}} \Big\{ L[\mathbf{x}(t_{N-1}), \mathbf{u}\{\mathbf{x}(t_{N-1}), t_{N-1}\}, t_{N-1}] + \int_{-\infty}^{\infty} \mathscr{C}^*[\xi, t_N] f_{\mathbf{x}}(\xi, t_N | \mathbf{x}(t_{N-1}), t_{N-1}) \, d\xi \Big\} \tag{13-64}$$

where $\mathscr{C}^*[\xi, t_N]$ is obtained as the result of the previous optimization, (13-58). The particular control function $\mathbf{u}[\cdot, t_{N-1}]$ that generates this minimum cost is then the optimal feedback control law function, $\mathbf{u}^*[\cdot, t_{N-1}]$.

By induction, we can obtain the general relationships needed to step backward in time to obtain the optimal control, known as the *stochastic dynamic programming* algorithm. At the general time $t_i$, the optimal control is obtained from the solution to

$$\mathscr{C}^*[\mathbf{x}(t_i), t_i] = \min_{\mathbf{u} \in \mathscr{U}} \Big\{ L[\mathbf{x}(t_i), \mathbf{u}\{\mathbf{x}(t_i), t_i\}, t_i] + \int_{-\infty}^{\infty} \mathscr{C}^*[\xi, t_{i+1}] f_{\mathbf{x}}(\xi, t_{i+1} | \mathbf{x}(t_i), t_i) \, d\xi \Big\} \tag{13-65}$$

or, equivalently, in terms of conditional expectations,

$$\mathscr{C}^*[\mathbf{x}(t_i), t_i] = \min_{\mathbf{u} \in \mathscr{U}} \{ L[\mathbf{x}(t_i), \mathbf{u}\{\mathbf{x}(t_i), t_i\}, t_i] + E\{\mathscr{C}^*[\mathbf{x}(t_{i+1}), t_{i+1}] | \mathbf{x}(t_i) = \mathbf{x}(t_i)\}\} \tag{13-66}$$

which is solved backwards in time from the terminal condition at $t_{N+1}$ of

$$\mathscr{C}^*[\mathbf{x}(t_{N+1}), t_{N+1}] = L_{\mathrm{f}}[\mathbf{x}(t_{N+1})] \tag{13-67}$$

Solving this backward to the initial time $t_0$ (or any time $t_i$ really), one obtains $\mathscr{C}^*[\mathbf{x}(t_0), t_0]$, the minimum expected cost to complete the process, or cost-to-go, from $\mathbf{x}(t_0)$ at time $t_0$, along with a specification in reverse time of the history of optimal control functions $\mathbf{u}^*[\cdot, t_0]$, $\mathbf{u}^*[\cdot, t_1], \ldots, \mathbf{u}^*[\cdot, t_N]$, i.e., a specification of $\mathbf{u}^*[\cdot, \cdot]$. If desired, one can evaluate the a priori mean total cost, averaged over all possible realizations of $\mathbf{x}(t_0, \cdot)$:

$$E\{\mathscr{C}^*[\mathbf{x}(t_0), t_0]\} = \int_{-\infty}^{\infty} \mathscr{C}^*[\xi, t_0] f_{\mathbf{x}(t_0)}(\xi)\, d\xi \tag{13-68}$$

Stochastic dynamic programming is critically dependent upon the *Markov* nature of the discrete-time $\mathbf{x}(\cdot,\cdot)$ process: this property allowed us to break out the integrations the way we did. Analogously, deterministic dynamic programming requires a complete *state* description to be able to break out the integrations in a similar manner. Such an algorithm can be used as an alternative to the maximum principle or calculus of variations for deriving deterministic optimal control laws. However, here we *had* to use an approach like dynamic programming, because there is no single optimal history of states and controls, but an entire family of possible trajectories due to the random process nature of the problem.

Conceptually, this method of stochastic dynamic programming is seen to be comprised of four basic steps: (1) Because of the Markov property of discrete-time $\mathbf{x}(\cdot,\cdot)$, the Chapman–Kolmogorov equation can be used to write the cost at time $t_i$ as

$$\mathscr{C}[\mathbf{x}(t_i), t_i] = L[\mathbf{x}(t_i), \mathbf{u}\{\mathbf{x}(t_i), t_i\}, t_i] + \int_{-\infty}^{\infty} \{\cdot\} f_{\mathbf{x}}(\xi, t_{i+1} | \mathbf{x}(t_i), t_i)\, d\xi \tag{13-69}$$

where the term $\{\cdot\}$ is recognized as $\mathscr{C}[\xi, t_{i+1}]$, the cost-to-go at time $t_{i+1}$ starting from $\mathbf{x}(t_{i+1}) = \xi$. (2) The control function $\mathbf{u}[\cdot,\cdot]$ is considered to be some *fixed* control function out of the admissible set of functions, $\mathscr{U}$; the optimal control function $\mathbf{u}^*[\cdot,\cdot]$ would be *one* such function. (3) No matter what $\mathbf{u}\{\mathbf{x}(t_i), t_i\}$ is used, in order to minimize $\mathscr{C}[\mathbf{x}(t_i), t_i]$, the value of $\mathscr{C}[\xi, t_{i+1}]$ in (13-69) must be the minimal $\mathscr{C}^*[\xi, t_{i+1}]$. (4) Then, to determine $\mathbf{u}^*[\cdot, t_i]$, we find the control function which yields minimal cost $\mathscr{C}[\mathbf{x}(t_i), t_i]$, denoted as $\mathscr{C}^*[\mathbf{x}(t_i), t_i]$, according to (13-65)–(13-67).

EXAMPLE 13.7 Consider a system described by a discrete-time linear state model driven by white Gaussian noise, with a quadratic cost criterion. This is a *unique* case in which a closed form solution is possible. Moreover, the control will be of the same form as for the corresponding deterministic optimal control problem: the same problem formulation, but with no dynamics driving noise or uncertainties.

Let the system be described by $n$-dimensional

$$\mathbf{x}(t_{i+1}) = \mathbf{\Phi}(t_{i+1}, t_i)\mathbf{x}(t_i) + \mathbf{B}_d(t_i)\mathbf{u}(t_i) + \mathbf{G}_d(t_i)\mathbf{w}_d(t_i)$$

where $\mathbf{w}_d(\cdot,\cdot)$ is $s$-vector zero-mean discrete-time white Gaussian noise with $E\{\mathbf{w}_d(t_i)\mathbf{w}_d^{\mathrm{T}}(t_i)\} = \mathbf{Q}_d(t_i)$. We desire to find the optimal $r$-vector function $\mathbf{u}^*$ as a function of perfectly measurable state $\mathbf{x}(t_i)$ and $t_i$ for all sample instants of interest, optimal in the sense that it minimizes a quadratic cost function given by (13-51), with "running cost" function $L$ given by

$$L[\mathbf{x}(t_j), \mathbf{u}(t_j), t_j] = \tfrac{1}{2}[\mathbf{x}^{\mathrm{T}}(t_j)\mathbf{X}(t_j)\mathbf{x}(t_j) + \mathbf{u}^{\mathrm{T}}(t_j)\mathbf{U}(t_j)\mathbf{u}(t_j)]$$

and with terminal cost function $L_f$ defined by

$$L_f[\mathbf{x}(t_{N+1})] = \tfrac{1}{2}\mathbf{x}^{\mathrm{T}}(t_{N+1})\mathbf{X}_f\mathbf{x}(t_{N+1})$$

where $\mathbf{X}(t_j)$ for all $j$ and $\mathbf{X}_f$ are $n$-by-$n$ positive semidefinite weighting matrices on state deviations and $\mathbf{U}(t_j)$ for all $t_j$ are $r$-by-$r$ positive definite weighting matrices on control. Special cases, such as $\mathbf{X}(t_j) \equiv \mathbf{0}$ for all $t_j$ with $\mathbf{X}_f \neq \mathbf{0}$, or nonzero $\mathbf{X}(t_j)$ and $\mathbf{X}_f = \mathbf{0}$, are then easily derived from this general problem. Using these definitions of $L$ and $L_f$, the costs $\mathscr{C}[\mathbf{x}(t_i), t_i]$ can be established backward in time. For the minimization in the dynamic programming algorithm, it will be assumed that the set of admissible controls $\mathscr{U}$ is the set of *all* functions of $(R^n \times T) \to R^r$, i.e., there is no constraint.

Thus, the optimization problem to be solved is (13-66),

$$\mathscr{C}^*[\mathbf{x}(t_i), t_i] = \min_{\mathbf{u}} \{\tfrac{1}{2}\mathbf{x}^T(t_i)\mathbf{X}(t_i)\mathbf{x}(t_i) + \mathbf{u}^T(t_i)\mathbf{U}(t_i)\mathbf{u}(t_i)$$
$$+ E[\mathscr{C}^*[\mathbf{x}(t_{i+1}), t_{i+1}] \,|\, \mathbf{x}(t_i) = \mathbf{x}(t_i)\}$$

to be solved backwards from the terminal condition (13-67),

$$\mathscr{C}^*[\mathbf{x}(t_{N+1}), t_{N+1}] = \tfrac{1}{2}\mathbf{x}^T(t_{N+1})\mathbf{X}_f\mathbf{x}(t_{N+1})$$

Now, with a considerable amount of hindsight, propose that this can be solved using a quadratic solution of the form

$$\mathscr{C}[\mathbf{x}(t_i), t_i] = \tfrac{1}{2}[\mathbf{x}^T(t_i)\mathbf{K}_c(t_i)\mathbf{x}(t_i) + g(t_i)]$$

where $\mathbf{K}_c(t_i)$ is assumed to be a symmetric $n$-by-$n$ matrix and $g(t_i)$ is a scalar, both to be determined. If this is to be the form of the solution, then from the terminal condition (13-67) it can be seen that

$$\mathbf{K}_c(t_{N+1}) = \mathbf{X}_f, \qquad g(t_{N+1}) = 0$$

are the appropriate terminal conditions.

Now the assumed form is substituted into the $\mathscr{C}^*[\mathbf{x}(t_i), t_i]$ equation derived from (13-66), in order to determine appropriate backward recursions for $\mathbf{K}_c(t_i)$ and $g(t_i)$. For convenience, define the new variable $\mathbf{x}^0(t_{i+1})$ as

$$\mathbf{x}^0(t_{i+1}) = \mathbf{\Phi}(t_{i+1}, t_i)\mathbf{x}(t_i) + \mathbf{B}_d(t_i)\mathbf{u}(t_i)$$

i.e., $\mathbf{x}^0(t_{i+1})$ is the known (deterministic) part of the state at time $t_{i+1}$, given the total state and control at time $t_i$, so that we can write the total state at time $t_{i+1}$ as

$$\mathbf{x}(t_{i+1}) = \mathbf{x}^0(t_{i+1}) + \mathbf{G}_d(t_i)\mathbf{w}_d(t_i)$$

If $\mathscr{C}^*[\mathbf{x}(t_i), t_i]$ is in fact of the assumed form, then the last term in (13-66) can be written in terms of this notation as

$$\begin{aligned}
&E\{\mathscr{C}^*[\mathbf{x}(t_{i+1}), t_{i+1}] \,|\, \mathbf{x}(t_i) = \mathbf{x}(t_i)\}\\
&\quad = E\{\tfrac{1}{2}[\mathbf{x}^T(t_{i+1})\mathbf{K}_c(t_{i+1})\mathbf{x}(t_{i+1}) + g(t_{i+1})] \,|\, \mathbf{x}(t_i) = \mathbf{x}(t_i)\}\\
&\quad = E\{\tfrac{1}{2}[\mathbf{x}^{0T}(t_{i+1})\mathbf{K}_c(t_{i+1})\mathbf{x}^0(t_{i+1}) + \mathbf{w}_d^T(t_i)\mathbf{G}_d^T(t_i)\mathbf{K}_c(t_{i+1})\mathbf{G}_d(t_i)\mathbf{w}_d(t_i) + g(t_{i+1}) \,|\, \mathbf{x}(t_i) = \mathbf{x}(t_i)\}\\
&\quad = \tfrac{1}{2}[\mathbf{x}^{0T}(t_{i+1})\mathbf{K}_c(t_{i+1})\mathbf{x}^0(t_{i+1}) + \mathrm{tr}\{\mathbf{G}_d(t_i)\mathbf{Q}_d(t_i)\mathbf{G}_d^T(t_i)\mathbf{K}_c(t_{i+1})\} + g(t_{i+1})
\end{aligned}$$

where, in the second equality, the two appropriate cross-terms have been removed since their expected values are zero, and the trace term in the last equality arises from

$$\begin{aligned}
E\{\mathbf{w}_d^T(t_i)\mathbf{G}_d^T(t_i)\mathbf{K}_c(t_{i+1})\mathbf{G}_d(t_i)\mathbf{w}_d(t_i) \,|\, \mathbf{x}(t_i) = \mathbf{x}(t_i)\} &= E\{\mathbf{w}_d^T(t_i)\mathbf{G}_d^T(t_i)\mathbf{K}_c(t_{i+1})\mathbf{G}_d(t_i)\mathbf{w}_d(t_i)\}\\
&= E\{\mathrm{tr}[\mathbf{G}_d(t_i)\mathbf{w}_d(t_i)\mathbf{w}_d^T(t_i)\mathbf{G}_d^T(t_i)\mathbf{K}_c(t_{i+1})]\}\\
&= \mathrm{tr}\{\mathbf{G}_d(t_i)E[\mathbf{w}_d(t_i)\mathbf{w}_d^T(t_i)]\mathbf{G}_d^T(t_i)\mathbf{K}_c(t_{i+1})\}
\end{aligned}$$

since $\mathbf{w}_d(\cdot,\cdot)$ is independent of $\mathbf{x}(t_i)$ and $\mathrm{tr}(\mathbf{AB}) = \mathrm{tr}(\mathbf{BA})$. Thus, the recursion for the optimal cost becomes

$$\begin{aligned}\mathscr{C}^*[\mathbf{x}(t_i), t_i] &= \min_{\mathbf{u}} \Big\{\tfrac{1}{2}[\mathbf{x}^{\mathrm{T}}(t_i)\mathbf{X}(t_i)\mathbf{x}(t_i) + \mathbf{u}^{\mathrm{T}}(t_i)\mathbf{U}(t_i)\mathbf{u}(t_i)] \\ &\quad + \tfrac{1}{2}\big[\mathbf{x}^{0\mathrm{T}}(t_{i+1})\mathbf{K}_{\mathrm{c}}(t_{i+1})\mathbf{x}^{0}(t_{i+1}) \\ &\quad + \mathrm{tr}\{\mathbf{G}_{\mathrm{d}}(t_i)\mathbf{Q}_{\mathrm{d}}(t_i)\mathbf{G}_{\mathrm{d}}^{\mathrm{T}}(t_i)\mathbf{K}_{\mathrm{c}}(t_{i+1})\} + g(t_{i+1})\big]\Big\} \\ &= \min_{\mathbf{u}} \Big\{\tfrac{1}{2}\big[\mathbf{x}^{\mathrm{T}}(t_i)[\mathbf{X}(t_i) + \mathbf{\Phi}^{\mathrm{T}}(t_{i+1}, t_i)\mathbf{K}_{\mathrm{c}}(t_{i+1})\mathbf{\Phi}(t_{i+1}, t_i)]\mathbf{x}(t_i) \\ &\quad + \mathbf{u}^{\mathrm{T}}(t_i)[\mathbf{U}(t_i) + \mathbf{B}_{\mathrm{d}}^{\mathrm{T}}(t_i)\mathbf{K}_{\mathrm{c}}(t_{i+1})\mathbf{B}_{\mathrm{d}}(t_i)]\mathbf{u}(t_i) \\ &\quad + 2\mathbf{x}^{\mathrm{T}}(t_i)\mathbf{\Phi}^{\mathrm{T}}(t_{i+1})\mathbf{K}_{\mathrm{c}}(t_{i+1}, t_i)\mathbf{B}_{\mathrm{d}}(t_i)\mathbf{u}(t_i) \\ &\quad + \mathrm{tr}[\mathbf{G}_{\mathrm{d}}(t_i)\mathbf{Q}_{\mathrm{d}}(t_i)\mathbf{G}_{\mathrm{d}}^{\mathrm{T}}(t_i)\mathbf{K}_{\mathrm{c}}(t_{i+1})] + g(t_{i+1})\big]\Big\}\end{aligned}$$

To perform the minimization, set the first derivative of the quantity in the $\{\cdot\}$ braces with respect to $\mathbf{u}(t_i)$ equal to zero, since $\mathscr{U}$ is unconstrained:

$$\mathbf{0}^{\mathrm{T}} = \frac{\partial\{\cdot\}}{\partial \mathbf{u}(t_i)} = \mathbf{u}^{\mathrm{T}}(t_i)[\mathbf{U}(t_i) + \mathbf{B}_{\mathrm{d}}^{\mathrm{T}}(t_i)\mathbf{K}_{\mathrm{c}}(t_{i+1})\mathbf{B}_{\mathrm{d}}(t_i)] + \mathbf{x}^{\mathrm{T}}(t_i)\mathbf{\Phi}^{\mathrm{T}}(t_{i+1}, t_i)\mathbf{K}_{\mathrm{c}}(t_{i+1})\mathbf{B}_{\mathrm{d}}(t_i)$$

to obtain a stationary point. From this equation we obtain

$$\mathbf{u}^*(t_i) = -[\mathbf{U}(t_i) + \mathbf{B}_{\mathrm{d}}^{\mathrm{T}}(t_i)\mathbf{K}_{\mathrm{c}}(t_{i+1})\mathbf{B}_{\mathrm{d}}(t_i)]^{-1}\mathbf{B}_{\mathrm{d}}^{\mathrm{T}}(t_i)\mathbf{K}_{\mathrm{c}}(t_{i+1})\mathbf{\Phi}(t_{i+1}, t_i)\mathbf{x}(t_i)$$

which is seen to be a linear transformation of $\mathbf{x}(t_i)$. Now look at the second derivative to determine the nature of this stationary point:

$$\frac{\partial^2\{\cdot\}}{\partial \mathbf{u}^2(t_i)} = [\mathbf{U}(t_i) + \mathbf{B}_{\mathrm{d}}^{\mathrm{T}}(t_i)\mathbf{K}_{\mathrm{c}}(t_{i+1})\mathbf{B}_{\mathrm{d}}(t_i)]$$

Thus, if we can show that this is a positive definite matrix (and we *will* be able to do so), then we are assured *both* that the inverse of

$$[\mathbf{U}(t_i) + \mathbf{B}_{\mathrm{d}}^{\mathrm{T}}(t_i)\mathbf{K}_{\mathrm{c}}(t_{i+1})\mathbf{B}_{\mathrm{d}}(t_i)]$$

exists as required to define $\mathbf{u}^*(t_i)$ *and* that the $\mathbf{u}^*(t_i)$ so generated is in fact an optimal, cost-*minimizing* control.

First, though, to evaluate the recursions for $\mathbf{K}_{\mathrm{c}}(t_i)$ and $g(t_i)$, substitute the $\mathbf{u}^*(t_i)$ expression back into the right hand side of the equation for $\mathscr{C}^*[\mathbf{x}(t_i), t_i]$, to obtain an equality between the assumed form,

$$\mathscr{C}^*[\mathbf{x}(t_i), t_i] = \tfrac{1}{2}[\mathbf{x}^{\mathrm{T}}(t_i)\mathbf{K}_{\mathrm{c}}(t_i)\mathbf{x}(t_i) + g(t_i)]$$

and the result of the substitution:

$$\begin{aligned}\mathscr{C}^*[\mathbf{x}(t_i), t_i] &= \tfrac{1}{2}\{\mathbf{x}^{\mathrm{T}}(t_i)[\mathbf{X}(t_i) + \mathbf{\Phi}^{\mathrm{T}}(t_{i+1}, t_i)\mathbf{K}_{\mathrm{c}}(t_{i+1})\mathbf{\Phi}(t_{i+1}, t_i)]\mathbf{x}(t_i) \\ &\quad - [\mathbf{x}^{\mathrm{T}}(t_i)\mathbf{\Phi}^{\mathrm{T}}(t_{i+1}, t_i)\mathbf{K}_{\mathrm{c}}(t_{i+1})\mathbf{B}_{\mathrm{d}}(t_i)][\mathbf{U}(t_i) + \mathbf{B}_{\mathrm{d}}^{\mathrm{T}}(t_i)\mathbf{K}_{\mathrm{c}}(t_{i+1})\mathbf{B}_{\mathrm{d}}(t_i)]^{-1} \\ &\quad \times [\mathbf{B}_{\mathrm{d}}^{\mathrm{T}}(t_i)\mathbf{K}_{\mathrm{c}}(t_{i+1})\mathbf{\Phi}(t_{i+1}, t_i)\mathbf{x}(t_i)] \\ &\quad + \mathrm{tr}[\mathbf{G}_{\mathrm{d}}(t_i)\mathbf{Q}_{\mathrm{d}}(t_i)\mathbf{G}_{\mathrm{d}}^{\mathrm{T}}(t_i)\mathbf{K}_{\mathrm{c}}(t_{i+1})] + g(t_{i+1})\}\end{aligned}$$

where the second term on the right hand side is actually the result of three such terms, two with a negative sign and one with a positive sign. The resulting equality must be valid for *all* $\mathbf{x}(t_i)$, so we can separately equate the terms that are quadratic in $\mathbf{x}(t_i)$ and those which are not functionally dependent on $\mathbf{x}(t_i)$. Doing this yields a discrete-time backward Riccati equation for $\mathbf{K}_{\mathrm{c}}(t_i)$ and a backward recursion for $g(t_i)$:

$$\begin{aligned}\mathbf{K}_{\mathrm{c}}(t_i) &= \mathbf{X}(t_i) + \mathbf{\Phi}^{\mathrm{T}}(t_{i+1}, t_i)\mathbf{K}_{\mathrm{c}}(t_{i+1})\mathbf{\Phi}(t_{i+1}, t_i) \\ &\quad - [\mathbf{\Phi}^{\mathrm{T}}(t_{i+1}, t_i)\mathbf{K}_{\mathrm{c}}(t_{i+1})\mathbf{B}_{\mathrm{d}}(t_i)][\mathbf{U}(t_i) + \mathbf{B}_{\mathrm{d}}^{\mathrm{T}}(t_i)\mathbf{K}_{\mathrm{c}}(t_{i+1})\mathbf{B}_{\mathrm{d}}(t_i)]^{-1} \\ &\quad \times [\mathbf{B}_{\mathrm{d}}^{\mathrm{T}}(t_i)\mathbf{K}_{\mathrm{c}}(t_{i+1})\mathbf{\Phi}(t_{i+1}, t_i)] \\ g(t_i) &= g(t_{i+1}) + \mathrm{tr}[\mathbf{G}_{\mathrm{d}}(t_i)\mathbf{Q}_{\mathrm{d}}(t_i)\mathbf{G}_{\mathrm{d}}^{\mathrm{T}}(t_i)\mathbf{K}_{\mathrm{c}}(t_{i+1})]\end{aligned}$$

Note that all of the statistical information about the effect of the dynamic noise is confined to the scalar $g(\cdot)$ function. Moreover, the result of setting $\mathbf{Q}_d(t_i) \equiv \mathbf{0}$ to remove this driving noise is that $g(t_i) = 0$ for all $t_i$, and the resulting cost and control are equivalent to the result for the corresponding deterministic optimal control problem: the cost-to-go is less in the deterministic case, as should be expected, but the optimal control function is *unaltered.*

Recall that we wanted to show that $[\mathbf{U}(t_i) + \mathbf{B}_d^T(t_i)\mathbf{K}_c(t_{i+1})\mathbf{B}_d(t_i)]$ is positive definite, to be assured of the existence of its inverse and the optimality of the stationary solution $\mathbf{u}^*(t_i)$. Since $\mathbf{U}(t_i)$ was itself assumed to be positive definite, the result is established if $[\mathbf{B}_d^T(t_i)\mathbf{K}_c(t_{i+1})\mathbf{B}_d(t_i)]$ can be shown to be positive semidefinite, which in turn is true if $\mathbf{K}_c(t_{i+1})$ is at worst positive semidefinite. First of all, at the terminal time, $\mathbf{K}_c(t_{N+1}) = \mathbf{X}_f$, which *is* symmetric and positive semidefinite. Thus, the term $[\mathbf{U}(t_N) + \mathbf{B}_d^T(t_N)\mathbf{K}_c(t_{N+1})\mathbf{B}_d(t_N)]$ is positive definite. To establish the result for $\mathbf{K}_c(t_N)$, note that since

$$\begin{aligned}\mathscr{C}^*[\mathbf{x}(t_N), t_N] &= E\{L[\mathbf{x}(t_N), \mathbf{u}^*(t_N), t_N] + L_f[\mathbf{x}(t_{N+1})] \,|\, \mathbf{x}(t_N) = \mathbf{x}(t_N)\} \\ &= E\{\tfrac{1}{2}[\mathbf{x}^T(t_N)\mathbf{X}(t_N)\mathbf{x}(t_N) + \mathbf{u}^{*T}(t_N)\mathbf{U}(t_N)\mathbf{u}^*(t_N) + \mathbf{x}^T(t_{N+1})\mathbf{X}_f\mathbf{x}(t_{N+1})] \,|\, \mathbf{x}(t_N) = \mathbf{x}(t_N)\}\end{aligned}$$

then $\mathscr{C}^*[\mathbf{x}(t_N), t_N]$ is, at worst, positive semidefinite because of the assumptions on $\mathbf{X}(t_N)$, $\mathbf{U}(t_N)$ and $\mathbf{X}_f$. Now, since this equals $\frac{1}{2}[\mathbf{x}^T(t_N)\mathbf{K}_c(t_N)\mathbf{x}(t_N) + g(t_N)]$, $\mathbf{K}_c(t_N) \geq \mathbf{0}$ and $g(t_N) \geq 0$. (Otherwise, negative values could be obtained.) By induction, $\mathbf{K}_c(t_i)$ is at worst positive semidefinite for all $t_i$, and $[\mathbf{U}(t_i) + \mathbf{B}_d^T(t_i)\mathbf{K}_c(t_{i+1})\mathbf{B}_d(t_i)]$ is positive definite for all $t_i$.

The result of this example is the optimal stochastic controller for the LQG problem, assuming perfect state knowledge at each sample time. The result is

$$\mathbf{u}^*(t_i) = -\mathbf{G}_c^*(t_i)\mathbf{x}(t_i)$$

a *linear state feedback* law in which the optimal controller gains $\mathbf{G}_c^*(t_i)$ are given by

$$\mathbf{G}_c^*(t_i) = [\mathbf{U}(t_i) + \mathbf{B}_d^T(t_i)\mathbf{K}_c(t_{i+1})\mathbf{B}_d(t_i)]^{-1}[\mathbf{B}_d^T(t_i)\mathbf{K}_c(t_{i+1})\mathbf{\Phi}(t_{i+1}, t_i)]$$

and $\mathbf{K}_c(t_{i+1})$ is found via the backward Riccati recursion derived earlier. This controller gain evaluation is identical to that of Section 13.3 (see (13-18), (13-19), (13-22), and (13-23) for comparison), and the block diagrams of this algorithm and its use in a sampled-data control system given in Fig. 13.9 are directly comparable to Figs. 13.4 and 13.5. ■

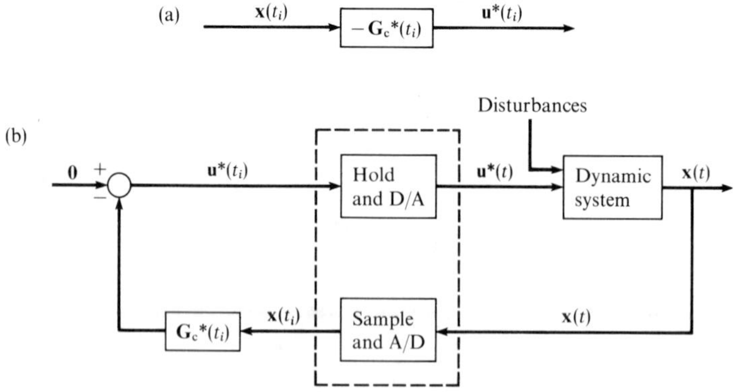


FIG. 13.9 Optimal stochastic controller for LQG problem with perfect state knowledge. (a) Feedback control generation (b) Sampled-data implementation.

The previous example demonstrated use of dynamic programming, (13-65)–(13-67), as a *conceptual tool* for deriving a useful stochastic controller under the linear-quadratic-Gaussian modeling assumptions. However, dynamic programming can also be considered a *computational algorithm* for generating feedback controllers for more general problems that do not lend themselves to such convenient controller solution forms. Unfortunately, this entails *discretizing state and control variables* if they are not naturally discrete-valued in the original problem definition. The dynamic programming algorithm then works backwards in time to generate, for every discrete value of $\mathbf{x}(t_i)$ at each $t_i$, the appropriate discrete-valued $\mathbf{u}^*[\mathbf{x}(t_i), t_i]$ and associated cost $\mathscr{C}^*[\mathbf{x}(t_i), t_i]$ by explicitly evaluating the costs $\mathscr{C}[\mathbf{x}(t_i), t_i]$ due to *each* discrete $\mathbf{u}[\mathbf{x}(t_i), t_i]$, and then choosing the control that yields the lowest cost. If $\mathbf{x}(t_i)$ is $n$-dimensional and each scalar component is discretized to $d_x$ levels, then $\mathbf{u}^*$ and $\mathscr{C}^*$ must be evaluated for $(d_x)^n$ separate state values at each sample time; maintaining a table with this many entries is a very harsh memory requirement. Moreover, if the $r$-dimensional $\mathbf{u}(t_i)$ is discretized to $d_u$ levels, each table entry requires $(d_u)^r$ separate cost computations! What has just been described is known in the literature as the "curse of dimensionality" [30] associated with dynamic programming. Because of it, this algorithm becomes infeasible for many practical problems. For example, if $d_x = d_u = 100$, $n = 8$, and $r = 2$, then $100^8 N = 10^{16} N$ table entries for $\mathbf{u}^*$ and $\mathscr{C}^*$ are required, where $N$ is the number of sample periods, and each entry would require $100^2 = 10^4$ separate cost evaluations! Nevertheless, computational dynamic programming has been used to generate some useful controllers.

EXAMPLE 13.8 This is a control problem in which the cost function is not quadratic and in which the set $\mathscr{U}$ of admissible controls is in fact a constrained set [12, 15, 32, 35, 36]. Consider the Apollo command module returning from the lunar mission and reentering the earth's atmosphere, as depicted in Fig. 13.10. Suppose we are interested in the control of the lateral motion of the vehicle, as displayed by the possible vehicle trajectory tracks in Fig. 13.10b, where the dashed line denotes the locus of possible landing sites, known as the vehicle "footprint."

Let us first describe how this lateral motion is controlled. The Apollo vehicle enters the atmosphere so that the top edge of its cone-shaped body is almost parallel to the free-stream velocity $V_\infty$, as shown in Fig. 13.11. Because the center of mass is purposely placed so as not to lie on the axis of symmetry of the cone, lift is generated. To obtain a limited degree of lateral maneuverability, the vehicle is rolled over so that some component of lift is in the horizontal plane. First, by knowing the vehicle state at the point of entry into the atmosphere (a predetermined window or threshold point) and the desired landing site, one knows the down-range distance from the threshold point to the landing site; this determines the amount of vertical force that must be maintained along the trajectory to reach the landing site. Looking at the vehicle head-on as in Fig. 13.12, if the vehicle generates a larger amount of lift force, we maintain the desired amount of vertical force by rolling the vehicle by the appropriate roll angle $\phi$, thereby generating a horizontal component of lift available for lateral maneuvering. Thus, the total lift is determined by vehicle and trajectory characteristics, and its vertical component and thus $|\phi|$ by down-range distance to the landing site: these are *not* at our disposal to change. Lateral control then entails deciding when to switch from $(+|\phi|)$ to $(-|\phi|)$ or vice versa.

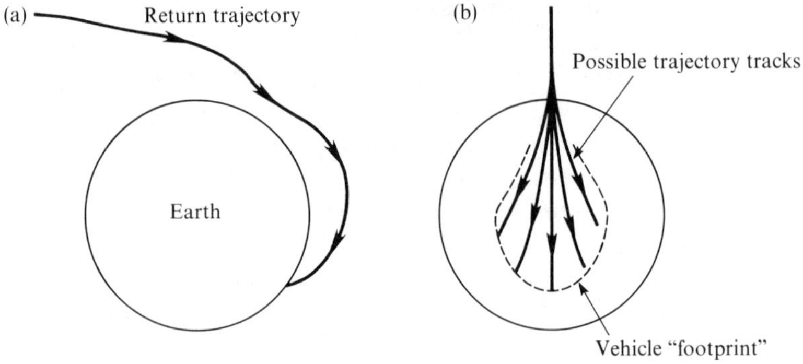


FIG. 13.10 Apollo command module returning to earth. (a) Side view of trajectory. (b) As seen from above.

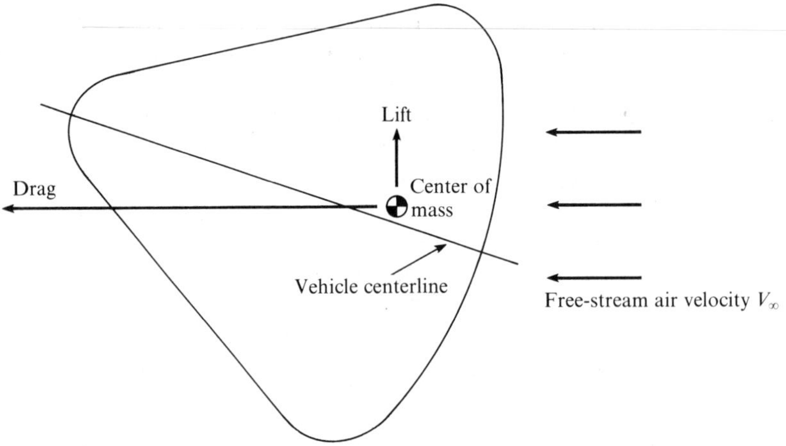


FIG. 13.11 Apollo vehicle reentry geometry.

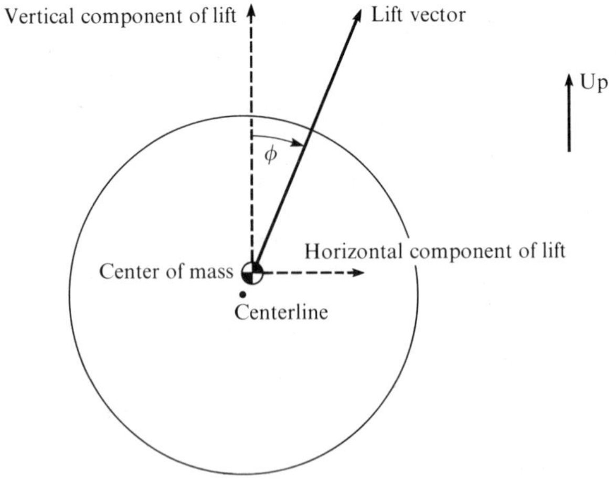


FIG. 13.12 Lift generation, head-on view.

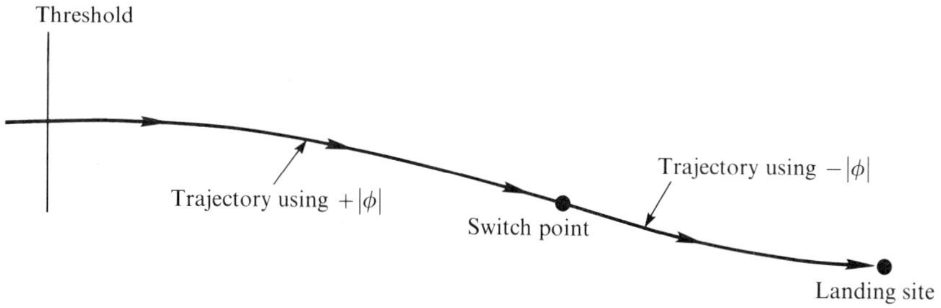


FIG. 13.13 Single-switch trajectory for deterministic problem.

One can envision a deterministic control problem of rolling to the right, and then determining when to roll left to reach the landing site exactly with a single switch. This is portrayed in Fig. 13.13, as seen from above. One might also consider using multiple switch points, but rolling the vehicle consumes attitude control rocket propellant, so a minimum number of rollovers (one for the deterministic case) is desirable.

Now consider the nondeterministic case. We want to minimize the cross-range error at the terminal point, while still trying to minimize the number of commanded rollovers due to fuel consumption considerations. Assume that the computer algorithm will update its information about once every 10 sec, so there are discrete control decision times, at which times we can command to roll to the left, roll to the right, or do nothing. Recalling that the roll angle *magnitude* is *not* under our control, one appropriate state variable for describing this problem is

$$x_1(t_i) = \begin{cases} +1 & \text{if vehicle is rolled to right at time } t_i \\ -1 & \text{if vehicle is rolled to left at time } t_i \end{cases}$$

This state then satisfies

$$\mathrm{x}_1(t_{i+1}) = \mathrm{x}_1(t_i) + 2\mathrm{u}(t_i)$$

where the admissible set of controls ($\mathscr{U}$) is then

$$u(t_i) = \begin{cases} +1 \quad \text{or} \quad 0 & \text{if} \quad x_1(t_i) = -1 \\ -1 \quad \text{or} \quad 0 & \text{if} \quad x_1(t_i) = +1 \end{cases}$$

so that $x_1(t_{i+1})$ can only assume values of $+1$ or $-1$. We will assume that the roll maneuver will be accomplished completely in less time than one sample period, 10 sec. There are a number of possible choices for a second state variable, but what was actually used was $x_2(t_i)$ defined as the cross-range error at the *final time* as *extrapolated linearly* from the present vehicle position and velocity, as shown in Fig. 13.14. The model of dynamics for this state is

$$\mathrm{x}_2(t_{i+1}) = \mathrm{x}_2(t_i) + a(t_i)\mathrm{x}_1(t_i) + b(t_i)\mathrm{u}(t_i) + \mathrm{w}_\mathrm{d}(t_i)$$

where $[a(t_i)x_1(t_i)]$ accounts for the effect of the roll state $x_1(t_i)$ on projected cross-range error over the next sample period; $[b(t_i)u(t_i)]$ is zero except when a rollover is commanded, at which point it accounts for the transient effects that are confined to that next sample period; and $\mathrm{w}_\mathrm{d}(\cdot,\cdot)$ accounts for random buffeting and model uncertainty, and is modeled as a zero-mean white Gaussian discrete-time noise with $E\{\mathrm{w}_\mathrm{d}(t_i)\mathrm{w}_\mathrm{d}(t_j)\} = Q_\mathrm{d}(t_i)\,\delta_{ij}$.

The objective is then to derive an optimal control law $\mathbf{u}^*[\cdot,\cdot]$, optimal in the sense that it minimizes a cost function composed of a terminal cost on cross-range error plus a penalty on the amount of rolling required:

$$\mathscr{C}[x_1(t_0), x_2(t_0), t_0] = E\left\{\sum_{i=0}^{N} |\mathrm{u}(t_i)| + \frac{1}{2} X_\mathrm{f}\mathrm{x}_2{}^2(t_{N+1}) \,\middle|\, \mathrm{x}_1(t_0) = x_1(t_0), \mathrm{x}_2(t_0) = x_2(t_0)\right\}$$

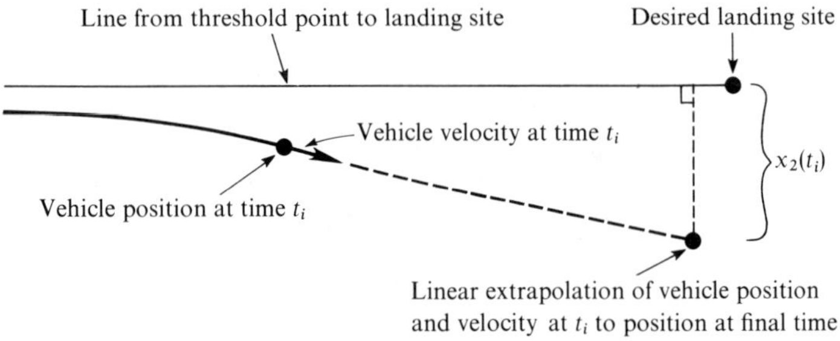


FIG. 13.14 Definition of $x_2(t_i)$.

Thus, by the dynamic programming algorithm, we want to find the solution to

$$\begin{aligned}\mathscr{C}^*[x_1(t_i), x_2(t_i), t_i] &= \min_{u \in \mathscr{U}} \left[|u(t_i)| + E\{\mathscr{C}^*[\mathrm{x}_1(t_{i+1}), \mathrm{x}_2(t_{i+1}), t_{i+1}] \,|\, \mathrm{x}_1(t_i) = x_1(t_i), \mathrm{x}_2(t_i) = x_2(t_i)\}\right] \\ &= \min_{u \in \mathscr{U}} \left[|u(t_i)| + E\{\mathscr{C}^*[\mathrm{x}_1(t_i) + 2u(t_i), \mathrm{x}_2(t_i) + a(t_i)\mathrm{x}_1(t_i) + b(t_i)u(t_i)\right. \\ &\quad \left. + \mathrm{w_d}(t_i), t_{i+1} \,|\, \mathrm{x}_1(t_i) = x_1(t_i), \mathrm{x}_2(t_i) = x_2(t_i)\}\right]\end{aligned}$$

But, $\mathrm{w_d}(t_i)$ is left as the only nondeterministic variable in the conditional expectation, so this can be written as

$$\begin{aligned}\mathscr{C}^*[x_1(t_i), x_2(t_i), t_i] = \min_{u \in \mathscr{U}} \Big\{ |u(t_i)| + \int_{-\infty}^{\infty} f_{\mathrm{w_d}(t_i)}(\xi) \\ \times \mathscr{C}^*[x_1(t_i) + 2u(t_i), x_2(t_i) + a(t_i)x_1(t_i) + b(t_i)u(t_i) + \xi, t_{i+1}]\, d\xi \Big\}\end{aligned}$$

where

$$f_{\mathrm{w_d}(t_i)}(\xi) = \frac{1}{\sqrt{2\pi Q_d(t_i)}} \exp\left\{-\frac{1}{2}\frac{\xi^2}{Q_d(t_i)}\right\}$$

For convenience, define the scalar function $\overline{\mathscr{C}^*}[\cdot, \cdot, \cdot]$ by

$$\overline{\mathscr{C}^*}[\alpha, \beta, t_i] = \int_{-\infty}^{\infty} f_{\mathrm{w_d}(t_i)}(\xi) \mathscr{C}^*[\alpha, \beta + \xi, t_{i+1}]\, d\xi$$

so that the dynamic programming algorithm can be written as

$$\mathscr{C}^*[x_1(t_i), x_2(t_i), t_i] = \min_{u \in \mathscr{U}} \{|u(t_i)| + \overline{\mathscr{C}^*}[x_1(t_i) + 2u(t_i), x_2(t_i) + a(t_i)x_1(t_i) + b(t_i)u(t_i), t_i]\}$$

Since there are only two possible values of $\alpha$ for $\overline{\mathscr{C}^*}[\alpha, \beta, t_i]$, recursion relations can be derived for the optimal cost in a rather simple fashion. If $x_1(t_i) = 1$, we have

$$\mathscr{C}^*[1, x_2(t_i), t_i] = \min_{u(t_i) = -1 \text{ or } 0} \{|u(t_i)| + \overline{\mathscr{C}^*}[1 + 2u(t_i), x_2(t_i) + a(t_i) + b(t_i)u(t_i), t_i]\}$$

while, if $x_1(t_i) = -1$, we have

$$\mathscr{C}^*[-1, x_2(t_i), t_i] = \min_{u(t_i) = 0 \text{ or } 1} \{|u(t_i)| + \overline{\mathscr{C}^*}[-1 + 2u(t_i), x_2(t_i) - a(t_i) + b(t_i)u(t_i), t_i]\}$$

These relations are then used to step backwards in time from the terminal conditions

$$\mathscr{C}^*[\pm 1, x_2(t_{N+1}), t_{N+1}] = \tfrac{1}{2} X_f x_2^2(t_{N+1})$$

By performing this propagation backward in time, the optimal control law $\mathbf{u}^*$ can be expressed as a function of $x_1(t_i)$, $x_2(t_i)$, and $t_i$ for all $i$. This entails discretizing $x_2(t_i)$, and for each of these

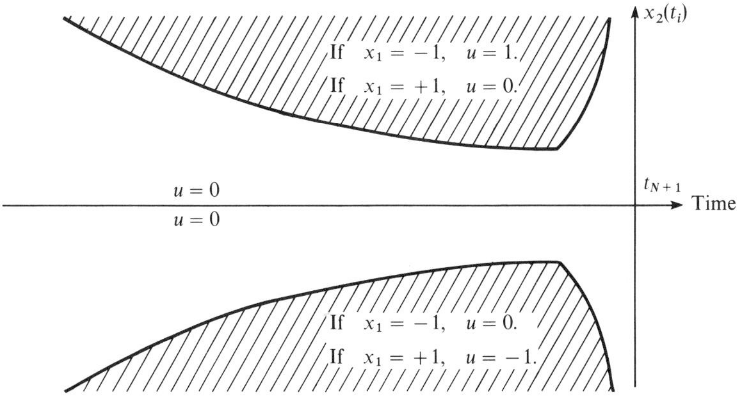


FIG. 13.15 Controller switching boundaries.

discrete values, evaluating both $\mathscr{C}^*[1, x_2(t_i), t_i]$ and $\mathscr{C}^*[-1, x_2(t_i), t_i]$. As seen before, each of these requires two cost evaluations (for each of two admissible $u(t_i)$ values), and selection of the minimal $\mathscr{C}^*$ and corresponding $u^*(t_i)$. When the calculations are performed, the result can be plotted with two switching boundaries, as in Fig. 13.15. In the center region, the cost is minimized by making no roll commands. In the upper region, the vehicle is "too far left": the terminal cross-range penalty is more significant than the cost of rolling over. Therefore, if the vehicle is in that region and is rolled to the left, i.e., if $x_1(t_i) = -1$, then $u(t_i) = 1$ and a rollover to the right is accomplished; if the vehicle is already rolled to the right, i.e., if $x_1(t_i) = 1$, no further command is made. The lower region is understood in the same manner. Note that the boundaries widen again near the end, because it does not pay to make corrections very late in the trajectory, just consuming fuel with little cross-range error benefit. The particular shape achieved is affected by choice of $X_f$; the larger $X_f$ is, the more important terminal errors are relative to the cost of rollovers and thus the narrower the middle region. In practice, $X_f$ is chosen so that the probability of requiring three or more rollovers is very small. Typical $x_2$ trajectories might appear as in Fig. 13.16, in which multiple rollovers are seen to be required on certain sample paths. ■

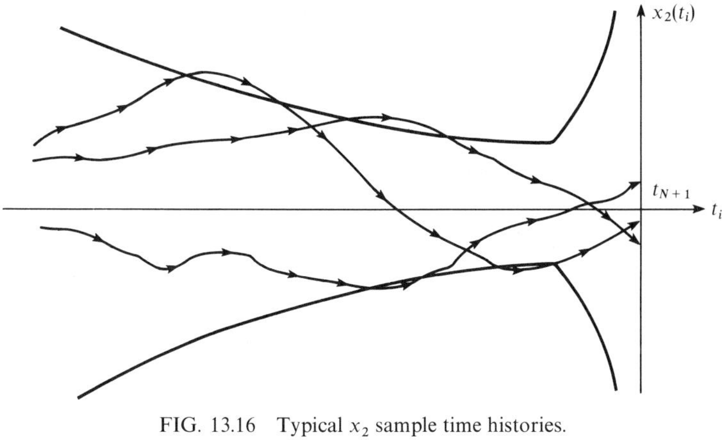


FIG. 13.16 Typical $x_2$ sample time histories.

The previous example employed a dynamics model in which driving noises and (at least some) states are fundamentally described as being able to assume values from a continuous range, with state discretization applied later in the dynamic programming algorithm itself. It is sometimes as natural, if not more so, to propose discrete-valued states and noises in the initial problem modeling [7]. For instance, in a resource allocation problem or communication relay network problem, states truly are discrete valued, and therefore noises in dynamics models must also be discrete valued so as not to yield propagated states that lie outside admissible values. Although the dynamics model, cost definition, and final dynamic programming algorithmic form will necessarily differ in appearance from previous developments, the basic concepts inherent in dynamic programming will remain unchanged.

Let $\mathbf{x}(t_i)$ denote the state vector at time $t_i$, which is typically discrete valued (it need not be for this development, though state discretization will usually be required for implementation of computational dynamic programming), before a decision is made at that time. Let $u(t_i)$ be a scalar, integer-valued control index used to label the various decisions available at time $t_i$, and let $\mathscr{U}$ indicate the constraint that certain decisions might not be available to the decision-maker at particular times and in certain states (to be demonstrated in examples to follow). For dynamics modeling, let $\boldsymbol{\phi}_k[\mathbf{x}(t_i), t_i]$ denote the $k$th member of a list ($k = 1, 2, \ldots, K$) of possible transformations that may operate on $\mathbf{x}(t_i)$ to yield the resulting state vector $\mathbf{x}(t_{i+1})$, and let $p[\boldsymbol{\phi}_k | \mathbf{x}(t_i), u(t_i), t_i]$ be the conditional probability that it will indeed be $\boldsymbol{\phi}_k$ that operates on $\mathbf{x}(t_i)$, given that the present system state is $\mathbf{x}(t_i)$ and that the decision-maker will select decision $u(t_i)$ at time $t_i$. The cost to the decision-maker of a transition from $\mathbf{x}(t_i)$ to $\mathbf{x}(t_{i+1})$ resulting from decision $u(t_i)$ at time $t_i$ is given by $L[\mathbf{x}(t_i), \mathbf{x}(t_{i+1}), u(t_i), t_i]$, and the cost at the final time associated with the system ending up in state $\mathbf{x}(t_{N+1})$ is $L_f[\mathbf{x}(t_{N+1})]$. For problems formulated in this manner, the appearance of the additional term $\mathbf{x}(t_{i+1})$ in the arguments of the running cost $L$ is not necessarily redundant or superfluous. For example, consider an altitude-channel aircraft collision avoidance problem: if the decision-maker decides to command the vehicle to climb ($u(t_i) = 1$, for sake of argument), then the cost is great for altitude $x(t_{i+1})$ values less than $x(t_i)$ (as due to wind buffeting) and perhaps zero for $x(t_{i+1}) > x(t_i)$; this would be reversed if he had decided instead to command a descent.

In terms of these definitions, the dynamic programming algorithm becomes

$$\begin{aligned}
&\mathscr{C}^*[\mathbf{x}(t_i), t_i] \\
&\quad = \min_{u \in \mathscr{U}} \left\{ \sum_{k=1}^{K} L[\mathbf{x}(t_i), \boldsymbol{\phi}_k\{\mathbf{x}(t_i), t_i\}, u(t_i), t_i] p[\boldsymbol{\phi}_k | \mathbf{x}(t_i), u(t_i), t_i] \right. \\
&\qquad \left. + \sum_{k=1}^{K} \mathscr{C}^*[\boldsymbol{\phi}_k\{\mathbf{x}(t_i), t_i\}, t_{i+1}] p[\boldsymbol{\phi}_k | \mathbf{x}(t_i), u(t_i), t_i] \right\} \qquad (13\text{-}70)
\end{aligned}$$

which is solved iteratively for $t_i = t_N, t_{N-1}, \ldots, t_0$, subject to the terminal condition

$$\mathscr{C}^*[\mathbf{x}(t_{N+1}), t_{N+1}] = L_f[\mathbf{x}(t_{N+1})] \tag{13-71}$$

The first summation in (13-70) is the expected cost that accrues over the next single sample period as a result of the decision made at time $t_i$. The second summation is the expected cost to complete the process from $t_{i+1}$ forward, assuming that all future decisions will also be optimal. Notice that Eqs. (13-70) and (13-71) are directly comparable to (13-65) or (13-66), and (13-67), respectively.

EXAMPLE 13.9 Consider the network portrayed in Fig. 13.17 [12]. We want to determine the minimum cost path from each leftmost node at time $t_0$ to the terminal plane at $t_{N+1} = t_3$. The numbers in circles at the terminal nodes denote the cost of ending up at that node; note that it is a multimodal terminal cost. Similarly, the number on each branch is the cost of using that branch.

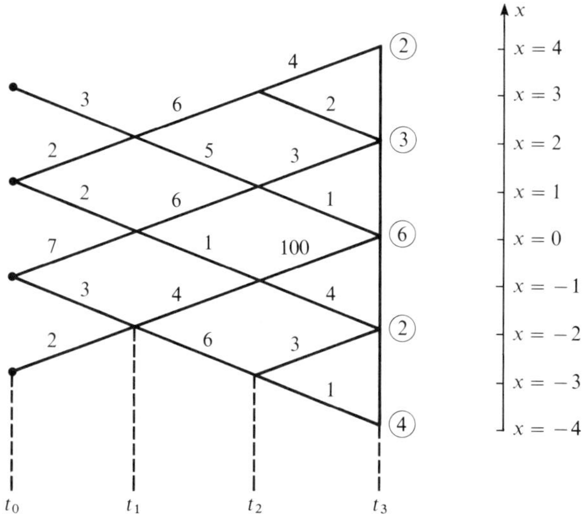


FIG. 13.17 Network to be traversed in Example 13.9. From Deyst [12].

Although this is a deterministic problem, it will be formulated in terms of (13-70) and (13-71), to allow direct comparison to the stochastic problem example to follow. The state $x(t_i)$ is the scalar that indicates the vertical ordinate of the node occupied at time $t_i$ before the decision $u(t_i)$ is made. Let the control decisions be

$$u(t_i) = 1: \quad \text{go upward}$$
$$u(t_i) = 2: \quad \text{go downward}$$

and notice that at some possible points in the state space (e.g., $x(t_0) = 3$), only one of these decisions is available, thereby specifying $\mathscr{U}$. There are only two possible state transformations that can result from decisions in this example,

$$\phi_1[x(t_i), t_i] = x(t_i) + 1, \qquad k = 1$$
$$\phi_2[x(t_i), t_i] = x(t_i) - 1, \qquad k = 2$$

Since the decisions in this example deterministically specify the succeeding state transformations, we have

$$p[\phi_1 | x(t_i), u(t_i) = 1, t_i] = 1 \qquad p[\phi_1 | x(t_i), u(t_i) = 2, t_i] = 0$$
$$p[\phi_2 | x(t_i), u(t_i) = 1, t_i] = 0 \qquad p[\phi_2 | x(t_i), u(t_i) = 2, t_i] = 1$$

For this problem, we have the simplification that $L[x(t_i), x(t_{i+1}), u(t_i), t_i]$ is not a function of the decision $u(t_i)$, and this quantity is specified by the numbers on the network branches; for example,

$$L[3, 2, 1, t_0] = L[3, 2, 2, t_0] = 3$$

The dynamic programming algorithm for this problem is the iterative solution to

$$\mathscr{C}^*[x(t_i), t_i] = \min_{u \in \mathscr{U}} \begin{cases} L[x(t_i), x(t_i) + 1, 1, t_i] + \mathscr{C}^*[x(t_i) + 1, t_{i+1}] & \text{if} \quad u(t_i) = 1 \\ L[x(t_i), x(t_i) - 1, 2, t_i] + \mathscr{C}^*[x(t_i) - 1, t_{i+1}] & \text{if} \quad u(t_i) = 2 \end{cases}$$

as worked backwards from $L_f[x(t_3)]$ as given in Fig. 13.17. Thus, we start at $t_i = t_2$. Consider first the node at $x(t_2) = 3$: we can decide to take the upward branch to incur a path cost of 4 and ter-

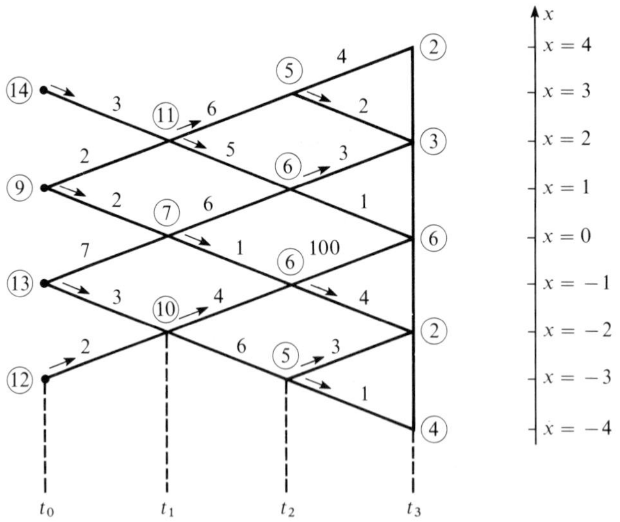


FIG. 13.18 Results of dynamic programming. From Deyst [12].

minal cost of 2, or the downward branch with costs of 2 and 3, respectively:

$$\mathscr{C}^*[x(t_2) = 3, t_2] = \min_{u \in \mathscr{U}} \begin{cases} 4 + 2 = 6 & \text{if} \quad u(t_2) = 1 \\ 2 + 3 = 5 & \text{if} \quad u(t_2) = 2 \end{cases}$$

Therefore, $\mathscr{C}^*[x(t_2) = 3, t_2] = 5$ and the optimal decision is to go downward, $u[x(t_2) = 3, t_2] = 2$. Continuing at $t_2$ yields the optimal costs-to-go encircled in Fig. 13.18 and the optimal decisions indicated by arrows.

Now we work backward to time $t_1$. Consider the node at $x(t_1) = 2$: we can decide to go upward, yielding the next single branch cost of 6 and *optimal* cost to complete the process from the resulting state $x(t_2) = 3$, *assuming* that we *will* make the optimal decision at time $t_2$, of 5; or we can decide to go downward with costs of 5 and 6:

$$\mathscr{C}^*[x(t_1) = 2, t_1] = \min_{u \in \mathscr{U}} \begin{cases} 6 + 5 = 11 & \text{if} \quad u(t_1) = 1 \\ 5 + 6 = 11 & \text{if} \quad u(t_1) = 2 \end{cases}$$

Here the optimal decision is not unique.

Progressing in this manner yields the results portrayed in Fig. 13.18, the optimal decision and associated cost-to-go for all possible state values at every stage. If we were free to choose the initial state at time $t_0$ at no extra cost, we would start at $x(t_0) = 1$, make downward decisions at each succeeding node, and incur a total cost of nine units. We have sequentially obtained the mappings $u^*[\cdot,\cdot]$ and $\mathscr{C}^*[\cdot,\cdot]$ as functions of the state space $R^1$ and the time set $T = \{t_0, t_1, t_2, t_3\}$, as tabulated below:

| | $u^*[\cdot,\cdot]$ | | | | | $\mathscr{C}^*[\cdot,\cdot]$ | | | |
|---|---|---|---|---|---|---|---|---|---|
| | $t_0$ | $t_1$ | $t_2$ | $t_3$ | | $t_0$ | $t_1$ | $t_2$ | $t_3$ |
| $x = 4$: | – | – | – | – | $x = 4$: | – | – | – | 2 |
| $x = 3$: | 2 | – | 2 | – | $x = 3$: | 14 | – | 5 | – |
| $x = 2$: | – | 1, 2 | – | – | $x = 2$: | – | 11 | – | 3 |
| $x = 1$: | 2 | – | 1 | – | $x = 1$: | 9 | – | 6 | – |
| $x = 0$: | – | 2 | – | – | $x = 0$: | – | 7 | – | 6 |
| $x = -1$: | 2 | – | 2 | – | $x = -1$: | 13 | – | 6 | – |
| $x = -2$: | – | 1 | – | – | $x = -2$: | – | 10 | – | 2 |
| $x = -3$: | 1 | – | 1, 2 | – | $x = -3$: | 12 | – | 5 | – |
| $x = -4$: | – | – | – | – | $x = -4$: | – | – | – | 4 |

■

EXAMPLE 13.10 Consider the same network and cost structure as in Fig. 13.17, but now let the problem be stochastic in nature. Assume that we no longer have perfect prediction of the outcomes of our decisions, but that

$$p[\phi_1 | x(t_i), u(t_i) = 1, t_i] = 0.8$$
$$p[\phi_2 | x(t_i), u(t_i) = 1, t_i] = 0.2$$
$$p[\phi_1 | x(t_i), u(t_i) = 2, t_i] = 0.3$$
$$p[\phi_2 | x(t_i), u(t_i) = 2, t_i] = 0.7$$

and that $\mathscr{U}$ still specifies that only the possible transition will occur for nodes from which a single branch emanates. Thus, decision $u(t_i) = 1$ corresponds to, "decide to go upward, with the result that an upward path will actually be taken with probability 0.8 and a downward path with probability 0.2," and similarly for $u(t_i) = 2$.

Applying the dynamic programming algorithm (13-70) at $x(t_2) = 3$ as in the previous example yields

$$\mathscr{C}^*[x(t_2) = 3, t_2] = \min_{u \in \mathscr{U}} \begin{cases} [(4)(0.8) + (2)(0.2)] + [(2)(0.8) + (3)(0.2)] = 5.8 & \text{if} \quad u(t_2) = 1 \\ [(4)(0.3) + (2)(0.7)] + [(2)(0.3) + (3)(0.7)] = 5.3 & \text{if} \quad u(t_2) = 2 \end{cases}$$

so that $\mathscr{C}^*[x(t_2) = 3, t_2] = 5.3$ and $u^*[x(t_2) = 3, t_2] = 2$, i.e., try to go downward. The full solution is presented in Fig. 13.19, where the arrows indicate the decision-maker's *preferred* direction: ↗ corresponds to $u(t_i) = 1$ and ↘ to $u(t_i) = 2$. The functions $u^*[\cdot,\cdot]$ and $\mathscr{C}^*[\cdot,\cdot]$ can again be completely tabulated with results taken from this figure.

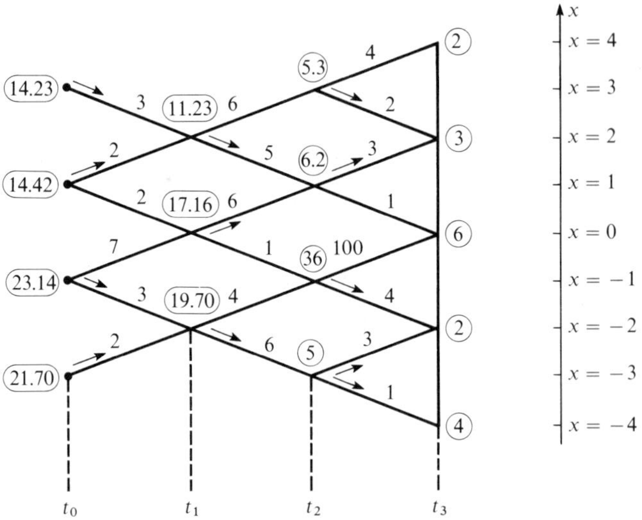


FIG. 13.19 Dynamic programming solution for stochastic controller. From Deyst [12].

Compare Figs. 13.18 and 13.19. Notice that the optimal decision function differs significantly in the two cases. This is an *important point*: direct extrapolations from deterministic optimal controllers do *not* always provide correct insights into stochastic control problem solutions, and the LQG problem is rather *unique* in that correct insights would be so provided. Moreover, there is not even a direct correspondence between points where the deterministic and stochastic optimal control problems have unique solutions necessarily, as seen at $x(t_1) = 2$. Finally, the value $x(t_0) = 3$, which was the least desirable initial state for the deterministic problem, now becomes the most desirable initial state for the stochastic version of the problem. The leftmost costs-to-go are costs that assume knowledge of $\mathrm{x}(t_0, \omega_j) = x(t_0)$. If desired, the a priori mean total cost can be evaluated as in (13-68):

$$E\{\mathscr{C}[\mathrm{x}(t_0), t_0]\} = \sum_{j=1}^{4} \mathscr{C}[x_j, t_0] p(\mathrm{x}(t_0) = x_j)$$

for $x_j$ taking on the values 3, 1, −1, and −3, provided one knows the required probabilities $p(\mathrm{x}(t_0) = x_j)$. ■

## 13.6 OPTIMAL STOCHASTIC CONTROL WITH NOISE-CORRUPTED MEASUREMENTS

Consider the control of a system described by the same dynamics models as in the previous section, (13-47) or the more general Markov discrete-time solutions to Itô stochastic differential equations (13-1) as viewed only at sample instants $t_i$. However, now assume perfect knowledge of the state $\mathbf{x}(t_i)$ at each sample time $t_i$ is *not* available from the system for use as an input to the controller. All that can be extracted from the system is the time history of measurements of the form (13-2), from $t_0$ to $t_i$ [1–3, 7–12, 19, 28, 29, 38, 40, 44]. Previously, we have often assumed arbitrarily that the first measurement is taken at time $t_1$, but this is easily modified to allow a measurement at $t_0$ as well. In this control application, we are more motivated to seek a measurement at $t_0$: without perfect knowledge of $\mathbf{x}(t_0)$ as assumed in the last section, it would be better to base the first control to apply at time $t_0$ on an *actual system output* such as $\mathbf{z}(t_0, \omega_k) = \mathbf{z}_0$ than to base it solely on a *mathematical model* such as the a priori density for $\mathbf{x}(t_0)$. Thus, we henceforth define the measurement history vector $\mathbf{Z}(t_i, \cdot)$ as

$$\mathbf{Z}(t_i) \triangleq [\mathbf{z}^{\mathrm{T}}(t_0), \mathbf{z}^{\mathrm{T}}(t_1), \ldots, \mathbf{z}^{\mathrm{T}}(t_i)]^{\mathrm{T}} \tag{13-72a}$$

and its realization similarly as

$$\mathbf{Z}_i \triangleq [\mathbf{z}_0^{\mathrm{T}}, \mathbf{z}_1^{\mathrm{T}}, \ldots, \mathbf{z}_i^{\mathrm{T}}]^{\mathrm{T}} \tag{13-72b}$$

The objective now is to determine an $r$-vector-valued control function $\mathbf{u}^*[\cdot,\cdot]$ as a function of the measurement history $\mathbf{Z}_i \in R^{(i+1)m}$ and time $t_i \in T$, to minimize a scalar cost-to-go function $\mathscr{C}[\cdot,\cdot]$, also a function of $R^{(i+1)m} \times T$. Of course, at time $t_i$ the controller can also remember the history of previous controls that it has applied, $\mathbf{u}(t_0), \mathbf{u}(t_1), \ldots, \mathbf{u}(t_{i-1})$. Since these inputs as well as the measured outputs $\mathbf{Z}_i$ provide information about the dynamic state of the system, it is convenient mathematically to denote the *information available to the controller at time* $t_i$ as

$$\mathbf{I}_i = \{\mathbf{z}_0^{\mathrm{T}}, [\mathbf{u}^{\mathrm{T}}(t_0), \mathbf{z}_1^{\mathrm{T}}], \ldots, [\mathbf{u}^{\mathrm{T}}(t_{i-1}), \mathbf{z}_i^{\mathrm{T}}]\}^{\mathrm{T}} \tag{13-73}$$

starting from $\mathbf{I}_0 = \mathbf{z}_0$. In (13-73) $\mathbf{I}_i$ can then be considered as a realization of the random vector $\mathbf{I}(t_i, \cdot)$ of dimension $\{i(m+r)+m\}$. In terms of this conceptualization, the objective is to determine an $r$-vector-valued control function $\mathbf{u}^*[\cdot,\cdot]$ as a function of the information $\mathbf{I}_i \in R^{i(m+r)+m}$ and time $t_i \in T$, to minimize a scalar cost-to-go function $\mathscr{C}[\cdot,\cdot]$, $R^{i(m+r)+m} \times T \to R^1$ defined by

$$\mathscr{C}[\mathbf{I}_i, t_i] \triangleq E\left\{ \sum_{j=1}^{N} L[\mathbf{x}(t_j), \mathbf{u}(t_j), t_j] + L_{\mathrm{f}}[\mathbf{x}(t_{N+1})] \,\middle|\, \mathbf{I}(t_i) = \mathbf{I}_i \right\} \tag{13-74}$$

which is similar to (13-51), except that now we condition upon knowledge of $\mathbf{I}_i$ instead of the $\mathbf{x}(t_i)$, which is no longer available. This objective will be accomplished in a manner analogous to that of the previous section.

First consider the cost function at the terminal time $t_{N+1}$. If we define a scalar function $\bar{L}_f$ by

$$\bar{L}_f[\mathbf{I}_{N+1}] \triangleq E\{L_f[\mathbf{x}(t_{N+1})] \,|\, \mathbf{I}(t_{N+1}) = \mathbf{I}_{N+1}\} \tag{13-75}$$

then (13-74) yields, for $i = N + 1$,

$$\mathscr{C}[\mathbf{I}_{N+1}, t_{N+1}] = E\{L_f[\mathbf{x}(t_{N+1})] \,|\, \mathbf{I}(t_{N+1}) = \mathbf{I}_{N+1}\} = \bar{L}_f[\mathbf{I}_{N+1}] \tag{13-76}$$

and similarly for $\mathscr{C}[\mathbf{I}(t_{N+1}), t_{N+1}]$. As discussed in conjunction with the comparable (13-53), we can also denote the result of (13-76) as $\mathscr{C}^*[\mathbf{I}_{N+1}, t_{N+1}]$.

Now step back to time $t_N$ to write

$$\begin{aligned}\mathscr{C}[\mathbf{I}_N, t_N] &= E\{L[\mathbf{x}(t_N), \mathbf{u}\{\mathbf{I}(t_N), t_N\}, t_N] + L_f[\mathbf{x}(t_{N+1})] \,|\, \mathbf{I}(t_N) = \mathbf{I}_N\} \\ &= \bar{L}[\mathbf{I}_N, \mathbf{u}\{\mathbf{I}_N, t_N\}, t_N] + E\{L_f[\mathbf{x}(t_{N+1})] \,|\, \mathbf{I}(t_N) = \mathbf{I}_N\}\end{aligned} \tag{13-77}$$

where we have defined the $\bar{L}$ function at $t_N$ via

$$\bar{L}[\mathbf{I}_N, \mathbf{u}\{\mathbf{I}_N, t_N\}, t_N] \triangleq E\{L[\mathbf{x}(t_N), \mathbf{u}\{\mathbf{I}(t_N), t_N\}, t_N] \,|\, \mathbf{I}(t_N) = \mathbf{I}_N\} \tag{13-78}$$

But, since

$$\mathbf{I}(t_{N+1}) = \{\mathbf{I}^{\mathrm{T}}(t_N), \mathbf{u}^{\mathrm{T}}(t_N), \mathbf{z}^{\mathrm{T}}(t_{N+1})\}^{\mathrm{T}}$$

we can write the last term in (13-77) via (12-22) of Volume 2 as

$$\begin{aligned}E\{L_f[\mathbf{x}(t_{N+1})] \,|\, \mathbf{I}(t_N) = \mathbf{I}_N\} &= E\{E\{L_f[\mathbf{x}(t_{N+1})] \,|\, \mathbf{I}(t_{N+1}) = \mathbf{I}(t_{N+1})\} \,|\, \mathbf{I}(t_N) = \mathbf{I}_N\} \\ &= E\{\bar{L}_f[\mathbf{I}(t_{N+1})] \,|\, \mathbf{I}(t_N) = \mathbf{I}_N\} \\ &= E\{\mathscr{C}^*[\mathbf{I}(t_{N+1}), t_{N+1}] \,|\, \mathbf{I}(t_N) = \mathbf{I}_N\}\end{aligned} \tag{13-79}$$

where the second equality uses (13-75), and the last equality invokes (13-76) and the discussion following it. Substituting back into (13-77) yields

$$\mathscr{C}[\mathbf{I}_N, t_N] = \bar{L}[\mathbf{I}_N, \mathbf{u}\{\mathbf{I}_N, t_N\}, t_N] + E\{\mathscr{C}^*[\mathbf{I}(t_{N+1}), t_{N+1}] \,|\, \mathbf{I}(t_N) = \mathbf{I}_N\} \tag{13-80}$$

The value of this cost is totally determined once we specify the particular control function $\mathbf{u}\{\cdot, t_N\}$ to be employed, and we wish to use the cost-minimizing control $\mathbf{u}^*\{\cdot, t_N\}$, which is generated by solving [2, 7, 20]:

$$\mathscr{C}^*[\mathbf{I}_N, t_N] = \min_{\mathbf{u} \in \mathscr{U}} \{\bar{L}[\mathbf{I}_N, \mathbf{u}\{\mathbf{I}_N, t_N\}, t_N] + E\{\mathscr{C}^*[\mathbf{I}(t_{N+1}), t_{N+1}] \,|\, \mathbf{I}(t_N) = \mathbf{I}_N\}\} \tag{13-81}$$

which is directly comparable to (13-58).

Stepping back to time $t_{N-1}$, we can write

$$\begin{aligned}\mathscr{C}[\mathbf{I}_{N-1}, t_{N-1}] &= E\{L[\mathbf{x}(t_{N-1}), \mathbf{u}\{\mathbf{I}(t_{N-1}), t_{N-1}\}, t_{N-1}] \\ &\quad + L[\mathbf{x}(t_N), \mathbf{u}\{\mathbf{I}(t_N), t_N\}, t_N] + L_f[\mathbf{x}(t_{N+1})] \,|\, \mathbf{I}(t_{N-1}) = \mathbf{I}_{N-1}\} \\ &= \bar{L}[\mathbf{I}_{N-1}, \mathbf{u}\{\mathbf{I}_{N-1}, t_{N-1}\}, t_{N-1}] + E\{L[\mathbf{x}(t_N), \mathbf{u}\{\mathbf{I}(t_N), t_N\}, t_N] \\ &\quad + L_f[\mathbf{x}(t_{N+1})] \,|\, \mathbf{I}(t_{N-1}) = \mathbf{I}_{N-1}\}\end{aligned} \tag{13-82}$$

where $\bar{L}$ is defined at $t_{N-1}$ as in (13-78), with $N$ replaced by $N-1$ throughout. As done at time $t_N$, we can evaluate the last expectation in (13-82) as

$$\begin{aligned} E\{\mathsf{L}+\mathsf{L}_{\mathrm{f}} | \mathbf{I}(t_{N-1}) = \mathbf{I}_{N-1}\} &= E\{E\{\mathsf{L}+\mathsf{L}_{\mathrm{f}} | \mathbf{I}(t_N) = \mathbf{I}(t_N)\} | \mathbf{I}(t_{N-1}) = \mathbf{I}_{N-1}\} \\ &= E\{\mathscr{C}[\mathbf{I}(t_N), t_N] | \mathbf{I}(t_{N-1}) = \mathbf{I}_{N-1}\} \end{aligned} \tag{13-83}$$

where the last equality uses the definition (13-74) for $j = N$. Thus, from (13-82) and (13-83),

$$\begin{aligned} \mathscr{C}[\mathbf{I}_{N-1}, t_{N-1}] = \bar{L}[\mathbf{I}_{N-1}, \mathbf{u}\{\mathbf{I}_{N-1}, t_{N-1}\}, t_{N-1}] \\ + E\{\mathscr{C}[\mathbf{I}(t_N), t_N] | \mathbf{I}(t_{N-1}) = \mathbf{I}_{N-1}\} \end{aligned} \tag{13-84}$$

Similar to the discussion following (13-63), the function $\mathscr{C}[\cdot, t_N]$ in the last term of this expression should always be $\mathscr{C}^*[\cdot, t_N]$ in order to minimize the overall cost, *independent* of the choice of the function $\mathbf{u}\{\cdot, t_{N-1}\}$. Then the cost minimizing control is found from solving

$$\begin{aligned} \mathscr{C}^*[\mathbf{I}_{N-1}, t_{N-1}] = \min_{\mathbf{u} \in \mathscr{U}} \{\bar{L}[\mathbf{I}_{N-1}, \mathbf{u}\{\mathbf{I}_{N-1}, t_{N-1}\}, t_{N-1}] \\ + E\{\mathscr{C}^*[\mathbf{I}(t_N), t_N] | \mathbf{I}(t_{N-1}) = \mathbf{I}_{N-1}\}\} \end{aligned} \tag{13-85}$$

By induction, the *stochastic dynamic programming recursion* to solve backwards in time is

$$\mathscr{C}^*[\mathbf{I}_i, t_i] = \min_{\mathbf{u} \in \mathscr{U}} \{\bar{L}[\mathbf{I}_i, \mathbf{u}\{\mathbf{I}_i, t_i\}, t_i] + E\{\mathscr{C}^*[\mathbf{I}(t_{i+1}), t_{i+1}] | \mathbf{I}(t_i) = \mathbf{I}_i\}\} \tag{13-86}$$

with a *terminal condition* at $t_{N+1}$ of

$$\mathscr{C}^*[\mathbf{I}_{N+1}, t_{N+1}] = \bar{L}_{\mathrm{f}}[\mathbf{I}_{N+1}] \tag{13-87}$$

where the $\bar{L}$ and $\bar{L}_{\mathrm{f}}$ functions are defined by

$$\bar{L}[\mathbf{I}_i, \mathbf{u}\{\mathbf{I}_i, t_i\}, t_i] \triangleq E\{L[\mathbf{x}(t_i), \mathbf{u}\{\mathbf{I}(t_i), t_i\}, t_i] | \mathbf{I}(t_i) = \mathbf{I}_i\} \tag{13-88a}$$

$$\bar{L}_{\mathrm{f}}[\mathbf{I}_{N+1}] \triangleq E\{L_{\mathrm{f}}[\mathbf{x}(t_{N+1})] | \mathbf{I}(t_{N+1}) = \mathbf{I}_{N+1}\} \tag{13-88b}$$

In a general problem in which $m$ (the measurement dimension), $r$ (the control dimension), and especially $N$ (the number of sample periods) can be large, the computational and memory requirements for calculating the $\mathbf{u}^*[\cdot,\cdot]$ function can readily become prohibitive. Even if $\mathbf{u}^*[\cdot,\cdot]$ can be so generated, it requires online storage of the entire growing-length $\mathbf{I}_i$ for all $i$ for eventual controller implementation, which is another significant drawback.

Thus, we are strongly motivated to seek a means of data reduction: we desire to express the information embodied in $\mathbf{I}(t_i, \omega_k) = \mathbf{I}_i$ by means of a random vector and particular realization having a smaller dimension, and ideally a dimension that does not grow with time. Such random vectors are known as *sufficient statistics* [11, 12, 40, 41]. A known function $\mathbf{S}_i$ of the random vector $\mathbf{I}(t_i, \cdot)$ is a *statistic* if $\mathbf{S}_i[\mathbf{I}(t_i, \cdot)]$ is itself a random variable mapping. The set of functions $\mathbf{S}_0[\cdot]$, $\mathbf{S}_1[\cdot], \ldots, \mathbf{S}_N[\cdot]$ constitute a set of *sufficient statistics* with respect to the problem under consideration if the right hand side term being minimized in (13-86) can be expressed solely as a function of $\mathbf{S}_i[\mathbf{I}_i]$, $\mathbf{u}(t_i)$, and $t_i$ for all $i$ and all admissible $\mathbf{u}(t_i)$. When the minimization is performed, it is clear that the optimal control function can be expressed as a function of $\mathbf{S}(t_i) \triangleq \mathbf{S}_i[\mathbf{I}(t_i)]$; its only dependence upon the information vector $\mathbf{I}(t_i)$ is through the sufficient statistic itself:

$$\mathbf{u}^*(t_i) = \mathbf{u}^*\{\mathbf{S}(t_i), t_i\} \triangleq \mathbf{u}^*\{\mathbf{S}_i[\mathbf{I}(t_i)], t_i\} \tag{13-89}$$

Therefore, *in terms of such sufficient statistics, the dynamic programming algorithm* is given by

$$\mathscr{C}^*[\mathbf{S}_i, t_i] = \min_{\mathbf{u} \in \mathscr{U}} \{\bar{L}'[\mathbf{S}_i, \mathbf{u}\{\mathbf{S}_i, t_i\}, t_i] + E\{\mathscr{C}^*[\mathbf{S}(t_{i+1}), t_{i+1}] \,|\, \mathbf{S}(t_i) = \mathbf{S}_i\}\} \tag{13-90}$$

solved backward recursively from the *terminal condition* at $t_{N+1}$ of

$$\mathscr{C}^*[\mathbf{S}_{N+1}, t_{N+1}] = \bar{L}_{\mathrm{f}}'[\mathbf{S}_{N+1}] \tag{13-91}$$

where the $\bar{L}'$ and $\bar{L}_{\mathrm{f}}'$ functions are defined by

$$\bar{L}'[\mathbf{S}_i, \mathbf{u}\{\mathbf{S}_i, t_i\}, t_i] \triangleq E\{L[\mathbf{x}(t_i), \mathbf{u}\{\mathbf{S}(t_i), t_i\}, t_i] \,|\, \mathbf{S}(t_i) = \mathbf{S}_i\} \tag{13-92a}$$

$$\bar{L}_{\mathrm{f}}'[\mathbf{S}_{N+1}] \triangleq E\{L_{\mathrm{f}}[\mathbf{x}(t_{N+1})] \,|\, \mathbf{S}(t_{N+1}) = \mathbf{S}_{N+1}\} \tag{13-92b}$$

Now a trivial case of a sufficient statistic is for $\mathbf{S}_i(\cdot)$ to be the appropriately dimensioned identity function that maps $\mathbf{I}(t_i)$ into itself for all $i$,

$$\mathbf{S}(t_i) \triangleq \mathbf{S}_i[\mathbf{I}(t_i)] = \mathbf{I}(t_i) \tag{13-93}$$

This is the trivial case since (13-90)–(13-92) become identical to (13-86)–(13-88), and no advantage is gained. However, if the sufficient statistic is characterized by fewer numbers than $\mathbf{I}(t_i)$, i.e., if $\mathbf{S}_i$ maps $R^{i(m+r)+m}$ into $R^q$ with $q < [i(m+r)+m]$ for all $i$ beyond some specified $i_0$ and particularly if $q$ does not grow with $i$, then the control law given by (13-89) might be significantly easier both to generate and to implement. One might consider the conditional density for the state at $t_i$ given the information vector $\mathbf{I}_i$ or, if possible, a finite-dimensional vector of parameters to characterize that density, as the sufficient statistic $\mathbf{S}(t_i)$.

To make this development less abstract, let us consider an important special case in which this conceptualization can be exploited fruitfully: the case in which adequate system models are in the form of *linear models driven by white Gaussian noise.* We do *not* yet impose the quadratic cost assumption of the LQG problem discussed in Section 13.3, but will see how far we can progress towards a useful controller form without this assumption. Thus, let the defining state equation be

$$\mathbf{x}(t_{i+1}) = \mathbf{\Phi}(t_{i+1}, t_i)\mathbf{x}(t_i) + \mathbf{B}_{\mathrm{d}}(t_i)\mathbf{u}(t_i) + \mathbf{G}_{\mathrm{d}}(t_i)\mathbf{w}_{\mathrm{d}}(t_i) \tag{13-94}$$

where $\mathbf{w}_d(\cdot,\cdot)$ is $s$-dimensional zero-mean white Gaussian discrete-time noise of covariance $\mathbf{Q}_d(t_i)$ for all $t_i$, independent of the initial condition $\mathbf{x}(t_0,\cdot)$, modeled as Gaussian and of mean $\hat{\mathbf{x}}_0$ and covariance $\mathbf{P}_0$. Equation (13-94) is viewed as an equivalent discrete-time model for a continuous-time system from which sampled-data measurements are available at $t_0, t_1, t_2$, etc. The available measurements are of the form

$$\mathbf{z}(t_i) = \mathbf{H}(t_i)\mathbf{x}(t_i) + \mathbf{v}(t_i) \tag{13-95}$$

where $\mathbf{v}(\cdot,\cdot)$ is zero-mean white Gaussian discrete-time noise of covariance $\mathbf{R}(t_i)$, and it is independent of both $\mathbf{x}(t_0,\cdot)$ and $\mathbf{w}_d(\cdot,\cdot)$. Now the objective is to find the optimal control function $\mathbf{u}^*[\cdot,\cdot]$ to minimize a cost-to-go function expressed as a function of the information vector and current time, and eventually as a function of an appropriate fixed-dimension sufficient statistic and current time. The total cost to be minimized from $t_0$ to final time $t_{N+1}$ is given by

$$J = E\left\{\sum_{i=0}^{N} L[\mathbf{x}(t_i), \mathbf{u}(t_i), t_i] + L_f[\mathbf{x}(t_{N+1})]\right\} \tag{13-96}$$

where the $L$ and $L_f$ functions are specified (Baire) functions. Thus we have the same problem formulation as in Section 13.3, except that (13-96) is a more general cost than (13-11).

In previous filtering developments associated with systems described by (13-94) and (13-95), we concentrated on the case of $\mathbf{u}(t_i) \equiv \mathbf{0}$, and then let $\mathbf{u}(t_i)$ be a known open-loop control for all $t_i$ and noted its presence only affected the conditional mean propagation in the associated Kalman filter (see Chapter 5 of Volume 1). Now we want to consider carefully state estimation when the system is forced by a feedback control law $\mathbf{u}^*[\cdot,\cdot]$ as a function of either $\mathbf{I}_i$ and $t_i$, or $\mathbf{S}_i$ and $t_i$. As we do so, the natural and desirable fixed-dimension sufficient statistic will become apparent.

For convenience, let us define two pseudostates [11, 12] for each $t_i$, "deterministic" $\mathbf{x}^d(t_i)$ and "stochastic" $\mathbf{x}^s(t_i)$, such that they satisfy

$$\mathbf{x}^d(t_{i+1}) = \mathbf{\Phi}(t_{i+1}, t_i)\mathbf{x}^d(t_i) + \mathbf{B}_d(t_i)\mathbf{u}(t_i) \tag{13-97a}$$

$$\mathbf{x}^s(t_{i+1}) = \mathbf{\Phi}(t_{i+1}, t_i)\mathbf{x}^s(t_i) + \mathbf{G}_d(t_i)\mathbf{w}_d(t_i) \tag{13-97b}$$

starting from initial conditions $\mathbf{x}^d(t_0) = \mathbf{0}$ and $\mathbf{x}^s(t_0) = \mathbf{x}(t_0)$. Note that $\mathbf{x}^d(t_{i+1})$ can be calculated exactly, deterministically at time $t_i$ after $\mathbf{u}^*(t_i)$ is calculated: the optimal control $\mathbf{u}^*[\cdot, t_i]$ is a function of $\mathbf{I}(t_i, \omega_k) = \mathbf{I}_i$ and is not known perfectly *a priori*, but once $\mathbf{u}^*[\mathbf{I}_i, t_i]$ is calculated at time $t_i$, this is a *known* input. Thus, $\mathbf{x}^d(t_i)$ is deterministic and contains all effects due to the control inputs, whereas $\mathbf{x}^s(t_i)$ embodies all of the randomness in the state description. Because we are considering a *linear* system description, the system state can be expressed as the sum of these two pseudostates:

$$\mathbf{x}(t_i) = \mathbf{x}^d(t_i) + \mathbf{x}^s(t_i) \qquad \text{for all} \quad t_i \tag{13-98}$$

Similarly, the measurement $\mathbf{z}(t_i)$ can be decomposed into two pseudo-measurements,

$$\mathbf{z}(t_i) = \mathbf{z}^{\mathrm{d}}(t_i) + \mathbf{z}^{\mathrm{s}}(t_i) \tag{13-99}$$

where

$$\mathbf{z}^{\mathrm{d}}(t_i) = \mathbf{H}(t_i)\mathbf{x}^{\mathrm{d}}(t_i) \tag{13-100a}$$

$$\mathbf{z}^{\mathrm{s}}(t_i) = \mathbf{H}(t_i)\mathbf{x}^{\mathrm{s}}(t_i) + \mathbf{v}(t_i) \tag{13-100b}$$

Thus $\mathbf{z}^{\mathrm{d}}(t_i)$ comprises that part of the measurement that we know exactly, while $\mathbf{z}^{\mathrm{s}}(t_i)$ is composed of that portion of the measurement which is not completely deterministic. Now the problem description is separated into the sum of two distinct parts.

Let us first consider estimation of $\mathbf{x}^{\mathrm{s}}(t_i)$. This can be accomplished conceptually by generating pseudomeasurements from available physical measurements via

$$\mathbf{z}^{\mathrm{s}}(t_i) = \mathbf{z}(t_i) - \mathbf{z}^{\mathrm{d}}(t_i) = \mathbf{z}(t_i) - \mathbf{H}(t_i)\mathbf{x}^{\mathrm{d}}(t_i) \tag{13-101}$$

These can be provided as input to a Kalman filter to estimate $\mathbf{x}^{\mathrm{s}}(t_i)$, and the error in this estimate would be

$$\mathbf{e}(t_i^-) \triangleq \mathbf{x}^{\mathrm{s}}(t_i) - \hat{\mathbf{x}}^{\mathrm{s}}(t_i^-) = \mathbf{x}(t_i) - \mathbf{x}^{\mathrm{d}}(t_i) - \hat{\mathbf{x}}^{\mathrm{s}}(t_i^-) \tag{13-102a}$$

$$\mathbf{e}(t_i^+) \triangleq \mathbf{x}^{\mathrm{s}}(t_i) - \hat{\mathbf{x}}^{\mathrm{s}}(t_i^+) = \mathbf{x}(t_i) - \mathbf{x}^{\mathrm{d}}(t_i) - \hat{\mathbf{x}}^{\mathrm{s}}(t_i^+) \tag{13-102b}$$

before and after measurement incorporation at time $t_i$, respectively. The statistics of these Gaussian errors would include

$$E\{\mathbf{e}(t_i^-)\} = \mathbf{0};\ E\{\mathbf{e}(t_i^-)\mathbf{e}^{\mathrm{T}}(t_i^-)\} = \mathbf{P}(t_i^-) \tag{13-103a}$$

$$E\{\mathbf{e}(t_i^+)\} = \mathbf{0};\ E\{\mathbf{e}(t_i^+)\mathbf{e}^{\mathrm{T}}(t_i^+)\} = \mathbf{P}(t_i^+) \tag{13-103b}$$

with covariances as calculated by the filter algorithm itself (assuming for now that the filter is not of reduced order or mismatched with the "truth model" description, as discussed in Chapter 6 of Volume 1). To perform such an estimation procedure, we could define a pseudomeasurement history in the same manner as the measurement history $\mathbf{Z}(t_i)$ was previously defined:

$$\mathbf{Z}^{\mathrm{s}}(t_i) \triangleq \begin{bmatrix} \mathbf{z}^{\mathrm{s}}(t_0) \\ \mathbf{z}^{\mathrm{s}}(t_1) \\ \vdots \\ \mathbf{z}^{\mathrm{s}}(t_i) \end{bmatrix} \tag{13-104}$$

From Section 5.4 of Volume 1, we know that $\mathbf{e}(t_i^+)$ and $\mathbf{Z}^{\mathrm{s}}(t_i)$ are independent random vectors.

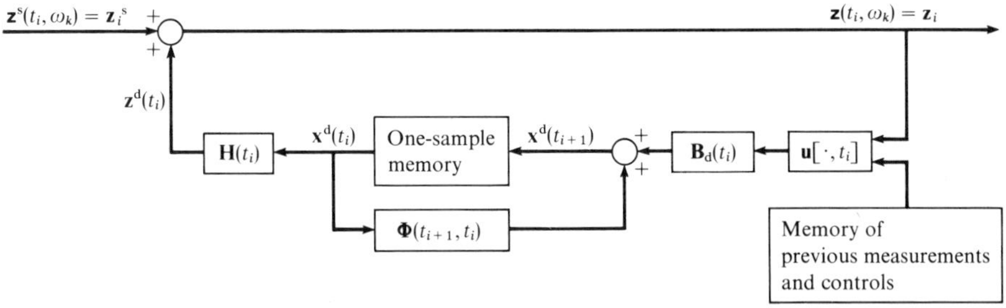


FIG. 13.20 $\mathbf{Z}(t_i)$ is a deterministic function of $\mathbf{Z}^s(t_i)$.

Now we want to show that $\mathbf{e}(t_i^+)$ is also independent of $\mathbf{Z}(t_i)$ itself in the current problem. To do so, assume that the control function $\mathbf{u}[\cdot, t_i]$ is a *prescribed* function of the information vector $\mathbf{I}_i$ (history of measurements through $t_i$ and of previous controls through $t_{i-1}$), so that the control applied at $t_i$ is

$$\mathbf{u}(t_i) = \mathbf{u}[\mathbf{I}_i, t_i] \tag{13-105}$$

With such a control function, it can be shown that $\mathbf{Z}(t_i)$ is a *deterministic function* of $\mathbf{Z}^s(t_i)$; in other words, given the entire pseudomeasurement history $\mathbf{Z}^s(t_i, \omega_k) = \mathbf{Z}_i^s$, we can generate the measurement history $\mathbf{Z}(t_i, \omega_k) = \mathbf{Z}_i$, and vice versa. This is demonstrated graphically in Fig. 13.20. Note that this conceptually depends on (1) $\mathbf{u}[\cdot, t_i]$ being defined for all $t_i$ and (2) the state dynamics and measurement models being linear so that superposition is valid as in (13-97)–(13-101). Also note that the indicated operations in the figure are well defined even at $t_0$ since $\mathbf{x}^d(t_0)$ is defined to be $\mathbf{0}$. Moreover, the block labeled "memory of previous measurements and controls" is a conceptual construction only, and actual memory of these histories will *not* be required in the eventual implementation. Therefore, since $\mathbf{e}(t_i^+)$ and $\mathbf{Z}^s(t_i)$ are independent and $\mathbf{Z}(t_i)$ is a deterministic function of $\mathbf{Z}^s(t_i)$, then $\mathbf{e}(t_i^+)$ and $\mathbf{Z}(t_i)$ are independent.

Now consider estimation of the total state $\mathbf{x}(t_i)$, defined through

$$\hat{\mathbf{x}}(t_i^-) = \mathbf{x}^d(t_i) + \hat{\mathbf{x}}^s(t_i^-) \tag{13-106a}$$

$$\hat{\mathbf{x}}(t_i^+) = \mathbf{x}^d(t_i) + \hat{\mathbf{x}}^s(t_i^+) \tag{13-106b}$$

By separately computing $\mathbf{x}^d(t_i)$ via (13-97a) and $\hat{\mathbf{x}}^s(t_i^-)$ and $\hat{\mathbf{x}}^s(t_i^+)$ via the pseudostate Kalman filter and adding the results, this state estimate is given by a standard Kalman filter algorithm, as expected. The measurement update for the state estimate at time $t_i$ is given by

$$\mathbf{K}(t_i) = \mathbf{P}(t_i^-)\mathbf{H}^T(t_i)[\mathbf{H}(t_i)\mathbf{P}(t_i^-)\mathbf{H}^T(t_i) + \mathbf{R}(t_i)]^{-1} \tag{13-107}$$

$$\hat{\mathbf{x}}(t_i^+) = \hat{\mathbf{x}}(t_i^-) + \mathbf{K}(t_i)[\mathbf{z}_i - \mathbf{H}(t_i)\hat{\mathbf{x}}(t_i^-)] \tag{13-108}$$

$$\mathbf{P}(t_i^+) = \mathbf{P}(t_i^-) - \mathbf{K}(t_i)\mathbf{H}(t_i)\mathbf{P}(t_i^-) \tag{13-109}$$

and the time propagation to the next sample time is

$$\hat{\mathbf{x}}(t_{i+1}^-) = \mathbf{\Phi}(t_{i+1}, t_i)\hat{\mathbf{x}}(t_i^+) + \mathbf{B}_d(t_i)\mathbf{u}(t_i) \tag{13-110}$$

$$\mathbf{P}(t_{i+1}^-) = \mathbf{\Phi}(t_{i+1}, t_i)\mathbf{P}(t_i^+)\mathbf{\Phi}^T(t_{i+1}, t_i) + \mathbf{G}_d(t_i)\mathbf{Q}_d(t_i)\mathbf{G}_d^T(t_i) \tag{13-111}$$

with $\mathbf{u}(t_i)$ given by (13-105). The appropriate initial conditions for this recursion are

$$\hat{\mathbf{x}}(t_0) = \hat{\mathbf{x}}_0 \tag{13-112a}$$

$$\mathbf{P}(t_0) = \mathbf{P}_0 \tag{13-112b}$$

As discussed previously, the time histories of (13-107), (13-109), and (13-111) can be precomputed and stored.

The error $\mathbf{e}(t_i^+)$ that was shown to be independent of $\mathbf{Z}(t_i)$ is also the error in this total state estimator:

$$\begin{aligned}\mathbf{e}(t_i^+) &= \mathbf{x}^s(t_i) - \hat{\mathbf{x}}^s(t_i^+) \\ &= [\mathbf{x}^d(t_i) + \mathbf{x}^s(t_i)] - [\mathbf{x}^d(t_i) + \hat{\mathbf{x}}^s(t_i^+)] \\ &= \mathbf{x}(t_i) - \hat{\mathbf{x}}(t_i^+)\end{aligned} \tag{13-113}$$

Furthermore, rearranging this yields

$$\mathbf{x}(t_i) = \hat{\mathbf{x}}(t_i^+) + \mathbf{e}(t_i^+) \tag{13-114}$$

where $\hat{\mathbf{x}}(t_i^+)$ is a *function* of $\mathbf{Z}(t_i)$ (or $\mathbf{I}(t_i)$), and $\mathbf{e}(t_i^+)$ is *independent* of $\mathbf{Z}(t_i)$ (or $\mathbf{I}(t_i)$) and is a Gaussian vector with mean zero and covariance $\mathbf{P}(t_i^+)$. The conditional density for $\mathbf{x}(t_i)$, conditioned on the entire time history of measurements and knowledge of applied controls $\mathbf{u}(t_0)$ through $\mathbf{u}(t_{i-1})$, is the Gaussian density

$$f_{\mathbf{x}(t_i)|\mathbf{I}(t_i)}(\boldsymbol{\xi}\,|\,\mathbf{I}_i) = \frac{1}{(2\pi)^{n/2}|\mathbf{P}(t_i^+)|^{1/2}} \exp\left\{-\frac{1}{2}[\boldsymbol{\xi} - \hat{\mathbf{x}}(t_i^+)]^T\mathbf{P}(t_i^+)^{-1}[\boldsymbol{\xi} - \hat{\mathbf{x}}(t_i^+)]\right\} \tag{13-115}$$

Note that $\mathbf{P}(t_i^+)$ is independent of $\mathbf{I}_i$ and that knowledge of $\hat{\mathbf{x}}(t_i^+)$ and $\mathbf{P}(t_i^+)$ is enough to define this density completely: $\hat{\mathbf{x}}(t_i^+)$ is a *sufficient statistic* for this estimation problem since it is sufficient to know its value, rather than that of $\mathbf{I}(t_i)$, in order to specify the density completely.

One last pertinent aspect of the estimation problem will be reviewed before considering the evaluation of the optimal control function. Recall the *new information* process $\mathbf{s}(\cdot,\cdot)$ defined in Section 5.4 of Volume 1 by

$$\mathbf{s}(t_i) \triangleq \mathbf{K}(t_i)[\mathbf{z}(t_i) - \mathbf{H}(t_i)\hat{\mathbf{x}}(t_i^-)] \tag{13-116}$$

which, for a particular information vector realization $\mathbf{I}(t_i, \omega_k) = \mathbf{I}_i$, is the term added to $\hat{\mathbf{x}}(t_i^-)$ to obtain $\hat{\mathbf{x}}(t_i^+)$, i.e., the filter gain $\mathbf{K}(t_i)$ times the residual. This $\mathbf{s}(t_i)$ was shown to be independent of the previous measurement history $\mathbf{Z}(t_{i-1})$

(and previous controls too), and the $\mathbf{s}(\cdot,\cdot)$ process was shown to be *white, Gaussian, zero-mean, and with covariance*

$$E\{\mathbf{s}(t_i)\mathbf{s}^{\mathrm{T}}(t_i)\} = \mathbf{K}(t_i)\mathbf{H}(t_i)\mathbf{P}(t_i^-) \tag{13-117}$$

which is recognized as the term subtracted from $\mathbf{P}(t_i^-)$ to yield $\mathbf{P}(t_i^+)$. Thus, we can conceptually write the density $f_{\mathbf{s}(t_i)|\mathbf{z}(t_{i-1})}(\boldsymbol{\xi}|\mathbf{Z}_{i-1}) = f_{\mathbf{s}(t_i)}(\boldsymbol{\xi})$ as Gaussian with mean zero and covariance as in (13-117); since $\mathbf{K}(t_i)\mathbf{H}(t_i)\mathbf{P}(t_i^-)$ is $n$-by-$n$ and of rank $m < n$ typically, a characteristic function description is actually more appropriate than a density function.

Now consider generating the optimal control law. The cost $\mathscr{C}^*[\mathbf{I}_i, t_i]$ is the minimum expected cost to complete the process from time $t_i$ forward, given the measurement history $\mathbf{Z}(t_i, \omega_k) = \mathbf{Z}_i$ that resulted from using an *admissible* (not necessarily optimal) control function $\mathbf{u}[\cdot, t_j]$ for $t_0 \le t_j \le t_{i-1}$ to generate the known values $\{\mathbf{u}(t_0), \mathbf{u}(t_1), \ldots, \mathbf{u}(t_{i-1})\}$, and assuming that the *optimal* admissible control function $\mathbf{u}^*[\cdot, t_j]$ will be used in the interval $t_i \le t_j \le t_N$. As usual, to generate $\mathscr{C}^*[\cdot,\cdot]$ and $\mathbf{u}^*[\cdot,\cdot]$ recursively, we start at the terminal time $t_{N+1}$ to write

$$\mathscr{C}^*[\mathbf{I}_{N+1}, t_{N+1}] = E\{L_{\mathrm{f}}[\mathbf{x}(t_{N+1})] | \mathbf{I}(t_{N+1}) = \mathbf{I}_{N+1}\} \tag{13-118}$$

as in (13-76), or (13-87) and (13-88b). Using the concept of a sufficient statistic, this can be written as

$$\begin{aligned}\mathscr{C}^*[\mathbf{I}_{N+1}, t_{N+1}] &= \int_{-\infty}^{\infty} L_{\mathrm{f}}(\boldsymbol{\xi}) f_{\mathbf{x}(t_{N+1})|\mathbf{I}(t_{N+1})}(\boldsymbol{\xi}|\mathbf{I}_{N+1})\, d\boldsymbol{\xi} \\ &= \int_{-\infty}^{\infty} L_{\mathrm{f}}(\boldsymbol{\xi}) f_{\mathbf{x}(t_{N+1})|\hat{\mathbf{x}}(t_{N+1}^+)}(\boldsymbol{\xi}|\hat{\mathbf{x}}(t_{N+1}^+))\, d\boldsymbol{\xi} \\ &\triangleq \bar{L}_{\mathrm{f}}'[\hat{\mathbf{x}}(t_{N+1}^+)]\end{aligned} \tag{13-119}$$

thereby establishing (13-91) and (13-92b) for this problem. The notation $f_{\mathbf{x}(t_{N+1})|\hat{\mathbf{x}}(t_{N+1}^+)}(\boldsymbol{\xi}|\hat{\mathbf{x}}(t_{N+1}^+))$ is meant to denote that the conditional density $f_{\mathbf{x}(t_{N+1})|\mathbf{I}(t_{N+1})}(\boldsymbol{\xi}|\mathbf{I}_{N+1})$ is totally specified by the sufficient statistic

$$\hat{\mathbf{x}}(t_{N+1}^+, \cdot) = \hat{\mathbf{x}}_{N+1}^+[\mathbf{I}(t_{N+1}, \cdot)] \tag{13-120}$$

EXAMPLE 13.11 To see a practical instance of (13-119), let the state be scalar and let $L_{\mathrm{f}}[\mathrm{x}(t_{N+1})] = \mathrm{x}^2(t_{N+1})$. Then, using (13-114),

$$\begin{aligned}\mathscr{C}^*[\mathbf{I}_{N+1}, t_{N+1}] &= E\{L_{\mathrm{f}}[\mathrm{x}(t_{N+1})] | \mathbf{I}(t_{N+1}) = \mathbf{I}_{N+1}\} \\ &= E\{L_{\mathrm{f}}[\hat{\mathrm{x}}(t_{N+1}^+) + \mathrm{e}(t_{N+1}^+)] | \mathbf{I}(t_{N+1}) = \mathbf{I}_{N+1}\} \\ &= E\{[\hat{\mathrm{x}}(t_{N+1}^+) + \mathrm{e}(t_{N+1}^+)]^2 | \mathbf{I}(t_{N+1}) = \mathbf{I}_{N+1}\} \\ &= \hat{x}^2(t_{N+1}^+) + P(t_{N+1}^+) \\ &\triangleq \bar{L}_{\mathrm{f}}'[\hat{x}(t_{N+1}^+)]\end{aligned}$$

Note that $\bar{L}_{\mathrm{f}}'[\cdot]$ typically differs from $L_{\mathrm{f}}[\cdot]$, and that $P(t_{N+1}^+)$ is not included in the arguments of $\bar{L}_{\mathrm{f}}'$ because it is a known, precomputable quantity independent of $\mathbf{I}_{N+1}$. ■

To emphasize the fact that $\mathscr{C}^*[\mathbf{I}_{N+1}, t_{N+1}]$ is a function of $\mathbf{I}_{N+1}$ only through the sufficient statistic realization $\hat{\mathbf{x}}(t_{N+1}^+)$, we can define $\mathscr{C}^*[\hat{\mathbf{x}}(t_{N+1}^+), t_{N+1}]$ as in (13-91) by making this notational change in (13-119):

$$\mathscr{C}^*[\hat{\mathbf{x}}(t_{N+1}^+), t_{N+1}] = \bar{L}_f'[\hat{\mathbf{x}}(t_{N+1}^+)] \tag{13-121}$$

Now go back one sample time. $\mathscr{C}[\mathbf{I}_N, t_N]$ is given in (13-77), where $\bar{L}$ is given by

$$\begin{aligned}\bar{L}[\mathbf{I}_N, \mathbf{u}(t_N), t_N] &= \int_{-\infty}^{\infty} L[\boldsymbol{\xi}, \mathbf{u}(t_N), t_N] f_{\mathbf{x}(t_N)|\mathbf{I}(t_N)}(\boldsymbol{\xi} | \mathbf{I}_N)\, d\boldsymbol{\xi} \\ &= \int_{-\infty}^{\infty} L[\boldsymbol{\xi}, \mathbf{u}(t_N), t_N] f_{\mathbf{x}(t_N)|\hat{\mathbf{x}}(t_N^+)}(\boldsymbol{\xi} | \hat{\mathbf{x}}(t_N^+))\, d\boldsymbol{\xi} \\ &\triangleq \bar{L}'[\hat{\mathbf{x}}(t_N^+), \mathbf{u}(t_N), t_N]\end{aligned} \tag{13-122}$$

using the same notation as in (13-119); this establishes (13-92a) for $t_i = t_N$. Now we can write (13-81) equivalently as

$$\mathscr{C}^*[\mathbf{I}_N, t_N] = \min_{\mathbf{u} \in \mathscr{U}} \{\bar{L}'[\hat{\mathbf{x}}(t_N^+), \mathbf{u}(t_N), t_N] + E\{L_f[\mathbf{x}(t_{N+1})] | \mathbf{I}(t_N) = \mathbf{I}_N\}\} \tag{13-123}$$

Consider the last term: The argument $\mathbf{x}(t_{N+1})$ in $L_f$ can be written as

$$\begin{aligned}\mathbf{x}(t_{N+1}) &= \hat{\mathbf{x}}(t_{N+1}^+) + \mathbf{e}(t_{N+1}^+) \\ &= \hat{\mathbf{x}}(t_{N+1}^-) + \mathbf{s}(t_{N+1}) + \mathbf{e}(t_{N+1}^+)\end{aligned} \tag{13-124}$$

using (13-114) and (13-108) combined with (13-116). Thus, $\mathbf{x}(t_{N+1})$ is the sum of three random variables that are mutually independent, with $\hat{\mathbf{x}}(t_{N+1}^-)$ dependent upon $\mathbf{Z}(t_N)$ and the control time history, whereas $\mathbf{s}(t_{N+1})$ and $\mathbf{e}(t_{N+1}^+)$ are both independent of these. It will now be shown that

$$E\{L_f[\mathbf{x}(t_{N+1})] | \mathbf{I}(t_N) = \mathbf{I}_N\} = \int_{-\infty}^{\infty} \bar{L}_f'[\hat{\mathbf{x}}(t_{N+1}^-) + \boldsymbol{\eta}] f_{\mathbf{s}(t_{N+1})}(\boldsymbol{\eta})\, d\boldsymbol{\eta} \tag{13-125}$$

where $\hat{\mathbf{x}}(t_{N+1}^-)$ is a *deterministic function* of the information vector realization $\mathbf{I}_N$ and chosen $\mathbf{u}(t_N)$, given by

$$\hat{\mathbf{x}}(t_{N+1}^-) = \boldsymbol{\Phi}(t_{N+1}, t_N)\hat{\mathbf{x}}(t_N^+) + \mathbf{B}_d(t_N)\mathbf{u}(t_N) \tag{13-126}$$

and where $\hat{\mathbf{x}}(t_N^+)$ is the sufficient statistic realization corresponding to $\mathbf{I}(t_N, \omega_k) = \mathbf{I}_N$. To do so [11, 12], we follow the same logical steps as used in (13-79):

$$\begin{aligned}E\{L_f[\mathbf{x}(t_{N+1})] | \mathbf{I}(t_N) = \mathbf{I}_N\} &= E\{E\{L_f[\mathbf{x}(t_{N+1})] | \mathbf{I}(t_{N+1}) = \mathbf{I}(t_{N+1})\} | \mathbf{I}(t_N) = \mathbf{I}_N\} \\ &= E\{\bar{L}_f[\mathbf{I}(t_{N+1})] | \mathbf{I}(t_N) = \mathbf{I}_N\} \\ &= E\{\bar{L}_f'[\hat{\mathbf{x}}(t_{N+1}^+)] | \mathbf{I}(t_N) = \mathbf{I}_N\}\end{aligned}$$

where the last equality invokes (13-119). Recognizing $\hat{\mathbf{x}}(t_{N+1}^+)$ as $[\hat{\mathbf{x}}(t_{N+1}^-) + \mathbf{s}(t_{N+1})]$ allows us to continue as

$$\begin{aligned}&E\{L_f[\mathbf{x}(t_{N+1})] | \mathbf{I}(t_N) = \mathbf{I}_N\} \\ &\quad = E\{\bar{L}_f'[\hat{\mathbf{x}}(t_{N+1}^-) + \mathbf{s}(t_{N+1})] | \mathbf{I}(t_N) = \mathbf{I}_N\} \\ &\quad \triangleq \int_{-\infty}^{\infty} \int_{-\infty}^{\infty} \bar{L}_f'[\boldsymbol{\xi} + \boldsymbol{\eta}] f_{\hat{\mathbf{x}}(t_{N+1}^-), \mathbf{s}(t_{N+1})|\mathbf{I}(t_N)}(\boldsymbol{\xi}, \boldsymbol{\eta} | \mathbf{I}_N)\, d\boldsymbol{\xi}\, d\boldsymbol{\eta}\end{aligned}$$

By Bayes' rule, the density in this expression becomes

$$f_{\hat{\mathbf{x}}^-,\mathbf{s}|\mathbf{I}} = f_{\hat{\mathbf{x}}^-|\mathbf{s},\mathbf{I}} f_{\mathbf{s}|\mathbf{I}} = f_{\hat{\mathbf{x}}^-|\mathbf{I}} f_{\mathbf{s}}$$

where the last equality is due to the independences mentioned below (13-124). Substituting this back in yields

$$\begin{aligned} E\{L_f[\mathbf{x}(t_{N+1})] \,|\, \mathbf{I}(t_N) = \mathbf{I}_N\} &= \int_{-\infty}^{\infty} \left[ \int_{-\infty}^{\infty} \bar{L}_f'[\boldsymbol{\xi} + \boldsymbol{\eta}] f_{\hat{\mathbf{x}}(t_{N+1}^-)|\mathbf{I}(t_N)}(\boldsymbol{\xi}\,|\,\mathbf{I}_N)\, d\boldsymbol{\xi} \right] f_{\mathbf{s}(t_{N+1})}(\boldsymbol{\eta})\, d\boldsymbol{\eta} \\ &= \int_{-\infty}^{\infty} \left[ \int_{-\infty}^{\infty} \bar{L}_f'[\boldsymbol{\xi} + \boldsymbol{\eta}] f_{\hat{\mathbf{x}}(t_{N+1}^-)|\hat{\mathbf{x}}(t_N^+)}(\boldsymbol{\xi}\,|\,\hat{\mathbf{x}}(t_N^+))\, d\boldsymbol{\xi} \right] f_{\mathbf{s}(t_{N+1})}(\boldsymbol{\eta})\, d\boldsymbol{\eta} \\ &= \int_{-\infty}^{\infty} \bar{L}_f'[\hat{\mathbf{x}}(t_{N+1}^-) + \boldsymbol{\eta}] f_{\mathbf{s}(t_{N+1})}(\boldsymbol{\eta})\, d\boldsymbol{\eta} \end{aligned}$$

thereby establishing (13-125) with $\hat{\mathbf{x}}(t_{N+1}^-)$ as given in (13-126).

Putting (13-125) into (13-123) yields $\mathscr{C}^*[\mathbf{I}_N, t_N]$ as

$$\mathscr{C}^*[\mathbf{I}_N, t_N] = \min_{\mathbf{u}\in\mathscr{U}} \left\{ \bar{L}'[\hat{\mathbf{x}}(t_N^+), \mathbf{u}(t_N), t_N] + \int_{-\infty}^{\infty} \mathscr{C}^*[\hat{\mathbf{x}}(t_{N+1}^-) + \boldsymbol{\eta}, t_{N+1}] f_{\mathbf{s}(t_{N+1})}(\boldsymbol{\eta})\, d\boldsymbol{\eta} \right\} \tag{13-127}$$

When this minimization is performed, since all terms are functions of $\mathbf{u}(t_N)$ and/or $\hat{\mathbf{x}}(t_N^+)$ and $t_N$, we obtain $\mathbf{u}^*$ as a function of $\hat{\mathbf{x}}(t_N^+)$ and $t_N$. Again to emphasize the fact that $\mathscr{C}^*[\mathbf{I}_N, t_N]$ is a function of only the sufficient statistic realization $\hat{\mathbf{x}}(t_N^+)$, we can replace $\mathscr{C}^*[\mathbf{I}_N, t_N]$ notationally by $\mathscr{C}^*[\hat{\mathbf{x}}(t_N^+), t_N]$ in (13-127).

By induction, we can write the general recursion to be solved to determine the optimal control function $\mathbf{u}^*[\cdot,\cdot]$ as

$$\mathscr{C}^*[\hat{\mathbf{x}}(t_i^+), t_i] = \min_{\mathbf{u}\in\mathscr{U}} \left\{ \bar{L}'[\hat{\mathbf{x}}(t_i^+), \mathbf{u}\{\hat{\mathbf{x}}(t_i^+), t_i\}, t_i] + \int_{-\infty}^{\infty} \mathscr{C}^*[\hat{\mathbf{x}}(t_{i+1}^-) + \boldsymbol{\eta}, t_{i+1}] f_{\mathbf{s}(t_{i+1})}(\boldsymbol{\eta})\, d\boldsymbol{\eta} \right\} \tag{13-128}$$

where $\hat{\mathbf{x}}(t_{i+1}^-)$ is given by (13-110), which is solved backwards in time from the terminal condition of

$$\mathscr{C}^*[\hat{\mathbf{x}}(t_{N+1}^+), t_{N+1}] = \bar{L}_f'[\hat{\mathbf{x}}(t_{N+1}^+)] \tag{13-129}$$

where $\bar{L}_f'$ and $\bar{L}'$ are defined as in (13-119) and (13-122), respectively, for general $t_i$ (see (13-92) also). In terms of conditional expectations, (13-128) can be written as

$$\mathscr{C}^*[\hat{\mathbf{x}}(t_i^+), t_i] = \min_{\mathbf{u} \in \mathscr{U}} \Big\{ \bar{L}'[\hat{\mathbf{x}}(t_i^+), \mathbf{u}\{\hat{\mathbf{x}}(t_i^+), t_i\}, t_i] + E\{\mathscr{C}^*[\hat{\mathbf{x}}(t_{i+1}^+), t_{i+1}] \mid \hat{\mathbf{x}}(t_i^+) = \hat{\mathbf{x}}(t_i^+)\} \Big\} \qquad (13\text{-}130)$$

where the random part of $\hat{\mathbf{x}}(t_{i+1}^+)$, conditioned on $\hat{\mathbf{x}}(t_i^+) = \hat{\mathbf{x}}(t_i^+)$, is $\mathbf{s}(t_{i+1})$. These results correspond directly to the general dynamic programming result for the case of exploiting sufficient statistics, (13-90)–(13-92).

Compare (13-129) and (13-130) to the dynamic programming algorithm for the case in which perfect knowledge of the entire state was assumed, (13-66) and (13-67). The structure of the results is the same except that:

(1) $\mathbf{x}(t_i)$ in (13-66) is replaced by $\hat{\mathbf{x}}(t_i^+)$ in (13-130);

(2) $L[\mathbf{x}(t_i), \mathbf{u}(t_i), t_i]$ in (13-66) is replaced by $\bar{L}'[\hat{\mathbf{x}}(t_i^+), \mathbf{u}(t_i), t_i]$ in (13-130), where

$$L[\mathbf{x}(t_i), \mathbf{u}(t_i), t_i] = E\{L[\mathbf{x}(t_i), \mathbf{u}\{\mathbf{x}(t_i), t_i\}, t_i] \mid \mathbf{x}(t_i) = \mathbf{x}(t_i)\} \qquad (13\text{-}131)$$

$$\bar{L}'[\hat{\mathbf{x}}(t_i^+), \mathbf{u}(t_i), t_i] = E\{L[\mathbf{x}(t_i), \mathbf{u}\{\hat{\mathbf{x}}(t_i^+), t_i\}, t_i] \mid \hat{\mathbf{x}}(t_i^+) = \hat{\mathbf{x}}(t_i^+)\} \qquad (13\text{-}132)$$

(3) $E\{\mathscr{C}^*[\mathbf{x}(t_{i+1}), t_{i+1}] \mid \mathbf{x}(t_i) = \mathbf{x}(t_i)\}$ of (13-66), where the expectation involves the effect of dynamics driving noise inputs, is replaced by $E\{\mathscr{C}^*[\hat{\mathbf{x}}(t_{i+1}^+), t_{i+1}] \mid \hat{\mathbf{x}}(t_i^+) = \hat{\mathbf{x}}(t_i^+)\}$ in (13-130), where the expectation is over the possible realizations of the new information $\mathbf{s}(t_{i+1})$, thereby incorporating uncertain knowledge of the state at $t_i$ and measurement uncertainties as well as dynamics driving noise between times $t_i$ and $t_{i+1}$;

(4) the terminal condition $L_f[\mathbf{x}(t_{N+1})]$ in (13-67) is replaced by $\bar{L}_f'[\hat{\mathbf{x}}(t_{N+1}^+)]$ in (13-129), where

$$L_f[\mathbf{x}(t_{N+1})] = E\{L_f[\mathbf{x}(t_{N+1})] \mid \mathbf{x}(t_{N+1}) = \mathbf{x}(t_{N+1})\} \qquad (13\text{-}133)$$

$$\bar{L}_f'[\hat{\mathbf{x}}(t_{N+1}^+)] = E\{L_f[\mathbf{x}(t_{N+1})] \mid \hat{\mathbf{x}}(t_{N+1}^+) = \hat{\mathbf{x}}(t_{N+1}^+)\} \qquad (13\text{-}134)$$

EXAMPLE 13.12 Consider the stochastic optimal controller for a problem described by a state dynamics model as in (13-94) with measurements modeled as in (13-95). Again we seek the optimal control function $\mathbf{u}^*[\cdot,\cdot]$ as a function of the sufficient statistic $\hat{\mathbf{x}}(t_i^+)$ and current time $t_i$, but now we impose the additional assumptions that the cost (13-96) is *quadratic* and that the set of admissible controls $\mathscr{U}$ is the set of *all* functions of $R^n \times T \rightarrow R^r$, i.e., there is *no constraint*. This example is analogous to Example 13.7 in which perfect state knowledge was assumed, and its result is the controller for the simplest LQG problem, discussed in Section 13.3.

For this problem, we will assume that the loss functions $L$ and $L_f$ in (13-96) are given by

$$L[\mathbf{x}(t_i), \mathbf{u}(t_i), t_i] = \tfrac{1}{2}[\mathbf{x}^T(t_i)\mathbf{X}(t_i)\mathbf{x}(t_i) + \mathbf{u}^T(t_i)\mathbf{U}(t_i)\mathbf{u}(t_i)]$$

$$L_f[\mathbf{x}(t_{N+1})] = \tfrac{1}{2}\mathbf{x}^T(t_{N+1})\mathbf{X}_f\mathbf{x}(t_{N+1})$$

where $\mathbf{X}(t_i)$ for all $i$ and $\mathbf{X}_f$ are $n$-by-$n$ positive semidefinite matrices and $\mathbf{U}(t_i)$ for all $t_i$ are $r$-by-$r$ positive definite matrices.

To generate the optimal controller, we will set up and explicitly solve the dynamic programming algorithm, (13-129) and (13-130). Consider the $\bar{L}'$ term in (13-130):

$$\bar{L}'[\hat{\mathbf{x}}(t_i^+), \mathbf{u}\{\hat{\mathbf{x}}(t_i^+), t_i\}, t_i] = \tfrac{1}{2}E\{\mathbf{x}^{\mathrm{T}}(t_i)\mathbf{X}(t_i)\mathbf{x}(t_i) + \mathbf{u}^{\mathrm{T}}(t_i)\mathbf{U}(t_i)\mathbf{u}(t_i) \mid \hat{\mathbf{x}}(t_i^+) = \hat{\mathbf{x}}(t_i^+)\}$$

where the expectation is implicitly conditioned on a specific choice of $\mathbf{u}\{\cdot, t_i\}$ as well as on knowledge of $\hat{\mathbf{x}}(t_i^+)$. Computing this expectation yields

$$\bar{L}'[\hat{\mathbf{x}}(t_i^+), \mathbf{u}(t_i), t_i] = \tfrac{1}{2}[\hat{\mathbf{x}}^{\mathrm{T}}(t_i^+)\mathbf{X}(t_i)\hat{\mathbf{x}}(t_i^+) + \operatorname{tr}\{\mathbf{X}(t_i)\mathbf{P}(t_i^+)\} + \mathbf{u}^{\mathrm{T}}(t_i)\mathbf{U}(t_i)\mathbf{u}(t_i)]$$

which is the same as the $L[\mathbf{x}(t_i), \mathbf{u}(t_i), t_i]$ in the case of perfect knowledge of the entire state, given in Example 13.7, except for $\hat{\mathbf{x}}(t_i^+)$ replacing $\mathbf{x}(t_i)$ and the appearance of the additional trace term, which is *not* a function of $\hat{\mathbf{x}}(t_i^+)$. Similarly, the $\bar{L}_{\mathrm{f}}'$ term in (13-129) can be expressed as

$$\bar{L}_{\mathrm{f}}'[\hat{\mathbf{x}}(t_{N+1}^+)] = \tfrac{1}{2}[\hat{\mathbf{x}}^{\mathrm{T}}(t_{N+1}^+)\mathbf{X}_{\mathrm{f}}\hat{\mathbf{x}}(t_{N+1}^+) + \operatorname{tr}\{\mathbf{X}_{\mathrm{f}}\mathbf{P}(t_{N+1}^+)\}]$$

and this is the same as $L_{\mathrm{f}}[\mathbf{x}(t_{N+1})]$ in Example 13.7, except for $\hat{\mathbf{x}}(t_{N+1}^+)$ replacing $\mathbf{x}(t_{N+1})$ and the appearance of the additional trace term, which is *not* a function of $\hat{\mathbf{x}}(t_{N+1}^+)$.

Thus, if there is no constraint imposed by $\mathscr{U}$, the recursion to be solved to determine the optimal control function is

$$\mathscr{C}^*[\hat{\mathbf{x}}(t_i^+), t_i] = \min_{\mathbf{u}} \{\tfrac{1}{2}[\hat{\mathbf{x}}^{\mathrm{T}}(t_i^+)\mathbf{X}(t_i)\hat{\mathbf{x}}(t_i^+) + \operatorname{tr}\{\mathbf{X}(t_i)\mathbf{P}(t_i^+)\}$$
$$+ \mathbf{u}^{\mathrm{T}}(t_i)\mathbf{U}(t_i)\mathbf{u}(t_i)] + E\{\mathscr{C}^*[\hat{\mathbf{x}}(t_{i+1}^+), t_{i+1}] \mid \hat{\mathbf{x}}(t_i^+) = \hat{\mathbf{x}}(t_i^+)\}\}$$

which is solved backwards from the terminal condition

$$\mathscr{C}^*[\hat{\mathbf{x}}(t_{N+1}^+), t_{N+1}] = \tfrac{1}{2}[\hat{\mathbf{x}}^{\mathrm{T}}(t_{N+1}^+)\mathbf{X}_{\mathrm{f}}\hat{\mathbf{x}}(t_{N+1}^+) + \operatorname{tr}\{\mathbf{X}_{\mathrm{f}}\mathbf{P}(t_{N+1}^+)\}]$$

As in Example 13.7, we attempt to solve this equation with an assumed quadratic form,

$$\mathscr{C}^*[\hat{\mathbf{x}}(t_i^+), t_i] = \tfrac{1}{2}[\hat{\mathbf{x}}^{\mathrm{T}}(t_i^+)\mathbf{K}_{\mathrm{c}}(t_i)\hat{\mathbf{x}}(t_i^+) + g(t_i)]$$

The same procedure is followed as used previously: terminal conditions and recursions for $\mathbf{K}_{\mathrm{c}}(t_i)$ and $g(t_i)$ are obtained by substitution of the assumed form into the equations, and $\mathbf{u}^*[\cdot, t_i]$ is determined for each $t_i$ by setting the first derivative of the quantity to be minimized equal to zero and verifying optimality with the second derivative. Performing these steps yields an optimal control as a linear function of $\hat{\mathbf{x}}(t_i^+)$, thereby providing a *linear feedback control*

$$\mathbf{u}^*[\hat{\mathbf{x}}(t_i^+), t_i] = -\mathbf{G}_{\mathrm{c}}^*(t_i)\hat{\mathbf{x}}(t_i^+)$$

where $\mathbf{G}_{\mathrm{c}}^*(t_i)$ is given by (13-23), based upon solving for $\mathbf{K}_{\mathrm{c}}$ via the backward Riccati equation (13-18) and terminal condition (13-19). The scalar $g(t_i)$, required only for cost evaluation and *not* for controller specification, is found by solving (13-20) backward from the terminal condition (13-21).

This controller is identical in structure to the result for Example 13.7,

$$\mathbf{u}^*[\mathbf{x}(t_i), t_i] = -\mathbf{G}_{\mathrm{c}}^*(t_i)\mathbf{x}(t_i)$$

using the *same* $\mathbf{G}_{\mathrm{c}}^*(t_i)$ and the *same* Riccati recursion and terminal condition for $\mathbf{K}_{\mathrm{c}}$. The scalar $g(t_i)$ in the current problem satisfies a different recursion relation and terminal condition than those valid for either the deterministic case ($g(t_i) \equiv 0$ for all $t_i$) or the stochastic case with perfect state knowledge, as in Example 13.7. Here the recursion can be written as

$$g(t_i) = g(t_{i+1}) + \operatorname{tr}\{\mathbf{X}(t_i)\mathbf{P}(t_i^+) + \boldsymbol{\Sigma}(t_{i+1})\mathbf{K}_{\mathrm{c}}(t_{i+\ +})\}$$

where $\boldsymbol{\Sigma}(t_{i+1})$ is the covariance of the new information random vector $\mathbf{s}(t_{i+1})$ as in (13-116) and (13-117),

$$\boldsymbol{\Sigma}(t_{i+1}) = \mathbf{K}(t_{i+1})\mathbf{H}(t_{i+1})\mathbf{P}(t_{i+1}^-)$$

Comparing the trace term in $g(t_i)$ to the corresponding term in the solution to Example 13.7, we see $\mathrm{tr}\{\boldsymbol{\Sigma}(t_{i+1})\mathbf{K}_c(t_{i+1})\}$ replacing $\mathrm{tr}\{[\mathbf{G}_d(t_i)\mathbf{Q}_d(t_i)\mathbf{G}_d^T(t_i)]\mathbf{K}_c(t_{i+1})\}$ and the appearance of the new term $\mathrm{tr}\{\mathbf{X}(t_i)\mathbf{P}(t_i^+)\}$. Similarly, the terminal condition of zero is replaced with $\mathrm{tr}\{\mathbf{X}_f\mathbf{P}(t_{N+1}^+)\}$, due to the additional uncertainty in this problem.

The controller we have just derived was discussed in detail in Section 13.3. ■

As in Section 13.5, we have developed an algorithm which can be used either as a theoretical or computational tool for deriving optimal stochastic controllers. If we restrict our attention to the LQG problem with unconstrained choice of $\mathbf{u}^*$ function, we are able to exploit it as a theoretical tool to obtain a useful closed form solution that is readily implemented for online use, as seen in Section 13.3. Viewed as a computational method, (13-86)–(13-88) are plagued by the "curse of dimensionality" because of the discretizations required to perform the optimizations in general. Distinct advantages are afforded by the use of sufficient statistics in (13-90)–(13-92) and the special case of (13-128)–(13-130), but even with this data reduction, the computational and memory burdens inherent in dynamic programming are formidable for general problems of practical significance.

## 13.7 SUMMARY

Section 13.2 presented the objectives and formulation of the basic stochastic control problem, and the structure of typical controllers that are potential problem solutions. It is assumed that systems to be controlled are often continuous time, with states well modeled as solutions to Itô stochastic differential equations as (13-1), from which sampled-data measurements of the form (13-2) are available. Continuous-time measurements could also be modeled as in (13-3) but further discussion of this case is deferred to later chapters. Linear system models driven by white Gaussian noise are an important special case, described by state differential equations (13-5) or equivalent discrete-time models (13-8), with sampled-data measurements modeled by (13-6) or possibly continuous-time measurements as in (13-7).

Section 13.3 introduced the concept of the performance index or *cost function* to be minimized by the controller: terminal and trajectory loss functions were discussed, as well as the *conditional expectation* of the *cost-to-go* as the natural cost to minimize at each stage of the problem. The objective was described as generating the cost-minimizing control function out of an admissible class of functions $\mathscr{U}$, expressed as a function of the current state and time if available, or of the measurement (or information) history or sufficient statistics for that history if the state is not available perfectly. To solve this problem, the *optimality principle* and *dynamic programming algorithm* were introduced, along with the *backward Kolmogorov equation* for assisting in the backward propagations inherent in these tools. Two potential structural properties of solutions to the dynamic programming algorithm, *certainty equivalence* and *separation*, were described as well.

Simultaneously, Section 13.3 discussed the optimal stochastic controller for the specific case of the LQG modeling assumptions (*linear* system model driven by white *Gaussian* noise, with a *quadratic* cost function) and unconstrained $\mathscr{U}$ being adequate. Cost functions of the form (13-11) were discussed in both mathematical and physical terms, and a designer's use of the LQG formulation was portrayed. This is a *tractable* problem, yielding a *readily synthesized, efficiently implemented, linear feedback* control law given by

$$\mathbf{u}^*[\hat{\mathbf{x}}(t_i^+), t_i] = -\mathbf{G}_c^*(t_i)\hat{\mathbf{x}}(t_i^+) \tag{13-22}$$

where $\hat{\mathbf{x}}(t_i^+)$ is generated by the Kalman filter associated with the given problem, and $\mathbf{G}_c^*(t_i)$ is given by (13-23) with $\mathbf{K}_c$ determined via the backward Riccati difference equation (13-18) and (13-19). Thus, the optimal LQG stochastic controller is the *cascade* of a *Kalman filter* with the optimal feedback *gain* of the corresponding *deterministic* optimal control problem. Here the conditional mean $\hat{\mathbf{x}}(t_i^+)$ is the sufficient statistic, and both the certainty equivalence and seperation properties pertain. The *synthesis* of practical sampled-data feedback controllers by this means is important, and the structure of the resulting controller algorithm is presented in Figs. 13.4 and 13.5.

Section 13.4 presented the fundamental *backward Kolmogorov equation* (13-33). Furthermore, its use in propagating costs backwards was illustrated in (13-41) and (13-42).

Section 13.5 presented optimal stochastic control assuming that perfect knowledge of the entire state is available. Although this is often an unrealistic simplification for practical problems, it provides useful insights into the solution for, and structure of, optimum control functions for the more practical problems of Section 13.6. For the general problem involving nonlinear system models (13-1) and (13-2), general cost (13-51), and a specification of admissible functions $\mathscr{U}$, the cost-minimizing control function $\mathbf{u}^*[\cdot,\cdot] \in \mathscr{U}$ is generated by solving the *stochastic dynamic programming* algorithm (13-65)–(13-67). This tool is based upon four basic steps: (1) because the discrete-time $\mathbf{x}(\cdot,\cdot)$ is *Markov*, the Chapman–Kolmogorov equation can be used to manipulate integrations to yield

$$\mathscr{C}[\mathbf{x}(t_i), t_i] = L[\mathbf{x}(t_i), \mathbf{u}\{\mathbf{x}(t_i), t_i\}, t_i] + E\{\mathscr{C}[\mathbf{x}(t_{i+1}), t_{i+1}] \,|\, \mathbf{x}(t_i) = \mathbf{x}(t_i)\} \tag{13-135}$$

as the cost to be minimized at time $t_i$; (2) the control function $\mathbf{u}\{\cdot, t_i\}$ is treated as *any fixed admissible* function; (3) *regardless* of the choice of $\mathbf{u}\{\cdot, t_i\}$, to minimize this $\mathscr{C}[\mathbf{x}(t_i), t_i]$, the $\mathscr{C}$ inside the expectation in (13-135) must be $\mathscr{C}^*$; (4) $\mathscr{C}[\mathbf{x}(t_i), t_i]$ is evaluated (conceptually) for every admissible choice of $\mathbf{u}\{\cdot, t_i\}$, and $\mathbf{u}^*\{\cdot, t_i\}$ is then the control that yields the minimal cost, denoted as $\mathscr{C}^*[\mathbf{x}(t_i), t_i]$, for each $\mathbf{x}(t_i)$. For the LQG problem with unconstrained $\mathscr{U}$, this algorithm yields the controller

$$\mathbf{u}^*(t_i) = -\mathbf{G}_c^*(t_i)\mathbf{x}(t_i) \tag{13-136}$$

as developed in Example 13.7. In the context of more general problems, dynamic programming can be used as a computational tool, inherently entailing discretizations of states and controls, and the associated "curse of dimensionality." If discretization is accomplished in the models before applying dynamic programming instead, then the algorithm can be expressed as in (13-70) and (13-71).

Finally, Section 13.6 developed optimal stochastic control under the more practical assumption that only incomplete, noise-corrupted measurements are available at the sample times. Thus, what is available to the controller at time $t_i$ is the *information vector* (13-73), composed of the time history of measurements up to $t_i$ and the past controls that have been applied. The natural cost to minimize is then the conditional expectation of the cost-to-go, given that information vector, (13-74). To solve this problem, one can use the *dynamic programming algorithm* (13-86)–(13-88). *Sufficient statistics* can sometimes be incorporated to assist in data reduction, yielding the general algorithm (13-89)–(13-92). An important illustration of this is given by the dynamic programming algorithm (13-128)–(13-130) for problems described by linear system models driven by white Gaussian noise. Additionally invoking the assumptions of quadratic cost criteria and unconstrained $\mathscr{U}$ as in Example 13.12 then yields the LQG stochastic controller portrayed in Section 13.3.

Because of the special *synthesis* capability associated with the LQG stochastic controller due to certainty equivalence, this particular controller form will be pursued in greater depth in Chapter 14.

## REFERENCES

1. Aoki, M., "Optimization of Stochastic Systems, Topics in Discrete-Time Systems." Academic Press, New York, 1967.
2. Åström, K. J., Optimal control of Markov processes with incomplete state information, *J. Math. Anal. Appl.* **10**, 174–205 (1965).
3. Åström, K. J., "Introduction to Stochastic Control Theory." Academic Press, New York, 1970.
4. Bellman, R., "Dynamic Programming." Princeton Univ. Press, Princeton, New Jersey, 1957.
5. Bellman, R., "Adaptive Control Processes, A Guided Tour." Princeton Univ. Press, Princeton, New Jersey, 1961.
6. Bellman, R., and Dreyfus, S. E., "Applied Dynamic Programming." Princeton Univ. Press, Princeton, New Jersey, 1962.
7. Bertsekas, D. P., "Dynamic Programming and Stochastic Control." Academic Press, New York, 1976.
8. Bertsekas, D., and Shreve, S., "Stochastic Optimal Control: The Discrete Time Case." Academic Press, New York, 1979.
9. Bryson, A. E., Jr., and Ho, Y., "Applied Optimal Control." Ginn (Blaisdell), Waltham, Massachusetts, 1969.
10. Davis, M. H. A., and Varaiya, P., Dynamic programming conditions for partially observable stochastic systems, *SIAM J. Control* **11** (2), pp. 226–261 (1973).
11. Deyst, J. J., Jr., Optimal Control in the Presence of Measurement Uncertainties, M. S. Thesis, MIT Exp. Astronom. Lab. Rep. TE-17, Cambridge, Massachusetts (January 1967).

12. Deyst, J. J., Jr., Estimation and Control of Stochastic Processes, unpublished course notes. MIT, Dept. of Aeronautics and Astronautics, Cambridge, Massachusetts (1970).
13. Dreyfus, S. E., Some types of optimal control of stochastic systems, *J. SIAM Control Ser. A* **2**, (1), 120–134 (1962).
14. Dreyfus, S. E., "Dynamic Programming and the Calculus of Variations." Academic Press, New York, 1965.
15. Edwards, R. M., Entry Monitoring System Ranging with Cross-Range Control, MSC Internal Note No. 69-FM-298, Mission Planning and Analysis Division, Manned Spacecraft Center, Houston, Texas (November 1969).
16. Elliot, R. J., The optimal control of a stochastic system, *SIAM J. Control*, **15**, (5), 756–778 (1977).
17. Fel'dbaum, A. A., "Optimal Control Systems." Academic Press, New York, 1967.
18. Fleming, W. H., Some Markovian optimization problems, *J. Math. Mech.* **12** (1), 131–140 (1963).
19. Fleming, W. H., Optimal control of partially observable diffusions, *SIAM J. Control* **6** (2), 194–214 (1968).
20. Fleming, W. H., and Nisio, N., On the existence of optimal stochastic controls, *J. Math. Mech.* **15** (5), 777–794 (1966).
21. Jacobson, D. H., and Mayne, D. Q., "Differential Dynamic Programming." Elsevier, Amsterdam 1970.
22. Joseph, P. D., and Tou, J. T., On linear control theory, *AIEE Trans. (Appl. Ind.)* 193–196 (September 1961).
23. Kozin, F., A survey of stability of stochastic systems, *Proc. Joint Automatic Control Conf., Stochastic Problems in Control, Ann Arbor, Michigan*, pp. 39–86 (June 1968).
24. Krasovskii, N. N., On optimal control in the presence of random disturbances, *Appl. Math. Mech.* **24** (1), 82–102 (1960).
25. Krasovskii, N. N., and Lidskii, E. A., Analytical design of controllers in systems with random attributes, I-III, *Automat. Rem. Control* **22** (9), 1021–1025 (1961); **22** (10), 1141–1146 (1961); **22** (11), 1289–1294 (1961).
26. Kushner, H. J., On the dynamical equations of conditional probability density functions with applications to stochastic control, *J. Math. Anal. Appl.* **8**, 332–344 (1964).
27. Kushner, H. J., "Stochastic Stability and Control." Academic Press, New York, 1967.
28. Kushner, H. J., "Introduction to Stochastic Control." Holt, New York, 1971.
29. Kwakernaak, H., and Sivan, R., "Linear Optimal Control Systems." Wiley, New York, 1972.
30. Larson, R. E., Survey of dynamic programming computational procedures, *IEEE Trans. Automat. Control* **AC-12** (6), 767–774 (1967).
31. Lidskii, E. A., Optimal control of systems with random properties, *Appl. Math. Mech.* **27** (1), 42–59 (1963).
32. Moseley, P. E., The Apollo Entry Guidance; A Review of the Mathematical Development and Its Operational Characteristics, TRW Note No. 69-FMT-791. Mission Planning and Analysis Division, Manned Spacecraft Center, Houston, Texas (December 1969).
33. Pontryagin, L. S., "The Mathematical Theory of Optimal Processes." Wiley, New York, 1962.
34. Potter, J. E., A Guidance-Navigation Separation Theorem. MIT Exp. Astronom. Lab. Rep. RE-11, Cambridge, Massachusetts (1964).
35. Project Apollo Entry Mission Plan-Apollo 8 Summary, MSC Internal Note No. 69-FM-130. Mission Planning and Analysis Division, Manned Spacecraft Center, Houston, Texas (May 1969).
36. Project Apollo Entry Mission Plan-Apollo 12 (Mission H-1), MSC Internal Note No. 69-FM-267. Mission Planning and Analysis Division, Manned Spacecraft Center, Houston, Texas (October 1969).

37. Shiryaev, A. N., Sequential analysis and controlled random processes (discrete time), *Kibernetika* **3**, 1–24 (1965).
38. Sorensen, H. W., An overview of filtering and control in dynamic systems, *in* "Control and Dynamic Systems: Advances in Theory and Applications" (C. T. Leondes, ed.), Vol. 12, pp. 1–61, Academic Press, New York, 1976.
39. Stratonovich, R. L., "Conditional Markov Processes and Their Application to the Theory of Optimal Control." Elsevier, Amsterdam, 1968.
40. Striebel, C., Sufficient statistics in the optimum control of stochastic systems, *J. Math. Anal. Appl.* **12**, 576–592 (1965).
41. Striebel, C., "Optimal Control of Discrete Time Stochastic Systems." Springer-Verlag, Berlin and New York, 1975.
42. Sworder, D., "Optimal Adaptive Control Systems." Academic Press, New York, 1966.
43. Wonham, W. M., Optimal stochastic control, *Proc. Joint Automatic Control Conf., Stochastic Problems in Control, Ann Arbor, Michigan* pp. 107–120 (June 1968).
44. Wonham, W. M., On the separation theorem of stochastic control, *SIAM J. Control* **6** (2), 312–326 (1968).
45. Wonham, W. M., Random differential equations in control theory, *in* "Probabilistic Methods in Applied Mathematics" (A. T. Bharucha-Reid, ed.), Vol. II. Academic Press, New York, 1970.

## PROBLEMS

**13.1** Consider the *scalar* time-invariant system description

$$\mathrm{x}(t_{i+1}) = \Phi\mathrm{x}(t_i) + B_{\mathrm{d}}\mathrm{u}(t_i) + G_{\mathrm{d}}\mathrm{w}_{\mathrm{d}}(t_i)$$

where $\mathrm{w}_{\mathrm{d}}(\cdot,\cdot)$ is zero-mean white Gaussian noise of variance $Q_{\mathrm{d}}$. Suppose you want to determine the control $u^*$ to minimize $\mathscr{C}[x(t_0), t_0]$ given by

$$\mathscr{C}[x(t_0), t_0] = E\left\{\frac{1}{2}\sum_{i=0}^{N}[X\mathrm{x}^2(t_i) + U\mathrm{u}^2(t_i)] + \frac{1}{2}X_{\mathrm{f}}\mathrm{x}^2(t_{N+1})\,\middle|\,\mathrm{x}(t_0) = x(t_0)\right\}$$

where the admissible controls are unconstrained and the value of the state $x(t_i)$ is assumed to be known perfectly for all $t_i$.

(a) Explicitly define $u^*[x(t_i), t_i]$ and the recursions for the scalars $K_{\mathrm{c}}(t_i)$ and $g(t_i)$ in the optimal cost functional

$$\mathscr{C}^*[x(t_i), t_i] = \tfrac{1}{2}[K_{\mathrm{c}}(t_i)x^2(t_i) + g(t_i)]$$

(b) Explicitly evaluate $K_{\mathrm{c}}(t_N)$, $g(t_N)$, $\mathscr{C}^*[x(t_N), t_N]$, and $u^*[x(t_N), t_N]$.

(c) If $U = 0$, what are $u^*(t_i)$ and $K_{\mathrm{c}}(t_i)$? Does this make sense conceptually?

(d) How do the results of the problem change if $x(t_i)$ is now known perfectly, but in fact only a noise-corrupted measurement of the state,

$$\mathrm{z}(t_i) = \mathrm{x}(t_i) + \mathrm{v}(t_i)$$

is available, where $\mathrm{v}(\cdot,\cdot)$ is white Gaussian noise of mean zero and variance $R$? Consider parts (a), (b), and (c) in light of this alteration.

**13.2** (a) Repeat the previous problem for the special case of $X_{\mathrm{f}} = 0$. What effect does this have on controller performance?

(b) Repeat the previous problem for the case of $X \equiv 0$ but $X_{\mathrm{f}} \neq 0$. What kind of behavior will such a controller exhibit?

(c) If $X_{\mathrm{f}} = X$, is the controller in steady state constant-gain performance for all time in part (a) of the previous problem? If not, what value of $X_{\mathrm{f}}$ does provide such performance? Under what conditions will the controller of part (d) of that problem be a constant-gain controller for all time? If these conditions are not met, when might it be reasonable to use a constant-gain *approximation* for all time, and how would these constant gains be evaluated?

**13.3** Consider the scalar system

$$x(t_{i+1}) = x(t_i) + w_d(t_i)$$

where $w_d(\cdot,\cdot)$ is a white Gaussian process noise with statistics

$$E\{w_d(t_i)\} = 0, \qquad E\{w_d(t_i)^2\} = \tfrac{1}{2}$$

The initial condition is described by means of a Gaussian random variable $x(t_1)$, independent of $w_d(\cdot,\cdot)$, with statistics

$$E\{x(t_1)\} = 1, \qquad E\{x(t_1)^2\} = 2$$

Two measurements are available, at times $t_1$ and $t_2$:

$$z(t_i) = x(t_i) + v(t_i), \qquad i = 1, 2$$

where $v(\cdot,\cdot)$ is a white Gaussian sequence, independent of $w_d(\cdot,\cdot)$ and $x(t_1)$, with statistics

$$E\{v(t_i)\} = 0, \qquad E\{v(t_i)^2\} = \tfrac{1}{4}$$

and let $z(t_1, \omega_k) = z_1$, $z(t_2, \omega_k) = z_2$. Obtain explicit estimation results: $\hat{x}(t_i^-)$, $\hat{x}(t_i^+)$, $P(t_i^-)$, $P(t_i^+)$, and $K(t_i)$ for times $t_1$ and $t_2$.

Now consider the same system, except that a control variable $u(t_i) = u\{\mathbf{Z}(t_i), t_i\}$ is added:

$$x(t_{i+1}) = x(t_i) + u(t_i) + w_d(t_i)$$

Find the control to apply at time $t_1$ in order to minimize the cost

$$J = E\{\tfrac{1}{2}x(t_1)^2 + \tfrac{1}{2}U u(t_1)^2 + \tfrac{1}{2}x(t_2)^2\}$$

where $U$ is a weighting constant to depict the tradeoff of driving the state to zero while not expending too much control energy. To see its effect, explicitly evaluate $u(t_1)$ for $U$ equal to three different values:

(a) $U = 0$,

(b) $U = 1$,

(c) $U = 10$.

**13.4** (a) Consider Example 13.2 along with an associated state dynamics model in which $\mathbf{\Phi} = \mathbf{B}_d = \mathbf{U} = \mathbf{I}$ and $\mathbf{X} = \mathbf{X}_f$ as given in that example. Generate the optimal control and cost-to-go functions for the last five sample periods before the final time, assuming perfect knowledge of the entire state at each sample time. Repeat for the case of $\mathbf{X} = \mathbf{X}_f = \mathbf{I}$, and show the difference in the way the controller expends its efforts.

(b) Repeat part (a), but under the assumption that only measurements of the form

$$z_1(t_i) = x_1(t_i) + v_1(t_i), \qquad z_2(t_i) = x_2(t_i) + v_2(t_i)$$

are available at each sample time, where $v_1(\cdot,\cdot)$ and $v_2(\cdot,\cdot)$ are independent noises, each being zero-mean white Gaussian discrete-time noise of variance $R = 1$. Repeat for the case of $v_1(\cdot,\cdot)$ having variance $R_1 = 1$ and $v_2(\cdot,\cdot)$ having variance $R_2 = 10$.

(c) Now let the state vector be composed of two scalars, position $p(t)$ and velocity $v(t)$, modeled by

$$\dot{p}(t) = v(t), \qquad \dot{v}(t) = a_{cmd}(t) + w(t)$$

where commanded acceleration $a_{cmd}$ is viewed as a control input to be generated via feedback. Establish the equivalent discrete-time model for this case, and thus define $\mathbf{\Phi}$, $\mathbf{B}_d$, and $Q_d$. Let $\mathbf{X} = \mathbf{X}_f$, as given in Example 13.2, let $U = 1$, and generate the optimal control and cost-to-go functions for the last five periods before the final time, assuming perfect knowledge of the entire state at each sample time. Repeat for the case of $\mathbf{X} = \mathbf{X}_f = \mathbf{I}$.

(d) Repeat part (c), but under the measurement assumptions of part (b).

(e) Repeat part (c), but under the assumption that only measurements of the form of $z_1(t_i)$ of part (b) are available. Repeat for $v_1(\cdot,\cdot)$ having variance of 10.

(f) Can this problem be solved based on measurements of $z_2(t_i)$ alone? Explain.

**13.5** Consider the scalar stochastic differential equation

$$dx(t) = f[x(t), t]\,dt + G[x(t), t]\,d\beta(t)$$

with $\beta(\cdot,\cdot)$ of diffusion $Q(t)$ for all $t$. Using properties (11-81)–(11-83) of the solution to this equation, namely

$$E\{[x(t + \Delta t) - x(t)] \,|\, x(t) = \rho\} = f[\rho, t] + o(\Delta t)$$
$$E\{[x(t + \Delta t) - x(t)]^2 \,|\, x(t) = \rho\} = G^2[\rho, t]Q(t) + o(\Delta t)$$
$$E\{[x(t + \Delta t) - x(t)]^k \,|\, x(t) = \rho\} = o(\Delta t), \qquad k > 2$$

where $[o(\Delta t)/(\Delta t)] \to 0$ as $\Delta t \to 0$, show that the transition density $f_x(\xi, t \,|\, \rho, t')$ satisfies the backward Kolmogorov equation

$$\partial f_x(\xi, t \,|\, \rho, t')/\partial t' = -f[\rho, t']\,\partial f_x(\xi, t \,|\, \rho, t')/\partial \rho - \tfrac{1}{2}G^2[\rho, t']Q(t')\,\partial^2 f_x(\xi, t \,|\, \rho, t')/\partial \rho^2$$

**13.6** Consider the scalar linear system investigated previously in Problem 11.17 (Volume 2):

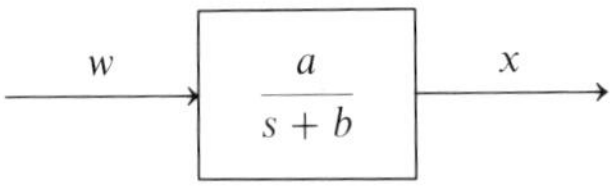


where $w(\cdot,\cdot)$ is white Gaussian noise. Show that the transition probability for $x(\cdot,\cdot)$ satisfies the backward Kolmogorov equation.

Define a terminal loss function as

$$L_f[x(t_f)] = x(t_f)^2$$

and the cost to complete the process from $x(t)$ at time $t$, to the final time $t_f$, as

$$\mathscr{C}[x(t), t] = E\{L_f[x(t_f)] \,|\, x(t) = x(t)\}$$

Derive an expression for $\mathscr{C}[\xi, t]$ for all $\xi$ and all $t \in [t_0, t_f]$.

**13.7** Consider the linear scalar system model

$$dx(t) = F(t)x(t)\,dt + G(t)\,d\beta(t)$$

where $\beta(\cdot,\cdot)$ is Brownian motion of diffusion $Q(t)$ for all $t$. Define the conditional characteristic function for $x(t)$ as

$$\phi_x(\mu, t \,|\, \rho, t') = E\{e^{j\mu x(t)} \,|\, x(t') = \rho\}$$
$$= \int_{-\infty}^{\infty} e^{j\mu\xi} f_x(\xi, t \,|\, \rho, t')\,d\xi$$

Using the backward Kolmogorov equation, derive the relation to express how this conditional characteristic function changes as the conditioning time is moved backward in time. Solve this equation by assuming a Gaussian solution. By so doing, also show how $E\{x(t) \,|\, x(t') = \rho\}$, $E\{[x(t)]^2 \,|\, x(t') = \rho\}$, and $E\{[x(t) - E[x(t) \,|\, x(t') = \rho]]^2 \,|\, x(t') = \rho\}$ change as $t'$ is moved backward in time.

**13.8** (a) In analogy to the derivation of (13-65)–(13-67), generate the deterministic dynamic programming algorithm.

(b) In analogy to the derivation of (13-70) and (13-71), develop the deterministic dynamic programming algorithm for discrete-valued states and controls.

(c) In analogy to the derivation of (13-86)–(13-92), develop the deterministic dynamic programming algorithm for the case of only *incomplete noise-corrupted* knowledge of the state being available at each sample time.

**13.9** In this chapter, we derive the optimal stochastic control law for a linear system and a quadratic cost criterion through the use of dynamic programming. This problem asks you to obtain the corresponding result for the deterministic case and thereby demonstrate the certainty equivalence property explicitly. That is to say, the optimal stochastic controller in the case of linear dynamics driven by white Gaussian noise and subject to quadratic cost should be of the same form as the optimal deterministic controller, with the state estimate $\hat{\mathbf{x}}(t_i^+)$ replacing the state $\mathbf{x}(t_i)$.

Consider the linear time-varying discrete system description

$$\mathbf{x}(t_{i+1}) = \mathbf{\Phi}(t_{i+1}, t_i)\mathbf{x}(t_i) + \mathbf{B}_d(t_i)\mathbf{u}(t_i)$$

and the cost functional

$$\mathscr{C}[\mathbf{x}(t_0), t_0] = \tfrac{1}{2}\mathbf{x}^T(t_{N+1})\mathbf{X}_f\mathbf{x}(t_{N+1}) + \frac{1}{2}\sum_{i=0}^{N}\{\mathbf{x}^T(t_i)\mathbf{X}(t_i)\mathbf{x}(t_i) + \mathbf{u}^T(t_i)\mathbf{U}(t_i)\mathbf{u}(t_i)\}$$

where it is assumed that $\mathbf{X}_f$ and $\mathbf{X}(t_i)$ are all symmetric and positive semidefinite, and that the $\mathbf{U}(t_i)$ are symmetric and positive definite. We further assume that the control $\mathbf{u}(t_i)$ is to be unconstrained and that $\mathbf{x}(t_0)$ is an arbitrary initial state.

(a) Show that the optimal control $\mathbf{u}^*(t_i)$ is related to the optimal state $\mathbf{x}^*(t_i)$ by

$$\mathbf{u}^*(t_i) = -[\mathbf{U}(t_i) + \mathbf{B}_d^T(t_i)\mathbf{K}_c(t_{i+1})\mathbf{B}_d(t_i)]^{-1}\mathbf{B}_d^T(t_i)\mathbf{K}_c(t_{i+1})\mathbf{\Phi}(t_{i+1}, t_i)\mathbf{x}^*(t_i)$$

where the matrix $\mathbf{K}_c(t_i)$ satisfies the recursive matrix Riccati difference equation

$$\begin{aligned}\mathbf{K}_c(t_i) = {} & \mathbf{X}(t_i) + \mathbf{\Phi}^T(t_{i+1}, t_i)\mathbf{K}_c(t_{i+1})\mathbf{\Phi}(t_{i+1}, t_i) \\ & - \mathbf{\Phi}^T(t_{i+1}, t_i)\mathbf{K}_c(t_{i+1})\mathbf{B}_d(t_i)[\mathbf{U}(t_i) + \mathbf{B}_d^T(t_i)\mathbf{K}_c(t_{i+1})\mathbf{B}_d(t_i)]^{-1}\mathbf{B}_d^T(t_i)\mathbf{K}_c(t_{i+1})\mathbf{\Phi}(t_{i+1}, t_i)\end{aligned}$$

which is solved backwards from the terminal condition

$$\mathbf{K}_c(t_{N+1}) = \mathbf{X}_f$$

Show this by means of dynamic programming.

(b) Show that the minimum cost can be expressed as

$$\mathscr{C}^*[\mathbf{x}^*(t_i), t_i] = \tfrac{1}{2}\mathbf{x}^{*T}(t_i)\mathbf{K}_c(t_i)\mathbf{x}^*(t_i)$$

where $\mathbf{x}^*(t_i)$ is the state along the optimal trajectory.

(c) Show that $\mathbf{K}_c(t_i)$ is a symmetric matrix.

(d) Show that $\mathbf{K}_c(t_i)$ is at least positive semidefinite.

**13.10** The discrete-time scalar process $\mathsf{x}(\cdot,\cdot)$ satisfies the equation

$$\mathsf{x}(t_{i+1}) = \mathsf{x}(t_i) + \mathsf{u}(t_i) + \mathsf{w}_d(t_i)$$

where

$$\mathsf{u}(t_i) \triangleq \text{control variable}$$

$$\mathsf{w}_d(\cdot,\cdot) \triangleq \text{independent process disturbance with}$$

$$E\{\mathsf{w}_d(t_i)\} = 0, \qquad E\{\mathsf{w}_d(t_i)\mathsf{w}_d(t_j)\} = 2\delta_{ij}$$

Consider the last step of the process from time $t_N$ to time $t_{N+1}$. The cost to make this step is

$$\mathscr{C}[\mathrm{x}(t_N), t_N] = E\{g[\mathrm{u}(t_N)] + \mathrm{x}(t_{N+1})^2 | \mathrm{x}(t_N) = \mathrm{x}(t_N, \cdot)\}$$

where $g$ is a step function on the control magnitude defined as follows:

$$g[u(t_N)] = \begin{cases} 0 & \text{if } |u(t_N)| < 1 \\ 1 & \text{if } |u(t_N)| \geq 1 \end{cases}$$

Find the optimal feedback control $u^*[x(t_N), t_N]$. Obtain the minimum cost as a function of $x(t_N)$. Plot these two functions versus $x(t_N)$.

**13.11** Solve the following dynamic programming problem for three stages ($t_i = t_0, t_1, t_2, t_3$). Let the state $\mathrm{x}(t_i)$ be a scalar that can assume values 1, 2, or 3 for all $t_i$. The possible state transitions are, using the notation of (13-70),

$$\phi_1(x) = x + 1, \qquad \phi_2(x) = x, \qquad \phi_3(x) = 3, \qquad \phi_4(x) = 1$$

Decision $u = 1$ is available in states $x = 1$ and $x = 2$, but not in $x = 3$. The consequences of this decision are specified by

$$p[\phi_1 | x, u = 1, t_i] = \tfrac{1}{3}, \qquad p[\phi_2 | x, u = 1, t_i] = \tfrac{1}{3}$$
$$p[\phi_3 | x, u = 1, t_i] = \tfrac{1}{3}, \qquad p[\phi_4 | x, u = 1, t_i] = 0$$

Decision $u = 2$ is available in all three states and is specified by

$$p[\phi_1 | x, u = 2, t_i] = 0, \qquad p[\phi_2 | x, u = 2, t_i] = \tfrac{1}{3}$$
$$p[\phi_3 | x, u = 2, t_i] = \tfrac{1}{3}, \qquad p[\phi_4 | x, u = 2, t_i] = \tfrac{1}{3}$$

Transition costs or losses are given by

$$L[x(t_i), x(t_{i+1}), u = 1, t_i] = 3|x(t_i) - x(t_{i+1})|$$
$$L[x(t_i), x(t_{i+1}), u = 2, t_i] = 4$$

and the terminal cost or loss function is

$$L_f[x(t_3)] = 10x(t_3)$$

The solution to this problem should be in the form of tables specifying the values of the functions $\mathscr{C}^*[x(t_i), t_i]$ and $u^*[x(t_i), t_i]$ for $x = 1, 2, 3$ and $t_i = t_0, t_1, t_2, t_3$.

**13.12** (a) Consider the control algorithm specified by (13-26)–(13-31), with use of (13-18)–(13-23). Show that this controller can also be expressed in the form

$$\mathbf{u}^*(t_i) = \mathbf{G}_{cx}(t_i)\mathbf{x}_c(t_i) + \mathbf{G}_{cz}(t_i)\mathbf{z}_i$$
$$\mathbf{x}_c(t_{i+1}) = \mathbf{\Phi}_c(t_{i+1}, t_i)\mathbf{x}_c(t_i) + \mathbf{B}_{cz}(t_i)\mathbf{z}_i$$

by letting the "controller state vector," $\mathbf{x}_c(t_i)$, be defined as $\hat{\mathbf{x}}(t_i^-)$. Obtain the four matrices of this form explicitly. This algorithm form is useful to provide the same input/output characteristics of (13-26)–(13-28), but minimizing the computational delay time between sampling the sensors to get $\mathbf{z}_i$ and producing the control command $\mathbf{u}^*(t_i)$. The state equation above can be calculated "in the background" before the next sample time, to provide a computed value of $[\mathbf{G}_{cx}(t_{i+1})\mathbf{x}_c(t_{i+1})]$ for use at that time. More will be said about this in the next chapter.

(b) To reduce computational delays even further, many controllers actually implement the suboptimal control law

$$\mathbf{u}(t_i) = -\mathbf{G}_c^*(t_i)\hat{\mathbf{x}}(t_i^-)$$

instead of (13-28). Repeat part (a) for this controller, and show the savings in computational delay. What are the disadvantages?

**13.13** Develop the appropriate modifications to the optimal stochastic controller of Example 13.12 in Section 13.6, due to each of the following extensions (considered individually):

(a) The dynamics driving noise $\mathbf{w}_d(\cdot,\cdot)$ in (13-94) and the measurement corruption noise $\mathbf{v}(\cdot,\cdot)$ in (13-95) are correlated according to

$$E\{\mathbf{w}_d(t_i)\mathbf{v}^T(t_j)\} = \mathbf{C}(t_i)\,\delta_{ij}$$

instead of being independent. (Recall Section 5.9 of Volume 1.)

(b) These noises are again assumed correlated, but instead according to

$$E\{\mathbf{w}_d(t_{i-1})\mathbf{v}^T(t_j)\} = \mathbf{C}(t_j)\,\delta_{ij}$$

(c) The measurement model (13-95) is replaced by a measurement that incorporates a direct feedthrough of control:

$$\mathbf{z}(t_i) = \mathbf{H}(t_i)\mathbf{x}(t_i) + \mathbf{D}_z(t_i)\mathbf{u}(t_i) + \mathbf{v}(t_i)$$

(d) The loss function of Example 13.12 is generalized to allow quadratic cross-terms involving $\mathbf{x}(t_i)$ and $\mathbf{u}(t_i)$:

$$L[\mathbf{x}(t_i), \mathbf{u}(t_i), t_i] = \tfrac{1}{2}[\mathbf{x}^T(t_i)\mathbf{X}(t_i)\mathbf{x}(t_i) + \mathbf{u}^T(t_i)\mathbf{U}(t_i)\mathbf{u}(t_i) + 2\mathbf{x}^T(t_i)\mathbf{S}(t_i)\mathbf{u}(t_i)]$$

**13.14** In Section 13.6, the extension of the dynamic programming algorithm based on full state knowledge at each $t_i$, (13-65)–(13-67), to the algorithm appropriate for the case of only incomplete noise-corrupted measurements being available at each $t_i$, (13-128)–(13-130), was made. In an analogous manner, develop the generalization of the dynamic programming algorithm for discrete states and controls, (13-70)–(13-71), to the case where only incomplete noise-corrupted measurements are available.

**13.15** In Example 13.12, it was stated that the dynamic programming algorithm for the LQG stochastic control problem could be solved by means of assuming a quadratic solution form,

$$\mathscr{C}^*[\hat{\mathbf{x}}(t_i^+), t_i] = \tfrac{1}{2}[\hat{\mathbf{x}}^T(t_i^+)\mathbf{K}_c(t_i)\hat{\mathbf{x}}(t_i^+) + g(t_i)]$$

substituting this form into the algorithm, and producing the backward recursions and terminal conditions $\mathbf{K}_c$ and $g$ would have to satisfy for the assumed form to be valid (in analogy to the development of Example 13.7). Perform this procedure and demonstrate the validity of the controller and cost-to-go function evaluations portrayed in Example 13.12.

CHAPTER 14

# Linear stochastic controller design and performance analysis

## 14.1 INTRODUCTION

This chapter expands upon the simple LQG optimal stochastic regulator introduced in the previous chapter, developing a systematic design approach based on LQG synthesis capability to produce prospective designs, to be evaluated in a tradeoff analysis of realistic performance capabilities versus computer loading. Discrete-time measurements are emphasized throughout, due to the importance of sampled-data controllers implemented as software in a digital computer. Section 14.2 first motivates the need for a generalized quadratic cost function in this sampled-data context, and then develops the corresponding optimal LQG controller.

A primary objective of feedback controller design is a stable closed-loop system, even if the true system varies from the design conditions used for controller generation. Sections 14.3–14.5 develop these issues, providing additional motivation to use LQG synthesis.

LQG synthesis can be used to develop more than simple regulators. Sections 14.6–14.10 treat trackers, nonzero setpoint controllers, controllers with prescribed disturbance rejection properties, PI (proportional plus integral) controllers, and command generator trackers, respectively, as designed via this methodology. A systematic approach to generating simplified, implementable linear sampled-data controller designs at appropriate sample frequencies by LQG synthesis, and an evaluation of realistic performance capabilities of each, are presented in Sections 14.11 and 14.12.

Finally, the case of continuous-time measurements is addressed in Section 14.13, concentrating on design of constant-gain controllers that can be readily implemented in analog fashion.

## 14.2 THE LQG STOCHASTIC REGULATOR

In Section 13.3 and Examples 13.7 and 13.12, we considered regulation of a system adequately described by a linear discrete-time state equation,

$$\mathbf{x}(t_{i+1}) = \mathbf{\Phi}(t_{i+1}, t_i)\mathbf{x}(t_i) + \mathbf{B}_\mathrm{d}(t_i)\mathbf{u}(t_i) + \mathbf{G}_\mathrm{d}(t_i)\mathbf{w}_\mathrm{d}(t_i) \tag{14-1}$$

viewed as an equivalent discrete-time model for a continuous-time system having sampled-data measurements; $\mathbf{w}_\mathrm{d}(\cdot,\cdot)$ is zero-mean white Gaussian discrete-time noise of covariance $\mathbf{Q}_\mathrm{d}(t_i)$ for all $t_i$, and assumed independent of the Gaussian initial $\mathbf{x}(t_0)$ with mean $\hat{\mathbf{x}}_0$ and covariance $\mathbf{P}_0$. As discussed in that chapter, we desire $p$ specified controlled variables, $\mathbf{y}_\mathrm{c}(t)$, to exhibit desirable characteristics; in the case of regulation, we wish these variables to assume values as close to zero as possible. If we let the controlled variables be related to the states by the linear expression

$$\mathbf{y}_\mathrm{c}(t_i) = \mathbf{C}(t_i)\mathbf{x}(t_i) \tag{14-2}$$

then we can seek the control function $\mathbf{u}^*[\cdot,\cdot]$ that minimizes the cost function

$$J = E\left\{\sum_{i=0}^{N} \frac{1}{2}[\mathbf{y}_\mathrm{c}^\mathrm{T}(t_i)\mathbf{Y}(t_i)\mathbf{y}_\mathrm{c}(t_i) + \mathbf{u}^\mathrm{T}(t_i)\mathbf{U}(t_i)\mathbf{u}(t_i)] + \frac{1}{2}\mathbf{y}_\mathrm{c}^\mathrm{T}(t_{N+1})\mathbf{Y}_\mathrm{f}\mathbf{y}_\mathrm{c}(t_{N+1})\right\} \tag{14-3}$$

with $\mathbf{Y}(t_i)$ for all $t_i$ and $\mathbf{Y}_\mathrm{f}$ naturally chosen as $p$-by-$p$ symmetric positive definite matrices (positive *definite* since we want a nonzero cost for any controlled variable nonzero values), and $\mathbf{U}(t_i)$ as $r$-by-$r$ symmetric matrices that are also usually positive definite (as explained in Section 13.3). For instance, a first choice in an iterative design of a stochastic controller might be to let $\mathbf{Y}(t_i)$, $\mathbf{Y}_\mathrm{f}$, and $\mathbf{U}(t_i)$ all be diagonal, with each diagonal term being the inverse of the square of the maximum allowed value of the associated component of $\mathbf{y}_\mathrm{c}(t_i)$ or $\mathbf{u}(t_i)$, as suggested in Example 13.3. In this manner, we are able to generate a cost function of the form depicted earlier,

$$J = E\left\{\sum_{i=0}^{N} \frac{1}{2}[\mathbf{x}^\mathrm{T}(t_i)\mathbf{X}(t_i)\mathbf{x}(t_i) + \mathbf{u}^\mathrm{T}(t_i)\mathbf{U}(t_i)\mathbf{u}(t_i)] + \frac{1}{2}\mathbf{x}^\mathrm{T}(t_{N+1})\mathbf{X}_\mathrm{f}\mathbf{x}(t_{N+1})\right\} \tag{14-4}$$

where

$$\mathbf{X}(t_i) = \mathbf{C}^\mathrm{T}(t_i)\mathbf{Y}(t_i)\mathbf{C}(t_i) \tag{14-5a}$$

$$\mathbf{X}_\mathrm{f} = \mathbf{C}^\mathrm{T}(t_{N+1})\mathbf{Y}_\mathrm{f}\mathbf{C}(t_{N+1}) \tag{14-5b}$$

are positive *semidefinite*: $n$-by-$n$ matrices of rank no greater than $p \le n$. The cost-minimizing controller has been shown to be

$$\mathbf{u}^*(t_i) = -\mathbf{G}_\mathrm{c}^*(t_i)\mathbf{x}(t_i) \tag{14-6}$$

if perfect knowledge of $\mathbf{x}(t_i)$ is assumed available at each sample time, or

$$\mathbf{u}^*(t_i) = -\mathbf{G}_c^*(t_i)\hat{\mathbf{x}}(t_i^+) \tag{14-7}$$

if only incomplete, noise-corrupted measurements are available and are processed by the appropriate Kalman filter [66]. These controller structures are displayed in Figs. 13.4, 13.5, and 13.9. The controller gains in these two cases are the same, and are equal to the gain of the associated optimal *deterministic* controller: *certainty equivalence* applies [12, 22, 37, 45, 53, 63, 100, 137, 161, 173, 176, 194]. For all three cases, these gains can be generated from the solution to the backward Riccati recursion

$$\mathbf{G}_c^*(t_i) = [\mathbf{U}(t_i) + \mathbf{B}_d^T(t_i)\mathbf{K}_c(t_{i+1})\mathbf{B}_d(t_i)]^{-1}[\mathbf{B}_d^T(t_i)\mathbf{K}_c(t_{i+1})\mathbf{\Phi}(t_{i+1}, t_i)] \tag{14-8}$$

$$\begin{aligned}\mathbf{K}_c(t_i) &= \mathbf{X}(t_i) + \mathbf{\Phi}^T(t_{i+1}, t_i)\mathbf{K}_c(t_{i+1})\mathbf{\Phi}(t_{i+1}, t_i) \\ &\quad - [\mathbf{\Phi}^T(t_{i+1}, t_i)\mathbf{K}_c(t_{i+1})\mathbf{B}_d(t_i)]\mathbf{G}_c^*(t_i) && \text{(14-9a)} \\ &= \mathbf{X}(t_i) + \mathbf{\Phi}^T(t_{i+1}, t_i)\mathbf{K}_c(t_{i+1})[\mathbf{\Phi}(t_{i+1}, t_i) - \mathbf{B}_d(t_i)\mathbf{G}_c^*(t_i)] && \text{(14-9b)}\end{aligned}$$

solved backwards from the terminal condition

$$\mathbf{K}_c(t_{N+1}) = \mathbf{X}_f \tag{14-10}$$

Note that a *sufficient*, though *not necessary*, condition for the inverse in (14-8) to *exist* and the cost-*minimizing* control to *exist* and be *unique*, is that $\mathbf{U}(t_i)$ is positive definite for all $t_i$ [57, 139, 195].

EXAMPLE 14.1 Consider the system described by the scalar time-invariant state equation

$$\mathsf{x}(t_{i+1}) = \Phi \mathsf{x}(t_i) + B_d u(t_i) + G_d \mathsf{w}_d(t_i)$$

with the state itself being the controlled variable, and assume we want to find the $\mathbf{u}^*[\cdot, t_i]$ for all $t_i$ that minimizes

$$J = E\left\{\sum_{i=0}^{N} \frac{1}{2}[X\mathsf{x}(t_i)^2 + U\mathsf{u}(t_i)^2] + \frac{1}{2} X_f \mathsf{x}(t_{N+1})^2\right\}$$

with $X$ and $U$ constant for all time. The optimal control is

$$u^*(t_i) = -G_c^*(t_i)x(t_i)$$

or

$$u^*(t_i) = -G_c^*(t_i)\hat{x}(t_i^+)$$

depending on what is available from the system at each sample time, where

$$G_c^*(t_i) = \frac{B_d K_c(t_{i+1})\Phi}{U + B_d^2 K_c(t_{i+1})}$$

and where $K_c(t_i)$ satisfies

$$K_c(t_i) = X + \Phi^2 K_c(t_{i+1}) - \frac{[B_d K_c(t_{i+1})\Phi]^2}{U + B_d{}^2 K_c(t_{i+1})}$$

$$K_c(t_{N+1}) = X_f$$

The $K_c(t_i)$ and gain $G_c{}^*(t_i)$ values generally have a terminal transient; working the Riccati equation backward in time, we can reach (under conditions to be specified later) a steady state value of $\bar{K}_c$ that satisfies $K_c(t_i) = K_c(t_{i+1}) = \bar{K}_c$:

$$\bar{K}_c = X + \Phi^2 \bar{K}_c - \frac{[B_d \bar{K}_c \Phi]^2}{U + B_d{}^2 \bar{K}_c}$$

i.e., the positive root of

$$(B_d{}^2)\bar{K}_c{}^2 + (U - \Phi^2 U - B_d{}^2 X)\bar{K}_c - (UX) = 0$$

Note that if $X_f = \bar{K}_c$, then $K_c(t_i)$ and $G_c{}^*(t_i)$ assume their steady state values $\bar{K}_c$ and $\bar{G}_c{}^*$ for all $t_i$. One might seek the steady state constant-gain control law

$$u(t_i) = -\bar{G}_c{}^* x(t_i) \qquad \text{or} \qquad u(t_i) = -\bar{G}_c{}^* \hat{x}(t_i{}^+)$$

to use for all time because of ease of implementation, accepting the performance degradation due to ignoring the terminal transient in $G_c{}^*(t_i)$ (and the initial transient in the Kalman filter for this problem, assuming stationary noise models so that a constant-gain filter is also achieved). Such design methodologies and evaluation of true performance of simplified designs will be developed in this chapter.

Consider the special case of $U \equiv 0$: physically, there is no penalty on the amount of control energy used; mathematically, this is the singular case in which $U$ is not positive definite as usually assumed. For this case,

$$G_c{}^*(t_i) = \frac{B_d \Phi K_c(t_{i+1})}{B_d{}^2 K_c(t_{i+1})} = \frac{\Phi}{B_d}$$

$$K_c(t_i) = X + \Phi^2 K_c(t_{i+1}) - \frac{[B_d \Phi K_c(t_{i+1})]^2}{B_d{}^2 K_c(t_{i+1})} = X$$

for $i = 0, 1, \ldots, N$, and $K_c(t_{N+1}) = X_f$. For the *deterministic* optimal control problem, this control is seen to drive the state $x(t_1)$ to zero regardless of the control energy required:

$$x(t_1) = \Phi x(t_0) + B_d u^*(t_0) = \Phi x(t_0) + B_d[-(\Phi/B_d)x(t_0)] = 0$$

The state is driven to zero in one sample period, and it stays there for all $t_i$, so the total cost is

$$J = \frac{1}{2} X x(t_0)^2 + \sum_{i=1}^{N} \frac{1}{2} X[x(t_i) = 0]^2 + \frac{1}{2} X_f[x(t_{N+1}) = 0]^2$$

$$= \tfrac{1}{2} X x(t_0)^2$$

as computed according to $\mathscr{C}^*[x(t_0), t_0] = \frac{1}{2}[K_c(t_0)x(t_0)^2 + g(t_0)]$, since $K_c(t_0) = X$ and $g(t_i) \equiv 0$ for the deterministic case. Note that the *closed-loop system eigenvalue* $(\Phi - B_d \bar{G}_c{}^*)$ is *zero* in this case. In the *stochastic* problem in which full state knowledge is assumed, the control

$$u(t_i) = -[\Phi/B_d]x(t_i)$$

then continually *attempts* to drive the state to zero in one sample period, but is generally prevented from doing so because of the presence of $w_d(t_i)$. The cost to complete the process from $t_0$ forward is then $\frac{1}{2}[Xx(t_0)^2 + g(t_0)]$ with $g(t_0)$ given as in Example 13.7:

$$g(t_0) = N[G_d{}^2 Q_d X] + G_d{}^2 Q_d X_f > 0$$

and the higher cost is due to the added uncertainty. If only noise-corrupted measurements are available, the control

$$u(t_i) = -[\Phi/B_d]\hat{x}(t_i{}^+)$$

again tries to drive the state to zero in a single period; here the cost-to-go at initial $t_0$ is again $\frac{1}{2}[Xx(t_0)^2 + g(t_0)]$, but with $g(t_0)$ given according to Example 13.12 and Eqs. (13-20) and (13-21) as

$$g(t_0) = \sum_{i=0}^{N} \{XP(t_i{}^+) + K(t_{i+1})HP(t_{i+1}^-)X\} + X_f P(t_{N+1}^+)$$

This cost is still higher because of the greater amount of uncertainty in this formulation.

Now consider the opposite limiting case of $U \to \infty$, which specifies infinite penalty for any finite control energy expenditure. It is then physically reasonable that

$$G_c{}^*(t_i) = \frac{B_d K_c(t_{i+1})\Phi}{U + B_d{}^2 K_c(t_{i+1})} \to 0$$

and thus no control is applied. This is the trivial case in which the closed-loop system eigenvalue equals the open-loop eigenvalue $\Phi$. Note that $G_c{}^*(t_i) \equiv 0$ is also the solution if $X = X_f = 0$ and $U$ is nonzero: if no penalty is placed on state deviations and any control application incurs a penalty, the optimal strategy is to use no control. ■

Computationally, (14-9a) can be made efficient by writing

$$\mathbf{A}_c(t_i) = \mathbf{B}_d{}^T(t_i)\mathbf{K}_c(t_{i+1})\mathbf{\Phi}(t_{i+1}, t_i) \qquad (14\text{-}11a)$$

$$\mathbf{G}_c{}^*(t_i) = [\mathbf{U}(t_i) + \mathbf{B}_d{}^T(t_i)\mathbf{K}_c(t_{i+1})\mathbf{B}_d(t_i)]^{-1}\mathbf{A}_c(t_i) \qquad (14\text{-}11b)$$

$$\mathbf{K}_c(t_i) = \mathbf{X}(t_i) + \mathbf{\Phi}^T(t_{i+1}, t_i)\mathbf{K}_c(t_{i+1})\mathbf{\Phi}(t_{i+1}, t_i) - \mathbf{A}_c^T(t_i)\mathbf{G}_c{}^*(t_i) \qquad (14\text{-}11c)$$

and recognizing $\mathbf{U}$, $\mathbf{B}_d{}^T\mathbf{K}_c\mathbf{B}_d$ and the inverse in (14-11b) and all three terms in (14-11c) to be symmetric, so that only the nonredundant (lower triangular) terms require explicit evaluation. An alternate, algebraically equivalent form of (14-9) is

$$\mathbf{K}_c(t_i) = [\mathbf{\Phi}(t_{i+1}, t_i) - \mathbf{B}_d(t_i)\mathbf{G}_c{}^*(t_i)]^T\mathbf{K}_c(t_{i+1})[\mathbf{\Phi}(t_{i+1}, t_i) - \mathbf{B}_d(t_i)\mathbf{G}_c{}^*(t_i)] + \mathbf{G}_c^{*T}(t_i)\mathbf{U}(t_i)\mathbf{G}_c{}^*(t_i) + \mathbf{X}(t_i) \qquad (14\text{-}12)$$

where again symmetry can be exploited for efficient computations.

As stated in Chapter 13, controllers of the form (14-6) or (14-7), with gains specified by (14-8)–(14-12), are the optimal stochastic controllers for the simplest form of LQG problem. Now we want to extend these results to a more general problem context [3, 5, 14, 38, 70, 105, 119, 189] and to motivate the need for such generalization in physical problems. First of all, let us generalize the quadratic cost function (14-4) to allow cross-terms between $\mathbf{x}(t_i)$ and $\mathbf{u}(t_i)$ for all $t_i$:

$$J = E\left\{\frac{1}{2}\mathbf{x}^{\mathrm{T}}(t_{N+1})\mathbf{X}_{\mathrm{f}}\mathbf{x}(t_{N+1}) + \sum_{i=0}^{N}\frac{1}{2}[\mathbf{x}^{\mathrm{T}}(t_i)\mathbf{X}(t_i)\mathbf{x}(t_i) + \mathbf{u}^{\mathrm{T}}(t_i)\mathbf{U}(t_i)\mathbf{u}(t_i) + 2\mathbf{x}^{\mathrm{T}}(t_i)\mathbf{S}(t_i)\mathbf{u}(t_i)]\right\} \tag{14-13a}$$

$$= E\left\{\frac{1}{2}\mathbf{x}^{\mathrm{T}}(t_{N+1})\mathbf{X}_{\mathrm{f}}\mathbf{x}(t_{N+1}) + \sum_{i=0}^{N}\frac{1}{2}\left(\begin{bmatrix}\mathbf{x}(t_i)\\ \mathbf{u}(t_i)\end{bmatrix}^{\mathrm{T}}\begin{bmatrix}\mathbf{X}(t_i) & \mathbf{S}(t_i)\\ \mathbf{S}^{\mathrm{T}}(t_i) & \mathbf{U}(t_i)\end{bmatrix}\begin{bmatrix}\mathbf{x}(t_i)\\ \mathbf{u}(t_i)\end{bmatrix}\right)\right\} \tag{14-13b}$$

with $\mathbf{X}_{\mathrm{f}}$ and $\mathbf{X}(t_i)$ symmetric positive semidefinite $n$-by-$n$ matrices, $\mathbf{U}(t_i)$ symmetric positive definite $r$-by-$r$ matrices, and $n$-by-$r$ $\mathbf{S}(t_i)$ chosen so that the symmetric composite matrix in the summation of (14-13b) is positive semidefinite (or, equivalently, so that $[\mathbf{X}(t_i) - \mathbf{S}(t_i)\mathbf{U}^{-1}(t_i)\mathbf{S}^{\mathrm{T}}(t_i)]$ is positive semidefinite) for all $t_i$. For such a cost criterion, the same derivation procedure as in Chapter 13 can be applied to yield the optimal controller again in the form of (14-6) or (14-7), but with the gain $\mathbf{G}_{\mathrm{c}}^*(t_i)$ specified by

$$\mathbf{G}_{\mathrm{c}}^*(t_i) = [\mathbf{U}(t_i) + \mathbf{B}_{\mathrm{d}}^{\mathrm{T}}(t_i)\mathbf{K}_{\mathrm{c}}(t_{i+1})\mathbf{B}_{\mathrm{d}}(t_i)]^{-1} \times [\mathbf{B}_{\mathrm{d}}^{\mathrm{T}}(t_i)\mathbf{K}_{\mathrm{c}}(t_{i+1})\mathbf{\Phi}(t_{i+1}, t_i) + \mathbf{S}^{\mathrm{T}}(t_i)] \tag{14-14}$$

where $\mathbf{K}_{\mathrm{c}}(t_i)$ satisfies the backward Riccati recursion, given equivalently as

$$\mathbf{K}_{\mathrm{c}}(t_i) = \mathbf{X}(t_i) + \mathbf{\Phi}^{\mathrm{T}}(t_{i+1}, t_i)\mathbf{K}_{\mathrm{c}}(t_{i+1})\mathbf{\Phi}(t_{i+1}, t_i) - [\mathbf{B}_{\mathrm{d}}^{\mathrm{T}}(t_i)\mathbf{K}_{\mathrm{c}}(t_{i+1})\mathbf{\Phi}(t_{i+1}, t_i) + \mathbf{S}^{\mathrm{T}}(t_i)]^{\mathrm{T}}\mathbf{G}_{\mathrm{c}}^*(t_i) \tag{14-15}$$

or

$$\mathbf{K}_{\mathrm{c}}(t_i) = [\mathbf{\Phi}(t_{i+1}, t_i) - \mathbf{B}_{\mathrm{d}}(t_i)\mathbf{G}_{\mathrm{c}}^*(t_i)]^{\mathrm{T}}\mathbf{K}_{\mathrm{c}}(t_{i+1})[\mathbf{\Phi}(t_{i+1}, t_i) - \mathbf{B}_{\mathrm{d}}(t_i)\mathbf{G}_{\mathrm{c}}^*(t_i)] + \mathbf{G}_{\mathrm{c}}^{*\mathrm{T}}(t_i)\mathbf{U}(t_i)\mathbf{G}_{\mathrm{c}}^*(t_i) + \mathbf{X}(t_i) - \mathbf{S}(t_i)\mathbf{G}_{\mathrm{c}}^*(t_i) - \mathbf{G}_{\mathrm{c}}^{*\mathrm{T}}(t_i)\mathbf{S}^{\mathrm{T}}(t_i) \tag{14-16}$$

solved backwards [93, 132, 160] from the terminal condition

$$\mathbf{K}_{\mathrm{c}}(t_{N+1}) = \mathbf{X}_{\mathrm{f}} \tag{14-17}$$

It can readily be seen that for the case of $\mathbf{S}(t_i) \equiv \mathbf{0}$, (14-14)–(14-16) reduce to (14-8), (14-9), and (14-12), respectively, as expected. However, let us motivate the need for $\mathbf{S}(t_i) \neq \mathbf{0}$. Consider the sampled-data control of a continuous-time system, using a zero-order hold interface such that

$$\mathbf{u}(t) = \mathbf{u}(t_i) \qquad \text{for all} \quad t \in [t_i, t_{i+1}) \tag{14-18}$$

Let the continuous-time state description be given by the usual linear stochastic differential equation

$$\mathbf{dx}(t) = \mathbf{F}(t)\mathbf{x}(t)\,dt + \mathbf{B}(t)\mathbf{u}(t)\,dt + \mathbf{G}(t)\,\mathbf{d}\boldsymbol{\beta}(t) \tag{14-19a}$$

with $\boldsymbol{\beta}(\cdot,\cdot)$ being Brownian motion of diffusion $\mathbf{Q}(t)$, or

$$\dot{\mathbf{x}}(t) = \mathbf{F}(t)\mathbf{x}(t) + \mathbf{B}(t)\mathbf{u}(t) + \mathbf{G}(t)\mathbf{w}(t) \tag{14-19b}$$

where $\mathbf{w}(\cdot,\cdot)$ is zero-mean white Gaussian noise of strength $\mathbf{Q}(t)$ for all $t$. Since we are interested in the behavior of the continuous-time system for *all* $t \in [t_0, t_{N+1}]$ and *not just* its behavior as seen only at the sample times $t_i$, let us assume that the performance objective of the control algorithm is to minimize an appropriate continuous-time quadratic cost of

$$J_c = E\left\{\frac{1}{2}\mathbf{x}^T(t_{N+1})\mathbf{X}_f\mathbf{x}(t_{N+1}) + \int_{t_0}^{t_{N+1}} \frac{1}{2}[\mathbf{x}^T(t)\mathbf{W}_{xx}(t)\mathbf{x}(t) + \mathbf{u}^T(t)\mathbf{W}_{uu}(t)\mathbf{u}(t) + 2\mathbf{x}^T(t)\mathbf{W}_{xu}(t)\mathbf{u}(t)]\,dt\right\} \tag{14-20a}$$

$$= E\left\{\frac{1}{2}\mathbf{x}^T(t_{N+1})\mathbf{X}_f\mathbf{x}(t_{N+1}) + \int_{t_0}^{t_{N+1}} \frac{1}{2}\left(\begin{bmatrix}\mathbf{x}(t)\\ \mathbf{u}(t)\end{bmatrix}^T \begin{bmatrix}\mathbf{W}_{xx}(t) & \mathbf{W}_{xu}(t)\\ \mathbf{W}_{xu}^T(t) & \mathbf{W}_{uu}(t)\end{bmatrix}\begin{bmatrix}\mathbf{x}(t)\\ \mathbf{u}(t)\end{bmatrix}\right)dt\right\} \tag{14-20b}$$

with $\mathbf{W}_{xx}(t)$ positive semidefinite and $\mathbf{W}_{uu}(t)$ positive definite for all $t$, and the composite matrix in the integrand of (14-20b) positive semidefinite for all $t$. We now show that this cost can be expressed equivalently in the form of (14-13). First divide the interval of interest, $[t_0, t_{N+1}]$, into $(N+1)$ control intervals (typically, but not necessarily, equal in length) to write

$$J_c = E\left\{\frac{1}{2}\mathbf{x}^T(t_{N+1})\mathbf{X}_f\mathbf{x}(t_{N+1}) + \sum_{i=0}^{N}\left(\int_{t_i}^{t_{i+1}} \frac{1}{2}[\mathbf{x}^T(t)\mathbf{W}_{xx}(t)\mathbf{x}(t) + \mathbf{u}^T(t)\mathbf{W}_{uu}(t)\mathbf{u}(t) + 2\mathbf{x}^T(t)\mathbf{W}_{xu}(t)\mathbf{u}(t)]\,dt\right)\right\} \tag{14-21}$$

Now for all $t \in [t_i, t_{i+1})$, $\mathbf{u}(t)$ is given by (14-18) and $\mathbf{x}(t)$ by

$$\mathbf{x}(t) = \mathbf{\Phi}(t,t_i)\mathbf{x}(t_i) + \bar{\mathbf{B}}(t,t_i)\mathbf{u}(t_i) + \int_{t_i}^{t} \mathbf{\Phi}(t,\tau)\mathbf{G}(\tau)\,\mathbf{d}\boldsymbol{\beta}(\tau) \tag{14-22}$$

$$\bar{\mathbf{B}}(t,t_i) \triangleq \int_{t_i}^{t} \mathbf{\Phi}(t,\tau)\mathbf{B}(\tau)\,d\tau \tag{14-23}$$

Substituting into (14-21) yields

$$J_c = E\left\{\frac{1}{2}\mathbf{x}^T(t_{N+1})\mathbf{X}_f\mathbf{x}(t_{N+1}) + \sum_{i=0}^{N}\frac{1}{2}[\mathbf{x}^T(t_i)\mathbf{X}(t_i)\mathbf{x}(t_i) + \mathbf{u}^T(t_i)\mathbf{U}(t_i)\mathbf{u}(t_i) + 2\mathbf{x}^T(t_i)\mathbf{S}(t_i)\mathbf{u}(t_i) + \mathsf{L}_r(t_i)]\right\} \tag{14-24}$$

where

$$\mathbf{X}(t_i) = \int_{t_i}^{t_{i+1}} \mathbf{\Phi}^{\mathrm{T}}(t, t_i)\mathbf{W}_{xx}(t)\mathbf{\Phi}(t, t_i)\, dt \tag{14-25a}$$

$$\mathbf{U}(t_i) = \int_{t_i}^{t_{i+1}} [\bar{\mathbf{B}}^{\mathrm{T}}(t, t_i)\mathbf{W}_{xx}(t)\bar{\mathbf{B}}(t, t_i) + \mathbf{W}_{uu}(t) + \bar{\mathbf{B}}^{\mathrm{T}}(t, t_i)\mathbf{W}_{xu}(t) + \mathbf{W}_{xu}^{\mathrm{T}}(t)\bar{\mathbf{B}}(t, t_i)]\, dt \tag{14-25b}$$

$$\mathbf{S}(t_i) = \int_{t_i}^{t_{i+1}} [\mathbf{\Phi}^{\mathrm{T}}(t, t_i)\mathbf{W}_{xx}(t)\bar{\mathbf{B}}(t, t_i) + \mathbf{\Phi}^{\mathrm{T}}(t, t_i)\mathbf{W}_{xu}(t)]\, dt \tag{14-25c}$$

$$\mathsf{L}_{\mathrm{r}}(t_i) = \int_{t_i}^{t_{i+1}} \left[\int_{t_i}^{t} \mathbf{\Phi}(t, \tau)\mathbf{G}(\tau)\,\mathbf{d\beta}(\tau)\right]^{\mathrm{T}} \mathbf{W}_{xx}(t) \left[\int_{t_i}^{t} \mathbf{\Phi}(t, \sigma)\mathbf{B}(\sigma)\,\mathbf{d\beta}(\sigma)\right] dt + \left\{\text{integrals of products of} \left[\int_{t_i}^{t} \mathbf{\Phi}(t, \tau)\mathbf{G}(\tau)\,\mathbf{d\beta}(\tau)\right] \text{with } \mathbf{x}(t_i) \text{ or } \mathbf{u}(t_i)\right\} \tag{14-25d}$$

First look specifically at (14-25c) and note that there are *two ways* in which the cross-terms in the generalized quadratic cost can arise: (1) *cross-terms in the continuous-time cost* (i.e., $\mathbf{W}_{xu}(t) \neq \mathbf{0}$ in (14-21)) yield nonzero $\mathbf{S}(t_i)$, and (2) even if $\mathbf{W}_{xu}(t) \equiv \mathbf{0}$ for all $t \in [t_0, t_{N+1}]$, $\mathbf{S}(t_i)$ is still generally nonzero due to the *objective of exerting desirable control influence on the state over the entire sample period* and not just at the sample times $t_0, t_1, \ldots, t_{N+1}$. This will be demonstrated physically in examples to follow.

Now consider the expectation of the residual loss function $\mathsf{L}_{\mathrm{r}}(t_i)$ of (14-24) that does not appear in the proposed cost (14-13); denote this residual cost as $J_{\mathrm{r}}(t_i) \triangleq E\{\mathsf{L}_{\mathrm{r}}(t_i)\}$. Since $[\int_{t_i}^{t} \mathbf{\Phi}(t, \tau)\mathbf{G}(\tau)\,\mathbf{d\beta}(\tau)]$ is zero-mean and uncorrelated with $\mathbf{x}(t_i)$ or $\mathbf{u}(t_i)$, only the leading term in (14-25d) will contribute to $J_{\mathrm{r}}(t_i)$, and we obtain

$$\begin{aligned}
J_{\mathrm{r}}(t_i) &= E\left\{\int_{t_i}^{t_{i+1}} \left[\int_{t_i}^{t} \mathbf{\Phi}(t, \tau)\mathbf{G}(\tau)\,\mathbf{d\beta}(\tau)\right]^{\mathrm{T}} \mathbf{W}_{xx}(t) \left[\int_{t_i}^{t} \mathbf{\Phi}(t, \sigma)\mathbf{G}(\sigma)\,\mathbf{d\beta}(\sigma)\right] dt\right\} \\
&= \int_{t_i}^{t_{i+1}} E\left\{\left[\int_{t_i}^{t} \mathbf{\Phi}(t, \tau)\mathbf{G}(\tau)\,\mathbf{d\beta}(\tau)\right]^{\mathrm{T}} \mathbf{W}_{xx}(t) \left[\int_{t_i}^{t} \mathbf{\Phi}(t, \sigma)\mathbf{G}(\sigma)\,\mathbf{d\beta}(\sigma)\right]\right\} dt \\
&= \int_{t_i}^{t_{i+1}} \mathrm{tr}\left(\mathbf{W}_{xx}(t) E\left\{\left[\int_{t_i}^{t} \mathbf{\Phi}(t, \sigma)\mathbf{G}(\sigma)\,\mathbf{d\beta}(\sigma)\right]\left[\int_{t_i}^{t} \mathbf{\Phi}(t, \tau)\mathbf{G}(\tau)\,\mathbf{d\beta}(\tau)\right]^{\mathrm{T}}\right\}\right) dt \\
&= \int_{t_i}^{t_{i+1}} \mathrm{tr}[\mathbf{W}_{xx}(t)\bar{\mathbf{Q}}(t, t_i)]\, dt
\end{aligned} \tag{14-26}$$

where

$$\bar{\mathbf{Q}}(t, t_i) \triangleq \int_{t_i}^{t} \mathbf{\Phi}(t, \tau)\mathbf{G}(\tau)\mathbf{Q}(\tau)\mathbf{G}^{\mathrm{T}}(\tau)\mathbf{\Phi}^{\mathrm{T}}(t, \tau)\, d\tau \tag{14-27}$$

This residual cost cannot be affected by the choice of control function $\mathbf{u}[\cdot,\cdot]$, and thus $J_r(t_i) \neq 0$ will have no bearing on the evaluation of the optimal control function $\mathbf{u}^*[\cdot,\cdot]$: the controller defined by (14-14)–(14-17) and either (14-6) or (14-7) is also optimal with respect to the criterion (14-24), or equivalently, (14-20). As in previous results, if we consider the deterministic optimal control problem, then $J_r(t_i) \equiv 0$ for all $t_i$, and the cost increases but the control function remains unaltered as we progress to the stochastic optimal control problems.

The *equivalent discrete-time problem* formulation for a sampled-data control problem described originally by (14-18)–(14-20) was expressed in terms of integral evaluations (14-23), (14-25), and (14-27). For efficient computation, these can be transformed by explicit differentiation into the corresponding differential equations to be integrated to yield the desired terms [115, 189]. Thus, the *differential equations to be integrated forward simultaneously from* $t_i$ *to* $t_{i+1}$ *to generate a discrete-time characterization of states and cost* (i.e., (14-1) with $\mathbf{G}_d(t_i) \equiv \mathbf{I}$ and (14-13)) *for a sampled-data control problem are*

$$\dot{\mathbf{\Phi}}(t, t_i) = \mathbf{F}(t)\mathbf{\Phi}(t, t_i) \tag{14-28a}$$

$$\dot{\bar{\mathbf{B}}}(t, t_i) = \mathbf{F}(t)\bar{\mathbf{B}}(t, t_i) + \mathbf{B}(t) \tag{14-28b}$$

$$\dot{\bar{\mathbf{Q}}}(t, t_i) = \mathbf{F}(t)\bar{\mathbf{Q}}(t, t_i) + \bar{\mathbf{Q}}(t, t_i)\mathbf{F}^{\mathrm{T}}(t) + \mathbf{G}(t)\mathbf{Q}(t)\mathbf{G}^{\mathrm{T}}(t) \tag{14-28c}$$

$$\dot{\bar{\mathbf{X}}}(t, t_i) = \mathbf{\Phi}^{\mathrm{T}}(t, t_i)\mathbf{W}_{xx}(t)\mathbf{\Phi}(t, t_i) \tag{14-28d}$$

$$\begin{aligned}\dot{\bar{\mathbf{U}}}(t, t_i) = {} & \bar{\mathbf{B}}^{\mathrm{T}}(t, t_i)\mathbf{W}_{xx}(t)\bar{\mathbf{B}}(t, t_i) + \mathbf{W}_{uu}(t) \\ & + \bar{\mathbf{B}}^{\mathrm{T}}(t, t_i)\mathbf{W}_{xu}(t) + \mathbf{W}_{xu}^{\mathrm{T}}(t)\bar{\mathbf{B}}(t, t_i)\end{aligned} \tag{14-28e}$$

$$\dot{\bar{\mathbf{S}}}(t, t_i) = \mathbf{\Phi}^{\mathrm{T}}(t, t_i)\mathbf{W}_{xx}(t)\bar{\mathbf{B}}(t, t_i) + \mathbf{\Phi}^{\mathrm{T}}(t, t_i)\mathbf{W}_{xu}(t) \tag{14-28f}$$

$$\dot{\bar{J}}_r(t, t_i) = \mathrm{tr}[\mathbf{W}_{xx}(t)\bar{\mathbf{Q}}(t, t_i)] \tag{14-28g}$$

starting from the initial conditions of $\mathbf{\Phi}(t_i, t_i) = \mathbf{I}$ and all others appropriately dimensioned zeros. Integration to the sample time $t_{i+1}$ yields the desired results as $\mathbf{\Phi}(t_{i+1}, t_i)$, $\mathbf{B}_d(t_i) = \bar{\mathbf{B}}(t_{i+1}, t_i)$, $\mathbf{Q}_d(t_i) = \bar{\mathbf{Q}}(t_{i+1}, t_i)$, $\mathbf{X}(t_i) = \bar{\mathbf{X}}(t_{i+1}, t_i)$, $\mathbf{U}(t_i) = \bar{\mathbf{U}}(t_{i+1}, t_i)$, $\mathbf{S}(t_i) = \bar{\mathbf{S}}(t_{i+1}, t_i)$, and $J_r(t_i) = \bar{J}_r(t_{i+1}, t_i)$. The numerical integration must be repeated for every sample period of interest, except in the case of a *time-invariant* system model driven by *stationary* noises and a cost with *constant* weighting matrices, in which an integration over one sample period suffices for all periods (a vast simplification).

EXAMPLE 14.2 Consider the system depicted in Fig. 14.1, an angular servo system for a gimbaled platform. This problem will form the context of numerous examples throughout this chapter. Here we consider the design of a sampled-data regulator assuming that $S(t_i) \equiv 0$ for all $t_i$, to which later results will be compared.

Assume that the dynamics of the gimbal angle $\theta(t)$ are well described by

$$\dot{\theta}(t) = -(1/T)\theta(t) + (1/T)\theta_{\mathrm{com}}(t) + \mathrm{w}(t)$$

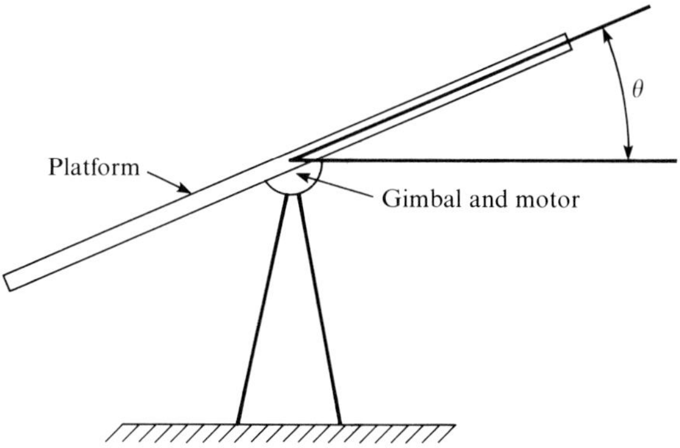


FIG. 14.1 Angular servo system for gimbaled platform.

where the commanded angle $\theta_{\text{com}}(t)$ is to be generated by a sampled-data controller (to be designed) such that

$$\theta_{\text{com}}(t) = \theta_{\text{com}}(t_i) \qquad \text{for all} \quad t \in [t_i, t_{i+1})$$

and stationary $\mathrm{w}(\cdot,\cdot)$ of strength (power spectral density level) $Q$ is meant to model both physical disturbances from the environment and the uncertainty associated with this simple model being a totally adequate description. A classical frequency domain model would depict the transfer function between commanded and delivered angle $\theta$ as

$$\frac{\theta(s)}{\theta_{\text{com}}(s)} = \frac{1}{Ts+1} = \frac{1/T}{s+(1/T)}$$

and similarly between noise input and $\theta$ output as $1/[s+(1/T)]$. Letting $\mathrm{x}(t) \triangleq \theta(t)$ and $u(t) \triangleq \theta_{\text{com}}(t)$, we have a time-invariant linear system model

$$\dot{\mathrm{x}}(t) = F\mathrm{x}(t) + Bu(t) + G\mathrm{w}(t)$$

with $F = -(1/T)$, $B = (1/T)$, and $G = 1$. The equivalent discrete-time model is

$$\begin{aligned} \mathrm{x}(t_{i+1}) &= [e^{-\Delta t/T}]\mathrm{x}(t_i) + \left[\int_{t_i}^{t_{i+1}} e^{-(t_{i+1}-\tau)/T}(1/T)\,d\tau\right]u(t_i) + \mathrm{w}_{\mathrm{d}}(t_i) \\ &= \underbrace{[e^{-\Delta t/T}]}_{\Phi}\mathrm{x}(t_i) + \underbrace{[1-e^{-\Delta t/T}]}_{B_{\mathrm{d}}}u(t_i) + \mathrm{w}_{\mathrm{d}}(t_i) \end{aligned}$$

where $\Delta t \triangleq [t_{i+1} - t_i]$ is chosen to be the fixed sample period of the digital controller algorithm, and $\mathrm{w}_{\mathrm{d}}(\cdot,\cdot)$ is zero-mean white Gaussian discrete-time noise of variance

$$Q_{\mathrm{d}} = \int_{t_i}^{t_{i+1}} e^{-2(t_{i+1}-\tau)/T} Q\,d\tau = QT[1-e^{-2\Delta t/T}]/2$$

For the sake of this example, let $T = 0.5$ sec and $\Delta t = 0.1$ sec (using an engineering rule of thumb of a sample frequency between 5 and 10 times the highest frequency signal of interest, rather than the minimum of twice that frequency specified by the Shannon sampling theorem), to yield $\Phi = 0.82$, $B_{\mathrm{d}} = 0.18$, and $Q_{\mathrm{d}} = 0.084Q$ to two significant figures.

If we want to generate the controller to minimize

$$J = E\left\{\sum_{i=0}^{N} \frac{1}{2}[X\mathrm{x}(t_i)^2 + U\mathrm{u}(t_i)^2] + \frac{1}{2}X_{\mathrm{f}}\mathrm{x}(t_{N+1})^2\right\}$$

assuming perfect state knowledge at each $t_i$, we can use (14-6) and (14-8)–(14-12) to synthesize a feedback controller given by

$$u^*(t_i) = -G_c^*(t_i)x(t_i)$$

where

$$G_c^*(t_i) = \frac{B_d K_c(t_{i+1})\Phi}{U + B_d^2 K_c(t_{i+1})} = \frac{0.15K_c(t_{i+1})}{U + 0.033K_c(t_{i+1})}$$

and

$$K_c(t_i) = X + \Phi^2 K_c(t_{i+1}) - \frac{[B_d K_c(t_{i+1})\Phi]^2}{U + B_d^2 K_c(t_{i+1})}$$

$$= X + 0.67K_c(t_{i+1}) - \frac{0.022K_c^2(t_{i+1})}{U + 0.033K_c(t_{i+1})}$$

with

$$K_c(t_{N+1}) = X_f$$

These gain evaluation and Riccati recursion equations are identical to those for the optimal deterministic controller for the same problem but with $\mathrm{w}(\cdot,\cdot)$ removed from the dynamics and no expectation operation being required in the cost definition.

Both $K_c$ and $G_c^*$ go through a terminal transient near $t_{N+1}$, but asymptotically approach steady state values $\bar{K}_c$ and $\overline{G_c^*}$ as one goes backwards in time. These values can be obtained as in Example 14.9 by solving for the $\bar{K}_c$ that satisfies $K_c(t_i) = K_c(t_{i+1}) = \bar{K}_c$ in the Riccati equation, i.e., the positive root of

$$(B_d^2)\bar{K}_c^2 + (U - \Phi^2 U - B_d^2 X)\bar{K}_c - (UX) = 0$$

In fact, $\bar{K}_c$ and $\overline{G_c^*}$ depend on the ratio $X/U$, and are independent of $X_f$. Letting $X = 1$ and $U$ take on a range of values yields the following steady state $\bar{K}_c$, gain $\overline{G_c^*}$, and closed-loop system eigenvalue or pole $(\Phi - B_d\overline{G_c^*})$:

| $U$: | 0 | 0.1 | 1 | 10 | $\infty$ |
|---|---|---|---|---|---|
| $\bar{K}_c$: | 1 $(=X)$ | 1.74 | 2.61 | 2.97 | – |
| $\overline{G_c^*}$: | 4.5 $(=\Phi/B_d)$ | 1.66 | 0.36 | 0.044 | 0 |
| $\Phi - B_d\overline{G_c^*}$: | 0 | 0.52 | 0.75 | 0.81 | 0.82 $(=\Phi)$ |

As in the previous example, when $U = 0$ (the singular case), $\bar{K}_c = X$, $\overline{G_c^*} = \Phi/B_d$, and the closed-loop eigenvalue is at zero. As $U$ increases, $\overline{G_c^*}$ decreases to zero in the limit as $U \to \infty$, the closed-loop eigenvalue goes from zero to the open-loop eigenvalue $\Phi$, and the $\frac{1}{2}\bar{K}_c x_0^2$ contribution to the optimal cost for process completion increases.

If only noise-corrupted measurements of the form

$$\mathrm{z}(t_i) = H\mathrm{x}(t_i) + \mathrm{v}(t_i)$$

are available at sample times instead of $x(t_i)$ itself, with $\mathrm{v}(\cdot,\cdot)$ zero-mean white Gaussian discrete-time noise of constant variance $R$ and independent of $\mathrm{w}(\cdot,\cdot)$, then the same $K_c(t_i)$ and $G_c^*(t_i)$ evaluations are used to generate

$$u^*(t_i) = -G_c^*(t_i)\hat{x}(t_i^+)$$

where $\hat{x}(t_i^+)$ is provided by the Kalman filter prescribed by

$$\hat{x}(t_i^+) = \hat{x}(t_i^-) + K(t_i)[z_i - H\hat{x}(t_i^-)]$$
$$\hat{x}(t_{i+1}^-) = 0.82\hat{x}(t_i^+) + 0.18[-G_c^*(t_i)\hat{x}(t_i^+)]$$

with

$$K(t_i) = \frac{P(t_i^-)H}{H^2P(t_i^-) + R}$$

$$P(t_i^+) = P(t_i^-) - K(t_i)HP(t_i^-)$$
$$P(t_{i+1}^-) = 0.67P(t_i^+) + 0.084Q$$

After an initial transient, the filter gain reaches a steady state value $K_{ss}$, which combined with the constant gain $\overline{G_c^*}$ valid before the terminal transient, yields a constant-gain controller for the intermediate times which can be written as

$$u(t_i) = -\overline{G_c^*}[I - K_{ss}H]\hat{x}(t_i^-) - \overline{G_c^*}K_{ss}z_i$$
$$\hat{x}(t_{i+1}^-) = [\Phi - B_d\overline{G_c^*}][I - K_{ss}H]\hat{x}(t_i^-) + [\Phi - B_d\overline{G_c^*}]z_i$$

Unlike the case where $x(t_i)$ is assumed known, this is a *dynamic compensator* operating on the available input $z_i$ and introducing another pole into the closed-loop system: the filter eigenvalue $[\Phi - \Phi K_{ss}H]$ as well as the original regulator eigenvalue $[\Phi - B_d\overline{G_c^*}]$. ■

EXAMPLE 14.3 Consider the same problem, but now assume that you desire to minimize the cost

$$J_c = E\left\{\frac{1}{2}\int_{t_0}^{t_{N+1}} [W_{xx}\mathrm{x}(t)^2 + W_{uu}\mathrm{u}(t)^2]\,dt + \frac{1}{2}X_f\mathrm{x}(t_{N+1})^2\right\}$$

where increasing $W_{xx}$ causes tighter control of the state *throughout* the interval $[t_0, t_{N+1}]$ instead of just at the sample times, increasing $W_{uu}$ decreases the amount of control energy you are willing to expend, and increasing $X_f$ decreases the terminal state deviations. (The ratio $W_{xx}/W_{uu}$ affects the steady state gains, and $X_f$ affects the magnitudes in the terminal transient.) For the discrete-time description of the cost, this will yield $S \neq 0$ even though $W_{xu} = 0$ in the $J_c$ above.

For this problem, Eq. (14-25) yields

$$X = W_{xx}T[1 - e^{-2\Delta t/T}]/2 = 0.082W_{xx}$$
$$U = W_{xx}\{\Delta t - 2T[1 - e^{-\Delta t/T}] + T[1 - e^{-2\Delta t/T}]/2\} + W_{uu}\Delta t$$
$$= 0.00115W_{xx} + 0.1W_{uu}$$
$$S = W_{xx}\{T[1 - e^{-\Delta t/T}] - T[1 - e^{-2\Delta t/T}]/2\} = 0.0082W_{xx}$$

The optimal controller is again given by (14-6) if the state is known at $t_i$, but with $G_c^*(t_i)$ given as (14-14):

$$G_c^*(t_i) = \frac{0.15K_c(t_{i+1}) + S}{U + 0.033K_c(t_{i+1})}$$

where, by (14-15) and (14-17),

$$K_c(t_i) = X + 0.67K_c(t_{i+1}) - \frac{[0.15K_c(t_{i+1}) + S]^2}{U + 0.033K_c(t_{i+1})}$$

$$K_c(t_{N+1}) = X_f$$

which are also identical to the corresponding deterministic optimal controller result.

Letting $W_{xx} = 1$, we can solve for the constant steady state gains and closed-loop eigenvalues for any $W_{uu}$, yielding

| $W_{uu}$: | 0 | 0.1 | 1 | 10 | $\infty$ |
|---|---|---|---|---|---|
| $\overline{G_c}^*$: | 9.09 | 3.15 | 1.15 | 0.236 | 0 |
| $\Phi - B_d\overline{G_c}^*$: | −0.82 | 0.25 | 0.61 | 0.78 | 0.82 |

Here the $\overline{G_c}^*$ is larger than the gain in the previous example corresponding to an $(X/U)$ ratio equivalent to the current problem's $(W_{xx}/W_{uu})$ ratio: greater control is being applied. Furthermore, the closed-loop system eigenvalues shown in this table are to the left of the corresponding result for the previous case. Also note that in the limit of $W_{uu} \to 0$, the closed-loop eigenvalue is at the mirror image about the imaginary axis of a complex plane ($z$ plane) plot of the open-loop eigenvalue, instead of the origin as when $U \to 0$ in that example: this controller can effect closed-loop pole locations that are unattainable by the previous form having $S = 0$.

As in the previous example, the same $G_c^*(t_i)$ is combined with (14-7) and the appropriate Kalman filter (which has not been altered from the previous example) if $x(t_i)$ is not known perfectly, by the certainty equivalence property. Once again, a dynamic compensator is generated, with closed-loop system poles at both $[\Phi - B_d\overline{G_c}^*]$ and $[\Phi - \Phi K_{ss}H]$ in steady state constant-gain form. ■

EXAMPLE 14.4 Consider the same example again, but assume that the control objective is modified. As before, we want to maintain small state deviations from zero, which is accomplished by incorporating the trajectory loss function $\frac{1}{2}\int_{t_0}^{t_{N+1}} W_{xx}\mathrm{x}(t)^2\,dt$ and the terminal loss function $\frac{1}{2}X_f\mathrm{x}(t_{N+1})^2$. However, instead of wanting to consider the design parameter $W_{uu}$ as a means of adjusting the amount of control energy applied, assume that you want a parameter for adjusting the amplitude of $\dot{\theta}$ more directly. For instance, the servo motor may have a maximum admissible angular rate, and we would want to be able to adjust a *single* such parameter if the controller based upon an initial choice of cost weighting parameters in fact violates this maximum rate.

Thus, we want to minimize

$$J_c = E\left\{\frac{1}{2}\int_{t_0}^{t_{N+1}} [W_{xx}\mathrm{x}(t)^2 + W_{\dot{x}\dot{x}}\dot{\mathrm{x}}(t)^2]\,dt + \frac{1}{2}X_f\mathrm{x}(t_{N+1})^2\right\}$$

$$= E\left\{\frac{1}{2}\int_{t_0}^{t_{N+1}} \left[W_{xx}\mathrm{x}(t)^2 + W_{\dot{x}\dot{x}}\left\{-\frac{1}{T}\mathrm{x}(t) + \frac{1}{T}\mathrm{u}(t) + \mathrm{w}(t)\right\}^2\right]dt + \frac{1}{2}X_f\mathrm{x}(t_{N+1})^2\right\}$$

$$= E\left\{\frac{1}{2}\int_{t_0}^{t_{N+1}} [\underbrace{\{W_{xx} + (W_{\dot{x}\dot{x}}/T^2)\}}_{W'_{xx}}\mathrm{x}(t)^2 + \underbrace{\{W_{\dot{x}\dot{x}}/T^2\}}_{W'_{uu}}\mathrm{u}(t)^2\right.$$

$$\left. + 2\underbrace{\{-W_{\dot{x}\dot{x}}/T^2\}}_{W'_{xu}}\mathrm{x}(t)\mathrm{u}(t) + \mathrm{w}(t)^2\}]\,dt + \frac{1}{2}X_f\mathrm{x}(t_{N+1})^2\right\}$$

In this expression we note that $E\{\mathrm{w}(t)^2\}$ is infinite and thus the cost is not finite due to the mathematical fictition of white noise $\mathrm{w}(\cdot,\cdot)$ appearing directly in our expression for $\dot{\mathrm{x}}(\cdot,\cdot)$. This difficulty can be removed by replacing $\mathrm{w}(\cdot,\cdot)$ with a finite-power noise $\mathrm{n}(\cdot,\cdot)$ with $\mathrm{n}(t)$ uncorrelated with $\mathrm{x}(t)$ or $\mathrm{u}(t)$: physically, a more realistic model. It is important that this term is not dependent on $\mathrm{x}(t)$ or $\mathrm{u}(t)$, and thus it will have no impact on determining the functional form of $u^*[\cdot,\cdot]$. This difficulty can also be removed by invoking certainty equivalence, and designing the deterministic optimal controller for the same problem and criterion, but with $\mathbf{w}(t)$ and the expectation operation removed from $J_c$.

Thus we have a case in which $W_{xu}$ is naturally nonzero. Equation (14-25) then yields

$$X = 0.082W'_{xx}, \qquad U = 0.00115W'_{xx} + 0.1W'_{uu} + 0.019W'_{xu}, \qquad S = 0.0082W'_{xx} + 0.091W_{xu}$$

or, in terms of $W_{xx}$ and $W_{\dot{x}\dot{x}}$,

$$X = 0.082W_{xx} + 0.021W_{\dot{x}\dot{x}}, \qquad U = 0.00115W_{xx} + 0.016W_{\dot{x}\dot{x}}, \qquad S = 0.0082W_{xx} - 0.043W_{\dot{x}\dot{x}}$$

With these evaluations of $X$, $U$, and $S$, the controller design and characteristics are as in the previous example. ■

In the context of a sampled-data optimal control problem, nonzero cross-terms between $\mathbf{x}(t_i)$ and $\mathbf{u}(t_i)$ in the cost function (14-13) are seen to arise naturally. Another way of formulating this is to say that the controlled variables of interest $\mathbf{y}_c(t_i)$ can be generalized from (14-2) to

$$\mathbf{y}_c(t_i) = \mathbf{C}(t_i)\mathbf{x}(t_i) + \mathbf{D}_y(t_i)\mathbf{u}(t_i) \tag{14-29}$$

In other words, the dynamic system composed of (14-1) and (14-29) is more likely to have a direct feedthrough of control inputs to the outputs of interest than is a corresponding continuous-time system model. Incorporating this generalized expression into the cost (14-3) yields

$$\begin{aligned} J &= E\left\{\sum_{i=0}^{N} \frac{1}{2}[\{\mathbf{x}^T(t_i)\mathbf{C}^T(t_i) + \mathbf{u}^T(t_i)\mathbf{D}_y^T(t_i)\}\mathbf{Y}(t_i)\{\mathbf{C}(t_i)\mathbf{x}(t_i) + \mathbf{D}_y(t_i)\mathbf{u}(t_i)\}\right. \\ &\qquad \left. + \mathbf{u}^T(t_i)\mathbf{U}(t_i)\mathbf{u}(t_i)] + \frac{1}{2}\mathbf{x}^T(t_{N+1})\mathbf{X}_f\mathbf{x}(t_{N+1})\right\} \\ &= E\left\{\frac{1}{2}\mathbf{x}^T(t_{N+1})\mathbf{X}_f\mathbf{x}(t_{N+1})\right. \\ &\qquad \left. + \sum_{i=0}^{N} \frac{1}{2}\left(\begin{bmatrix}\mathbf{x}(t_i)\\ \mathbf{u}(t_i)\end{bmatrix}^T \begin{bmatrix}\mathbf{C}^T(t_i)\mathbf{Y}(t_i)\mathbf{C}(t_i) & \mathbf{C}^T(t_i)\mathbf{Y}(t_i)\mathbf{D}_y(t_i)\\ \mathbf{D}_y^T(t_i)\mathbf{Y}(t_i)\mathbf{C}(t_i) & \mathbf{D}_y^T(t_i)\mathbf{Y}(t_i)\mathbf{D}_y(t_i) + \mathbf{U}(t_i)\end{bmatrix} \begin{bmatrix}\mathbf{x}(t_i)\\ \mathbf{u}(t_i)\end{bmatrix}\right)\right\} \end{aligned} \tag{14-30}$$

which is recognized to be of the form of (14-13b). Thus, for the regulator design problem, generalized output expressions such as (14-29) are inherently addressed by the generalized cost (14-13), so no additional controller modification is required to handle (14-29); later, when we extend beyond regulators to PI (proportional plus integral) and other controller forms, (14-29) will again be considered.

In similar fashion, the model for discrete-time noise-corrupted measurements can be extended from

$$\mathbf{z}(t_i) = \mathbf{H}(t_i)\mathbf{x}(t_i) + \mathbf{v}(t_i) \tag{14-31}$$

with $\mathbf{v}(\cdot,\cdot)$ zero-mean white Gaussian discrete-time noise of covariance $\mathbf{R}(t_i)$ for all $t_i$, typically assumed independent of $\mathbf{x}(t_0,\cdot)$ and $\mathbf{w}_d(\cdot,\cdot)$, to

$$\mathbf{z}(t_i) = \mathbf{H}(t_i)\mathbf{x}(t_i) + \mathbf{D}_z(t_i)\mathbf{u}(t_i) + \mathbf{v}(t_i) \tag{14-32}$$

The only effect this has is to modify the residual in the Kalman filter to yield an update equation of

$$\hat{\mathbf{x}}(t_i^+) = \hat{\mathbf{x}}(t_i^-) + \mathbf{K}(t_i)[\mathbf{z}_i - \mathbf{H}(t_i)\hat{\mathbf{x}}(t_i^-) - \mathbf{D}_z(t_i)\mathbf{u}(t_i)] \tag{14-33}$$

With all of the appropriate extensions just developed, we are still afforded the *synthesis capability* seen in the previous chapter for the simplest LQG problem: the stochastic LQG optimal controller is given by (14-6) or (14-7), with $\mathbf{G}_c^*(t_i)$ equal to that of the associated deterministic LQ optimal state regulator [3, 5, 38, 83, 168, 189]. Thus we are motivated to investigate deterministic LQ optimal state regulator theory, using full state feedback, as a means of generating appropriate controller structures for more general problems, and eventually to exploit this structure by replacing $\mathbf{x}(t_i)$ with $\hat{\mathbf{x}}(t_i^+)$ from the appropriate Kalman filter, to generate the stochastic optimal controller. This will be pursued in later sections.

First, however, we must consider *stability* of the resulting closed-loop system. Classical controller design, especially frequency-domain design of constant-gain controllers for time-invariant system and stationary noise models, typically entails ensuring "sufficient" stability first and then seeking desirable performance from the stabilized system with any remaining design freedom. In contrast, we have been concentrating on the desired performance objective, despite the *criticality of overall system stability as a prerequisite to attaining that performance.* Therefore, we must assess the stabilizing effect of the controllers we are synthesizing, not only under the design conditions assumed but also under the physically important case of mismatch between assumed models and actual systems. In the case of *full-state feedback* controllers, this will provide strong *additional motivation* to exploiting LQG controller design, while simultaneously raising an important concern for the proper design of controllers that employ filters to generate $\hat{\mathbf{x}}(t_i^+)$ when $\mathbf{x}(t_i)$ is not known perfectly.

## 14.3 STABILITY

In this section some fundamental concepts of stability are presented. Since we are most interested in the sampled-data control of continuous-time systems, emphasis is concentrated on stability for continuous-time system models, although discrete-time analogies are evident. Deterministic system stability is considered first to facilitate understanding of basic ideas, and stochastic stability is treated as an extension of these ideas. This is necessarily a brief overview of a vast field; it suffices to set the context of stability properties to be presented subsequently, but the reader is also referred to [16, 28, 36, 48, 51, 55, 69, 84, 87–89, 92, 96, 103, 130, 143, 146, 174, 190–193] for more extensive presentations.

Deterministic system stability is a qualitative property associated with the solution to differential or difference equations, and one that can often be investigated without explicitly solving the equations themselves. Primarily, we want to characterize the asymptotic behavior of solutions as the independent variable time grows without bound: whether the solutions tend to grow indefinitely or reach a limit point or limiting set (and, if so, if the limits are unique or change with perturbations to the inputs), whether there are conditions on initial conditions and/or controls that can assure uniform boundedness of the solutions, whether certain perturbations in the system (as initial conditions, inputs, or parameters describing the system model itself) will not cause excessive changes in the output, etc.

Let us start by considering the deterministic state differential equation

$$\dot{\mathbf{x}}(t) = \mathbf{f}[\mathbf{x}(t), \mathbf{u}(t), t]; \qquad \mathbf{x}(t_0) = \mathbf{x}_0 \tag{14-34}$$

There are two fundamental categories of stability, zero-input stability and bounded input/bounded output stability. *Zero-input stability* concepts concentrate on the autonomous system described by

$$\dot{\mathbf{x}}(t) = \mathbf{f}[\mathbf{x}(t), t]; \qquad \mathbf{x}(t_0) = \mathbf{x}_0 \tag{14-35}$$

and ask under what conditions the transient solution will die out (completely, or stay "small," etc.), to leave just a given nominal solution. On the other hand, *bounded input/bounded output (BIBO) stability* concepts focus on the forced solutions corresponding to

$$\dot{\mathbf{x}}(t) = \mathbf{f}[\mathbf{x}(t), \mathbf{u}(t), t]; \qquad \mathbf{x}(t_0) = \mathbf{0} \quad \text{(typically)} \tag{14-36}$$

and ask under what conditions you are assured of achieving a bounded output from the solution to the given differential equation when a bounded input is applied.

*Zero-Input Stability for General Nonlinear System Models*

Consider a *nominal solution* $\mathbf{x}_n(\cdot)$ to (14-35), i.e., the function $\mathbf{x}_n(t)$ for all $t$ of interest satisfying

$$\dot{\mathbf{x}}_n(t) = \mathbf{f}[\mathbf{x}_n(t), t]; \qquad \mathbf{x}_n(t_0) = \mathbf{x}_0 \tag{14-35$'$}$$

(If $\mathbf{x}_n(t) \equiv \mathbf{x}_n =$ constant for all $t$, then we call this particular nominal an *equilibrium state.*) This particular nominal solution to the differential equation (14-35) is *stable in the sense of Lyapunov* if, for any $t_0$ and any $\varepsilon > 0$, there exists a $\delta(\varepsilon, t_0)$ which may depend on $\varepsilon$ and possibly $t_0$ as well, such that

$$\|\mathbf{x}(t_0) - \mathbf{x}_n(t_0)\| \leq \delta \tag{14-37a}$$

implies that

$$\|\mathbf{x}(t) - \mathbf{x}_n(t)\| < \varepsilon \tag{14-37b}$$

for all $t \geq t_0$, where $\|\cdot\|$ denotes an appropriate norm, such as Euclidean length of a vector. This says that, by choosing the initial state "close enough" to the nominal solution initial condition, the state $\mathbf{x}(t)$ will not depart from $\mathbf{x}_n(t)$ "too much." This is a very weak form of stability: if you start out within a "$\delta$-ball" of the nominal, you are assured only of staying within an "$\varepsilon$-ball" of that nominal. This does not say that $\mathbf{x}(t)$ will necessarily converge to the nominal, and $\varepsilon$ may well be considerably larger than $\delta$. Moreover, so far we are speaking only of stability "in the small," applying only in the infinitesimal region about the nominal solution. Notice that, for general nonlinear system models, we must speak of the *stability of a given nominal solution* and cannot simply talk of stability of the system model itself, as we can for linear system models.

The nominal solution $\mathbf{x}_n(t)$ to (14-35) is *asymptotically stable* if it is not only stable in the sense of Lyapunov, but also there exists a scalar $\alpha(t_0)$ for all $t_0$, whose value may depend on $t_0$, such that

$$\|\mathbf{x}(t_0) - \mathbf{x}_n(t_0)\| < \alpha \tag{14-38a}$$

implies that

$$\|\mathbf{x}(t) - \mathbf{x}_n(t)\| \to 0 \qquad \text{as} \quad t \to \infty \tag{14-38b}$$

This is a stronger form of stability, in that choosing an initial deviation small enough assures you that the solution $\mathbf{x}(t)$ will in fact converge to the nominal $\mathbf{x}_n(t)$. However, this is still "stability in the small."

The descriptor *uniformly* basically states that it does not matter what absolute time it is. A nominal solution is uniformly stable in the sense of Lyapunov if $\delta$ in (14-37a) is not a function of $t_0$, and $\mathbf{x}_n(t)$ is uniformly asymptotically stable if $\alpha$ in (14-38a) is not a function of $t_0$. This is an issue of concern for time-varying systems since their characteristics change in time, but not an issue for a time-invariant system described by $\dot{\mathbf{x}}(t) = \mathbf{f}[\mathbf{x}(t)]$.

Now consider going beyond the concept of stability in the small. The nominal solution $\mathbf{x}_n(t)$ to (14-35) is said to be *asymptotically stable in the large* (*ASIL*), or *globally asymptotically stable*, if it is not only stable in the sense of Lyapunov, but also (14-38b) is true for *any* initial $\mathbf{x}(t_0)$. This basically states that no matter what initial state is chosen, the solution converges to the nominal. It is generally difficult to prove.

Finally, the nominal solution $\mathbf{x}_n(t)$ to (14-35) is *uniformly asymptotically stable in the large* (*UASIL*) or *uniformly globally asymptotically stable*, if it is not only stable in the sense of Lyapunov, but also (14-38b) is true for *any* $\mathbf{x}(t_0)$ and *any* $t_0$. It does not matter what the initial state is or what initial time is chosen (again, important for time-varying systems), the solution will converge to $\mathbf{x}_n(t)$ asymptotically. *All* other solutions would eventually converge to that nominal solution. This is the strongest form of stability. Yet, notice that we still do not distinguish between the case in which transient deviations from the nominal die out quickly and the case of a more sluggish response in which the rate of convergence is very slow: stability is just a requisite threshold, beyond which additional performance requirements are naturally levied.

*Establishing Zero-Input Stability for General Nonlinear System Models*

To establish zero-input stability of a given solution to (14-35) by the *second method of Lyapunov* (the "first" method deals only with small perturbations by considering linearized perturbation equations), one seeks a scalar "Lyapunov" function that will meet certain sufficient conditions for the various types of stability just discussed, without resorting to explicit solution of (14-35) itself. Let $V[\cdot,\cdot]$ be a scalar-valued function of $\mathbf{x} \in \mathscr{S} \subset R^n$ and $t \in T$ that is continuous with respect to $\mathbf{x}$ and $t$ and has first partial derivatives with respect to both that are also continuous in $\mathbf{x}$ and $t$. $V[\cdot,\cdot]$ is a *Lyapunov function* for (14-35) on the set $\mathscr{S}$ if $V$ is bounded for $\|\mathbf{x}\|$ bounded and if $\dot{V}$, the time rate of change of $V$ along the solutions to the differential equation given by

$$\dot{V}[\mathbf{x},t] = \sum_{i=1}^{n} \left( \frac{\partial V[\mathbf{x},t]}{\partial x_i} f_i[\mathbf{x},t] \right) + \frac{\partial V[\mathbf{x},t]}{\partial t} \tag{14-39}$$

satisfies

$$\dot{V}[\mathbf{x},t] \leq W(\mathbf{x}) \leq 0 \tag{14-40}$$

for all $x \in \mathscr{S}$ and some continuous, nonpositive $W(\cdot)$. A scalar function $V[\cdot,\cdot]$ is called *positive definite* if

$$V[\mathbf{0},t] = 0 \tag{14-41a}$$

and if there exists a continuous nondecreasing scalar-valued function $V_1$ of a scalar argument, such that

$$V_1(0) = 0 \tag{14-41b}$$

and

$$V[\mathbf{x},t] \geq V_1(\|\mathbf{x}\|) > 0 \qquad \text{for} \quad \|\mathbf{x}\| \neq 0 \tag{14-41c}$$

Furthermore, $V[\cdot,\cdot]$ is termed *decrescent* if there exists a scalar function $V_2$ of a scalar argument that is continuous and nondecreasing, such that

$$V_2(0) = 0 \tag{14-42a}$$

$$V[\mathbf{x},t] \leq V_2(\|\mathbf{x}\|) \tag{14-42b}$$

The basic stability theorems [48, 96, 103] state that, if $V[\cdot,\cdot]$ is a Lyapunov function for (14-35) on $\mathscr{S}$ that is positive definite and decrescent and if $\mathbf{f}[\mathbf{0},t] = \mathbf{0}$, then the null solution is uniformly stable in the sense of Lyapunov on the set $\mathscr{S}$. If, in addition, $\{-W(\mathbf{x})\}$ as given in (14-40) is positive definite, then the null solution is uniformly asymptotically stable on the set $\mathscr{S}$; this can be relaxed to $\dot{V}[\mathbf{x},t] \leq 0$, and $\dot{\mathbf{V}}[\mathbf{x},t]$ is not identically zero along any system solution other than the zero solution. Finally, if $\mathscr{S}$ is the entire state space $R^n$ and $V[\mathbf{x},t] \to \infty$ as $\|\mathbf{x}\| \to \infty$, then the null solution is uniformly asymptotically stable in the large. Lyapunov functions are not unique, and some Lyapunov functions can provide more meaningful stability results than others. However, in the general nonlinear case, finding even a single adequate Lyapunov function is often a difficult, if not intractable, task. There are sufficient conditions for instability of solutions as well [103].

*Zero-Input Stability for Linear System Models*

Let us now restrict our attention to linear system models described by a special case of (14-35), namely

$$\dot{\mathbf{x}}(t) = \mathbf{F}(t)\mathbf{x}(t); \qquad \mathbf{x}(t_0) = \mathbf{x}_0 \tag{14-43}$$

Here we no longer need to consider a particular nominal solution: stability about one nominal is the same as about any other. Since a solution $\mathbf{x}(t)$ and a nominal solution $\mathbf{x}_n(t)$ both satisfy (14-43), then the deviation from the nominal, $[\mathbf{x}(t) - \mathbf{x}_n(t)]$, also satisfies (14-43), so we confine our attention to stability of the zero solution, and we can speak in terms of *stability of the system model* instead of stability of a particular nominal solution. Similarly, "in the small" versus "in the large" is no longer an issue. For instance, asymptotic stability and asymptotic stability in the large (or global asymptotic stability) are one in the same.

A form of zero-input stability defined for linear systems, and which will be of use to us subsequently, is exponential stability. The system model defined by (14-43) is termed (uniformly) *exponentially stable* [16] if there exist positive constants $\beta$ and $\gamma$ such that

$$\|\mathbf{x}(t)\| \le \beta e^{-\gamma(t-t_0)}\|\mathbf{x}(t_0)\| \tag{14-44a}$$

for all $t \ge t_0$ of interest, starting from any initial state $\mathbf{x}(t_0)$ at any initial time $t_0$; or, equivalently, if

$$\|\mathbf{\Phi}(t, t_0)\| \le \beta e^{-\gamma(t-t_0)} \tag{14-44b}$$

where the norm of a matrix is defined appropriately, such as $\|\mathbf{A}\|^2 = \sum\sum_{ij} A_{ij}^2$. Thus, if the system model is exponentially stable, the state solution converges to zero exponentially, no matter what the initial state or time. Although this says more about rates of convergence, we still do not distinguish between "good" and "bad" values of $\gamma$ when claiming exponential stability. Sufficient conditions for exponential stability will be discussed later, in conjunction with BIBO stability.

*Zero-Input Stability for Linear Time-Invariant System Models*

Confining our attention to the class of time-invariant forms of linear system models, as

$$\dot{\mathbf{x}}(t) = \mathbf{F}\mathbf{x}(t); \qquad \mathbf{x}(t_0) = \mathbf{x}_0 \tag{14-45}$$

we find that stability is determined by the *eigenvalues* of $\mathbf{F}$. The system model (14-45) is *stable in the sense of Lyapunov* if and only if all eigenvalues of $\mathbf{F}$ have *nonpositive* real parts and, to any eigenvalue on the imaginary axis with multiplicity $\mu$, there correspond exactly $\mu$ eigenvectors of $\mathbf{F}$ (this is always valid if $\mu = 1$). For time-invariant linear system models, Laplace transforms and frequency domain interpretations can be employed to state this more classically as: The system poles must be in the left-half plane or on the imaginary axis of a Laplace $s$-plane plot, with similar additional constraints for multiple poles on the imaginary axis. For instance, an undamped second order linear oscillator with a single complex conjugate pair of poles on the imaginary axis is stable in the sense of Lyapunov: the state solution does not converge to zero, but is confined to a region of state space defined via the maximum amplitudes of sinusoidal oscillations of the two state components.

The system model (14-45) is *asymptotically stable* if and only if all eigenvalues of $\mathbf{F}$ have *strictly negative* real parts. Stated classically, the system poles must lie strictly in the left-half $s$ plane. (For linear time-invariant discrete-time or discretized systems, the eigenvalues of $\mathbf{\Phi}$ must be strictly less than one, so that we have a dynamics model that is a "contraction mapping," or, the system poles must lie strictly within the unit circle drawn on a $z$-transform complex plane plot of poles and zeros.) Thus, the undamped oscillator that was found to be stable in the sense of Lyapunov is not asymptotically stable.

For time-invariant linear system models, *asymptotic stability implies, and is implied by, exponential stability*. As in the case of the more general time-varying linear system models, stability of any form "in the small" and "in the large" are identical concepts. Moreover, by time invariance of the model, all stability results are true uniformly. Therefore, these two types of stability are the strongest stability properties for a linear time-invariant system model, *both equivalent to uniform asymptotic stability in the large.*

Another means of demonstrating stability (as of a proposed closed-loop system) that is closely associated with the previous stability theorems is to choose as a Lyapunov function the general quadratic

$$V[\mathbf{x}(t)] = \mathbf{x}^{\mathrm{T}}(t)\mathbf{K}_v\mathbf{x}(t) \tag{14-46}$$

with $\mathbf{K}_v$ symmetric and positive definite, so that

$$\begin{aligned} \dot{V}[\mathbf{x}(t)] &= \dot{\mathbf{x}}^{\mathrm{T}}(t)\mathbf{K}_v\mathbf{x}(t) + \mathbf{x}^{\mathrm{T}}(t)\mathbf{K}_v\dot{\mathbf{x}}(t) \\ &= \mathbf{x}^{\mathrm{T}}(t)\{\mathbf{F}^{\mathrm{T}}\mathbf{K}_v + \mathbf{K}_v\mathbf{F}\}\mathbf{x}(t) \\ &= -\mathbf{x}^{\mathrm{T}}(t)\mathbf{K}_{\dot{v}}\mathbf{x}(t) \end{aligned} \tag{14-47}$$

where $\mathbf{K}_{\dot{v}}$ is given by the *Lyapunov equation* [43]

$$\mathbf{K}_{\dot{v}} + \mathbf{F}^{\mathrm{T}}\mathbf{K}_v + \mathbf{K}_v\mathbf{F} = \mathbf{0} \tag{14-48}$$

If one chooses an arbitrary positive semidefinite $\mathbf{K}_{\dot{v}}$ and solves (14-48) for $\mathbf{K}_v$ (a unique solution exists if no two eigenvalues of the system have real parts which add to zero), then positive definiteness of $\mathbf{K}_v$ is both necessary and sufficient for asymptotic stability.

It is useful for certain stability characterizations to define the *stable and unstable subspaces* of real Euclidean state space $R^n$ associated with a time-invariant linear system model (14-45). If $\mathbf{F}$ has $n$ distinct eigenvalues, the stable subspace is the real linear subspace of $R^n$ spanned by the eigenvectors of $\mathbf{F}$ corresponding to eigenvalues with *strictly negative* real parts; if $\mathbf{F}$ has any repeated eigenvalues, this generalizes to the real subspace formed as the direct sum of null spaces [35] corresponding to eigenvalues with strictly negative real parts. The unstable subspace is the span of eigenvectors or direct sum of null spaces, respectively, corresponding to eigenvalues with nonnegative real parts. Then $R^n$ is the direct sum of stable and unstable subspaces.

### *Zero-Input Stability: Implications about Stability of Nonlinear Solutions from Linearized System Stability*

Assume that you are given a nonlinear system model as (14-34) with a nominal solution $\mathbf{x}_n(t)$ for all $t$, and you design a linear perturbation controller based upon the linearized equations

$$\delta\dot{\mathbf{x}}(t) = \mathbf{F}(t)\,\delta\mathbf{x}(t) + \mathbf{B}(t)\,\delta\mathbf{u}(t) \tag{14-49}$$

as expanded about the nominal state $\mathbf{x}_n(t)$ and control $\mathbf{u}_n(t)$ for all time $t$:

$$\mathbf{F}(t) = \left.\frac{\partial\mathbf{f}[\mathbf{x},\mathbf{u},t]}{\partial\mathbf{x}}\right|_{\mathbf{x}=\mathbf{x}_n(t),\ \mathbf{u}=\mathbf{u}_n(t)} \tag{14-50a}$$

$$\mathbf{B}(t) = \left.\frac{\partial\mathbf{f}[\mathbf{x},\mathbf{u},t]}{\partial\mathbf{u}}\right|_{\mathbf{x}=\mathbf{x}_n(t),\ \mathbf{u}=\mathbf{u}_n(t)} \tag{14-50b}$$

If the perturbation controller yields a stable closed-loop linearized system model, what can be said about stability of solutions to the original nonlinear system model as modified by the linear feedback controller? To answer that in the current context, consider the time-invariant system described by $\dot{\mathbf{x}}(t) = \mathbf{f}[\mathbf{x}(t)]$ and suppose this differential equation has an equilibrium solution $\mathbf{x}_n$, that $\mathbf{f}[\mathbf{x}]$ possesses partial derivatives with respect to the components of $\mathbf{x}$ at the point $\mathbf{x} = \mathbf{x}_n$, and that $\mathbf{f}$ and its partials are continuous (this precludes some important nonlinearities such as saturation or dead zones). Then, if the time-invariant linear perturbation system

$$\delta\dot{\mathbf{x}}(t) = \left.\frac{\partial\mathbf{f}[\mathbf{x}]}{\partial\mathbf{x}}\right|_{\mathbf{x}=\mathbf{x}_n} \delta\mathbf{x}(t) \tag{14-51}$$

is *asymptotically stable*, $\mathbf{x}(t) = \mathbf{x}_n$ is an *asymptotically stable (in the small) solution* [92, 143] of $\dot{\mathbf{x}}(t) = \mathbf{f}[\mathbf{x}(t)]$. As expected, one cannot claim anything about stability in the large for the nonlinear system solutions based only on linear perturbation system stability, but we can gain positive implications for "sufficiently small" deviations about the nominal. Of course, if the linearized system is unstable, nonlinear system solutions are also. Note that if $\mathbf{F}$ has any eigenvalues with zero real parts (i.e., linearized system poles on the imaginary axis of the $s$ plane, which correspond to discretized system $\mathbf{\Phi}$ eigenvalues or poles lying on the unit circle in the $z$ plane), then no conclusions can be made about stability of nonlinear system solutions.

*The BIBO Stability for General Nonlinear System Models*

A system model is termed *bounded-input/bounded-output stable*, or *BIBO stable*, if for every input within a sufficiently small bound, the system output is bounded. To be more precise, a system described by (14-36) is BIBO stable if, for any $t_0$ and some finite bound $B$, there exists a $U(B, t_0)$ such that all controls $\mathbf{u}(\cdot)$ defined on $[t_0, \infty)$ such that

$$\|\mathbf{u}(t)\| \leq U \qquad \text{for all} \quad t \in [t_0, \infty) \tag{14-52}$$

the solution $\mathbf{x}(t)$ to the differential equation satisfies

$$\|\mathbf{x}(t)\| \leq B \qquad \text{for all} \quad t \in [t_0, \infty) \tag{14-53}$$

If the bounds do not depend on $t_0$, the system is *uniformly BIBO stable.*

*The BIBO Stability for Linear System Models*

A system described by the linear model

$$\dot{\mathbf{x}}(t) = \mathbf{F}(t)\mathbf{x}(t) + \mathbf{B}(t)\mathbf{u}(t) \tag{14-54a}$$

$$\mathbf{y}_c(t) = \mathbf{C}(t)\mathbf{x}(t) + \mathbf{D}_y(t)\mathbf{u}(t) \tag{14-54b}$$

is said to be *uniformly BIBO stable* if, for all $t_0$ and for $\mathbf{x}(t_0) = \mathbf{0}$, *every* bounded input defined on $[t_0, \infty)$ gives rise to a bounded response on $[t_0, \infty)$. Stated more precisely, if there exists a constant $B$ independent of $t_0$ such that, for all $t_0$,

$$\mathbf{x}(t_0) = \mathbf{0} \tag{14-55a}$$

$$\|\mathbf{u}(t)\| \leq 1 \qquad \text{for all} \quad t \in [t_0, \infty) \tag{14-55b}$$

(where 1 in (14-55b) is chosen arbitrarily) imply that

$$\|\mathbf{y}(t)\| \leq B \qquad \text{for all} \quad t \in [t_0, \infty) \tag{14-56}$$

*Sufficient conditions* for uniform BIBO stability are as follows. If $\mathbf{F}(t)$ is bounded for all $t \in (-\infty, \infty)$, the system model (14-54) with $\mathbf{B}(t) = \mathbf{C}(t) \equiv \mathbf{I}$ and $\mathbf{D}_y(t) \equiv \mathbf{0}$ is uniformly BIBO stable. Moreover, if $\|\mathbf{u}(t)\|$ is not only bounded but also converges to zero as $t$ grows without bound, then $\|\mathbf{x}(t)\| \to 0$ also. If $\mathbf{F}(t)$ and $\mathbf{B}(t)$ are bounded on $(-\infty, \infty)$, and if there exist positive constants

$\varepsilon$ and $\delta$ (each independent of $t_0$) such that the controllability Gramian $\mathbf{W}(t_0, t_0 + \delta)$ is positive definite:

$$\mathbf{W}(t_0, t_0 + \delta) \triangleq \int_{t_0}^{t_0+\delta} \mathbf{\Phi}(t_0, \tau)\mathbf{B}(\tau)\mathbf{B}^{\mathrm{T}}(\tau)\mathbf{\Phi}^{\mathrm{T}}(t_0, \tau)\, d\tau \geq \varepsilon\mathbf{I} \tag{14-57}$$

then the system model (14-54) with $\mathbf{C}(t) \equiv \mathbf{I}$ and $\mathbf{D}_y(t) \equiv \mathbf{0}$ is uniformly BIBO stable. Finally, if $\mathbf{F}(t)$, $\mathbf{B}(t)$, and $\mathbf{C}(t)$ are all bounded on $(-\infty, \infty)$, and there exist positive constants $\varepsilon$, $\delta$, $\varepsilon'$, and $\delta'$ such that (14-57) is true and also the observability Gramian $\mathbf{M}(t_0, t_0 + \delta')$ is positive definite:

$$\mathbf{M}(t_0, t_0 + \delta') \triangleq \int_{t_0}^{t_0+\delta'} \mathbf{\Phi}^{\mathrm{T}}(\tau, t_0)\mathbf{H}^{\mathrm{T}}(\tau)\mathbf{H}(\tau)\mathbf{\Phi}(\tau, t_0)\, d\tau \geq \varepsilon'\mathbf{I} \tag{14-58}$$

then the system model (14-54) with $\mathbf{D}_y(t) \equiv \mathbf{0}$ is uniformly BIBO stable.

The sufficient conditions just stated are also the sufficient conditions for *exponential stability* of the system modeled by $\dot{\mathbf{x}}(t) = \mathbf{F}(t)\mathbf{x}(t)$. Thus, for *linear* system models, uniform BIBO stability and uniform exponential stability imply, and are implied by, each other. Such a link between zero-input and BIBO stabilities does not exist for general nonlinear system models.

### *The BIBO and Asymptotic Stabilities for Linear Time-Invariant System Models*

As stated before, for time-invariant linear system models, exponential and asymptotic stabilities are equivalent, and "uniformly" is no longer an issue. Therefore, the sufficient conditions for BIBO stability are either as just stated or the fact that all eigenvalues of $\mathbf{F}$ have strictly negative real parts.

A word of caution against extrapolating time-invariant system model stability concepts to the time-varying case is in order. For the model (14-54a) with $\mathbf{F}$ and $\mathbf{B}$ constant, BIBO and asymptotic (exponential) stability criteria require the system poles (eigenvalues of $\mathbf{F}$) to lie in the left-half complex plane. However, for more general time-varying system models, it is *not valid* to consider $\mathbf{F}(t)$ for all $t \in [t_0, \infty)$ and require the eigenvalues of each such $\mathbf{F}(t)$ to have strictly negative real parts. In other words, it is improper to look at a "frozen" system model at each time $t$ and apply the stability criteria for time-invariant system models. In numerous practical time-varying systems, the impulse response of the "frozen" system will be bounded at some times and unbounded at others; for instance, a ground-launched missile will have an unstable and uncontrollable "frozen" description at time of launch, but the stability and controllability improve as the velocity increases. Other physically important systems periodically go through intervals of apparent stability and instability of the "frozen" system description. We want bounded output response, and there is *no* direct relationship between this behavior and the position of poles of the "frozen" model. In fact, there are cases in which an "apparently" stable system, i.e., one in which the eigenvalues of $\mathbf{F}(t)$ for all $t \in (-\infty, \infty)$ stay in the left-half plane, will yield unbounded output in response to certain bounded inputs. There are even examples of *homogeneous* linear system models of this type that produce unbounded outputs.

*Stochastic Stability* [84, 87–89, 193]

As we have seen, deterministic system stability primarily characterizes behavior of solutions to differential or difference equations as time grows without bound. Once such *convergence* properties are defined for deterministic models, they can be extended conceptually to stochastic models simply by employing any of the convergence concepts discussed in Chapter 4 of Volume 1: *convergence almost surely* (*with probability one*), *mean square convergence*, and *convergence in probability*. For instance, if a deterministic equation solution were to reach a given limit point asymptotically, we might ask whether all sample solutions (except possibly a set of samples having total probability of zero) of the corresponding stochastic equation driven by noises and uncertainty converge to a limit point, and, if so, whether it is the same limit point as when stochastic effects are absent. Instead, we might ask whether the mean squared deviation from that limit point converges to zero, or at least to a specifiable bound if not zero. Or, we might ask whether the probability that deviations from a limit point value take on any finite or infinite nonzero magnitude goes to zero asymptotically. Thus, any deterministic stability conceptualization can yield at least three correponding interpretations for the stochastic case.

The extensions just described are properly concerned with process *sample* behavior on the interval $[t_0, \infty)$. Nevertheless, a considerable amount of investigation has been devoted to establishing stability properties of the *moments* as well as the distributions of the solution processes, and the interrelationship of these generally weaker results to stability of process samples themselves [84, 88].

## 14.4 STABILITY OF LQG REGULATORS

In order to state the stability results for discrete-time LQG stochastic optimal regulators in a precise and convenient fashion [92, 105, 196], we must review some system model structural characteristics from Volume 1 [115] and introduce some new related characteristics as well [16, 35, 67, 92]. First consider deterministic linear time-varying system representations corresponding to (14-1) and (14-31) with noises and uncertainties removed. Such a system model is termed *completely controllable* (from the points of entry of $\mathbf{u}(t_i)$) if the state of the system can be transferred from the zero state at any initial time $t_0$ to any terminal state $\mathbf{x}(t_f) = \mathbf{x}_f$ within some finite time $(t_f - t_0)$. This is true if and only if the range space of the $n$-by-$n$ discrete-time controllability Gramian

$$\mathbf{W}_D(t_0, t_f) \triangleq \sum_{i=1}^{f} \mathbf{\Phi}(t_0, t_i)\mathbf{B}_d(t_{i-1})\mathbf{B}_d^T(t_{i-1})\mathbf{\Phi}^T(t_0, t_i) \tag{14-59}$$

is all of $R^n$ or, equivalently, if $\mathbf{W}_D(t_0, t_f)$ is of full rank $n$ or is positive definite, for some value of $t_f$. (Note that Eq. (14-59) corresponds directly to the controllability Gramian (14-57) for continuous-time systems.) The system model is called *completely observable* (from the points of output of $\mathbf{z}(t_i)$) if, for each $t_0$ there exists a $t_f < \infty$ such that the output response $\mathbf{z}(t_i; t_0, \mathbf{x}_0, \mathbf{u})$ has the property that

$$\mathbf{z}(t_i; t_0, \mathbf{x}_0, \mathbf{u}) = \mathbf{z}(t_i; t_0, \mathbf{x}_0', \mathbf{u}) \qquad \text{for all} \quad t_i \in [t_0, t_f] \tag{14-60}$$

for all choices of $\mathbf{u}$ function over the interval $[t_0, t_f]$ implies that $\mathbf{x}_0 = \mathbf{x}_0'$. The model is called *completely reconstructible* if, for each $t_f$ there exists a $t_0$ with $-\infty < t_0 < t_f$, such that the same condition is true. It has been argued that this is a more natural property to seek in a model than observability: that you can *always* generate the present state value, i.e., $\mathbf{x}(t_f)$ as derivable from $\mathbf{x}_0$ once it is established, from measurements extending *into the past from any current time* $t_f$, rather than being able to do so from measurements extending into the future from any given $t_0$. The former of these two properties is possessed by a model if and only if the null space of the $n$-by-$n$ *observability (reconstructibility) Gramian*

$$\mathbf{M}_D(t_0, t_f) \triangleq \sum_{i=1}^{f} \mathbf{\Phi}^T(t_i, t_0)\mathbf{H}^T(t_i)\mathbf{H}(t_i)\mathbf{\Phi}(t_i, t_0) \tag{14-61}$$

is all of $R^n$ or, equivalently, if $\mathbf{M}_D(t_0, t_f)$ is of full rank $n$ or is positive definite, for some value of $t_f$, given any $t_0$; the latter corresponds to the same criterion but for some value of $t_0$, given any $t_f$. (Equation (14-61) is analogous to the observability Gramian (14-58) for continuous-time system models.)

Now consider deterministic linear *time-invariant* system models. The concept of (*complete*) *controllability* is the same, but it is established more easily by verifying that

$$\mathbf{W}_{DTI} \triangleq [\mathbf{B}_d \,\vdots\, \mathbf{\Phi}\mathbf{B}_d \,\vdots\, \cdots \,\vdots\, \mathbf{\Phi}^{n-1}\mathbf{B}_d] \tag{14-62}$$

is of full rank $n$ for the discrete-time case, or that

$$\mathbf{W}_{TI} \triangleq [\mathbf{B} \,\vdots\, \mathbf{F}\mathbf{B} \,\vdots\, \cdots \,\vdots\, \mathbf{F}^{n-1}\mathbf{B}] \tag{14-63}$$

is of rank $n$ in the continuous-time case. It is convenient here to define the *controllable subspace* [16, 35, 92] as the linear subspace of states that are reachable from the zero state in a finite time, which is then the span of the columns of either controllability matrix, (14-62) or (14-63). In a number of applications, it is useful to consider a weaker condition than controllability, called stabilizability. A linear time-invariant system model is *stabilizable* [92] if its unstable subspace is contained in its controllable subspace. If a model has this property, then any vector in the unstable subspace is also in the controllable subspace, so that at least all unstable modes can be completely controlled. By its definition, stabilizability is seen to be implied by either complete controllability or asymptotic stability.

*Reconstructibility* and (*complete*) *observability* are also the same concepts as for the time-varying case, but they are *indistinguishable* for time-invariant system models. Furthermore, they are verified more easily by the full rank of either

$$\mathbf{M}_{\mathrm{DTI}} \triangleq [\mathbf{H}^{\mathrm{T}} \vdots \boldsymbol{\Phi}^{\mathrm{T}}\mathbf{H}^{\mathrm{T}} \vdots \cdots \vdots (\boldsymbol{\Phi}^{\mathrm{T}})^{n-1}\mathbf{H}^{\mathrm{T}}] \tag{14-64}$$

in the discrete-time case, or

$$\mathbf{M}_{\mathrm{TI}} \triangleq [\mathbf{H}^{\mathrm{T}} \vdots \mathbf{F}^{\mathrm{T}}\mathbf{H}^{\mathrm{T}} \vdots \cdots \vdots (\mathbf{F}^{\mathrm{T}})^{n-1}\mathbf{H}^{\mathrm{T}}] \tag{14-65}$$

for continuous-time models. Here it is also convenient to define the *unobservable subspace* [16, 35, 92], or unreconstructible subspace, as the linear subspace of states $\mathbf{x}_0$ for which $\mathbf{z}(t_i; t_0, \mathbf{x}_0, \mathbf{0}) = \mathbf{0}$ for all $t_i \geq t_0$, so that it is not possible to distinguish between states in this subspace by looking just at the output; this turns out to be the null space of either observability matrix, (14-64) or (14-65). As previously, it is useful to consider a condition that is weaker than observability, namely detectability. A linear time-invariant system model is *detectable* [92] if its unobservable subspace is contained in its stable subspace: the zero-input response of the system model state in the unobservable subspace converges to zero; i.e., any modes that are not observable are assured to be stable modes. It can readily be seen that either complete observability or asymptotic stability will imply detectability.

Now consider a stochastic representation composed of discrete-time dynamics (14-1) or continuous-time dynamics (14-19), and discrete-time measurements (14-31) or (14-32). (Continuous-time noise-corrupted measurements will also be considered later.) As described in Section 5.8 of Volume 1, the system model is *stochastically controllable* (from the points of entry of the dynamics driving noise) if there exist positive numbers $\alpha$ and $\beta$, $0 < \alpha < \beta < \infty$, and a positive integer $N$ such that, for all $i \geq N$,

$$\alpha\mathbf{I} \leq \sum_{j=i-N+1}^{i} \boldsymbol{\Phi}(t_i, t_j)\mathbf{G}_{\mathrm{d}}(t_{j-1})\mathbf{Q}_{\mathrm{d}}(t_{j-1})\mathbf{G}_{\mathrm{d}}{}^{\mathrm{T}}(t_{j-1})\boldsymbol{\Phi}^{\mathrm{T}}(t_i, t_j) \leq \beta\mathbf{I} \tag{14-66}$$

or if there exist such an $\alpha$ and $\beta$ and a time interval $\Delta t$ such that, for all $t \geq t_0 + \Delta t$,

$$\alpha\mathbf{I} \leq \int_{t-\Delta t}^{t} \boldsymbol{\Phi}(t,\tau)\mathbf{G}(\tau)\mathbf{Q}(\tau)\mathbf{G}^{\mathrm{T}}(\tau)\boldsymbol{\Phi}^{\mathrm{T}}(t,\tau)\,d\tau \leq \beta\mathbf{I} \tag{14-67}$$

where $\mathbf{M}_1 \leq \mathbf{M}_2$ denotes that $(\mathbf{M}_2 - \mathbf{M}_1)$ is positive semidefinite. Analogously, the model is *stochastically observable* if

$$\alpha\mathbf{I} \leq \sum_{j=i-N+1}^{i} \boldsymbol{\Phi}^{\mathrm{T}}(t_j, t_i)\mathbf{H}^{\mathrm{T}}(t_j)\mathbf{R}^{-1}(t_j)\mathbf{H}(t_j)\boldsymbol{\Phi}(t_j, t_i) \leq \beta\mathbf{I} \tag{14-68}$$

for the case of discrete-time measurements, and if

$$\alpha\mathbf{I} \leq \int_{t-\Delta t}^{t} \boldsymbol{\Phi}^{\mathrm{T}}(\tau, t)\mathbf{H}^{\mathrm{T}}(\tau)\mathbf{R}_{\mathrm{c}}^{-1}(\tau)\mathbf{H}(\tau)\boldsymbol{\Phi}(\tau, t)\,d\tau \leq \beta\mathbf{I} \tag{14-69}$$

for continuous-time measurements being available [166]. Equations (14-66)–(14-69) without the appearance of $\mathbf{Q}_d$, $\mathbf{Q}$, $\mathbf{R}^{-1}$, or $\mathbf{R}_c^{-1}$, are the conditions for *uniformly* complete controllability and observability: generally stricter conditions than complete controllability and observability in that the Gramians are bounded above as well as below, but equivalent conditions in the time-invariant special case.

With appropriate model characteristics thus defined, we can investigate the stability of LQG stochastic controllers, looking first at the deterministic optimal full-state feedback controller, then the Kalman filter, and finally their cascade combination. Consider the deterministic LQ optimal regulator based upon a dynamics model of (14-1) with all uncertainties removed,

$$\mathbf{x}(t_{i+1}) = \mathbf{\Phi}(t_{i+1}, t_i)\mathbf{x}(t_i) + \mathbf{B}_d(t_i)\mathbf{u}(t_i) \tag{14-70}$$

assuming that the full state $\mathbf{x}(t_i)$ is available perfectly at each sample time, and designed to minimize the quadratic cost function (14-13) with the expectation operation removed, with $\mathbf{X}_f \geq \mathbf{0}$, $\mathbf{X}(t_i) \geq \mathbf{0}$, $\mathbf{U}(t_i) > \mathbf{0}$, and $\mathbf{S}(t_i)$ chosen so that the composite matrix in the summation of (14-13b) is positive semidefinite (or, equivalently, such that $[\mathbf{X}(t_i) - \mathbf{S}(t_i)\mathbf{U}^{-1}(t_i)\mathbf{S}^T(t_i)] \geq \mathbf{0}$) for all $t_i$, as assumed previously. Furthermore, assume that $\mathbf{\Phi}(t_{i+1}, t_i)$, $\mathbf{B}_d(t_i)$, $\mathbf{X}(t_i)$, $\mathbf{U}(t_i)$, and $\mathbf{S}(t_i)$ for all $t_i$, and $\mathbf{X}_f$, are all bounded, and consider the backward propagation of (14-14)–(14-17) to generate the optimal control law (14-6). Definitive stability statements can be made about the optimal controller once it has achieved its steady state condition sufficiently far from the terminal time $t_{N+1}$. This will be particularly useful for time-invariant system descriptions, for which steady state controllers of the form (14-6) simply entail premultiplication of $\mathbf{x}(t_i)$ by constant gains; this simple form causes it to be considered for implementation for *all* time in a finite interval that is long compared to the terminal transient period, ignoring the time-varying gains of the truly optimal controller.

First consider the more general time-varying system description. If (14-70) is either *completely controllable* or *exponentially stable*, then the solution $\mathbf{K}_c(t_i)$ to the backward Riccati equation (14-14)–(14-16) with $\mathbf{K}_c(t_{N+1}) = \mathbf{0}$ converges to a positive semidefinite bounded steady state solution sequence $\bar{\mathbf{K}}_c(t_i)$ for all $t_i$ as $t_i \to -\infty$, and this $\bar{\mathbf{K}}_c(t_i)$ for all $t_i$ is itself a solution to the backward equation. Basically, once existence and uniqueness of solutions in the region local to final time $t_{N+1}$ are established, complete controllability enables $\mathbf{K}_c(t_i)$ to be bounded from above, and this backward running sequence is also nondecreasing (shown by relating $\frac{1}{2}\mathbf{x}^T(t_i)\mathbf{K}_c(t_i)\mathbf{x}(t_i)$ to cost evaluations), and thus the sequence must have a limit, which is $\bar{\mathbf{K}}_c(t_i)$. Further claims can be made if we strengthen our assumptions to the system described by (14-70) and the artificial output equation

$$\mathbf{y}_a(t_i) = \sqrt{\mathbf{X}(t_i)}^T\mathbf{x}(t_i) \tag{14-71}$$

being *uniformly completely controllable and uniformly completely reconstructible.*

The last assumption states that (14-68) is satisfied with $\mathbf{H}(t_j) = \sqrt{\mathbf{X}(t_j)}^{\mathrm{T}}$ and $\mathbf{R}^{-1}(t_j) = \mathbf{I}$, i.e.,

$$\alpha \mathbf{I} \le \sum_{j=i-N+1}^{i} \mathbf{\Phi}^{\mathrm{T}}(t_j, t_i)\mathbf{X}(t_j)\mathbf{\Phi}(t_j, t_i) \le \beta \mathbf{I} \tag{14-72}$$

and is equivalent to (14-1) and (14-2) being uniformly completely reconstructible if $\mathbf{Y}(t_i)$ in (14-3) is assumed positive *definite* for all $t_i$. Under either *this more restrictive assumption*, or again the *exponential stability* of (14-70), then (1) $\mathbf{K}_c(t_i)$ converges to $\bar{\mathbf{K}}_c(t_i)$ as $t_i \to -\infty$, *regardless* of the terminal condition $\mathbf{K}(t_{N+1}) = \mathbf{X}_f \ge \mathbf{0}$, (2) the steady state optimal control law

$$\mathbf{u}(t_i) = -\bar{\mathbf{G}}_c^*(t_i)\mathbf{x}(t_i) \tag{14-73a}$$

$$\begin{aligned}\bar{\mathbf{G}}_c^*(t_i) = {} & [\mathbf{U}(t_i) + \mathbf{B}_d^{\mathrm{T}}(t_i)\bar{\mathbf{K}}_c(t_{i+1})\mathbf{B}_d(t_i)]^{-1} \\ & \times [\mathbf{B}_d^{\mathrm{T}}(t_i)\bar{\mathbf{K}}_c(t_{i+1})\mathbf{\Phi}(t_{i+1}, t_i) + \mathbf{S}^{\mathrm{T}}(t_i)]\end{aligned} \tag{14-73b}$$

in fact minimizes the limiting cost

$$\begin{aligned}\lim J = \lim_{j \to -\infty} & \left\{ \frac{1}{2} \mathbf{x}^{\mathrm{T}}(t_{N+1})\mathbf{X}_f \mathbf{x}(t_{N+1}) \right. \\ & \left. + \sum_{i=j}^{N} \frac{1}{2} \left( \begin{bmatrix} \mathbf{x}(t_i) \\ \mathbf{u}(t_i) \end{bmatrix}^{\mathrm{T}} \begin{bmatrix} \mathbf{X}(t_i) & \mathbf{S}(t_i) \\ \mathbf{S}^{\mathrm{T}}(t_i) & \mathbf{U}(t_i) \end{bmatrix} \begin{bmatrix} \mathbf{x}(t_i) \\ \mathbf{u}(t_i) \end{bmatrix} \right) \right\}\end{aligned} \tag{14-74}$$

for all $\mathbf{X}_f \ge \mathbf{0}$, and that minimal value is the finite and positive definite $\frac{1}{2}\mathbf{x}^{\mathrm{T}}(t_j)\bar{\mathbf{K}}_c(t_j)\mathbf{x}(t_j)]$ at the initial limiting $t_j$ (or we can replace $t_j$ by $t_0$ and write the limit as $N \to \infty$ instead), and (3) *this steady state optimal control law causes the closed-loop system representation*

$$\begin{aligned}\mathbf{x}(t_{i+1}) &= \mathbf{\Phi}(t_{i+1}, t_i)\mathbf{x}(t_i) + \mathbf{B}_d(t_i)[-\bar{\mathbf{G}}_c^*(t_i)\mathbf{x}(t_i)] \\ &= [\mathbf{\Phi}(t_{i+1}, t_i) - \mathbf{B}_d(t_i)\bar{\mathbf{G}}_c^*(t_i)]\mathbf{x}(t_i)\end{aligned} \tag{14-75}$$

*to be exponentially stable* [38, 65, 92]. Thus, by synthesizing the deterministic optimal LQ regulator under these assumptions, we are *assured* of an exponentially stabilizing full-state feedback law.

Now let us confine attention to the case of a time-invariant system model with constant cost function matrices, and maintain the bounded assumptions on constant $\mathbf{X}_f \ge \mathbf{0}$, $\mathbf{X} \ge \mathbf{0}$, $\mathbf{U} > \mathbf{0}$, and $[\mathbf{X} - \mathbf{S}\mathbf{U}^{-1}\mathbf{S}^{\mathrm{T}}] \ge \mathbf{0}$. If the system model

$$\mathbf{x}(t_{i+1}) = \mathbf{\Phi}\mathbf{x}(t_i) + \mathbf{B}_d\mathbf{u}(t_i) \tag{14-76}$$

is *stabilizable*, then the solution to the discrete backward Riccati recursion (14-14)–(14-16) with $\mathbf{K}_c(t_{N+1}) = \mathbf{0}$ converges to a *constant* steady state solution $\bar{\mathbf{K}}_c$ as $t_i \to -\infty$, where this $\bar{\mathbf{K}}_c$ satisfies the *algebraic* Riccati equation

$$\bar{\mathbf{K}}_c = \mathbf{X} + \mathbf{\Phi}^{\mathrm{T}}\bar{\mathbf{K}}_c\mathbf{\Phi} - [\mathbf{B}_d^{\mathrm{T}}\bar{\mathbf{K}}_c\mathbf{\Phi} + \mathbf{S}^{\mathrm{T}}]^{\mathrm{T}}\bar{\mathbf{G}}_c^* \tag{14-77a}$$

or equivalently,

$$\bar{\mathbf{K}}_c = [\mathbf{\Phi} - \mathbf{B}_d\bar{\mathbf{G}}_c^*]^T\bar{\mathbf{K}}_c[\mathbf{\Phi} - \mathbf{B}_d\bar{\mathbf{G}}_c^*] + \bar{\mathbf{G}}_c^{*T}\mathbf{U}\bar{\mathbf{G}}_c^* + \mathbf{X} - \mathbf{S}\bar{\mathbf{G}}_c^* - \bar{\mathbf{G}}_c^{*T}\mathbf{S}^T \tag{14-77b}$$

where

$$\bar{\mathbf{G}}_c^* = [\mathbf{U} + \mathbf{B}_d^T\bar{\mathbf{K}}_c\mathbf{B}_d]^{-1}[\mathbf{B}_d^T\bar{\mathbf{K}}_c\mathbf{\Phi} + \mathbf{S}^T] \tag{14-78}$$

and the steady state control law

$$\mathbf{u}(t_i) = -\bar{\mathbf{G}}_c^*\mathbf{x}(t_i) \tag{14-79}$$

is then *time invariant*. If we additionally assume that the system model composed of (14-76) and the artificial output (recall discussion below Eq. (14-72))

$$\mathbf{y}_a(t_i) = \sqrt{\mathbf{X}}^T\mathbf{x}(t_i) \tag{14-80}$$

is both *stabilizable and detectable*, then the preceding results are valid, and also (1) the solution to the Riccati recursion converges to $\bar{\mathbf{K}}_c$ for *any* $\mathbf{K}_c(t_{N+1}) = \mathbf{X}_f \geq \mathbf{0}$ (existence and uniqueness are guaranteed), (2) the constant-gain steady state law (14-79) *minimizes* (14-74) for $\mathbf{X}_f \geq \mathbf{0}$ and this minimum value is $[\frac{1}{2}\mathbf{x}_0{}^T\bar{\mathbf{K}}_c\mathbf{x}_0]$, and (3) this control law causes the closed-loop system model

$$\mathbf{x}(t_{i+1}) = [\mathbf{\Phi} - \mathbf{B}_d\bar{\mathbf{G}}_c^*]\mathbf{x}(t_i) \tag{14-81}$$

to be *asymptotically stable*; i.e., all of the eigenvalues of $[\mathbf{\Phi} - \mathbf{B}_d\bar{\mathbf{G}}_c^*]$ have magnitude strictly less than one [92]. These eigenvalues can be characterized still further: let $z_j$, $j = 1, 2, \ldots, n' \leq n$ denote the roots of

$$\det\left\{\begin{bmatrix} z\mathbf{I} - [\mathbf{\Phi} - \mathbf{B}_d\mathbf{U}^{-1}\mathbf{S}^T] & \mathbf{B}_d\mathbf{U}^{-1}\mathbf{B}_d^T \\ -[\mathbf{X} + \mathbf{S}\bar{\mathbf{G}}_c^*] & z^{-1}\mathbf{I} - \mathbf{\Phi}^T \end{bmatrix}\right\} = 0 \tag{14-82}$$

that have magnitude strictly less than one; these $z_j$ constitute $n'$ eigenvalues of the closed-loop system model (14-81), and the remaining $(n - n')$ are at the origin (an extension of results in [92]). If the detectability assumption is strengthened to *observability*, these results are valid, and also $\bar{\mathbf{K}}_c$ is assured to be *positive definite* as well and $[\frac{1}{2}\mathbf{x}(t_i)^T\bar{\mathbf{K}}_c\mathbf{x}(t_i)]$ can serve as a Lyapunov function for establishing stability.

To put these results in perspective, consider an *arbitrary* linear time-invariant control law

$$\mathbf{u}(t_i) = -\mathbf{G}_c\mathbf{x}(t_i) \tag{14-83}$$

applied to a system modeled as (14-76). If and only if the model is *stabilizable*, there *does exist* a constant $\mathbf{G}_c$ such that the closed-loop system is *asymptotically stable*: the eigenvalues of $[\mathbf{\Phi} - \mathbf{B}_d\mathbf{G}_c]$ are all of magnitude strictly less than one [92]. Unfortunately, how to find such a $\mathbf{G}_c$ is not specified by such an existence result [201]. What the LQ full-state regulator provides is a systematic means of *synthesizing* a control law with this desirable stability property, if the system model (14-76) and (14-80) is both stabilizable and detectable.

Similarly, if and only if the model (14-76) has the more restrictive property of *complete controllability*, then the poles of the closed-loop system resulting from use of (14-83), i.e., the eigenvalues of $[\mathbf{\Phi} - \mathbf{B}_d\mathbf{G}_c]$, *can be arbitrarily located* in the complex plane (with the restriction that complex values must appear in complex conjugate pairs) by choosing $\mathbf{G}_c$ suitably [15, 92]. Thus, an alternative design procedure to LQ optimal regulator synthesis can entail finding a $\mathbf{G}_c$ so as to place the closed-loop system poles in desired locations as specified by the designer. For instance, for a controllable system model it is possible to select $\mathbf{G}_c$ so as to force all the eigenvalues of $[\mathbf{\Phi} - \mathbf{B}_d\mathbf{G}_c]$ to the origin. By the Cayley–Hamilton theorem, every matrix satisfies its own characteristic equation, so $[\mathbf{\Phi} - \mathbf{B}_d\mathbf{G}_c]^n = \mathbf{0}$ under these conditions ($[\mathbf{\Phi} - \mathbf{B}_d\mathbf{G}_c]$ is then termed "nilpotent with index $n$"), and

$$\mathbf{x}(t_n) = [\mathbf{\Phi} - \mathbf{B}_d\mathbf{G}_c]^n\mathbf{x}_0 = \mathbf{0} \tag{14-84}$$

Thus, *any* initial state $\mathbf{x}_0$ can be reduced to $\mathbf{0}$ in at most $n$ sample periods, a desirable characteristic known as *state deadbeat response*.

Under the assumptions of complete *controllability and observability* of the linear model (14-76) and (14-80), the optimal regulator, designed to minimize the quadratic cost (14-74) with bounded and constant $\mathbf{X} \geq \mathbf{0}$, $\mathbf{U} > \mathbf{0}$, and $[\mathbf{X} - \mathbf{S}\mathbf{U}^{-1}\mathbf{S}^T] \geq \mathbf{0}$ (i.e., the composite matrix in (14-74) is positive semidefinite), is the *time-invariant, asymptotically stabilizing* law (14-78) and (14-79), where $\bar{\mathbf{K}}_c$ is the *unique, positive definite, real, symmetric* solution to the steady state discrete Riccati equation (14-77). The interrelationships of appropriate choices for $\mathbf{X}$, $\mathbf{U}$, and $\mathbf{S}$ and the desired placement of closed-loop system poles have been studied, as have the loci of poles of $[\mathbf{\Phi} - \mathbf{B}_d\bar{\mathbf{G}}_c{}^*]$ for the case of $\mathbf{U} = \mu\mathbf{U}_0$ and $\mathbf{U}_0 > \mathbf{0}$ in the limits as $\mu \to 0$ (the case of "cheap" control) or $\mu \to \infty$, basically by manipulation of (14-82) [29, 44, 50, 54, 91, 92, 147, 159, 167, 188]. A classical result [92] considers the case of a scalar input/scalar controlled output system described by

$$\mathbf{x}(t_{i+1}) = \mathbf{\Phi}\mathbf{x}(t_i) + \mathbf{B}_d u(t_i) \tag{14-85a}$$

$$y_c(t_i) = \mathbf{C}\mathbf{x}(t_i) \tag{14-85b}$$

or, taking the *z-transform* [23, 27, 92, 140, 179] of these, defined by

$$\mathbf{x}(z) \triangleq \sum_{i=0}^{\infty} \mathbf{x}(t_i)z^{-i} \tag{14-85c}$$

we get

$$z\mathbf{x}(z) - z\mathbf{x}_0 = \mathbf{\Phi}\mathbf{x}(z) + \mathbf{B}_d u(z) \tag{14-85d}$$

$$y_c(z) = \mathbf{C}\mathbf{x}(z)$$

and solving as

$$y_c(z) = \mathbf{C}[(z\mathbf{I} - \mathbf{\Phi})^{-1}z]\mathbf{x}_0 + [\mathbf{C}(z\mathbf{I} - \mathbf{\Phi})^{-1}\mathbf{B}_d]u(z) \tag{14-85e}$$

to yield an open-loop $z$-transform transfer function between input $u$ and output $y_c$ of

$$\mathbf{C}[z\mathbf{I} - \mathbf{\Phi}]^{-1}\mathbf{B}_d = \frac{Kz^{\mathscr{s}-\mathscr{q}}\prod_{i=1}^{\mathscr{q}}(z - z_i)}{z^{n-\mathscr{p}}\prod_{j=1}^{\mathscr{p}}(z - p_j)}, \qquad K \neq 0 \tag{14-85f}$$

where the $z_i$ and $p_j$ are, respectively, the $\mathscr{q}$ zeros and $\mathscr{p}$ poles (eigenvalues) of the open-loop system with nonzero values, with $n \geq \mathscr{p} \geq \mathscr{q}$ and $(n-1) \geq \mathscr{s} \geq \mathscr{q}$. For such a system model, the closed-loop system resulting from applying the LQ optimal regulator designed to minimize a cost (14-74) with $\mathbf{S} = \mathbf{0}$, $U = \mu$, and $Y = 1$ so that $\mathbf{X} = \mathbf{C}^{\mathrm{T}}\mathbf{C}$, has the properties that (1) $(n - \mathscr{p})$ of the closed-loop poles are at the origin for all $\mu$, (2) as $\mu \to 0$, $\mathscr{q}$ closed-loop poles approach the locations $\tilde{z}_i$ related to open-loop zeros by

$$\tilde{z}_i = \begin{cases} z_i & \text{if } |z_i| \leq 1 \\ 1/z_i & \text{if } |z_i| > 1 \end{cases} \tag{14-86}$$

and the remaining $(\mathscr{p} - \mathscr{q})$ poles go to the origin, and (3) as $\mu \to \infty$, the $\mathscr{p}$ nonzero closed-loop eigenvalues converge to the locations $\tilde{p}_j$ related to open-loop poles by

$$\tilde{p}_j = \begin{cases} p_j & \text{if } |p_j| \leq 1 \\ 1/p_j & \text{if } |p_j| > 1 \end{cases} \tag{14-87}$$

(Analogous results can be obtained [92] for multiple input/multiple output system descriptions if we consider the poles and zeros of $\det\{\mathbf{C}[z\mathbf{I} - \mathbf{\Phi}]^{-1}\mathbf{B}_d\}$.) Thus, even in the extreme cases of $\mu \to 0$ or $\mu \to \infty$, the closed-loop system is precluded from ever being unstable. Note that for the case of *all* open-loop zeros within the unit circle, i.e., the case of a *minimum-phase* open-loop system model, in the limit as $\mu \to 0$, the $\mathscr{q}$ closed-loop poles that converge to $\tilde{z}_i = z_i$ are *canceled* by the open-loop zeros at those locations. This causes closed-loop system behavior to be dictated by the $(n - \mathscr{q})$ poles at the origin, namely *output deadbeat response* in which the outputs can be driven to zero in at most $(n - \mathscr{q})$ sample periods, but at which time the internal system state is generally not completely quiescent. However, an open-loop system that is not minimum phase yields very different characteristics since there are fewer than $\mathscr{q}$ pole-zero cancellations; the need to treat nonminimum phase open-loop systems as a unique special case will recur in later discussions as well. It should also be pointed out that we have been considering the steady state $\bar{K}_c$ and $\bar{G}_c{}^*$ in the limit as $\mu \to 0$, and that these need *not* be obtainable merely by substituting $\mathbf{U} \equiv \mathbf{0}$ into the Riccati difference equation and letting it reach steady state (reversing the order of taking the limits).

Now consider only the Kalman filter portion of the LQG stochastic optimal regulator. Conditions for stability have already been established in Chapter 5 of Volume 1: If the system model upon which the Kalman filter is based is stochastically controllable (from the points of entry of dynamics noise) and stochastically observable (from the points of extraction of measurements), then the filter is uniformly asymptotically globally stable. In fact, this can be viewed as a *duality* result corresponding to the controller stability characteristics just stated [22, 92]. Given a system model

$$\mathbf{x}(t_{i+1}) = \mathbf{\Phi}(t_{i+1}, t_i)\mathbf{x}(t_i) + \mathbf{B}_d(t_i)\mathbf{u}(t_i) \tag{14-88a}$$

$$\mathbf{y}_c(t_i) = \mathbf{C}(t_i)\mathbf{x}(t_i) \tag{14-88b}$$

its *dual* with respect to sample time $t^*$ is

$$\mathbf{x}^d(t_{i+1}) = \mathbf{\Phi}^T(t^* - t_{i-1}, t^* - t_i)\mathbf{x}^d(t_i) + \mathbf{C}^T(t^* - t_i)\mathbf{u}^d(t_i) \tag{14-89a}$$

$$\mathbf{y}^d(t_i) = \mathbf{B}_d{}^T(t^* - t_i)\mathbf{x}^d(t_i) \tag{14-89b}$$

and the original system is itself the dual of (14-89). The importance of duals lies in the fact that a system model is completely controllable, completely reconstructible, or exponentially stable if and only if its dual is completely reconstructible, completely controllable, or exponentially stable, respectively. For the time-invariant case, it is additionally true that a model is stabilizable or detectable if and only if its dual is, respectively, detectable or stabilizable. To exploit duality for the usual case of a Kalman filter developed under the assumption that dynamics driving noise and measurement noise are uncorrelated, let the cross-weighting $\mathbf{S}(t_i) \equiv \mathbf{0}$ for all $t_i$. Further assume that $n$-by-$n$ $[\mathbf{G}_d(t_i)\mathbf{Q}_d(t_i)\mathbf{G}_d{}^T(t_i)] \geq \mathbf{0}$ is of rank $s$ so that $s$-by-$s$ $\mathbf{Q}_d(t_i) > \mathbf{0}$ can be generated. Then, if we make the correspondences between the optimal LQ regulator and Kalman filter as indicated in the first six entries of Table 14.1 for all $t_i \leq t_N$ (note we assume $\mathbf{C}^T(t_{N+1})\mathbf{Y}(t_{N+1})\mathbf{C}(t_{N+1})$ is zero; otherwise, this could be

TABLE 14.1

*Duality Relationships between Optimal LQ Regulator and Kalman Filter*

| Optimal LQ regulator | Kalman filter |
|---|---|
| By design choice: | |
| $\mathbf{\Phi}(t_{i+1}, t_i)$ | $\mathbf{\Phi}^T(t^* - t_{i-1}, t^* - t_i)$ |
| $\mathbf{B}_d(t_i)$ | $\mathbf{H}^T(t^* - t_i)$ |
| $\mathbf{C}(t_{i+1})$ | $\mathbf{G}_d{}^T(t^* - t_i)$ |
| $\mathbf{Y}(t_{i+1}) > \mathbf{0}$ | $\mathbf{Q}_d(t^* - t_i) > \mathbf{0}$ |
| $\mathbf{U}(t_i) > \mathbf{0}$ | $\mathbf{R}(t^* - t_i) > \mathbf{0}$ |
| $\mathbf{X}_f \geq \mathbf{0}$ | $\mathbf{P}_0 \geq \mathbf{0}$ |
| By resulting computations: | |
| $\mathbf{K}_c(t_{i+1})$ | $\mathbf{P}[(t^* - t_i)^-]$ |
| $\mathbf{G}_c{}^*(t_i)$ | $\mathbf{K}^T(t^* - t_i)\mathbf{\Phi}^T(t^* - t_{i-1}, t^* - t_i)$ |

subtracted from $\mathbf{X}_f$ in the table), then the last two correspondences in the table result by direct computation, and the closed-loop system state for the regulator problem,

$$\mathbf{x}(t_{i+1}) = [\mathbf{\Phi}(t_{i+1}, t_i) - \mathbf{B}_d(t_i)\mathbf{G}_c^*(t_i)]\mathbf{x}(t_i) \tag{14-90}$$

and the unforced estimation error before measurement incorporation for the Kalman filter problem,

$$\mathbf{e}(t_{i+1}) = [\mathbf{\Phi}(t_{i+1}, t_i) - \{\mathbf{\Phi}(t_{i+1}, t_i)\mathbf{K}(t_i)\}\mathbf{H}(t_i)]\mathbf{e}(t_i^-) \tag{14-91}$$

are dual with respect to $t^* = (t_0 + t_N)$.

Duality can be used advantageously in a number of ways. For instance, computer software tools as developed for Kalman filter design can be used to design LQ optimal regulators, and vice versa. Of specific interest to us here is that duality can be invoked to establish additional stability results for the Kalman filter. Thus, assume the system model is given by (14-1) and (14-31) with $\mathbf{\Phi}(t_{i+1}, t_i)$, $\mathbf{B}_d(t_i)$, $\mathbf{G}_d(t_i)$, $\mathbf{H}(t_i)$, $\mathbf{Q}_d(t_i)$, and $\mathbf{R}(t_i)$ bounded for all $t_i$; $[\mathbf{G}_d(t_i)\mathbf{Q}_d(t_i)\mathbf{G}_d^T(t_i)]$ is positive semidefinite (with $\mathbf{Q}_d(t_i) > \mathbf{0}$ corresponding to $\mathbf{Y}(t_i) > \mathbf{0}$ in the regulator problem), $\mathbf{R}(t_i)$ is positive definite, and $E\{\mathbf{w}_d(t_i)\mathbf{v}^T(t_j)\} = \mathbf{0}$ for all $t_i$ and $t_j$. Then, if the system model is either *completely reconstructible* or *exponentially stable*, then the error covariance solution $\mathbf{P}(t_i^-)$ to the forward Riccati equation (13-29)–(13-31) with $\mathbf{P}_0 = \mathbf{0}$ converges to a positive semidefinite bounded steady state solution sequence $\bar{P}(t_i^-)$ for all $t_i$ as $t_i \to \infty$, and this $\bar{P}(t_i^-)$ for all $t_i$ is itself a solution to the forward Riccati recursion. If we strengthen the assumptions to the model being *uniformly completely reconstructible and uniformly completely controllable* from the points of entry of $\mathbf{w}_d(\cdot,\cdot)$, or again assume the model is *exponentially stable*, then (1) $\mathbf{P}(t_i^-)$ converges to $\bar{\mathbf{P}}(t_i^-)$ as $t_i \to \infty$, regardless of the initial condition $\mathbf{P}_0 \geq \mathbf{0}$, (2) the steady state filter corresponding to gain calculations based on $\bar{\mathbf{P}}(t_i^-)$ in fact minimizes the limiting mean square error cost

$$\lim J = \lim_{j\to\infty} E\{\tfrac{1}{2}\mathbf{e}^T(t_j^-)\mathbf{W}(t_j)\mathbf{e}(t_j^-)\} \tag{14-92}$$

for all $\mathbf{W}(t_j) \geq \mathbf{0}$ and all $\mathbf{P}_0 \geq \mathbf{0}$, and that minimal value is the finite and positive definite $[\frac{1}{2}\operatorname{tr}\{\bar{\mathbf{P}}(t_j^-)\mathbf{W}(t_j)\}]$, and (3) this steady state Kalman filter is *exponentially stable.*

Restricting attention to time-invariant system and stationary noise models, we again can use duality to attain stability results corresponding to previous regulator properties. If the system model is *stabilizable and detectable*, then (1) the solution $\mathbf{P}(t_i^-)$ to the forward Riccati recursion converges to a *constant* steady state solution $\bar{\mathbf{P}}$ as $t_i \to \infty$ for any $\mathbf{P}_0 \geq \mathbf{0}$, where $\bar{\mathbf{P}}$ is the solution to the algebraic Riccati equation resulting from setting $\mathbf{P}(t_{i+1}^-) = \mathbf{P}(t_i^-) = \bar{\mathbf{P}}$; (2) the steady state Kalman filter is a *constant-gain* ($\bar{\mathbf{K}} = \bar{\mathbf{P}}\mathbf{H}^T[\mathbf{H}\bar{\mathbf{P}}\mathbf{H}^T + \mathbf{R}]^{-1}$), *time-invariant* algorithm that *minimizes* (14-92) for all $\mathbf{P}_0 \geq \mathbf{0}$ and this minimum value is $[\frac{1}{2}\operatorname{tr}\{\bar{\mathbf{P}}\mathbf{W}\}]$; and (3) this constant-gain steady state filter is *asymptotically stable* regardless of choice of $\mathbf{Q}_d > \mathbf{0}$ and $\mathbf{R} > \mathbf{0}$; and the roots $z_j$, $j = 1, 2, \ldots, n' \leq n$, of

$$\det\left\{\begin{bmatrix} z\mathbf{I} - \mathbf{\Phi}^T & \mathbf{H}^T\mathbf{R}^{-1}\mathbf{H} \\ -\mathbf{G}_d\mathbf{Q}_d\mathbf{G}_d{}^T & z^{-1}\mathbf{I} - \mathbf{\Phi} \end{bmatrix}\right\} = 0 \tag{14-93}$$

that have magnitude strictly less than one constitute the $n'$ eigenvalues of the (unforced) estimation error system model, with the remaining $(n - n')$ poles at the origin [92]. If the stabilizability assumption is strengthened to controllability, $\bar{\mathbf{P}}$ is additionally assured to be positive definite.

As for the regulator case, these results can be compared to those for an *arbitrary time-invariant full-order observer* [92, 108, 109, 126, 127, 181, 182, 185, 186, 203] for a system modeled by time-invariant (14-1) and (14-32), given by

$$\tilde{\mathbf{x}}(t_{i+1}^-) = \mathbf{\Phi}\tilde{\mathbf{x}}(t_i^-) + \mathbf{B}_d\mathbf{u}(t_i) + \tilde{\mathbf{K}}[\mathbf{z}_i - \mathbf{H}\tilde{\mathbf{x}}(t_i^-) - \mathbf{D}_z\mathbf{u}(t_i)] \tag{14-94}$$

with $\tilde{\mathbf{K}}$ not necessarily equal to $\mathbf{\Phi}$ times the steady state Kalman filter gain. If and only if the system model is *detectable*, there *does exist* a constant $\tilde{\mathbf{K}}$ such that the observer is *asymptotically stable*; if the model is detectable and stabilizable, the Kalman filter provides a systematic means of generating a specific $\tilde{\mathbf{K}} = \{\mathbf{\Phi}\bar{\mathbf{K}}\}$ that does yield such stability. If and only if the system model is *completely reconstructible* (*observable*), then the observer poles, i.e., the eigenvalues of $[\mathbf{\Phi} - \tilde{\mathbf{K}}\mathbf{H}]$, *can be arbitrarily located* in the complex plane, with the only restriction being that complex values must appear in complex conjugate pairs, by choosing $\tilde{\mathbf{K}}$ suitably. Under these conditions, observers can be designed by searching for a $\tilde{\mathbf{K}}$ to provide desirable observer pole locations, rather than using a steady state Kalman gain. For instance, $\tilde{\mathbf{K}}$ can be selected so as to force all eigenvalues of $[\mathbf{\Phi} - \tilde{\mathbf{K}}\mathbf{H}]$ to the origin, yielding a "deadbeat state observer" in which the unforced reconstruction error (corresponding to $\mathbf{e}(t_i^-)$ with $\mathbf{w}_d \equiv \mathbf{0}$ and $\mathbf{v} \equiv \mathbf{0}$) can be reduced to zero in at most $n$ sample periods. Analogous to previous results, these two design techniques can be interrelated and the loci of eigenvalues of $[\mathbf{\Phi} - \{\mathbf{\Phi}\bar{\mathbf{K}}\}\mathbf{H}]$ can be studied for $\mathbf{R} = \mu\mathbf{R}_0$ in the limits as $\mu \to 0$ or $\mu \to \infty$.

Now consider the LQG stochastic optimal controller formed by cascading the Kalman filter with the corresponding LQ optimal deterministic state regulator. Separately designing a stable filter and stabilizing regulator will yield a *stable* closed-loop system when the resulting LQG stochastic optimal controller is applied to the original system model, and thus the appropriate form of sufficient conditions is the *intersection of sufficient conditions* for stability of the two components. Thus, if the system model composed of (14-1), (14-31) or (14-32), and (14-71) with bounded defining matrices and standard positivity assumptions ($\mathbf{U}(t_i) > \mathbf{0}$ and $\mathbf{R}(t_i) > \mathbf{0}$; $\mathbf{X}(t_i) \geq \mathbf{0}$, $[\mathbf{X}(t_i) - \mathbf{S}(t_i)\mathbf{U}^{-1}(t_i)\mathbf{S}^T(t_i)] \geq \mathbf{0}$, $\mathbf{X}_f \geq \mathbf{0}$; $[\mathbf{G}_d(t_i)\mathbf{Q}_d(t_i)\mathbf{G}_d{}^T(t_i)] \geq \mathbf{0}$, $E\{\mathbf{w}_d(t_i)\mathbf{v}^T(t_j)\} = \mathbf{0}$, $\mathbf{P}_0 \geq \mathbf{0}$) is either *exponentially stable*, or *uniformly completely controllable* from the points of entry of *both* $\mathbf{u}(t_i)$ and $\mathbf{w}_d(t_i)$ and *uniformly completely reconstructible* from the points of extraction of *both* $\mathbf{y}_a(t_i)$ and $\mathbf{z}(t_i)$, then the closed-loop representation created by applying the LQG stochastic optimal regulator to the original system model is *exponentially stable.* In the time-invariant case, stabilizability with respect to $\mathbf{u}(t_i)$ and $\mathbf{w}_d(t_i)$ and detectability with respect to $\mathbf{y}_a(t_i)$ and $\mathbf{z}(t_i)$ is a sufficient set of conditions for the *asymptotic stability* of the closed-loop system model.

Due to computational considerations (to be discussed later in more detail), what is often considered for implementation is not the law given by (14-7), but a suboptimal

$$\mathbf{u}(t_i) = -\mathbf{G}_c{}^*(t_i)\hat{\mathbf{x}}(t_i{}^-) \tag{14-95}$$

with the same sufficient conditions for stability. This can be related directly to the generic cascading of an *arbitrary* full-state observer and *arbitrary* state feedback law to yield

$$\mathbf{u}(t_i) = -\mathbf{G}_c(t_i)\tilde{\mathbf{x}}(t_i{}^-) \tag{14-96a}$$

$$\begin{aligned}\tilde{\mathbf{x}}(t_{i+1}^-) = {} & \mathbf{\Phi}(t_{i+1}, t_i)\tilde{\mathbf{x}}(t_i{}^-) + \mathbf{B}_d(t_i)\mathbf{u}(t_i)\\ & + \tilde{\mathbf{K}}(t_i)[\mathbf{z}_i - \mathbf{H}(t_i)\tilde{\mathbf{x}}(t_i{}^-) - \mathbf{D}_z(t_i)\mathbf{u}(t_i)]\end{aligned} \tag{14-96b}$$

with gains $\mathbf{G}_c(t_i)$ and $\tilde{\mathbf{K}}_c(t_i)$ chosen according to some appropriate criteria [92, 120, 184]. Sufficient conditions for the *existence* of gain matrices $\mathbf{G}_c(t_i)$ and $\tilde{\mathbf{K}}(t_i)$ for all $t_i \in [t_0, t_N]$ such that the closed-loop system model is *exponentially stable* are either *exponential stability* of the open-loop system model, or its being *uniformly completely controllable* with respect to $\mathbf{u}(t_i)$ and *uniformly completely reconstructible* with respect to $\mathbf{z}(t_i)$. In the *time-invariant* system case, the eigenvalues of the closed-loop system simply consist of the regulator poles and observer poles, i.e., the eigenvalues of $[\mathbf{\Phi} - \mathbf{B}_d\mathbf{G}_c]$ and $[\mathbf{\Phi} - \tilde{\mathbf{K}}\mathbf{H}]$, respectively. If and only if the open-loop system model is both *stabilizable* with respect to $\mathbf{u}(t_i)$ and *detectable* with respect to $\mathbf{z}(t_i)$, there do exist gains $\mathbf{G}_c$ and $\tilde{\mathbf{K}}$ that can provide asymptotic closed-loop stability; making the stronger assumptions of *complete controllability and complete observability* allow the poles of both the regulator and observer to be placed arbitrarily (with the complex conjugate pair restriction). Assigning all of them to the origin yields the "output feedback state deadbeat control system," able to reduce the initial state to zero in at most $2n$ steps in the absence of noises and uncertainties.

## 14.5 STABILITY ROBUSTNESS OF LQG REGULATORS

Notice the careful couching of all stability claims in the previous section. All were based upon the *unrealistic* premise that the system model, upon which are based the regulator, filter, and cascade thereof, is a totally adequate depiction of the real-world system to be controlled. The closed-loop system under consideration was consistently the original system *model* (14-1) with appropriate feedback controller supplying the $\mathbf{u}(t_i)$. Since no model is perfect, it is important to ask how stabilizing properties of a given *fixed* controller function are affected by variations in the real-world system from design conditions upon which the controller was based. This is the important issue of *stability robustness*, specification of finite "regions" of models about a nominal model (not merely infinitesimal regions as in small sensitivity analyses) for which stability is preserved. It addresses all possible variations from design conditions: (1) *parameter variations*, in which the basic model structure is adequate, but defining matrices $\mathbf{\Phi}$, $\mathbf{B}_d$, $\mathbf{G}_d$, $\mathbf{H}$, $\mathbf{D}_z$, $\mathbf{Q}_d$, $\mathbf{R}$, and $\mathbf{P}_0$ might differ from assumed values; (2) *order reduction of the design model* from a higher-dimensional linear "truth model" to produce a computationally advantageous controller, but one which ignores certain modes of system performance (this is *always* an issue, since there are no $n$-dimensional physical systems, only $n$-dimensional models deemed to be adequate representations of those systems); and (3) *inadequacy of the linearity assumption itself*, caused by knowingly or unknowingly ignoring nonlinear effects in order to generate an LQG model that allows controller synthesis. These same variations will cause other important *performance degradations*, and these will be assessed subsequently, but the potential loss of closed-loop system stability is such a critical issue that it should be investigated first.

To understand some of the basic underlying concepts, consider a *time-invariant* linear system representation. For single-input/single-output models of this form, classical frequency domain methods and associated graphical depictions of system characteristics such as Bode and Nyquist plots can readily display minimum changes in the frequency response that will yield instability [23, 125]. Gain margin and phase margin [23, 92, 125, 150] are typical specifications, though they do not completely describe the stability margin characteristics (e.g., on a Nyquist plot, they correspond only to vertical and horizontal displacements from the origin, rather than shifts in arbitrary directions). For either single-input/single-output or multiple-input/multiple-output system models [13, 110–112, 114, 116, 144–146], the robustness issue centers on the properties of loop gain and return difference mappings [40, 92, 110, 154]. These are understood in terms of Fig. 14.2 of a typical system configuration in which $\mathbf{G}_1(z)$ can be considered an equivalent discrete-time model of the system to be controlled and $\mathbf{G}_2(z)$ can represent the controller itself. The $z$-transform transfer functions as shown can be related to state models, as in (14-85), and are readily extended to multiple-input/multiple-output cases and direct feedthroughs into outputs ($\mathbf{y}_c = \mathbf{Cx} + \mathbf{D}_y\mathbf{u}$ simply yields a transfer function of $[\mathbf{C}(z\mathbf{I} - \mathbf{\Phi})^{-1}\mathbf{B}_d + \mathbf{D}_y]$ in (14-85e)). Conceptually, if we cut the "loop" of interconnections as

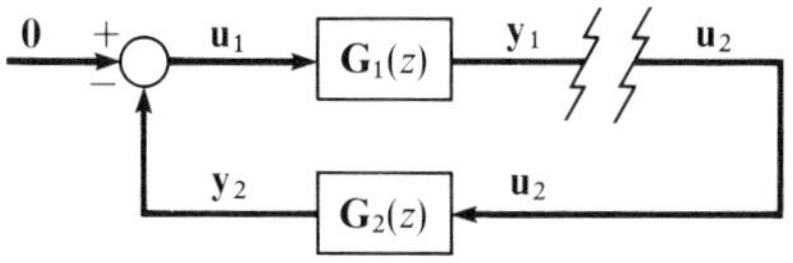


FIG. 14.2 Loop gain and return difference.

shown, and inject an input $\mathbf{u}_2(t)$, then the "returned variable" $\mathbf{y}_1(t)$ from the system can be written as

$$\mathbf{y}_1(z) = -\mathbf{G}_1(z)\mathbf{G}_2(z)\mathbf{u}_2(z) \triangleq -\mathbf{G}_L(z)\mathbf{u}_2(z) \tag{14-97}$$

in terms of the *loop gain* $\mathbf{G}_L(z)$,

$$\mathbf{G}_L(z) = \mathbf{G}_1(z)\mathbf{G}_2(z) \tag{14-98}$$

The difference between the injected $\mathbf{u}_2(t)$ and returned $\mathbf{y}_1(t)$ is then given by

$$\mathbf{u}_2(z) - \mathbf{y}_1(z) = [\mathbf{I} + \mathbf{G}_L(z)]\mathbf{u}_2(z) \tag{14-99}$$

in terms of the *return difference transformation* $[\mathbf{I} + \mathbf{G}_L(z)]$. Note that cutting the loop at different locations yields different loop gains; for instance, cutting it just to the right of the summing junction yields a loop gain of $[\mathbf{G}_2(z)\mathbf{G}_1(z)]$ and a corresponding change to the return difference. Thus, a relationship of primary interest for studying a loop that is cut at some particular point is given by (14-99), which is often written in terms of new variables **e** and **r** as

$$\mathbf{r}(z) = [\mathbf{I} + \mathbf{G}_L(z)]\mathbf{e}(z) \tag{14-100}$$

as diagrammed in Fig. 14.3a. To consider robustness, assume that the nominal system as just described is stable (in some sense). We are then interested in whether the closed-loop system retains this stability property subject to additive or multiplicative alterations in the loop gain causing $\mathbf{G}_L(z)$ to become $\{\mathbf{G}_L(z) + \Delta\mathbf{G}(z)\}$ or $\{\mathbf{G}_L(z)[\mathbf{I} + \Delta\mathbf{G}(z)]\}$, as depicted in Figs. 14.3b and 14.3c, respectively. From small-scale perturbation analysis [10, 30, 90, 92, 135, 136], it can be shown that the system is insensitive to additive alterations as in Fig. 14.3b and also to

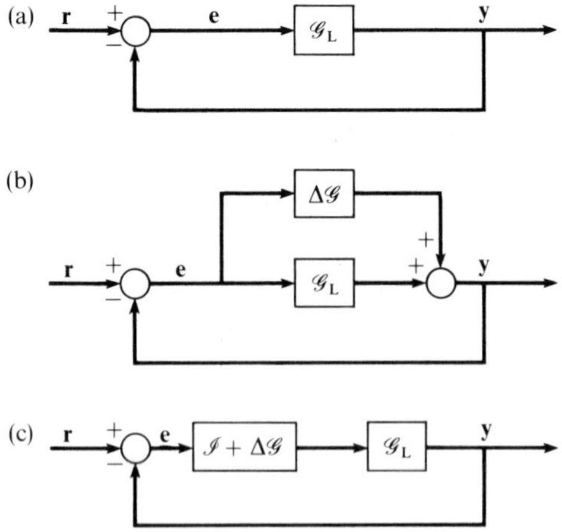


FIG. 14.3 Basic closed-loop system representation and alterations. (a) Nominal system representation. (b) System with additive alterations. (c) System with multiplicative alterations.

dynamics disturbances over a frequency band of interest if the return differences $[\mathbf{I} + \mathbf{G}_L(\exp\{2\pi j\omega/\omega_s\})]$ is small (in the sense of simple magnitude for single input/single output descriptions, or an appropriate norm more generally) for all $\omega$ in that frequency band, where $\omega_s$ is the fixed sampling frequency $(2\pi/\Delta t$, where $\Delta t$ is the sample period). But we are interested in more than just *small* perturbations here, so we must proceed further.

Notice that although Fig. 14.3 was motivated by $z$-transform equations (14-97)–(14-100), the diagrams themselves are expressed in terms of the *transformations* (*operators*) $\mathscr{G}_L, \Delta\mathscr{G}$, and $\mathscr{I}$ that are *represented by matrix multiplications* $\mathbf{G}_L(z)$, $\Delta\mathbf{G}(z)$, and $\mathbf{I}$, respectively. These diagrams, and Eqs. (14-97)–(14-100) themselves, can be understood in this more general operator context, which is the standard context for studying BIBO stability [36, 153, 190, 192]. Thus, (14-100) becomes

$$(\mathscr{I} + \mathscr{G}_L)\mathbf{e} = \mathbf{r} \tag{14-101}$$

and thus the return difference operator $(\mathscr{I} + \mathscr{G}_L)$ maps the function $\mathbf{e}$ into the function $\mathbf{r}$. To be more precise, let $\mathscr{B}$ be some (Banach) space of functions $\mathbf{u}(\cdot)$ that map the time set $T$ into a finite-dimensional vector space, as $R^r$, with an appropriate norm $(\|\cdot\|_{\mathscr{B}})$ associated with it to specify the "length" of these functions appropriately; let $\mathscr{B}_e$ be the extension of this space defined by

$$\mathscr{B}_e = \{\mathbf{u}(\cdot) : \mathscr{P}_\tau \mathbf{u} \in \mathscr{B} \text{ for all } \tau \in T\} \tag{14-102a}$$

where

$$(\mathscr{P}_\tau \mathbf{u})(t) = \begin{cases} \mathbf{u}(t) & \text{for} \quad t \le \tau \\ \mathbf{0} & \text{for} \quad t > \tau \end{cases} \tag{14-102b}$$

$\mathscr{I}$ is the identity mapping from $\mathscr{B}_e$ into itself, and $\mathscr{G}_L$ also maps $\mathscr{B}_e$ into itself. If $\mathscr{P}_\tau \mathscr{G} \mathscr{P}_\tau = \mathscr{P}_\tau \mathscr{G}$ for all $\tau$, $\mathscr{G}$ is termed causal, and this corresponds physically to a nonanticipative system, i.e., one that does not respond to inputs before they arrive. If a system model loop gain is given by the *causal linear operator* $\mathscr{G}_L$, then the closed-loop system model of Fig. 14.3a is *BIBO stable* if the mapping $(\mathscr{I} + \mathscr{G}_L)^{-1}$ from $\mathscr{B}_e$ into itself *exists*, is *causal*, and is a *bounded linear operator* when restricted to $\mathscr{B} \subset \mathscr{B}_e$: for each $\mathbf{r} \in \mathscr{B}_e$ there exists a unique causally related $\mathbf{e} \in \mathscr{B}_e$ satisfying (14-101), and also for each $\mathbf{r} \in \mathscr{B} \subset \mathscr{B}_e$, the corresponding $\mathbf{e}$ and $(\mathbf{r} - \mathbf{e}) = \mathbf{y}$ are also elements of $\mathscr{B} \subset \mathscr{B}_e$. Notice that this formulation admits infinite-dimensional as well as finite-dimensional state space representations, and therefore can address continuous-time systems with time delays and other important applications.

An important result for robustness proven in [153] is as follows. Let $\mathscr{A}$ be a causal linear operator mapping $\mathscr{B}_e$ into $\mathscr{B}_e$, and assume $\mathscr{A}^{-1}$ exists, is causal, and is bounded when restricted to $\mathscr{B} \subset \mathscr{B}_e$. Then if $\Delta\mathscr{A}$ is a causal linear operator from $\mathscr{B}_e$ into $\mathscr{B}_e$ that is bounded when restricted to $\mathscr{B}$, and if

$$\|\mathscr{A}^{-1}\Delta\mathscr{A}\|_{\mathscr{B}} < 1 \tag{14-103}$$

then $(\mathscr{A} + \Delta\mathscr{A})^{-1}$ exists as a mapping from $\mathscr{B}_e$ into $\mathscr{B}_e$, is causal, and is bounded when restricted to $\mathscr{B} \subset \mathscr{B}_e$. (Moreover, the linearity of $\Delta\mathscr{A}$ is not essential.) Thus, if the system represented by Fig. 14.3a is stable, then the system with additive alterations as in Fig. 14.3b remains stable provided

$$\|(\mathscr{I} + \mathscr{G}_L)^{-1}\Delta\mathscr{G}\|_{\mathscr{B}} < 1 \tag{14-104a}$$

and the system with multiplicative alterations as in Fig. 14.3c remains stable provided

$$\|[\mathscr{I} - (\mathscr{I} + \mathscr{G}_L)^{-1}]\Delta\mathscr{G}\|_{\mathscr{B}} < 1 \tag{14-104b}$$

Note that (14-104a) follows directly from (14-103), and that

$$\begin{aligned}
&[\mathscr{I} + \mathscr{G}_L(\mathscr{I} + \Delta\mathscr{G})]\mathbf{e} \\
&\quad = [\mathscr{I} + \mathscr{G}_L + (\mathscr{I} + \mathscr{G}_L)\Delta\mathscr{G} - \Delta\mathscr{G}]\mathbf{e} \\
&\quad = [(\mathscr{I} + \mathscr{G}_L)\{\mathscr{I} + \Delta\mathscr{G} - (\mathscr{I} + \mathscr{G}_L)^{-1}\Delta\mathscr{G}\}]\mathbf{e} \\
&\quad = [(\mathscr{I} + \mathscr{G}_L)\{\mathscr{I} + [\mathscr{I} - (\mathscr{I} + \mathscr{G}_L)^{-1}]\Delta\mathscr{G}\}]\mathbf{e} \\
&\quad = [(\mathscr{I} + \mathscr{G}_L)[\mathscr{I} - (\mathscr{I} + \mathscr{G}_L)^{-1}]^{-1}\{[\mathscr{I} - (\mathscr{I} + \mathscr{G}_L)^{-1}]^{-1} + \Delta\mathscr{G}\}]\mathbf{e}
\end{aligned}$$

and (14-104b) guarantees the invertibility of $\{[\mathscr{I} - (\mathscr{I} + \mathscr{G}_L)^{-1}]^{-1} + \Delta\mathscr{G}\}$ and thus of $[\mathscr{I} + \mathscr{G}_L(I + \Delta\mathscr{G})]$ as desired. Since $\|\mathscr{A}^{-1}\Delta\mathscr{A}\|_{\mathscr{B}} \le \|\mathscr{A}^{-1}\|_{\mathscr{B}}\|\Delta\mathscr{A}\|_{\mathscr{B}}$, a sufficient condition for (14-103) is

$$\|\Delta\mathscr{A}\|_{\mathscr{B}} < 1/\|\mathscr{A}^{-1}\|_{\mathscr{B}} \tag{14-105}$$

and thus a *sufficient condition for stability with additive alterations* $\Delta\mathscr{G}$ is

$$\|\Delta\mathscr{G}\|_{\mathscr{B}} < 1/\|(\mathscr{I} + \mathscr{G}_L)^{-1}\|_{\mathscr{B}} \tag{14-106a}$$

and, for *multiplicative alterations* $\Delta\mathscr{G}$ is

$$\|\Delta\mathscr{G}\|_{\mathscr{B}} < 1/\|\mathscr{I} - (\mathscr{I} + \mathscr{G}_L)^{-1}\|_{\mathscr{B}} \tag{14-106b}$$

At this point, it is apparent that these robustness results are intimately tied to a basic result of numerical analysis that states: If an $r$-by-$r$ matrix $\mathbf{A}$ is invertible, then $(\mathbf{A} + \Delta\mathbf{A})$ is invertible for all $\Delta\mathbf{A}$ such that

$$\sigma_{\max}(\Delta\mathbf{A}) = \|\Delta\mathbf{A}\| < \frac{1}{\|\mathbf{A}^{-1}\|} = \sigma_{\min}(\mathbf{A}) \tag{14-107}$$

where the spectral norm $\|\cdot\|$ for $r$-by-$r$ matrix $\mathbf{A}$ is defined by

$$\|\mathbf{A}\| = \max_{\|\mathbf{x}\|=1} \|\mathbf{A}\mathbf{x}\| \tag{14-108a}$$

where for any $\mathbf{y}$ in complex $r$-dimensional space

$$\|\mathbf{y}\| = \left(\sum_{i=1}^{r} |y_i|^2\right)^{1/2} = (\mathbf{y}^*\mathbf{y})^{1/2} \tag{14-108b}$$

and ( )* denotes a conjugate transpose, and $\sigma_{\max}$ and $\sigma_{\min}$ are the maximum and minimum *singular values* of their matrix arguments. *Singular value decomposition* (*SVD*) is an important and practical tool [80, 172] in numerical linear algebra that can be applied to nonsquare matrices in general; the special case of the SVD of an $r$-by-$r$ nonsingular matrix $\mathbf{A}$ is given by

$$\mathbf{A} = \mathbf{U}\mathbf{\Sigma}\mathbf{V}^* \tag{14-109}$$

where $\mathbf{U}$ and $\mathbf{V}$ are unitary $r$-by-$r$ matrices ($\mathbf{U}^*\mathbf{U} = \mathbf{I}$, $\mathbf{V}^*\mathbf{V} = \mathbf{I}$) and $\mathbf{\Sigma}$ is a diagonal matrix with diagonal terms as the singular values $\sigma_i$, the nonnegative square roots of the eigenvalues of $\mathbf{A}^*\mathbf{A}$, often arrayed in descending order: $\sigma_{\max} = \sigma_1 \geq \sigma_2 \geq \cdots \geq \sigma_r = \sigma_{\min}$. It should be noted that other valid matrix norms exist besides (14-108); for instance the Frobenius norm [171]

$$\|\mathbf{A}\| = \left( \sum_{i=j}^{r} \sum_{j=1}^{r} |A_{ij}|^2 \right)^{1/2} = (\mathrm{tr}\, \mathbf{A}^*\mathbf{A})^{1/2} \tag{14-108$'$}$$

is a more pessimistic norm that is greater than or equal to the spectral norm, that also satisfies the desirable consistency condition $\|\mathbf{AB}\| \leq \|\mathbf{A}\|\,\|\mathbf{B}\|$, and that is often used for computations because of its simpler evaluation.

For *time-invariant* cases, in which the various operators in (14-106) are representable as convolution summations, this linkage to singular value decompositions can be exploited to yield practical tests for robustness. Consider the operator $\mathscr{G}$ defined by the convolution summation

$$(\mathscr{G}\mathbf{u})(t_i) = \sum_{k=0}^{\infty} \mathbf{G}(t_i - t_k)\mathbf{u}(t_k) \tag{14-110}$$

where the elements of the impulse response matrix are assumed to be absolutely summable ($\sum_{k=0}^{\infty} |G_{ij}(t_k)| < \infty$), with operator norm $\|\mathscr{G}\|$ (called $l_2{}^r$ norm) of $[\sum_{k=0}^{\infty} \|\mathbf{G}(t_k)\|^2]^{1/2}$ with $\|\mathbf{G}(t_k)\|$ as defined previously. Then, as a result of the discrete form of *Parseval's theorem* for a real scalar sequence $\{f(k)\}$ with $\sum_{k=0}^{\infty} f^2(k) < \infty$, namely,

$$\sum_{k=0}^{\infty} f^2(k) = \frac{1}{2\pi j} \oint F(z)F(z^{-1})z^{-1}\, dz \tag{14-111}$$

where the line integral is taken around the unit circle and evaluated by residues [23, 27], it can be shown that

$$\|\mathscr{G}\|_{l_2{}^r} = \sigma^*_{\max} \tag{14-112a}$$

where

$$\begin{aligned} \sigma^*_{\max} &= \max_{0 \leq \omega < \omega_s/2} \sigma_{\max}[\mathbf{G}(\exp\{2\pi j\omega/\omega_s\})] \\ &= \max_{0 \leq \omega < \omega_s/2} \max_{1 \leq i \leq r} \sigma_i[\mathbf{G}(\exp\{2\pi j\omega/\omega_s\})] \end{aligned} \tag{14-112b}$$

and $\sigma_i[\mathbf{G}(\exp\{2\pi j\omega/\omega_s\})]$ denotes the $i$th singular value of the transfer function $\mathbf{G}(z)$ corresponding to $\mathscr{G}$, evaluated at $z = \exp\{2\pi j\omega/\omega_s\}$ (somewhere on the upper half of the unit circle in the complex $z$ plane), and $\omega_s$ is the fixed sampling frequency, $2\pi/\Delta t$. It is also assumed that the sampling rate $\omega_s$ is chosen to be at least twice the frequency of any signal content of interest for proper representation in the equivalent discrete-time model, as dictated by the Shannon sampling theorem. (Classically, this corresponds to all poles and zeros of the underlying continuous-time system lying within the "primary strip" of the $s$ plane between $j\omega_s/2$ and $-j\omega_s/2$ that gets mapped into the $z$ plane. If $\omega_s$ is improperly chosen so that singularities lie outside this primary strip, "folding" occurs in which the primary strip corresponding to the discrete-time representation has superimposed on it the singularities of all strips in the $s$ plane of width $j\omega_s$ and centered at $j[\omega_s/2 \pm N\omega_s]$ for integer $N$: a misrepresentation known as "aliasing.") Combining (14-106), (14-107), and (14-112) yields the *sufficient condition for stability with additive alterations* $\Delta\mathscr{G}$ represented as $\mathbf{\Delta G}(z)$ given by

$$\sigma_{\max}[\mathbf{\Delta G}(\exp\{2\pi j\omega/\omega_s\})] < \sigma_{\min}[\mathbf{I} + \mathbf{G}_L(\exp\{2\pi j\omega/\omega_s\})] \tag{14-113a}$$

for all $\omega \in [0, \omega_s/2)$, and for *multiplicative alterations* by

$$\sigma_{\max}[\mathbf{\Delta G}(\exp\{2\pi j\omega/\omega_s\})] < \sigma_{\min}[\mathbf{I} + \mathbf{G}_L^{-1}(\exp\{2\pi j\omega/\omega_s\})] \tag{14-113b}$$

for all $\omega \in [0, \omega_s/2)$. Note that (14-113b) has used

$$\sigma_{\min}[\mathbf{I} + \mathbf{A}^{-1}] = 1/\|(\mathbf{I} + \mathbf{A}^{-1})^{-1}\|$$

and the matrix inversion lemma to write, for invertible $\mathbf{A}$,

$$(\mathbf{I} + \mathbf{A}^{-1})^{-1} = \mathbf{I} - (\mathbf{I} + \mathbf{A})^{-1} \tag{14-114}$$

The minimum singular values in the strict inequalities of (14-113) can be readily computed and plotted as a function of $\omega$, and such a plot plays the same role as the classical graphical presentations play for establishing robustness properties of single input/single output models, as discussed earlier. Since classical gain and phase margins correspond to multiplicative alterations, $\sigma_{\min}[\mathbf{I} + \mathbf{G}_L^{-1}(\exp\{2\pi j\omega/\omega_s\})]$ is a particularly useful measure. Although this generally cannot be expressed directly in terms of the minimum singular value in (14-113a), Eq. (14-114) can be used to show that

$$\|(\mathbf{I} + \mathbf{A}^{-1})^{-1}\| - \|(\mathbf{I} + \mathbf{A})^{-1}\| \le 1 \tag{14-115}$$

and thus these singular values are related by the inequality [97, 128]

$$\sigma_{\min}(\mathbf{I} + \mathbf{A}^{-1}) \ge \frac{\sigma_{\min}(\mathbf{I} + \mathbf{A})}{1 + \sigma_{\min}(\mathbf{I} + \mathbf{A})} \tag{14-116}$$

This will be used to great advantage subsequently.

We want to establish and compare the robustness characteristics of the optimal LQ full-state feedback law (14-6) in steady state as given by (14-79), the optimal LQG stochastic controller (14-7) in steady state, and the suboptimal law (14-95). Since the signals extracted from the physical system differ in dimension among these cases ($n$ versus $m$), the appropriate point to cut the loop for those comparisons is at the point where $\mathbf{u}(t_i) \in R^r$ leaves any of the controllers and enters the physical system. This point is marked on Figs. 14.4–14.6. (Note that *very misleading* conclusions can be drawn if we artificially break the loop just to the right of the identity matrix in the forward path of each controller [41].) In the measurement models for Figs. 14.5 and 14.6, it is assumed that $\mathbf{D}_z = \mathbf{0}$ (see (14-32)). The $z$-transform representation of an equivalent discrete-time model for the real system was developed in (14-85), and the analogous representation for the Kalman filter is found by taking the $z$-transform of

$$\hat{\mathbf{x}}(t_{i+1}^{+}) = \mathbf{\Phi}\hat{\mathbf{x}}(t_i^{+}) + \mathbf{B}_{\mathrm{d}}\mathbf{u}(t_i) + \mathbf{K}\{\mathbf{z}_{i+1} - \mathbf{H}[\mathbf{\Phi}\hat{\mathbf{x}}(t_i^{+}) + \mathbf{B}_{\mathrm{d}}\mathbf{u}(t_i)]\} \tag{14-117}$$

for Fig. 14.5, and for Fig. 14.6:

$$\hat{\mathbf{x}}(t_{i+1}^{-}) = \mathbf{\Phi}[\hat{\mathbf{x}}(t_i^{-}) + \mathbf{K}\{\mathbf{z}_i - \mathbf{H}\hat{\mathbf{x}}(t_i^{-})\}] + \mathbf{B}_{\mathrm{d}}\mathbf{u}(t_i) \tag{14-118}$$

Finally, note that the three controllers have been drawn in a manner similar to Figs. 13.3 and 13.5, foretelling controller structures in which an input command $\mathbf{u}_{\mathrm{com}}$ need not be zero and in which constant matrices $\mathbf{I}$ and $\bar{\mathbf{G}}_{\mathrm{c}}^*$ might be time-varying matrices or even dynamic compensators: These forms will be discussed later.

For the *full-state LQ optimal steady state feedback law* of Fig. 14.4, certain *robustness guarantees* can be made: it is inherently a robust design, though generally not as robust as continuous-time full-state feedback laws. The following result has been proven [134, 149–151] for the case of $\mathbf{S} \equiv \mathbf{0}$ in (14-13), i.e., for the simplest regulator as discussed in Chapter 13, and assuming that $\mathbf{U}$ is diagonal with diagonal elements $U_{ii} > 0$ for $i = 1, 2, \ldots, r$. For concise statement of results, parameters $a_1, a_2, \ldots, a_r$ are defined by

$$a_i \triangleq \sqrt{U_{ii}/[U_{ii} + \lambda_{\max}(\mathbf{B}_{\mathrm{d}}^{\mathrm{T}}\bar{\mathbf{K}}_{\mathrm{c}}\mathbf{B}_{\mathrm{d}})]} \tag{14-119}$$

where $\lambda_{\max}(\cdot)$ indicates the maximum eigenvalue of a matrix. Let a linear time-invariant finite-gain scalar input/scalar output device with $z$-transfer function $L_i(z)$, $i = 1, 2, \ldots, r$, be placed on each of the $r$ paths at the loop break point in Fig. 14.4. Then the closed-loop system is stable if the $z$-transform Nyquist locus of every $L_i(z)$ in the complex plane, $\{L_i(\exp[j\theta]), \theta \in [0, \pi]\}$, lies entirely in the circle of radius $[a_i/(1 - a_i^2)]$ centered at the point $[1/(1 - a_i^2) + j0]$ in

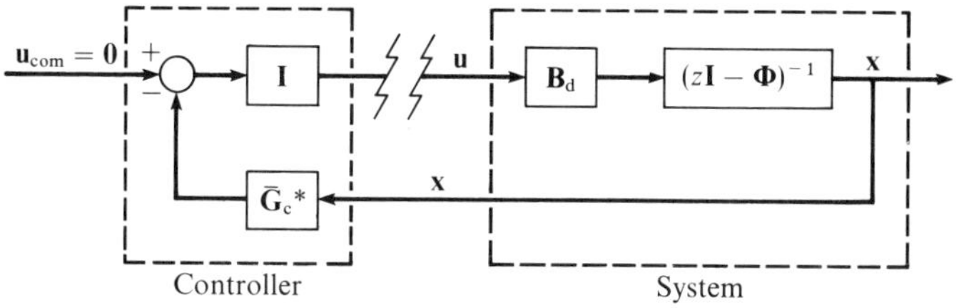


FIG. 14.4 Full-state feedback law $\mathbf{u}(t_i) = -\bar{\mathbf{G}}_{\mathrm{c}}^*\mathbf{x}(t_i)$.

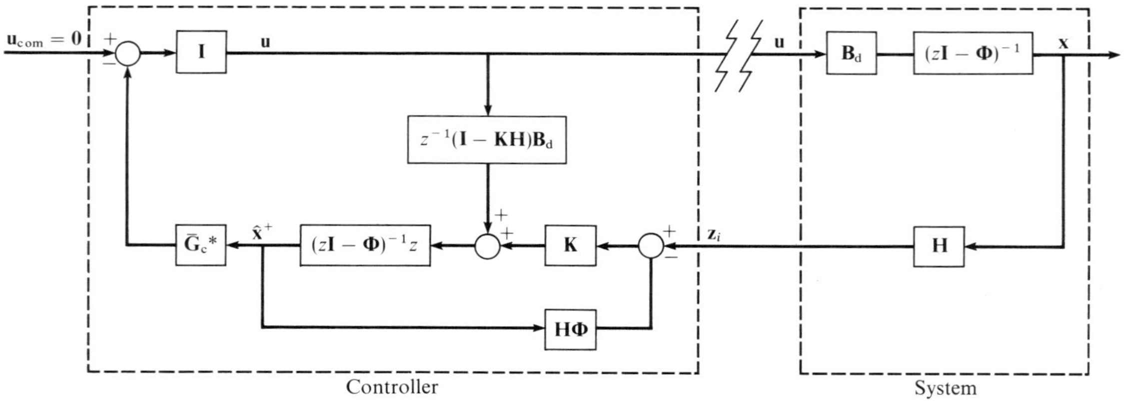


FIG. 14.5 Optimal LQG law $\mathbf{u}(t_i) = -\bar{\mathbf{G}}_c{}^*\hat{\mathbf{x}}(t_i{}^+)$.

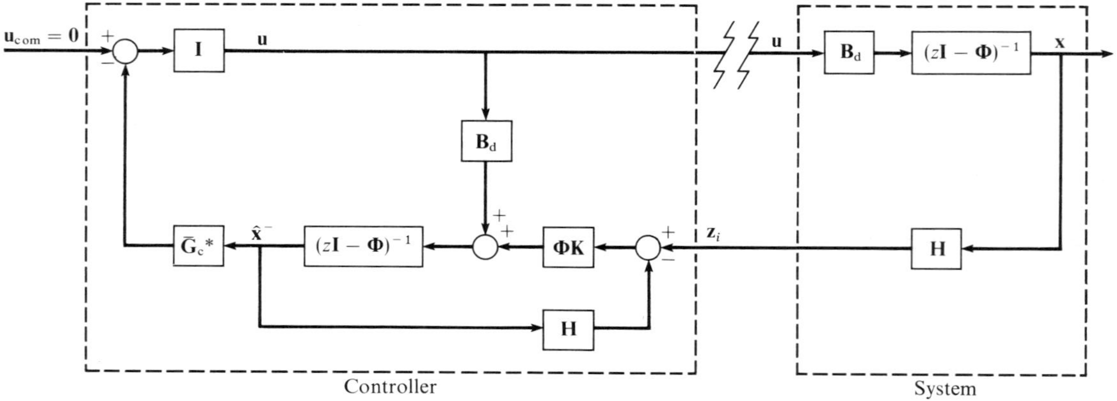


FIG. 14.6 Suboptimal law $\mathbf{u}(t_i) = -\bar{\mathbf{G}}_c{}^*\hat{\mathbf{x}}(t_i{}^-)$.

the complex plane. As a special case corresponding to specification of *simultaneous minimal gain-margins*, let $L_i(z)$ be a simple gain $k_i$ for all $i$; then stability is preserved if

$$\frac{1}{1+a_i} \leq k_i \leq \frac{1}{1-a_i}, \qquad i = 1, 2, \ldots, r \tag{14-120a}$$

On the other hand, if a pure phase shift of angle $\phi_i$ is inserted into each channel by letting $L_i(z) = e^{j\phi_i}$, then stability is preserved under the *simultaneous minimal phase-margin* conditions

$$|\phi_i| \leq \arcsin(a_i/2), \qquad i = 1, 2, \ldots, r \tag{14-120b}$$

Moreover, when $\lambda_{\max}(\mathbf{B}_d{}^T\bar{\mathbf{K}}_c\mathbf{B}_d) \ll U_{ii}$ for all $i$, which will be the case in the limit as the sample period goes to zero, then $a_i \rightarrow 1$, and (14-120a) yields a gain margin that goes to *infinity* and a gain reduction tolerance that goes to *at least* 50% in each control input channel, and (14-120b) yields a phase margin that approaches *at least* $\pm 60°$.

These limiting characteristics correspond directly to the stability margins of steady state LQ optimal state feedback laws for the case of continuous perfect measurement of state, and time-invariant models. (These will be studied in more detail in Section 14.13.) For the continuous-time case, (14-101)–(14-106) are still valid, but (14-110) is replaced by

$$(\mathscr{G}\mathbf{u})(t) = \int_0^\infty \mathbf{G}(t-\tau)\mathbf{u}(\tau)\,d\tau \tag{14-121}$$

where the elements of the impulse response function matrix $\mathbf{G}$ are assumed to be absolutely integrable; also the $\mathscr{L}_2{}^r$ norm of the operator $\mathscr{G}$ defined as $[\int_0^\infty \|\mathbf{G}(t)\|^2\,dt]^{1/2}$ is equal to $\sigma_{\max}^*$ by Parseval's theorem in continuous form, where

$$\sigma_{\max}^* = \max_{\omega \geq 0}\ \max_{1 \leq i \leq r}\ \sigma_i[\mathbf{G}(j\omega)] \tag{14-122}$$

and $\sigma_i[\mathbf{G}(j\omega)]$ denotes the $i$th singular value of the Laplace transfer function $\mathbf{G}(s)$ corresponding to $\mathscr{G}$, evaluated at $s = j\omega$ (somewhere on the positive imaginary axis in the complex $s$ plane). Then the corresponding *sufficient condition for stability with additive alterations* $\Delta\mathscr{G}$ represented by $\Delta\mathbf{G}(s)$ is

$$\sigma_{\max}[\Delta\mathbf{G}(j\omega)] < \sigma_{\min}[\mathbf{I} + \mathbf{G}_L(j\omega)], \qquad \omega > 0 \tag{14-123a}$$

and *for multiplicative alterations*,

$$\sigma_{\max}[\Delta\mathbf{G}(j\omega)] < \sigma_{\min}[\mathbf{I} + \mathbf{G}_L^{-1}(j\omega)], \qquad \omega > 0 \tag{14-123b}$$

Consider the LQ state regulator for $\dot{\mathbf{x}}(t) = \mathbf{F}\mathbf{x}(t) + \mathbf{B}\mathbf{u}(t)$ choosing $\mathbf{W}_{xx} = \mathbf{C}^T\mathbf{C}$ and $\mathbf{W}_{uu} = \mathbf{I}$ in (14-20) and requiring that $\mathbf{W}_{xu} \equiv \mathbf{0}$; also assume the system with the given state equation and artificial output $\mathbf{C}\mathbf{x}(t)$ is completely observable and controllable. It will be shown in Section 14.13 that the steady state LQ optimal regulator is then

$$\mathbf{u}^*(t) = -\mathbf{B}^{\mathrm{T}}\bar{\mathbf{K}}_{\mathrm{c}}\mathbf{x}(t) \tag{14-124a}$$

where $\bar{\mathbf{K}}_{\mathrm{c}}$ is the unique positive definite solution to the algebraic Riccati equation

$$\mathbf{0} = \mathbf{F}^{\mathrm{T}}\bar{\mathbf{K}}_{\mathrm{c}} + \bar{\mathbf{K}}_{\mathrm{c}}\mathbf{F} + \mathbf{C}^{\mathrm{T}}\mathbf{C} - \bar{\mathbf{K}}_{\mathrm{c}}\mathbf{B}\mathbf{B}^{\mathrm{T}}\bar{\mathbf{K}}_{\mathrm{c}} \tag{14-124b}$$

From the equality [68] for the return difference matrix for this loop,

$$[\mathbf{I}+\mathbf{G}_{\mathrm{L}}(-s)]^{\mathrm{T}}[\mathbf{I}+\mathbf{G}_{\mathrm{L}}(s)]=\mathbf{I}+[\mathbf{C}(-s\mathbf{I}-\mathbf{F})^{-1}\mathbf{B}]^{\mathrm{T}}[\mathbf{C}(s\mathbf{I}-\mathbf{F})^{-1}\mathbf{B}]\geq\mathbf{I} \tag{14-125}$$

it is readily shown that $\sigma_{\min}[\mathbf{I} + \mathbf{G}_{\mathrm{L}}(j\omega)] \geq 1$ for all $\omega > 0$, and thus by (14-123a), the closed-loop system is very robust to additive perturbations. Furthermore, (14-116) then yields [128], that $\sigma_{\min}[\mathbf{I} + \mathbf{G}_{\mathrm{L}}^{-1}(j\omega)] \geq \frac{1}{2}$ so that stability is preserved for multiplicative alterations $\Delta\mathscr{G}$ such that

$$\sigma_{\max}[\Delta G(j\omega)] < 1/2 \qquad \text{for all} \quad \omega \geq 0 \tag{14-126}$$

as indicated by (14-120) in the limit as $a_i \to 1$.

All of the preceding robustness characteristics are *guaranteed* properties for general constant-gain LQ *state feedback* laws: individual designs can achieve even better characteristics, as can be evaluated through (14-113) or (14-123). However, care must be taken in interpreting these results. First, inserting pure gain or pure phase shift into the loop to yield gain and phase margins corresponds to displacements of a Nyquist plot locus in only two of an infinite number of possible directions as discussed previously, and these do not adequately describe stability under simultaneous gain and phase changes. Second, these types of multiplicative alteration do *not address the adequacy of the linear model of the system*: ignored states or nonlinearities are not included in $\Delta\mathscr{G}$, and these can seriously affect stability. For instance, an $n$th order linear model might be an adequate representation of structural bending properties within some frequency range, but a stable and robust controller based on this model may not provide the "guaranteed margins" because of the effect of higher order modes outside that model's frequency realm of adequacy. There are always unmodeled dynamics in a real system that will contribute significant phase and attenuation at a sufficiently high frequency, but effects on stability are not serious if the unmodeled effects are well separated in frequency from the states in the design model and if the loop transfer function bandwidth is not too high [56]. Thus, good model generation before the LQ regulator is designed, and purposeful limiting of bandwidth during or after the design procedure, are usually sought [168]; but in some applications such as control of large-scale space structures with many closely spaced flexure modes, this cannot be accomplished with complete success.

An important point is that when a filter or observer is cascaded with the full-state feedback law, there are *no* guaranteed stability margins [39, 41, 124]. It is the dependence of the filter or observer upon an internal system model to generate a state estimate that degrades the robustness properties of the resulting feedback control law. As a result, it is particularly important to evaluate (14-113) or (14-123) to display the actual robustness of a proposed design.

It is in fact possible to generate an observer to cascade with the full-state feedback law that *recovers some of the robustness properties* usually lost with filters or observers in the loop [29, 31, 41, 90, 92, 102]. Let the system model be completely observable and controllable. Consider the control law depicted in Fig. 14.6: what is desired is that the return difference matrix $[\mathbf{I} + \mathbf{G}_L(z)]$ of this law duplicate the return difference of the full-state feedback law of Fig. 14.4. It can be readily shown that this is achieved if an observer gain $\tilde{\mathbf{K}} = \mathbf{\Phi K}$ can be found that satisfies

$$\mathbf{\Phi K}[\mathbf{I} + \mathbf{H}(z\mathbf{I} - \mathbf{\Phi})^{-1}\mathbf{\Phi K}]^{-1} = \mathbf{B}_d[\mathbf{H}(z\mathbf{I} - \mathbf{\Phi})^{-1}\mathbf{B}_d]^{-1} \tag{14-127}$$

If $\mathbf{K}(q)$, parameterized as a function of a scalar $q$, were selected such that

$$\lim_{q \to \infty} \mathbf{\Phi K}/q = \mathbf{B}_d \mathbf{W} \tag{14-128}$$

for any nonsingular $m$-by-$m$ $\mathbf{W}$, then (14-127) is satisfied asymptotically as $q \to \infty$, as is verified by direct substitution. Thus, we choose (for finite $q$)

$$\mathbf{K} = q\mathbf{\Phi}^{-1}\mathbf{B}_d\mathbf{W} \tag{14-129}$$

One particular choice of $\mathbf{W}$ is motivated by considering a dual state equation

$$\mathbf{x}^d(t_{i+1}) = \mathbf{\Phi}^T\mathbf{x}^d(t_i) + \mathbf{H}^T\mathbf{v}(t_i) \tag{14-130a}$$

where $\mathbf{v}(t_i) = -\mathbf{K}^T\mathbf{\Phi}^T\mathbf{x}_d(t_i)$, and $\mathbf{K}^T$ is selected so as to minimize $[\sum_i \mathbf{y}^d(t_i)^T\mathbf{y}^d(t_i)]$ and $\mathbf{y}^d(t_i)$ is an artificial output generated as

$$\mathbf{y}^d(t_i) = \mathbf{D}^d\mathbf{x}^d(t_i) \tag{14-130b}$$

The solution to this control problem for the dual system is

$$\mathbf{K}^T = [\mathbf{D}^d\mathbf{H}^T]^{-1}\mathbf{D}^d \to \mathbf{K} = \mathbf{D}^{dT}[\mathbf{H}\mathbf{D}^{dT}]^{-1} \tag{14-131a}$$

provided the indicated inverse exists; with this $\mathbf{K}$, the observer's unforced error equation becomes

$$\mathbf{e}(t_{i+1}^-) = [\mathbf{\Phi} - \mathbf{\Phi K H}]\mathbf{e}(t_i^-) = [\mathbf{\Phi} - \mathbf{\Phi}\mathbf{D}^{dT}(\mathbf{H}\mathbf{D}^{dT})^{-1}\mathbf{H}]\mathbf{e}(t_i^-) \tag{14-131b}$$

and $m$ eigenvalues of this closed-loop system are assigned to the origin, and the remaining $(n - m)$ to the $(n - m)$ invariant zeros of the system (14-130) [29, 33, 113, 131]. Furthermore, since $\mathbf{HK} = \mathbf{I}$ for this choice of $\mathbf{K}$, the forced state estimation error equation obeys

$$\begin{aligned}\mathbf{He}(t_{i+1}^+) &= \mathbf{H}\{(\mathbf{I} - \mathbf{KH})[\mathbf{\Phi e}(t_i^+) + \mathbf{G}_d\mathbf{w}_d(t_i)] - \mathbf{Kv}(t_{i+1})\} \\ &= -\mathbf{v}(t_{i+1})\end{aligned} \tag{14-131c}$$

which has the physical interpretation of no filtering of the measurement noise taking place. If we choose $\mathbf{D}^d = \mathbf{B}_d^T\mathbf{\Phi}^{-T}$, then

$$\mathbf{K} = \mathbf{\Phi}^{-1}\mathbf{B}_d(\mathbf{H}\mathbf{\Phi}^{-1}\mathbf{B}_d)^{-1} \tag{14-131d}$$

thereby motivating a choice of $\mathbf{W} = (\mathbf{H}\mathbf{\Phi}^{-1}\mathbf{B}_d)^{-1}$ in (14-129). Other choices of $\mathbf{W}$ are also possible and warrant further investigation. See also Problem 14.16h.

In the continuous measurement case, an analogous development leads to equations similar to (14-127) and (14-128), namely [41],

$$\mathbf{K}[\mathbf{I} + \mathbf{H}(s\mathbf{I} - \mathbf{F})^{-1}\mathbf{K}] = \mathbf{B}[\mathbf{H}(s\mathbf{I} - \mathbf{F})^{-1}\mathbf{B}]^{-1} \tag{14-132}$$

as satisfied asymptotically by $\mathbf{K}$ selected such that

$$\lim_{q\to\infty} \frac{\mathbf{K}}{q} = \mathbf{B}\mathbf{W} \tag{14-133}$$

If the system model is *minimum phase*, then this can in fact be solved by means of a *Kalman filter* in which the dynamic noise strength matrix $[\mathbf{G}\mathbf{Q}\mathbf{G}^{\mathrm{T}}]$ has been altered by adding the term $[q^2\mathbf{B}\mathbf{V}\mathbf{B}^{\mathrm{T}}]$ for any positive definite symmetric matrix $\mathbf{V}$. Thus, in this case, by tuning the filter by addition of white pseudonoise *of an appropriate covariance kernel structure* (adding it at the entry point of $\mathbf{u}$ itself), robustness properties are recovered; this is *not* true for *arbitrary* tuning of the filter via pseudonoise addition. In a dual procedure [90, 92], the state feedback gains $\bar{\mathbf{G}}_c$ are adjusted appropriately instead to account for state estimates replacing perfectly known state variables.

## 14.6 THE LQG SYNTHESIS OF TRACKERS

To this point, we have been considering the design of controllers intended to regulate controlled variables $\mathbf{y}_c$, defined by (14-2) (or (14-29)) for a system modeled by state equation (14-1), to *zero* for all time. Now we wish to extend the LQG synthesis process to the design of controllers that instead cause $\mathbf{y}_c$ to track some *target* or *reference variable*, $\mathbf{y}_r$ [92], generated as the output of a shaping filter,

$$\mathbf{x}_r(t_{i+1}) = \mathbf{\Phi}_r(t_{i+1}, t_i)\mathbf{x}_r(t_i) + \mathbf{G}_{dr}(t_i)\mathbf{w}_{dr}(t_i) \tag{14-134a}$$

$$\mathbf{y}_r(t_i) = \mathbf{C}_r(t_i)\mathbf{x}_r(t_i) \tag{14-134b}$$

with $\mathbf{w}_{dr}(\cdot,\cdot)$ a discrete-time, zero-mean, white Gaussian noise with autocorrelation kernel

$$E\{\mathbf{w}_{dr}(t_i)\mathbf{w}_{dr}^{\mathrm{T}}(t_j)\} = \mathbf{Q}_{dr}(t_i)\,\delta_{ij} \tag{14-135}$$

Furthermore, $\mathbf{w}_{dr}(\cdot,\cdot)$ is typically assumed independent of $\mathbf{w}_d(\cdot,\cdot)$ of autocorrelation kernel $\mathbf{Q}_d(t_i)\,\delta_{ij}$ in (14-1), $\mathbf{x}_r(t_0)$ is usually independent of $\mathbf{x}(t_0)$ of (14-1), and the dynamics driving noises are independent of the state initial conditions. The reference variable process might truly be a stochastic process, as in the case of tracking airborne targets whose dynamics are modeled stochastically, or we might want to consider the "trivial" case in which $\mathbf{w}_{dr} \equiv \mathbf{0}$, such as in desiring to follow some commanded reference signal that can be modeled as the output of an undriven linear system.

To keep the tracking error

$$\mathbf{e}(t_i) = \mathbf{y}_c(t_i) - \mathbf{y}_r(t_i) \tag{14-136}$$

small, while restricting inputs from excessive amplitudes or energy, an appropriate cost function to minimize is

$$\begin{aligned} J = E\Big\{&\frac{1}{2}[\mathbf{y}_c(t_{N+1}) - \mathbf{y}_r(t_{N+1})]^T\mathbf{Y}_f[\mathbf{y}_c(t_{N+1}) - \mathbf{y}_r(t_{N+1})] \\ &+ \sum_{i=0}^{N} \frac{1}{2}\big([\mathbf{y}_c(t_i) - \mathbf{y}_r(t_i)]^T\mathbf{Y}(t_i)[\mathbf{y}_c(t_i) - \mathbf{y}_r(t_i)] \\ &+ \mathbf{u}^T(t_i)\mathbf{U}(t_i)\mathbf{u}(t_i) + 2[\mathbf{y}_c(t_i) - \mathbf{y}_r(t_i)]^T\mathbf{S}_y(t_i)\mathbf{u}(t_i)\big)\Big\} \end{aligned} \tag{14-137}$$

with $\mathbf{Y}_f > \mathbf{0}$, $\mathbf{Y}(t_i) > \mathbf{0}$, and $\mathbf{U}(t_i) > \mathbf{0}$ for all $t_i$, and $\mathbf{S}_y(t_i)$ allowed to be nonzero for the reasons cited in Section 14.2 (such as exerting control over an interval, and replacing (14-2) by (14-29) as via (14-30)).

To generate the desired tracker by LQG controller synthesis, first express the problem description in terms of the augmented state process

$$\mathbf{x}_a(\cdot,\cdot) \triangleq \begin{bmatrix} \mathbf{x}(\cdot,\cdot) \\ \mathbf{x}_r(\cdot,\cdot) \end{bmatrix} \tag{14-138}$$

as

$$\begin{aligned} \begin{bmatrix} \mathbf{x}(t_{i+1}) \\ \mathbf{x}_r(t_{i+1}) \end{bmatrix} &= \begin{bmatrix} \mathbf{\Phi}(t_{i+1},t_i) & \mathbf{0} \\ \mathbf{0} & \mathbf{\Phi}_r(t_{i+1},t_i) \end{bmatrix}\begin{bmatrix} \mathbf{x}(t_i) \\ \mathbf{x}_r(t_i) \end{bmatrix} + \begin{bmatrix} \mathbf{B}_d(t_i) \\ \mathbf{0} \end{bmatrix}\mathbf{u}(t_i) \\ &+ \begin{bmatrix} \mathbf{G}_d(t_i) & \mathbf{0} \\ \mathbf{0} & \mathbf{G}_{dr}(t_i) \end{bmatrix}\begin{bmatrix} \mathbf{w}_d(t_i) \\ \mathbf{w}_{dr}(t_i) \end{bmatrix} \end{aligned} \tag{14-139}$$

$$E\left\{\begin{bmatrix} \mathbf{w}_d(t_i) \\ \mathbf{w}_{dr}(t_i) \end{bmatrix}[\mathbf{w}_d^T(t_j) \quad \mathbf{w}_{dr}^T(t_j)]\right\} = \begin{bmatrix} \mathbf{Q}_d(t_i) & \mathbf{0} \\ \mathbf{0} & \mathbf{Q}_{dr}(t_i) \end{bmatrix}\delta_{ij} \tag{14-140}$$

with augmented system controlled variable, to be regulated to zero, of

$$\mathbf{e}(t_i) = [\mathbf{C}(t_i) \quad -\mathbf{C}_r(t_i)]\begin{bmatrix} \mathbf{x}(t_i) \\ \mathbf{x}_r(t_i) \end{bmatrix} \tag{14-141}$$

which is to be accomplished by minimization of the quadratic cost

$$\begin{aligned} J = E\Big\{&\frac{1}{2}\mathbf{x}_a^T(t_{N+1})\mathbf{X}_{fa}\mathbf{x}_a(t_{N+1}) \\ &+ \sum_{i=0}^{N}\frac{1}{2}\left(\begin{bmatrix} \mathbf{x}_a(t_i) \\ \mathbf{u}(t_i) \end{bmatrix}^T\begin{bmatrix} \mathbf{X}_a(t_i) & \mathbf{S}_a(t_i) \\ \mathbf{S}_a^T(t_i) & \mathbf{U}(t_i) \end{bmatrix}\begin{bmatrix} \mathbf{x}_a(t_i) \\ \mathbf{u}(t_i) \end{bmatrix}\right)\Big\} \end{aligned} \tag{14-142}$$

where

$$\mathbf{X}_{\mathrm{fa}} = \mathbf{C}_{\mathrm{a}}^{\mathrm{T}}(t_{N+1})\mathbf{Y}_{\mathrm{f}}\mathbf{C}_{\mathrm{a}}(t_{N+1}) \tag{14-143a}$$

$$\mathbf{X}_{\mathrm{a}}(t_i) = \mathbf{C}_{\mathrm{a}}^{\mathrm{T}}(t_i)\mathbf{Y}(t_i)\mathbf{C}_{\mathrm{a}}(t_i) \tag{14-143b}$$

$$\mathbf{S}_{\mathrm{a}}(t_i) = \mathbf{C}_{\mathrm{a}}^{\mathrm{T}}(t_i)\mathbf{S}_y(t_i) \tag{14-143c}$$

with $\mathbf{C}_{\mathrm{a}}(t_i) \triangleq [\mathbf{C}(t_i) \vdots -\mathbf{C}_{\mathrm{r}}(t_i)]$ as in (14-141). Here we have $\mathbf{X}_{\mathrm{fa}} \geq \mathbf{0}$, $\mathbf{X}_{\mathrm{a}}(t_i) \geq \mathbf{0}$, $\mathbf{U}(t_i) > \mathbf{0}$, and $\mathbf{S}_{\mathrm{a}}(t_i)$ chosen so that the composite matrix in (14-142) is positive semidefinite. Thus, in terms of this *augmented system description*, we simply have an LQG stochastic *regulator* problem as described by (14-1), (14-2), and (14-13), for which the solution is either (14-6) or (14-7), with gains calculated via (14-14)–(14-17).

First we design the deterministic LQ optimal regulator, under the simplifying assumption that we have *perfect* knowledge of *both* $\mathbf{x}(t_i)$ *and* $\mathbf{x}_{\mathrm{r}}(t_i)$ for all time $t_i$. Note that this is especially unrealistic since we seldom have access to $\mathbf{x}_{\mathrm{r}}(t_i)$: the best we can hope for is noise-free measurements of the reference *variables* $\mathbf{y}_{\mathrm{r}}(t_i)$ to be tracked, and even that does not happen very often. Applying (14-14)–(14-17) to the augmented system and partitioning as in (14-138), we obtain the full-state feedback law of

$$\begin{aligned}\mathbf{u}^*(t_i) &= -[\mathbf{G}_{\mathrm{c1}}^*(t_i) \quad \mathbf{G}_{\mathrm{c2}}^*(t_i)]\begin{bmatrix}\mathbf{x}(t_i)\\ \mathbf{x}_{\mathrm{r}}(t_i)\end{bmatrix}\\ &= -\mathbf{G}_{\mathrm{c1}}^*(t_i)\mathbf{x}(t_i) - \mathbf{G}_{\mathrm{c2}}^*(t_i)\mathbf{x}_{\mathrm{r}}(t_i)\end{aligned} \tag{14-144}$$

where, by (14-14) and (14-15) written explicitly, but without time arguments (all are $t_i$ except for $t_{i+1}$ on $\mathbf{K}_{\mathrm{ca}}$ on the right hand sides),

$$\begin{aligned}[\mathbf{G}_{\mathrm{c1}}^* \quad \mathbf{G}_{\mathrm{c2}}^*] &= \left\{\mathbf{U} + [\mathbf{B}_{\mathrm{d}}^{\mathrm{T}} \quad \mathbf{0}]\begin{bmatrix}\mathbf{K}_{\mathrm{c11}} & \mathbf{K}_{\mathrm{c12}}\\ \mathbf{K}_{\mathrm{c12}}^{\mathrm{T}} & \mathbf{K}_{\mathrm{c22}}\end{bmatrix}\begin{bmatrix}\mathbf{B}_{\mathrm{d}}\\ \mathbf{0}\end{bmatrix}\right\}^{-1}\\ &\quad\times\left\{[\mathbf{B}_{\mathrm{d}}^{\mathrm{T}} \quad \mathbf{0}]\begin{bmatrix}\mathbf{K}_{\mathrm{c11}} & \mathbf{K}_{\mathrm{c12}}\\ \mathbf{K}_{\mathrm{c12}}^{\mathrm{T}} & \mathbf{K}_{\mathrm{c22}}\end{bmatrix}\begin{bmatrix}\mathbf{\Phi} & \mathbf{0}\\ \mathbf{0} & \mathbf{\Phi}_{\mathrm{r}}\end{bmatrix} + [\mathbf{S}_y^{\mathrm{T}}\mathbf{C} \quad -\mathbf{S}_y^{\mathrm{T}}\mathbf{C}_{\mathrm{r}}]\right\}\\ &= \{\mathbf{U} + \mathbf{B}_{\mathrm{d}}^{\mathrm{T}}\mathbf{K}_{\mathrm{c11}}\mathbf{B}_{\mathrm{d}}\}^{-1}\{\mathbf{B}_{\mathrm{d}}^{\mathrm{T}}\mathbf{K}_{\mathrm{c11}}\mathbf{\Phi} + \mathbf{S}_y^{\mathrm{T}}\mathbf{C} \vdots \mathbf{B}_{\mathrm{d}}^{\mathrm{T}}\mathbf{K}_{\mathrm{c12}}\mathbf{\Phi}_{\mathrm{r}} \quad -\mathbf{S}_y^{\mathrm{T}}\mathbf{C}_{\mathrm{r}}]\end{aligned} \tag{14-145a}$$

$$\begin{aligned}\begin{bmatrix}\mathbf{K}_{\mathrm{c11}} & \mathbf{K}_{\mathrm{c12}}\\ \mathbf{K}_{\mathrm{c12}}^{\mathrm{T}} & \mathbf{K}_{\mathrm{c22}}\end{bmatrix} &= \begin{bmatrix}\mathbf{C}^{\mathrm{T}}\\ -\mathbf{C}_{\mathrm{r}}^{\mathrm{T}}\end{bmatrix}\mathbf{Y}[\mathbf{C} \quad -\mathbf{C}_{\mathrm{r}}] + \begin{bmatrix}\mathbf{\Phi}^{\mathrm{T}} & \mathbf{0}\\ \mathbf{0} & \mathbf{\Phi}_{\mathrm{r}}^{\mathrm{T}}\end{bmatrix}\begin{bmatrix}\mathbf{K}_{\mathrm{c11}} & \mathbf{K}_{\mathrm{c12}}\\ \mathbf{K}_{\mathrm{c12}}^{\mathrm{T}} & \mathbf{K}_{\mathrm{c22}}\end{bmatrix}\begin{bmatrix}\mathbf{\Phi} & \mathbf{0}\\ \mathbf{0} & \mathbf{\Phi}_{\mathrm{r}}\end{bmatrix}\\ &\quad - \begin{bmatrix}\mathbf{\Phi}^{\mathrm{T}}\mathbf{K}_{\mathrm{c11}}\mathbf{B}_{\mathrm{d}} + \mathbf{C}^{\mathrm{T}}\mathbf{S}_y\\ \mathbf{\Phi}_{\mathrm{r}}^{\mathrm{T}}\mathbf{K}_{\mathrm{c12}}^{\mathrm{T}}\mathbf{B}_{\mathrm{d}} - \mathbf{C}_{\mathrm{r}}^{\mathrm{T}}\mathbf{S}_y\end{bmatrix}[\mathbf{G}_{\mathrm{c1}}^* \quad \mathbf{G}_{\mathrm{c2}}^*]\end{aligned} \tag{14-145b}$$

The motivation for carrying this out in detail will be apparent once we express the separate partitions of these relations explicitly. Thus, we obtain

$$\mathbf{G}_{c1}^*(t_i) = [\mathbf{U}(t_i) + \mathbf{B}_d^T(t_i)\mathbf{K}_{c11}(t_{i+1})\mathbf{B}_d(t_i)]^{-1} \times [\mathbf{B}_d^T(t_i)\mathbf{K}_{c11}(t_{i+1})\mathbf{\Phi}(t_{i+1}, t_i) + \mathbf{S}_y^T(t_i)\mathbf{C}(t_i)] \quad (14\text{-}146a)$$

$$\mathbf{G}_{c2}^*(t_i) = [\mathbf{U}(t_i) + \mathbf{B}_d^T(t_i)\mathbf{K}_{c11}(t_{i+1})\mathbf{B}_d(t_i)]^{-1} \times [\mathbf{B}_d^T(t_i)\mathbf{K}_{c12}(t_{i+1})\mathbf{\Phi}_r(t_{i+1}, t_i) - \mathbf{S}_y^T(t_i)\mathbf{C}_r(t_i)] \quad (14\text{-}146b)$$

with

$$\mathbf{K}_{c11}(t_i) = \mathbf{C}^T\mathbf{Y}\mathbf{C} + \mathbf{\Phi}^T\mathbf{K}_{c11}\mathbf{\Phi} - [\mathbf{\Phi}^T\mathbf{K}_{c11}\mathbf{B}_d + \mathbf{C}^T\mathbf{S}_y]\mathbf{G}_{c1}^*$$
$$\mathbf{K}_{c11}(t_{N+1}) = \mathbf{C}^T(t_{N+1})\mathbf{Y}_f\mathbf{C}(t_{N+1}) \quad (14\text{-}147a)$$

$$\mathbf{K}_{c12}(t_i) = -\mathbf{C}^T\mathbf{Y}\mathbf{C}_r + \mathbf{\Phi}^T\mathbf{K}_{c12}\mathbf{\Phi}_r - \mathbf{G}_{c1}^{*T}[\mathbf{B}_d^T\mathbf{K}_{c12}\mathbf{\Phi}_r - \mathbf{S}_y^T\mathbf{C}_r]$$
$$\mathbf{K}_{c12}(t_{N+1}) = -\mathbf{C}^T(t_{N+1})\mathbf{Y}_f\mathbf{C}_r(t_{N+1}) \quad (14\text{-}147b)$$

$$\mathbf{K}_{c22}(t_i) = \mathbf{C}_r^T\mathbf{Y}\mathbf{C}_r + \mathbf{\Phi}_r^T\mathbf{K}_{c22}\mathbf{\Phi}_r - \mathbf{G}_{c2}^{*T}[\mathbf{B}_d^T\mathbf{K}_{c12}\mathbf{\Phi}_r - \mathbf{S}_y^T\mathbf{C}_r]$$
$$\mathbf{K}_{c22}(t_{N+1}) = \mathbf{C}_r^T(t_{N+1})\mathbf{Y}_f\mathbf{C}_r(t_{N+1}) \quad (14\text{-}147c)$$

Note the structure of this result. First of all, consider the *feedback* connection from $\mathbf{x}(t_i)$ to $\mathbf{u}(t_i)$, the first term in (14-144) with gain expressed by (14-146a), able to be evaluated by solving *only* the backward Riccati equation (14-147a). This term is identical to the simpler regulator solution for the unaugmented system description, given by (14-6) and (14-14)–(14-17) with the appropriate identifications: (14-143) with the subscript a removed. Thus, $\mathbf{K}_{c11}$ and $\mathbf{G}_{c1}^*$ are *independent* of the properties of the reference variable to be tracked, and they are found by solving the corresponding *deterministic* LQ optimal *regulator* problem with the *reference variable omitted*. Second, consider the *feedforward* connection from $\mathbf{x}_r(t_i)$ to $\mathbf{u}(t_i)$. The gain $\mathbf{G}_{c2}^*$ *does* depend on the reference variable properties, as seen from (14-146b) and (14-147b) with $\mathbf{G}_{c1}^*$ supplied from prior computations for the feedback path. However, $\mathbf{K}_{c22}(t_i)$ given by (14-147c) *need not be evaluated* in order to generate this feedforward gain [92].

The structure of this tracker control law is displayed in Fig. 14.7, in which the block labeled "controlled system" is composed of the underlying continuous-time system, with a zero-order hold and D/A conversion on the "front end" and sampling and A/D conversion to generate the $\mathbf{x}(t_i)$ output. Note that this is meant to represent a "snapshot" of variables at sample time $t_i$; for instance, $\mathbf{u}^*(t_i)$ as generated in the figure will affect the *future* state $\mathbf{x}(t_{i+1})$ and not the state $\mathbf{x}(t_i)$ currently being output from the system. Also, the dashed portion of the diagram corresponds to *artificial* constructions to indicate the relationship of variables.

If we restrict our attention to *time-invariant system and shaping filter models* (driven by stationary noises, but this has no effect on the deterministic LQ optimal regulator design) and *constant cost matrices* in (14-137), then we can seek the steady state *constant* gains $\bar{\mathbf{G}}_{c1}^*$ and $\bar{\mathbf{G}}_{c2}^*$ to be implemented for *all time*

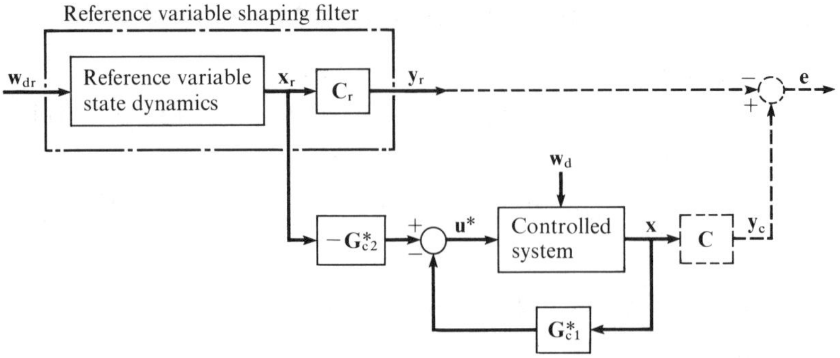


FIG. 14.7 Structure of tracker control law.

as a purposeful simplification to the optimal solution requiring considerably less computation. However, from Fig. 14.7 it can be seen that the augmented system configuration is *not* controllable with respect to the points of entry of the input $\mathbf{u}(t_i)$: $\mathbf{u}(t_i)$ cannot control the shaping filter states. Therefore, to employ this synthesis, we must *require* $\mathbf{\Phi}_r$ to correspond to an *asymptotically stable* system model: all of its eigenvalues must lie *strictly within* the unit circle on the complex plane, and not on the unit circle itself. If poles were to lie on or outside the unit circle, then the augmented system model would have modes that are simultaneously unstable, uncontrollable, and reconstructible. We are then not assured of the existence of a constant steady state solution $\bar{\mathbf{K}}_{ca}$, or of a bounded cost, in the limit as $t_0 \to -\infty$. For this reason, it is important *not* to formulate a nonzero setpoint controller simply as a tracker of this form, with reference variable shaping filter composed of undriven integrators outputting the desired setpoint $\mathbf{y}_r =$ constant. Such a nonzero setpoint controller will be developed properly as a special case in the next section.

EXAMPLE 14.5 Recall the angular servo problem of Examples 14.2–14.4, in which the equivalent discrete-time model for gimbal angle $x$ is

$$\begin{aligned} x(t_{i+1}) &= [e^{-\Delta t/T}]x(t_i) + [1 - e^{-\Delta t/T}]u(t_i) + w_d(t_i) \\ &= [0.82]x(t_i) + [0.18]u(t_i) + w_d(t_i) \end{aligned}$$

with $\Delta t = 0.1$ sec, $T = 0.5$ sec, and $Q_d = 0.084Q$ with $Q$ being the strength of the continuous white Gaussian noise in the original differential equation for the servo angle. Now we desire the servo to *track a target* (reference variable) described by a first order Gauss–Markov process,

$$\dot{x}_r(t) = -[1/T_r]x_r(t) + w_r(t); \qquad y_r(t) = x_r(t)$$

with $w_r(\cdot,\cdot)$ zero-mean white Gaussian noise with autocorrelation kernel $E\{w_r(t)w_r(t+\tau)\} = [2\sigma_r^2/T_r]\,\delta(\tau)$, and $\sigma_r$ is the rms value of $x_r(\cdot,\cdot)$. The equivalent discrete-time model for this target is

$$x_r(t_{i+1}) = [e^{-\Delta t/T_r}]x_r(t_i) + w_{dr}(t_i); \qquad y_r(t_i) = x_r(t_i)$$

where $E\{w_{dr}(t_i)w_{dr}(t_j)\} = Q_{dr}\,\delta_{ij}$ and $Q_{dr}$ is evaluated as $\sigma_r^2[1 - e^{-2\Delta t/T_r}]$. Let $T_r = 0.2$ sec so that $\Phi_r = e^{-\Delta t/T_r} = 0.61$ and $Q_{dr} = 0.63\sigma_r^2$.

Let us design the controller to minimize

$$J = E\left\{\sum_{i=0}^{N} \frac{1}{2}\left[Y\mathrm{e}(t_i)^2 + U\mathrm{u}(t_i)^2\right] + \frac{1}{2} Y_{\mathrm{f}}\mathrm{e}(t_{N+1})^2\right\}$$

where $\mathrm{e}(t_i) = [\mathrm{y}_c(t_i) - \mathrm{y}_r(t_i)] = [\mathrm{x}(t_i) - \mathrm{x}_r(t_i)]$ and $X = Y$ for this example. Note this is the simplest case in which $S \equiv 0$. The solution is

$$u^*(t_i) = -G_{c1}^*(t_i)x(t_i) - G_{c2}^*(t_i)x_r(t_i)$$

where the first term involving feedback gain $G_{c1}^*(t_i)$ is identical to the deterministic optimal LQ regulator gain for the problem omitting the reference variable; by (14-146a) and (14-147a),

$$G_{c1}^*(t_i) = \frac{0.15K_{c1}(t_{i+1})}{U + 0.033K_{c11}(t_{i+1})}$$

$$K_{c11}(t_i) = X + 0.67K_{c11}(t_{i+1}) - \frac{0.022K_{c11}^2(t_{i+1})}{U + 0.033K_{c11}(t_{i+1})}$$

as developed in Example 14.2. Once $K_{c11}$ is computed for all time, $K_{c12}$ can be evaluated (or the $K_{c12}$ can be evaluated simultaneously) according to (14-146b) and (14-147b):

$$\begin{aligned} K_{c12}(t_i) &= -[1]Y[1] + [0.82]K_{c12}(t_{i+1})[0.61] \\ &\quad - \frac{0.15K_{c11}(t_{i+1})}{U + 0.033K_{c11}(t_{i+1})}\{[0.18]K_{c12}(t_{i+1})[0.61]\} \\ &= -X + 0.50K_{c12}(t_{i+1}) - \frac{0.016[K_{c11}(t_{i+1})][K_{c12}(t_{i+1})]}{U + 0.033K_{c11}(t_{i+1})} \end{aligned}$$

$$G_{c2}^*(t_i) = \frac{[0.18]K_{c12}(t_{i+1})[0.61]}{U + 0.033K_{c11}(t_{i+1})} = \frac{0.11K_{c12}(t_{i+1})}{U + 0.033K_{c11}(t_{i+1})}$$

As in Example 14.2, we can generate steady state constant gains for a suboptimal control law

$$u(t_i) = -\bar{G}_{c1}^* x(t_i) - \bar{G}_{c2}^* x_r(t_i)$$

If we let $X = 1$, we obtain the $\bar{K}_{c11}$, $\bar{G}_{c1}^*$ closed-loop system eigenvalue $[\Phi - B_d\bar{G}_{c1}^*]$, $\bar{K}_{c12}$, and $\bar{G}_{c2}^*$ values shown in the accompanying table as a function of chosen $U$ parameter.

| $U$: | 0 | 0.1 | 1 | 10 | $\infty$ |
|---|---|---|---|---|---|
| $\bar{K}_{c11}$: | 1 | 1.74 | 2.61 | 2.97 | — |
| $\bar{G}_{c1}^*$: | 4.5 | 1.66 | 0.36 | 0.044 | 0 |
| $\Phi - \mathbf{B}_d\bar{G}_{c1}^*$: | 0 | 0.52 | 0.75 | 0.81 | 0.82 |
| $\bar{K}_{c12}$: | −1.00 | −1.47 | −1.85 | −1.98 | −2.00 |
| $\bar{G}_{c2}^*$: | −3.33 | −1.03 | −0.19 | −0.022 | 0 |

The first three rows of this table are identical to the result in Example 14.2. As $U$ increases, $\bar{G}_{c2}^*$ decreases in magnitude as expected, and in the limit of $U \to \infty$, both $\bar{G}_{c1}^*$ and $\bar{G}_{c2}^*$ go to zero, leaving the system uncontrolled.

If the reference variable dynamics model is changed, the $G_{c2}^*(t_i)$ feedforward gain history is altered, but the feedback connection, i.e., the $G_{c1}^*(t_i)$ time history, is unchanged. To show this, suppose the target were again modeled as a first order Gauss–Markov process, but with $T_r = 0.3$ sec, so that $\Phi_r = e^{-1/3} = 0.72$. Then the $K_{c11}$ and $G_{c1}^*$ evaluations are as given previously, but (14-146b) and (14-147b) yield

$$K_{c12}(t_i) = -X + 0.58K_{c12}(t_{i+1}) - \frac{0.019[K_{c11}(t_{i+1})][K_{c12}(t_{i+1})]}{U + 0.033K_{c11}(t_{i+1})}$$

$$G_{c2}^*(t_i) = \frac{0.13K_{c12}(t_{i+1})}{U + 0.033K_{c11}(t_{i+1})}$$

In steady state operation, we obtain $\bar{K}_{c11}$, $\bar{G}_{c1}^*$, and $[\Phi - B_d\bar{G}_{c1}^*]$ results as in the preceding table, but with $\bar{K}_{c12}$ and $\bar{G}_{c2}^*$ replaced by

| $U$: | 0 | 0.1 | 1 | 10 | $\infty$ |
|---|---|---|---|---|---|
| $\bar{K}_{c12}$: | −1.00 | −1.58 | −2.15 | −2.35 | −2.38 |
| $\bar{G}_{c2}^*$: | −3.93 | −1.31 | −0.26 | −0.030 | 0 |

With the target correlation time increased from 0.2 to 0.3 sec, the controller is "told" the current reference variable value will persist longer into the future, and thus it can afford to put a larger magnitude gain on it to generate a control to minimize tracking error. ■

Now assume that only noise-corrupted measurements are available from the controlled system, as

$$\mathbf{z}(t_i) = \mathbf{H}(t_i)\mathbf{x}(t_i) + \mathbf{v}(t_i) \tag{14-148a}$$

and only noise-corrupted measurements of the reference variable are available as well,

$$\mathbf{z}_r(t_i) = \mathbf{y}_r(t_i) + \mathbf{v}_r(t_i) = \mathbf{C}_r(t_i)\mathbf{x}_r(t_i) + \mathbf{v}_r(t_i) \tag{14-148b}$$

In terms of the augmented state, these can be written as

$$\begin{bmatrix} \mathbf{z}(t_i) \\ \mathbf{z}_r(t_i) \end{bmatrix} = \begin{bmatrix} \mathbf{H}(t_i) & \mathbf{0} \\ \mathbf{0} & \mathbf{C}_r(t_i) \end{bmatrix} \begin{bmatrix} \mathbf{x}(t_i) \\ \mathbf{x}_r(t_i) \end{bmatrix} + \begin{bmatrix} \mathbf{v}(t_i) \\ \mathbf{v}_r(t_i) \end{bmatrix} \tag{14-148c}$$

By LQG controller synthesis for this augmented system, the Kalman filter for (14-139) and (14-148c) is to be cascaded with (14-144). *If* $\mathbf{x}(t_0)$ and $\mathbf{x}_r(t_0)$ are uncorrelated, $\mathbf{w}_d(\cdot,\cdot)$ and $\mathbf{w}_{dr}(\cdot,\cdot)$ are uncorrelated as in (14-140), and $\mathbf{v}(\cdot,\cdot)$ and $\mathbf{v}_r(\cdot,\cdot)$ are uncorrelated in (14-148c), then this augmented system filter *decouples into two totally independent filters.* One of them inputs $\mathbf{z}(t_i)$ and $\mathbf{u}(t_{i-1})$ to generate an estimate of the controlled system state $\hat{\mathbf{x}}(t_i^+)$, and the other inputs $\mathbf{z}_r(t_i)$ to provide the reference variable state estimate $\hat{\mathbf{x}}_r(t_i^+)$. The optimal stochastic controller is then

$$\mathbf{u}^*(t_i) = -\mathbf{G}_{c1}^*(t_i)\hat{\mathbf{x}}(t_i^+) - \mathbf{G}_{c2}^*(t_i)\hat{\mathbf{x}}_r(t_i^+) \tag{14-149}$$

where again the first term would be identical to the optimal stochastic *regulator* for the problem omitting the reference variable shaping filter model. The structure of this controller is indicated in Fig. 14.8; as previously, dashed lines

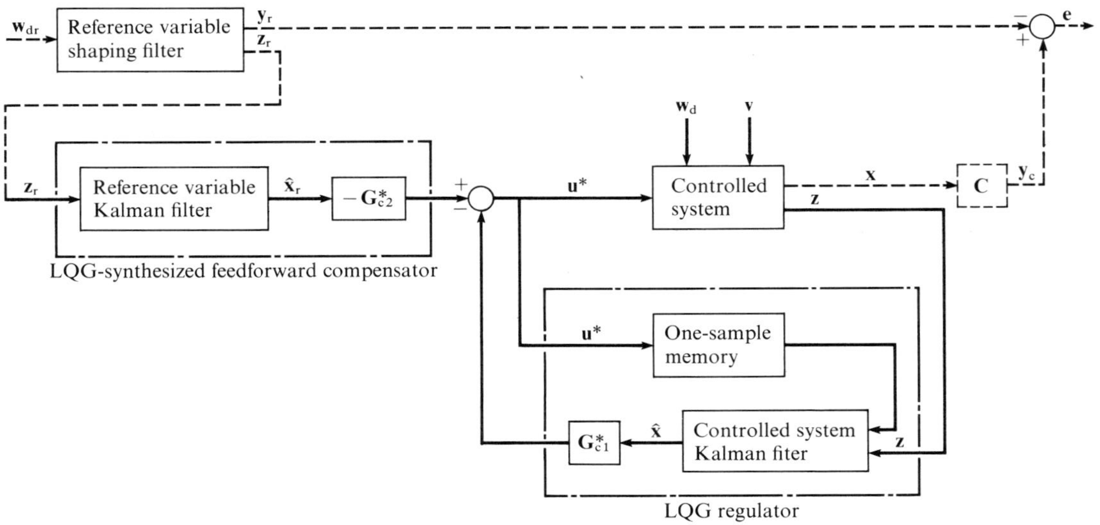


FIG. 14.8 Structure of tracker that inputs noise-corrupted measurements.

represent variables that are not physically accessible. For computational reasons, the suboptimal

$$\mathbf{u}(t_i) = -\mathbf{G}_{c1}^*(t_i)\hat{\mathbf{x}}(t_i^-) - \mathbf{G}_{c2}^*(t_i)\hat{\mathbf{x}}_r(t_i^-) \tag{14-150}$$

is often implemented, and its structure can be discerned from the same figure.

EXAMPLE 14.6 If in the previous example only noise-corrupted measurements of $x(t_i)$ and and $x_r(t_i)$ were available, and *if* $E\{\mathbf{x}(t_0)\mathbf{x}_r(t_0)\} = E\{\mathbf{x}(t_0)\}E\{\mathbf{x}_r(t_0)\}$, $E\{\mathbf{w}_d(t_i)\mathbf{w}_{dr}(t_j)\} = 0$ and $E\{v(t_i)v_r(t_j)\} = 0$ for all $t_i$ and $t_j$, then the optimal controller is given by (14-149) with the gains as specified in that example, $\hat{x}(t_i^+)$ provided by the filter of Example 14.2, and $\hat{x}_r(t_i^+)$ generated by a separate Kalman filter:

$$\hat{x}_r(t_i^+) = \hat{x}_r(t_i^-) + K_r(t_i)[z_r(t_i) - C_r(t_i)\hat{x}_r(t_i^-)]$$
$$\hat{x}_r(t_{i+1}^-) = 0.61\hat{x}(t_i^+)$$

with

$$K_r(t_i) = \frac{P_r(t_i^-)C_r(t_i)}{C_r^2(t_i)P_r(t_i^-) + R_r(t_i)}$$

$$P_r(t_i^+) = P_r(t_i^-) - K_r(t_i)C_r(t_i)P_r(t_i^-)$$
$$P_r(t_{i+1}^-) = 0.37P_r(t_i^+) + 0.63\sigma_r^2$$

Without the uncorrelatedness assumptions, the full-state controller still decomposes as in Fig. 14.7, but the augmented system Kalman filter does *not* decouple to yield the structure of Fig. 14.8. ■

## 14.7 NONZERO SETPOINT CONTROLLERS

In the last section, it was mentioned that nonzero setpoint controllers should not be considered simply as trackers in which the reference variables are constants in time, since we are not assured of the existence of a constant steady state solution $\bar{\mathbf{K}}_{ca}$ to the augmented system Riccati equation when reference variable system eigenvalues have magnitude of one. Here we consider the *time-invariant case* and generate a prospective design [92] via deterministic LQ regulator theory, as part of the overall LQG synthesis procedure. Some of its properties will motivate investigation of PI designs, as accomplished in Section 14.9, as an alternative solution.

Given the deterministic time-invariant system model

$$\mathbf{x}(t_{i+1}) = \mathbf{\Phi}\mathbf{x}(t_i) + \mathbf{B}_d\mathbf{u}(t_i); \qquad \mathbf{x}(t_0) = \text{given} \tag{14-151}$$

with the controlled variable

$$\mathbf{y}_c(t_i) = \mathbf{C}\mathbf{x}(t_i) + \mathbf{D}_y\mathbf{u}(t_i) \tag{14-152}$$

it is desired to maintain the controlled variable $\mathbf{y}_c(t_i)$ at a *nonzero* equilibrium value $\mathbf{y}_d$ *with zero steady state error*. Note the explicit inclusion of $\mathbf{D}_y$ in (14-152), since we will be performing more than just LQ synthesis, in which nonzero $\mathbf{D}_y$ is easily imbedded in the cost by (14-30). To maintain $\mathbf{y}_c(t_i)$ at the desired setpoint $\mathbf{y}_d$, an equilibrium solution must be found to yield $\mathbf{x}(t_i) = \mathbf{x}_0$ for all $t_i$, such that $\mathbf{y}_c(t_i) = \mathbf{y}_d$. The *nominal control* $\mathbf{u}_0$ to hold the system at that equilibrium point is found as the solution to

$$\mathbf{x}_0 = \mathbf{\Phi}\mathbf{x}_0 + \mathbf{B}_d\mathbf{u}_0 \tag{14-153a}$$

$$\mathbf{y}_d = \mathbf{C}\mathbf{x}_0 + \mathbf{D}_y\mathbf{u}_0 \tag{14-153b}$$

or

$$\begin{bmatrix} (\mathbf{\Phi} - \mathbf{I}) & \mathbf{B}_d \\ \mathbf{C} & \mathbf{D}_y \end{bmatrix} \begin{bmatrix} \mathbf{x}_0 \\ \mathbf{u}_0 \end{bmatrix} = \begin{bmatrix} \mathbf{0} \\ \mathbf{y}_d \end{bmatrix} \tag{14-154}$$

Given $\mathbf{y}_d$, we must solve (14-154) for $\mathbf{x}_0$ and $\mathbf{u}_0$; to assure these exist and are unique for every $\mathbf{y}_d$, assume that the composite matrix in (14-154) is *square* (implying that the dimension $r$ of $\mathbf{u}(t_i)$ equals the dimension $p$ of $\mathbf{y}_c(t_i)$) and *invertible* (there exists a matrix $\mathbf{A}^{-1}$ such that $\mathbf{A}\mathbf{A}^{-1} = \mathbf{A}^{-1}\mathbf{A} = \mathbf{I}$ for $\mathbf{A}$ given in (14-154)), with

$$\begin{bmatrix} (\mathbf{\Phi} - \mathbf{I}) & \mathbf{B}_d \\ \mathbf{C} & \mathbf{D}_y \end{bmatrix}^{-1} = \begin{bmatrix} \mathbf{\Pi}_{11} & \mathbf{\Pi}_{12} \\ \mathbf{\Pi}_{21} & \mathbf{\Pi}_{22} \end{bmatrix} \tag{14-155}$$

Then, (14-154) can be solved as

$$\begin{bmatrix} \mathbf{x}_0 \\ \mathbf{u}_0 \end{bmatrix} = \begin{bmatrix} (\mathbf{\Phi} - \mathbf{I}) & \mathbf{B}_d \\ \mathbf{C} & \mathbf{D}_y \end{bmatrix}^{-1} \begin{bmatrix} \mathbf{0} \\ \mathbf{y}_d \end{bmatrix} = \begin{bmatrix} \mathbf{\Pi}_{11} & \mathbf{\Pi}_{12} \\ \mathbf{\Pi}_{21} & \mathbf{\Pi}_{22} \end{bmatrix} \begin{bmatrix} \mathbf{0} \\ \mathbf{y}_d \end{bmatrix} \tag{14-156a}$$

or

$$\mathbf{x}_0 = \mathbf{\Pi}_{12}\mathbf{y}_d, \qquad \mathbf{u}_0 = \mathbf{\Pi}_{22}\mathbf{y}_d \tag{14-156b}$$

Setting $r = p$ is not overly restrictive: if the dimension of $\mathbf{y}_c$ were less than that of $\mathbf{u}$, one has additional design freedom and can increase the dimension of $\mathbf{y}_c$ to accomplish additional objectives, or the right inverse can be used; if the dimension of $\mathbf{y}_c$ is greater than that of $\mathbf{u}$, in general there is no solution, but one can obtain an approximate result with the left inverse. Before proceeding, right and left inverses are developed further [175].

If $p < r$, which is to say that the number of controlled outputs $\mathbf{y}_c$ is less than the number of controls $\mathbf{u}$, then the solution is nonunique. Assume that the rank of the composite matrix in (14-154) is $(n + p)$; any redundant components of $\mathbf{y}_c$ can be eliminated to achieve this condition. Then (14-154) is of the form $\mathbf{A}\mathbf{x} = \mathbf{b}$ to be solved for $\mathbf{x}$, given $\mathbf{A}$ and $\mathbf{b}$, with $\dim(\mathbf{b}) < \dim(\mathbf{x})$; the nonunique solution can be expressed in terms of the *right inverse* $\mathbf{A}^R$ as $\mathbf{x} = \mathbf{A}^R\mathbf{b}$, where $\mathbf{A}\mathbf{A}^R = \mathbf{I}$ and $\mathbf{I}$ is the $\dim(\mathbf{b})$-by-$\dim(\mathbf{b})$ identity. However, there is a *unique* minimum norm solution, a solution whose generalized norm or length, $[\mathbf{x}^T\mathbf{W}\mathbf{x}]^{1/2}$, is minimal; it is given by $\mathbf{x} = \mathbf{A}^{RMW}\mathbf{b}$ where $\mathbf{A}^{RMW}$ is the minimal weight right inverse, or weighted pseudoinverse,

$$\mathbf{A}^{RMW} = \mathbf{W}^{-1}\mathbf{A}^T\{\mathbf{A}\mathbf{W}^{-1}\mathbf{A}^T\}^{-1}; \qquad \mathbf{W} > \mathbf{0} \tag{14-157}$$

and note $\mathbf{A}\mathbf{A}^{\mathrm{RMW}} = \mathbf{A}\mathbf{W}^{-1}\mathbf{A}^{\mathrm{T}}\{\mathbf{A}\mathbf{W}^{-1}\mathbf{A}^{\mathrm{T}}\}^{-1} = \mathbf{I}$, but generally $\mathbf{A}^{\mathrm{RMW}}\mathbf{A} \neq \mathbf{I}$. Choice of $\mathbf{W}$ other than $\mathbf{I}$ specifies that you want to minimize some components of $\mathbf{x}$ more than others: you are really minimizing $[\mathbf{x}^{\mathrm{T}}\sqrt{\mathbf{W}}][\sqrt{\mathbf{W}}^{\mathrm{T}}\mathbf{x}]$, i.e., the Euclidean length of $\sqrt{\mathbf{W}}^{\mathrm{T}}\mathbf{x}$ in transformed coordinates. If $\mathbf{W}$ is diagonal and, for physical reasons in a particular problem, $x_i$ is to be minimized at the expense of other components, then $W_{ii}$ should be made the largest weighting value (see Problem 14.24). Thus, the solution in this case can be written as in (14-156), but with (14-155) replaced by

$$\begin{bmatrix} \mathbf{\Pi}_{11} & \mathbf{\Pi}_{12} \\ \mathbf{\Pi}_{21} & \mathbf{\Pi}_{22} \end{bmatrix} = \begin{bmatrix} (\mathbf{\Phi} - \mathbf{I}) & \mathbf{B}_{\mathrm{d}} \\ \mathbf{C} & \mathbf{D}_y \end{bmatrix}^{\mathrm{R}} \tag{14-158}$$

with $\mathbf{A}^{\mathrm{R}}$ typically evaluated as $\mathbf{A}^{\mathrm{RMW}}$.

Finally, if $p > r$ so that there are more controlled outputs than controls, then (14-154) is overdetermined and an exact solution does not exist in general. Assuming the rank of the composite matrix in (14-154) to be $(n + r)$, we can write the nonunique approximations, *none* of which are true solutions to $\mathbf{A}\mathbf{x} = \mathbf{b}$, as $\mathbf{x} = \mathbf{A}^{\mathrm{L}}\mathbf{b}$ in terms of the *left inverse* $\mathbf{A}^{\mathrm{L}}$, where $\mathbf{A}^{\mathrm{L}}\mathbf{A} = \mathbf{I}$ and $\mathbf{I}$ is the dim($\mathbf{x}$)-by-dim($\mathbf{x}$) identity. As in the previous case, there is a *unique* minimum norm approximation of this form such that the generalized distance between $\mathbf{A}\mathbf{x}$ and $\mathbf{b}$, $[(\mathbf{b} - \mathbf{A}\mathbf{x})^{\mathrm{T}}\mathbf{W}(\mathbf{b} - \mathbf{A}\mathbf{x})]^{1/2}$, is minimized; it is given by $\mathbf{x} = \mathbf{A}^{\mathrm{LMW}}\mathbf{b}$, where $\mathbf{A}^{\mathrm{LMW}}$ is the minimal weighted left inverse or weighted pseudoinverse

$$\mathbf{A}^{\mathrm{LMW}} = \{\mathbf{A}^{\mathrm{T}}\mathbf{W}\mathbf{A}\}^{-1}\mathbf{A}^{\mathrm{T}}\mathbf{W}; \qquad \mathbf{W} > \mathbf{0} \tag{14-159}$$

Again note $\mathbf{A}^{\mathrm{LMW}}\mathbf{A} = \mathbf{I}$ but generally $\mathbf{A}\mathbf{A}^{\mathrm{LMW}} \neq \mathbf{I}$, and that $\mathbf{W}$ other than $\mathbf{I}$ is used to specify desired components to be especially close approximations. Unlike the case of the right inverse, the "solution" $\mathbf{x} = \mathbf{A}^{\mathrm{L}}\mathbf{b}$ is only an approximation and substituting it back into the original equation $\mathbf{A}\mathbf{x} = \mathbf{b}$ will generally *not* yield a valid equality. The approximation for this case is as given in (14-156), but with (14-155) replaced by

$$\begin{bmatrix} \mathbf{\Pi}_{11} & \mathbf{\Pi}_{12} \\ \mathbf{\Pi}_{21} & \mathbf{\Pi}_{22} \end{bmatrix} = \begin{bmatrix} (\mathbf{\Phi} - \mathbf{I}) & \mathbf{B}_{\mathrm{d}} \\ \mathbf{C} & \mathbf{D}_y \end{bmatrix}^{\mathrm{L}} \tag{14-160}$$

with $\mathbf{A}^{\mathrm{L}}$ typically evaluated as $\mathbf{A}^{\mathrm{LMW}}$.

Having solved for $\mathbf{x}_0$ and $\mathbf{u}_0$, we can define the perturbation variables

$$\delta\mathbf{x}(t_i) \triangleq \mathbf{x}(t_i) - \mathbf{x}_0 = \mathbf{x}(t_i) - \mathbf{\Pi}_{12}\mathbf{y}_{\mathrm{d}} \tag{14-161a}$$

$$\delta\mathbf{u}(t_i) \triangleq \mathbf{u}(t_i) - \mathbf{u}_0 = \mathbf{u}(t_i) - \mathbf{\Pi}_{22}\mathbf{y}_{\mathrm{d}} \tag{14-161b}$$

$$\delta\mathbf{y}_{\mathrm{c}}(t_i) \triangleq \mathbf{y}_{\mathrm{c}}(t_i) - \mathbf{y}_{\mathrm{d}} \tag{14-161c}$$

which satisfy

$$\delta\mathbf{x}(t_{i+1}) = \mathbf{\Phi}\,\delta\mathbf{x}(t_i) + \mathbf{B}_{\mathrm{d}}\,\delta\mathbf{u}(t_i); \qquad \delta\mathbf{x}(t_0) = \mathbf{x}(t_0) - \mathbf{x}_0 \tag{14-162a}$$

$$\delta\mathbf{y}_{\mathrm{c}}(t_i) = \mathbf{C}\,\delta\mathbf{x}(t_i) + \mathbf{D}_y\,\delta\mathbf{u}(t_i) \tag{14-162b}$$

In terms of these perturbation variables, a *perturbation LQ regulator* can be designed to minimize the cost criterion

$$J = \sum_{i=0}^{N} \frac{1}{2}\left(\begin{bmatrix}\boldsymbol{\delta}\mathbf{x}(t_i)\\ \boldsymbol{\delta}\mathbf{u}(t_i)\end{bmatrix}^{\mathrm{T}}\begin{bmatrix}\mathbf{X} & \mathbf{S}\\ \mathbf{S}^{\mathrm{T}} & \mathbf{U}\end{bmatrix}\begin{bmatrix}\boldsymbol{\delta}\mathbf{x}(t_i)\\ \boldsymbol{\delta}\mathbf{u}(t_i)\end{bmatrix}\right) + \frac{1}{2}\boldsymbol{\delta}\mathbf{x}^{\mathrm{T}}(t_{N+1})\mathbf{X}_{\mathrm{f}}\,\boldsymbol{\delta}\mathbf{x}(t_{N+1}) \tag{14-163}$$

according to (14-30). Then the solution is

$$\boldsymbol{\delta}\mathbf{u}^*(t_i) = -\mathbf{G}_{\mathrm{c}}^*(t_i)\,\boldsymbol{\delta}\mathbf{x}(t_i) \tag{14-164}$$

with $\mathbf{G}_{\mathrm{c}}^*(t_i)$ defined by (14-14)–(14-17). By appropriate choice of cost matrices, this controller will maintain desirably small excursions in $\boldsymbol{\delta}\mathbf{y}_{\mathrm{c}}(t_i)$, i.e., small deviations in $\mathbf{y}_{\mathrm{c}}(t_i)$ from $\mathbf{y}_{\mathrm{d}}$, while not exerting too much perturbation control $\boldsymbol{\delta}\mathbf{u}(t_i)$ to do so. Expressing it in terms of the original variables, we have

$$\mathbf{u}^*(t_i) - \mathbf{u}_0 = -\mathbf{G}_{\mathrm{c}}^*(t_i)[\mathbf{x}(t_i) - \mathbf{x}_0] \tag{14-165}$$

which can be rearranged into one of two convenient manners,

$$\mathbf{u}^*(t_i) = \mathbf{u}_0 - \mathbf{G}_{\mathrm{c}}^*(t_i)[\mathbf{x}(t_i) - \mathbf{x}_0] \tag{14-166a}$$

or

$$\mathbf{u}^*(t_i) = -\mathbf{G}_{\mathrm{c}}^*(t_i)\mathbf{x}(t_i) + [\boldsymbol{\Pi}_{22} + \mathbf{G}_{\mathrm{c}}^*(t_i)\boldsymbol{\Pi}_{12}]\mathbf{y}_{\mathrm{d}} \tag{14-166b}$$

These two forms are diagrammed in Fig. 14.9. In Fig. 14.9a, the values of $\mathbf{u}_0$ and $\mathbf{x}_0$ are typically considered as being calculated offline according to (14-156); this particular structure provides insights into the design of controllers for nonlinear systems as well, as will be seen in Chapter 15. Figure 14.9b has the structure of the generic controller of Fig. 13.3, with the LQ *regulator gain* $\mathbf{G}_{\mathrm{c}}^*(t_i)$ (obtained independent of any setpoint considerations) providing feedback compensation and $[\boldsymbol{\Pi}_{22} + \mathbf{G}_{\mathrm{c}}^*(t_i)\boldsymbol{\Pi}_{12}]$ constituting the feedforward compensation. This is particularly useful if we used the steady state constant gain

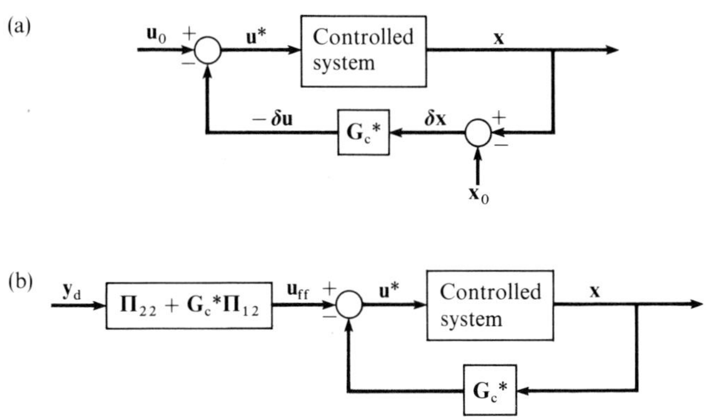


FIG. 14.9 Structure of nonzero setpoint controller. (a) Block diagram of Eq. (14-166a). (b) Block diagram of Eq. (14-166b).

$\bar{\mathbf{G}}_c{}^*$ for all time. An important point is that, as in the previous section, the *LQ optimal regulator* is a basic building block of the overall controller structure.

EXAMPLE 14.7 Again consider the angular servo problem, but assume we want to generate a controller that will achieve a gimbal angle equal to some commanded value $y_d$, where $y_c(t_i) = x(t_i)$, so $C = 1$ and $D_y = 0$. To do so, we must evaluate (14-155) as

$$\begin{bmatrix} \Pi_{11} & \Pi_{12} \\ \Pi_{21} & \Pi_{22} \end{bmatrix} = \begin{bmatrix} (\Phi - 1) & B_d \\ 1 & 0 \end{bmatrix}^{-1} = \begin{bmatrix} 0 & 1 \\ \dfrac{1}{B_d} & \dfrac{-(\Phi - 1)}{B_d} \end{bmatrix} = \begin{bmatrix} 0 & 1 \\ 5.6 & 1 \end{bmatrix}$$

and thus, by (14-156b), $x_0 = y_d$ and $u_0 = y_d$. The desired controller is given by (14-162) as

$$\begin{aligned} u^*(t_i) &= y_d - G_c{}^*(t_i)[x(t_i) - y_d] \\ &= -G_c{}^*(t_i)x(t_i) + [1 + G_c{}^*(t_i)]y_d \end{aligned}$$

with $G_c{}^*(t_i)$ given as in Example 14.2. ■

This controller is also readily extended to the case in which only noise-corrupted measurements are available from the system: a Kalman filter is put in cascade with the gain $\mathbf{G}_c{}^*(t_i)$ in Fig. 14.9b. For Fig. 14.9a, a *perturbation state Kalman filter* of the same structure could generate $\widehat{\delta \mathbf{x}}(t_i{}^+)$ by processing perturbation measurements

$$\delta \mathbf{z}(t_i) = \mathbf{z}(t_i) - \mathbf{z}_0 \tag{14-167}$$

where $\mathbf{z}_0$ are the noise-free measurements that would be generated from the system at equilibrium

$$\mathbf{z}_0 = \mathbf{H}\mathbf{x}_0 + \mathbf{D}_z\mathbf{u}_0 = [\mathbf{H}\mathbf{\Pi}_{12} + \mathbf{D}_z\mathbf{\Pi}_{22}]\mathbf{y}_d \tag{14-168}$$

This will also be developed more fully in Chapter 15.

Although this design achieves an asymptotically stable and robust closed-loop system *model* in which the controlled output $\mathbf{y}_c(t_i)$ attains the desired $\mathbf{y}_d$ with zero error in steady state, it suffers from a serious drawback. The steady state value of the true $\mathbf{y}_c(t_i)$ will *not* equal $\mathbf{y}_d$ if there are *any errors in modeling the actual physical system*, which there always will be. We would prefer a controller than maintains desired steady state system output despite modeling errors, with an asymptotically stable and robust closed loop. This, in part, motivates the "PI" designs of Section 14.9, as will some considerations in the next section.

## 14.8 REJECTION OF TIME-CORRELATED DISTURBANCES

Consider regulation or tracking in the face of time-correlated dynamics disturbances instead of, or in addition to, white dynamics noise [32, 58–61, 92]. Let the system of interest be described by

$$\mathbf{x}(t_{i+1}) = \mathbf{\Phi}(t_{i+1}, t_i)\mathbf{x}(t_i) + \mathbf{B}_d(t_i)\mathbf{u}(t_i) + \mathbf{n}_d(t_i) + \mathbf{w}_d(t_i) \tag{14-169}$$

where $\mathbf{n}_d(\cdot,\cdot)$ is time-correlated Gaussian noise, generated as the output of a shaping filter,

$$\mathbf{x}_n(t_{i+1}) = \mathbf{\Phi}_n(t_{i+1}, t_i)\mathbf{x}_n(t_i) + \mathbf{G}_{dn}(t_i)\mathbf{w}_{dn}(t_i) \tag{14-170a}$$

$$\mathbf{n}_d(t_i) = \mathbf{H}_n(t_i)\mathbf{x}_n(t_i) + \mathbf{D}_n(t_i)\mathbf{w}_{dn}(t_i) \tag{14-170b}$$

and $\mathbf{w}_{dn}(\cdot,\cdot)$ is discrete-time zero-mean white Gaussian noise with autocorrelation kernel $E\{\mathbf{w}_{dn}(t_i)\mathbf{w}_{dn}^T(t_j)\} = \mathbf{Q}_{dn}(t_i)\,\delta_{ij}$ with $\mathbf{Q}_{dn}(t_i) \geq \mathbf{0}$ for all $t_i$ (including the special case of $\mathbf{Q}_{dn} \equiv \mathbf{0}$); the initial conditions and driving noises of the original physical system model and shaping filter are all typically assumed pairwise independent. Now assume that we want to regulate $\mathbf{y}_c(t_i)$ as defined by (14-2) or (14-29) to zero; this encompasses the tracking case if we let (14-169) be an augmented system model. Thus, we want to minimize a cost of the form of (14-30), and therefore as also given by (14-13b), despite the influence of the disturbances $\mathbf{n}_d(t_i)$ in (14-169).

The LQG controller synthesis can be used if we express the problem equivalently in terms of the augmented state process

$$\mathbf{x}_a(\cdot,\cdot) \triangleq \begin{bmatrix} \mathbf{x}(\cdot,\cdot) \\ \mathbf{x}_n(\cdot,\cdot) \end{bmatrix} \tag{14-171}$$

similar to what was done in Section 14.6. We obtain

$$\begin{bmatrix} \mathbf{x}(t_{i+1}) \\ \mathbf{x}_n(t_{i+1}) \end{bmatrix} = \begin{bmatrix} \mathbf{\Phi}(t_{i+1}, t_i) & \mathbf{H}_n(t_i) \\ \mathbf{0} & \mathbf{\Phi}_n(t_{i+1}, t_i) \end{bmatrix} \begin{bmatrix} \mathbf{x}(t_i) \\ \mathbf{x}_n(t_i) \end{bmatrix} + \begin{bmatrix} \mathbf{B}_d(t_i) \\ \mathbf{0} \end{bmatrix} \mathbf{u}(t_i)$$

$$+ \begin{bmatrix} \mathbf{G}_d(t_i) & \mathbf{D}_n(t_i) \\ \mathbf{0} & \mathbf{G}_{dn}(t_i) \end{bmatrix} \begin{bmatrix} \mathbf{w}_d(t_i) \\ \mathbf{w}_{dn}(t_i) \end{bmatrix} \tag{14-172}$$

$$E\left\{\begin{bmatrix} \mathbf{w}_d(t_i) \\ \mathbf{w}_{dn}(t_i) \end{bmatrix} \begin{bmatrix} \mathbf{w}_d^T(t_j) & \mathbf{w}_{dn}^T(t_j) \end{bmatrix}\right\} = \begin{bmatrix} \mathbf{Q}_d(t_i) & \mathbf{0} \\ \mathbf{0} & \mathbf{Q}_{dn}(t_i) \end{bmatrix} \delta_{ij} \tag{14-173}$$

$$\mathbf{y}_c(t_i) = \begin{bmatrix} \mathbf{C}(t_i) & \mathbf{0} \end{bmatrix} \begin{bmatrix} \mathbf{x}(t_i) \\ \mathbf{x}_n(t_i) \end{bmatrix} \tag{14-174}$$

where the regulation of $\mathbf{y}_c(t_i)$ is accomplished by generating the control to minimize the cost as given by (14-142) and (14-143), but with $\mathbf{C}_a(t_i) = [\mathbf{C}(t_i) \,\vdots\, \mathbf{0}]$ as in (14-174). Note that a term of $[\mathbf{D}_y(t_i)\mathbf{u}(t_i)]$ added to (14-174) can be incorporated readily, via (14-130).

The first step is to generate the deterministic optimal LQ regulator, unrealistically assuming perfect knowledge of $\mathbf{x}(t_i)$ and $\mathbf{x}_n(t_i)$. As in Section 14.6, we apply (14-14)–(14-17) to the augmented system description and partition the result according to (14-171), to yield

$$\begin{aligned} \mathbf{u}^*(t_i) &= -\begin{bmatrix} \mathbf{G}_{c1}^*(t_i) & \mathbf{G}_{c2}^*(t_i) \end{bmatrix} \begin{bmatrix} \mathbf{x}(t_i) \\ \mathbf{x}_n(t_i) \end{bmatrix} \\ &= -\mathbf{G}_{c1}^*(t_i)\mathbf{x}(t_i) - \mathbf{G}_{c2}^*(t_i)\mathbf{x}_n(t_i) \end{aligned} \tag{14-175}$$

where

$$\mathbf{G}_{c1}^{*}(t_i) = [\mathbf{U}(t_i) + \mathbf{B}_d^{\mathrm{T}}(t_i)\mathbf{K}_{c11}(t_{i+1})\mathbf{B}_d(t_i)]^{-1} \times [\mathbf{B}_d^{\mathrm{T}}(t_i)\mathbf{K}_{c11}(t_{i+1})\boldsymbol{\Phi}(t_{i+1}, t_i) + \mathbf{S}_y^{\mathrm{T}}(t_i)\mathbf{C}(t_i)] \tag{14-176a}$$

$$\mathbf{G}_{c2}^{*}(t_i) = [\mathbf{U}(t_i) + \mathbf{B}_d^{\mathrm{T}}(t_i)\mathbf{K}_{c11}(t_{i+1})\mathbf{B}_d(t_i)]^{-1} \times [\mathbf{B}_d^{\mathrm{T}}(t_i)\mathbf{K}_{c12}(t_{i+1})\boldsymbol{\Phi}_n(t_{i+1}, t_i) + \mathbf{B}_d^{\mathrm{T}}(t_i)\mathbf{K}_{c11}(t_{i+1})\mathbf{H}_n(t_i)] \tag{14-176b}$$

with the partitions $\mathbf{K}_{c11}$ and $\mathbf{K}_{c12}$ given by the backward recursions

$$\begin{aligned} \mathbf{K}_{c11}(t_i) &= \mathbf{C}^{\mathrm{T}}\mathbf{Y}\mathbf{C} + \boldsymbol{\Phi}^{\mathrm{T}}\mathbf{K}_{c11}\boldsymbol{\Phi} - [\boldsymbol{\Phi}^{\mathrm{T}}\mathbf{K}_{c11}\mathbf{B}_d + \mathbf{C}^{\mathrm{T}}\mathbf{S}_y]\mathbf{G}_{c1}^{*} \\ \mathbf{K}_{c11}(t_{N+1}) &= \mathbf{C}^{\mathrm{T}}(t_{N+1})\mathbf{Y}_f\mathbf{C}(t_{N+1}) \end{aligned} \tag{14-177a}$$

$$\begin{aligned} \mathbf{K}_{c12}(t_i) &= \boldsymbol{\Phi}^{\mathrm{T}}\mathbf{K}_{c11}\mathbf{H}_n + \boldsymbol{\Phi}^{\mathrm{T}}\mathbf{K}_{c12}\boldsymbol{\Phi}_n \\ &\quad - \mathbf{G}_{c1}^{*\mathrm{T}}[\mathbf{B}_d^{\mathrm{T}}\mathbf{K}_{c12}\boldsymbol{\Phi}_n + \mathbf{B}_d^{\mathrm{T}}\mathbf{K}_{c11}\mathbf{H}_n] \\ \mathbf{K}_{c11}(t_{N+1}) &= \mathbf{0} \end{aligned} \tag{14-177b}$$

using the same time index convention as in (14-147). The third partition, $\mathbf{K}_{c22}(t_i)$, can also be evaluated by

$$\begin{aligned} \mathbf{K}_{c22}(t_i) &= \mathbf{H}_n^{\mathrm{T}}\mathbf{K}_{c11}\mathbf{H}_n + \boldsymbol{\Phi}_n^{\mathrm{T}}\mathbf{K}_{c12}^{\mathrm{T}}\mathbf{H}_n + \mathbf{H}_n^{\mathrm{T}}\mathbf{K}_{c12}\boldsymbol{\Phi}_n + \boldsymbol{\Phi}_n^{\mathrm{T}}\mathbf{K}_{c22}\boldsymbol{\Phi}_n \\ &\quad - \mathbf{G}_{c2}^{*\mathrm{T}}[\mathbf{B}_d\mathbf{K}_{c12}\boldsymbol{\Phi}_n + \mathbf{B}_d^{\mathrm{T}}\mathbf{K}_{c11}\mathbf{H}_n] \\ \mathbf{K}_{c22}(t_{N+1}) &= \mathbf{0} \end{aligned} \tag{14-177c}$$

As in Section 14.6, $\mathbf{K}_{c11}$ and $\mathbf{G}_{c1}^{*}$ can be generated *alone* via (14-176a) and (14-177a) to provide definition of the feedback connection from $\mathbf{x}(t_i)$ to $\mathbf{u}^{*}(t_i)$: they are *independent* of the properties of the disturbances $\mathbf{n}_d(\cdot,\cdot)$, and are found by solving the *deterministic* LQ optimal *regulator* problem with the disturbances totally *omitted*. The *feedforward* connection is defined by $\mathbf{G}_{c2}^{*}$ and $\mathbf{K}_{c12}$ computed via (14-176b) and (14-177b), with $\mathbf{K}_{c11}$ values supplied from the solution to (14-177a): they *do* depend on the disturbance properties, but the $\mathbf{K}_{c22}$ partition given by (14-177c) *need not be computed* [92]. This full-state feedback controller structure is shown in Fig. 14.10, using the same block diagram conventions as in Fig. 14.7.

If we restrict our attention to *time-invariant* system and shaping filter models and *constant* cost matrices, then we can implement the *simplified* controller composed of steady state *constant* gains $\bar{\mathbf{G}}_{c1}^{*}$ and $\bar{\mathbf{G}}_{c2}^{*}$ for all time. Similar to the case in Section 14.6, the shaping filter states are not controllable by $\mathbf{u}(t_i)$ and so we must assume that all eigenvalues of $\boldsymbol{\Phi}_n$ are of magnitude *strictly* less than one. Otherwise, the augmented system model would include modes that are simultaneously unstable, uncontrollable, and reconstructible, thereby not meeting the sufficient conditions for assuming the existence of constant $\bar{\mathbf{K}}_{ca}$ or finite cost for the augmented system LQ problem. Constant disturbances, generated as outputs of integrators (with corresponding $\boldsymbol{\Phi}_n$ eigenvalues at unity) will be discussed as a special case.

Although the full-state feedback control law decomposes as in Fig. 14.10, the Kalman filter in the LQG optimal stochastic controller generally does *not*

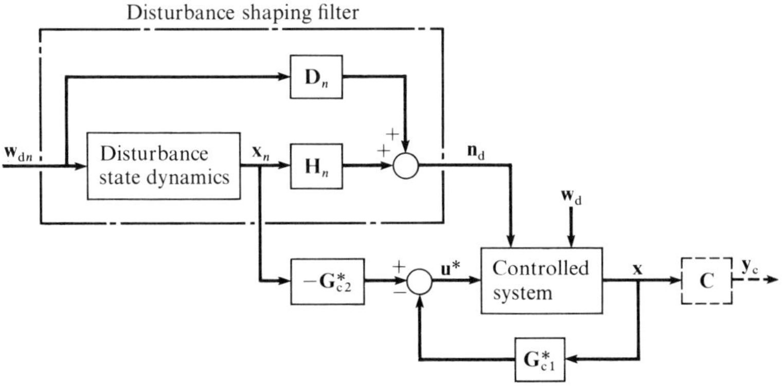


FIG. 14.10 Structure of full-state feedback disturbance rejection controller.

partition into two separate filters. Note that the augmented system state transition matrix in (14-172) is not block diagonal and that, even if the augmented system noise covariance $\mathbf{Q}_{da}(t_i)$ is block diagonal as in (14-173), $[\mathbf{G}_{da}(t_i)\mathbf{Q}_{da}(t_i)\mathbf{G}_{da}^T(t_i)]$ generally is not. Because of this intercoupling, the overall stochastic controller is *not* just the LQG optimal regulator with a separate filter/control-gain combination as a feedforward compensator, as seen earlier in Fig. 14.9 for trackers.

EXAMPLE 14.8 Recall the gimbaled platform problem of previous examples, but now let the system dynamics be described by

$$\dot{x}(t) = -(1/T)x(t) + (1/T)u(t) + n(t)$$

where $n(\cdot,\cdot)$ is a disturbance to be rejected, modeled as a stationary first order Gauss–Markov process, the steady state output of the shaping filter:

$$\dot{x}_n(t) = -(1/T_n)x_n(t) + w_n(t), \qquad n(t) = x_n(t)$$

with $E\{w_n(t)w_n(t+\tau)\} = Q_n\delta(\tau)$, and $Q_n = [2\sigma_n{}^2/T_n]$ where $\sigma_n$ = rms value of $n(\cdot,\cdot)$. The augmented continuous-time model is then

$$\begin{bmatrix} \dot{x}(t) \\ \dot{x}_n(t) \end{bmatrix} = \begin{bmatrix} -1/T & 1 \\ 0 & -1/T_n \end{bmatrix}\begin{bmatrix} x(t) \\ x_n(t) \end{bmatrix} + \begin{bmatrix} 1/T \\ 0 \end{bmatrix} u(t) + \begin{bmatrix} 0 \\ 1 \end{bmatrix} w_n(t)$$

$$y_c(t_i) = \begin{bmatrix} 1 & 0 \end{bmatrix}\begin{bmatrix} x(t) \\ x_n(t) \end{bmatrix}$$

The equivalent discrete-time model is given by

$$\begin{bmatrix} x(t_{i+1}) \\ x_n(t_{i+1}) \end{bmatrix} = \begin{bmatrix} e^{-\Delta t/T} & (TT_n)(e^{-\Delta t/T_n} - e^{-\Delta t/T})/(T_n - T) \\ 0 & e^{-\Delta t/T_n} \end{bmatrix}\begin{bmatrix} x(t_i) \\ x_n(t_i) \end{bmatrix}$$

$$+ \begin{bmatrix} (1 - e^{-\Delta t/T}) \\ 0 \end{bmatrix} u(t_i) + \begin{bmatrix} w_{da1}(t_i) \\ w_{da2}(t_i) \end{bmatrix}$$

$$y_c(t_i) = \begin{bmatrix} 1 & 0 \end{bmatrix}\begin{bmatrix} x(t_i) \\ x_n(t_i) \end{bmatrix}$$

where $\mathbf{w}_{da}(\cdot,\cdot)$ is zero-mean white Gaussian discrete-time noise of covariance

$$E\{\mathbf{w}_{da}(t_i)\mathbf{w}_{da}^T(t_i)\} = \int_{t_i}^{t_{i+1}} \mathbf{\Phi}_a(t_{i+1}-\tau)\mathbf{G}_a Q_n \mathbf{G}_a^T \mathbf{\Phi}_a^T(t_{i+1}-\tau)\,d\tau$$

$$\triangleq \mathbf{Q}_{da} = \begin{bmatrix} Q_{da11} & Q_{da12} \\ Q_{da12} & Q_{da22} \end{bmatrix}$$

As before, let $T = 0.5$ sec and $\Delta t = 0.1$ sec, and let the disturbance be described by $T_n = 1.0$ sec, so that the bandwidth of the disturbance is only half the bandpass of the system, and thus the disturbance would not be well modeled as white. Putting these numbers in yields

$$\begin{bmatrix} \mathrm{x}(t_{i+1}) \\ \mathrm{x}_n(t_{i+1}) \end{bmatrix} = \begin{bmatrix} 0.82 & 0.086 \\ 0 & 0.905 \end{bmatrix} \begin{bmatrix} \mathrm{x}(t_i) \\ \mathrm{x}_n(t_i) \end{bmatrix} + \begin{bmatrix} 0.18 \\ 0 \end{bmatrix} \mathrm{u}(t_i) + \begin{bmatrix} \mathrm{w}_{da1}(t_i) \\ \mathrm{w}_{da2}(t_i) \end{bmatrix}$$

with

$$\mathbf{Q}_{da} = \sigma_n^2 \begin{bmatrix} 0.00053 & 0.0085 \\ 0.0085 & 0.18 \end{bmatrix} \qquad \text{(not diagonal)}$$

Let us design the controller to minimize

$$J = E\left\{ \sum_{i=0}^{N} \frac{1}{2}[X\mathrm{x}(t_i)^2 + U\mathrm{u}(t_i)^2] + \frac{1}{2} X_f \mathrm{x}(t_{N+1})^2 \right\}$$

which is the simplest case with $S \equiv 0$. The solution is

$$u^*(t_i) = -G_{c1}^*(t_i)x(t_i) - G_{c2}^*(t_i)x_n(t_i)$$

where $G_{c1}^*(t_i)$ is the deterministic optimal LQ *regulator* gain for the problem omitting the disturbance: computation via (14-176a) and (14-177a) is identical to that of Example 14.2. With $K_{c11}$ so calculated, $K_{c12}$ is computed via (14-176b) and (14-177b) as

$$K_{c12}(t_i) = 0.0705K_{c11} + 0.74K_{c12} - \frac{0.0023K_{c11}^2 + 0.024K_{c11}K_{c12}}{U + 0.033K_{c11}}$$

$$G_{c2}^*(t_i) = \frac{0.0155K_{c11} + 0.16K_{c12}}{U + 0.033K_{c11}}$$

where all terms on the right hand side have time arguments of $t_{i+1}$.

In steady state operation, we obtain $\bar{K}_{c11}$, $\bar{G}_{c1}^*$, and $[\Phi - B_d\bar{G}_{c1}^*]$ results as in Examples 14.2 and 14.5, for $X = 1$ and $U$ assuming a range of values. The corresponding $\bar{K}_{c12}$ and $\bar{G}_{c2}^*$ values are as in the following table:

| $U$: | 0 | 0.1 | 1 | 10 | $\infty$ |
|---|---|---|---|---|---|
| $\bar{K}_{c12}$: | 0.00081 | 0.15 | 0.53 | 0.78 | — |
| $\bar{G}_{c2}^*$: | 0.47 | 0.32 | 0.12 | 0.017 | 0 |

When system and shaping filter states are not accessible perfectly, a two-state Kalman filter can be generated to provide $\hat{x}(t_i^+)$ and $\hat{x}_n(t_i^+)$ to replace $x(t_i)$ and $x_n(t_i)$, respectively. However, this filter does not decompose into two one-state filters. ■

Now let us consider the special case of rejecting *constant disturbances*, instead of stochastic processes which are zero-mean in steady state or deterministic inputs which converge asymptotically to zero [58–61, 92]. As part of the LQG synthesis procedure, consider the deterministic time-invariant model

$$\mathbf{x}(t_{i+1}) = \mathbf{\Phi x}(t_i) + \mathbf{B}_d \mathbf{u}(t_i) + \mathbf{d} \tag{14-178a}$$

$$\mathbf{y}_c(t_i) = \mathbf{C x}(t_i) \tag{14-178b}$$

and assume it is desired to regulate the controlled variable $\mathbf{y}_c(t_i)$ to zero despite the constant disturbance $\mathbf{d}$, typically of *unknown* magnitude. First, to motivate such a problem, let us consider a typical way in which $\mathbf{d}$ arises. Consider the linearization of a nonlinear time-invariant system model

$$\mathbf{x}(t_{i+1}) = \boldsymbol{\phi}[\mathbf{x}(t_i), \mathbf{u}(t_i), \mathbf{a}] \tag{14-179}$$

where $\mathbf{a}$ is a constant parameter vector of nominal value $\mathbf{a}_0$, but whose true value may differ from this by an amount $\boldsymbol{\delta}\mathbf{a} = [\mathbf{a} - \mathbf{a}_0]$. A nominal state and control for equilibrium conditions can be established by solving $\mathbf{x}(t_{i+1}) - \mathbf{x}(t_i) = \mathbf{0}$, or

$$\boldsymbol{\phi}[\mathbf{x}_0, \mathbf{u}_0, \mathbf{a}_0] - \mathbf{x}_0 = \mathbf{0} \tag{14-180}$$

to yield the desired $\mathbf{x}_0$ and $\mathbf{u}_0$. However, the equilibrium conditions are changed if $\mathbf{a} \neq \mathbf{a}_0$:

$$\boldsymbol{\phi}[\mathbf{x}_0, \mathbf{u}_0, \mathbf{a}] - \mathbf{x}_0 \neq \mathbf{0} \qquad \text{for} \quad \mathbf{a} \neq \mathbf{a}_0 \tag{14-181}$$

Now consider the linearization about the equilibrium conditions via Taylor series:

$$\boldsymbol{\delta}\mathbf{x}(t_{i+1}) = \left[\frac{\partial \boldsymbol{\phi}}{\partial \mathbf{x}}\right]\bigg|_{\mathbf{x}_0, \mathbf{u}_0, \mathbf{a}_0} \boldsymbol{\delta}\mathbf{x}(t_i) + \left[\frac{\partial \boldsymbol{\phi}}{\partial \mathbf{u}}\right]\bigg|_{\mathbf{x}_0, \mathbf{u}_0, \mathbf{a}_0} \boldsymbol{\delta}\mathbf{u}(t_i) + \left[\frac{\partial \boldsymbol{\phi}}{\partial \mathbf{a}}\right]\bigg|_{\mathbf{x}_0, \mathbf{u}_0, \mathbf{a}_0} \boldsymbol{\delta}\mathbf{a} + \cdots \tag{14-182}$$

The partials in this expression are constant, precomputable, and totally known, but $\boldsymbol{\delta}\mathbf{a}$ is *unknown and constant*. One might neglect the higher order terms in (14-182), and choose a quadratic cost criterion in part to justify being able to neglect them, to generate a perturbation control law designed to keep $\boldsymbol{\delta}\mathbf{x}(t_i)$ and $\boldsymbol{\delta}\mathbf{u}(t_i)$ small for all $t_i$. This gives rise to a linear time-invariant model as proposed in (14-178), with

$$\mathbf{d} = \left[\frac{\partial \boldsymbol{\phi}}{\partial \mathbf{a}}\right]\bigg|_{\mathbf{x}_0, \mathbf{u}_0, \mathbf{a}_0} \boldsymbol{\delta}\mathbf{a} \tag{14-183}$$

Let us show that straightforward LQ optimal regulator synthesis is not adequate for achieving the control objective. Consider trying to use the quadratic cost

$$J = \sum_{i=0}^{\infty} \frac{1}{2} [\mathbf{x}^T(t_i)\mathbf{X}\mathbf{x}(t_i) + \mathbf{u}^T(t_i)\mathbf{U}\mathbf{u}(t_i)] \tag{14-184}$$

in an effort to generate a steady state constant-gain controller. To prevent the cost from growing unbounded, we require

$$\mathbf{x}_{ss} = \lim_{i \to \infty} \mathbf{x}(t_i) = \mathbf{0} \tag{14-185}$$

and thus in steady state, (14-178a) becomes

$$\mathbf{0} = \mathbf{\Phi 0} + \mathbf{B}_d \mathbf{u}_{ss} + \mathbf{d} = \mathbf{B}_d \mathbf{u}_{ss} + \mathbf{d} \tag{14-186}$$

But this implies that the cost (14-184) will grow unbounded, since $\mathbf{u}_{ss} \neq \mathbf{0}$. Of course, one could use

$$J = \sum_{i=0}^{\infty} \frac{1}{2} [\mathbf{x}^T(t_i)\mathbf{X}\mathbf{x}(t_i) + \{\mathbf{u}(t_i) - \mathbf{u}_{ss}\}^T \mathbf{U} \{\mathbf{u}(t_i) - \mathbf{u}_{ss}\}] \tag{14-187}$$

once $\mathbf{u}_{ss}$ is generated as a solution to (14-186). However, this is a valid approach *only* if $\mathbf{d}$ is *known* a priori, which it typically is *not*, and so this form of controller is generally inadequate.

From classical control theory, we expect that by *adding an integral to the proportional control* afforded by $[-\mathbf{G}_c^*(t_i)\mathbf{x}(t_i)]$, we shall be able to handle both constant disturbances and the nonzero setpoints of the previous section, with zero steady state error in the deterministic case. In the next section, we formulate appropriate augmented system LQG optimal regulator problems to allow synthesis of such controllers.

## 14.9 THE LQG SYNTHESIS OF PI CONTROLLERS

Consider the following deterministic control problem with all states assumed accessible, which was motivated by the two previous sections, as part of an eventual stochastic control problem addressed by certainty equivalence. Let a system be described by the *time-invariant* model

$$\mathbf{x}(t_{i+1}) = \mathbf{\Phi x}(t_i) + \mathbf{B}_d \mathbf{u}(t_i) + \mathbf{d}; \qquad \mathbf{x}(t_0) = \text{given} \tag{14-188a}$$

$$\mathbf{y}_c(t_i) = \mathbf{C}\mathbf{x}(t_i) + \mathbf{D}_y \mathbf{u}(t_i) \tag{14-188b}$$

where $\mathbf{x}(t_i) \in R^n$, $\mathbf{u}(t_i) \in R^r$, and $\mathbf{y}_c(t_i) \in R^p$, and noting that (14-188b) specifically admits direct feedthrough. Assume that it is desired to generate a controller that will maintain $\mathbf{y}_c(t_i)$ at a nonzero desired value $\mathbf{y}_d$ (that may change its value infrequently) with zero steady state error, without being told the value of the disturbance $\mathbf{d}$, i.e., despite unmodeled constant disturbances. This is known as achieving the "type-1 property" [21, 155].

The nominal control $\mathbf{u}_0$ to hold the system at the desired condition when $\mathbf{d} = \mathbf{0}$ was given in Section 14.7 by (14-156), and perturbations about the nominal were described by (14-161) and (14-162). As discussed in that section, one can often assume that the number of controls equals the number of controlled outputs ($r = p$) and that the inverse in (14-156) exists; if this is not the case, (14-158) or (14-160) can replace that inverse. Merely designing an LQ optimal regulator for perturbations from this nominal equilibrium is often insufficient, because steady state $\mathbf{y}_c(t_i)$ will not equal $\mathbf{y}_d$ if there are modeling errors (causing miscalculated $[\mathbf{\Pi}_{22} + \mathbf{G}_c{}^*\mathbf{\Pi}_{12}]$ in (14-166b)) or unmodeled disturbances ($\mathbf{d}$ in (14-188a)) in the system representation. One seeks a controller structure that includes accepting the sensed error between desired $\mathbf{y}_d$ and the achieved $\mathbf{y}_c(t_i)$, and generating the appropriate control to keep this small. Initially, $\mathbf{y}_c(t_i)$ will be assumed to be known perfectly, computed as in (14-188b), and later it will be replaced by

$$\hat{\mathbf{y}}_c(t_i^+) = \mathbf{C}\hat{\mathbf{x}}(t_i^+) + \mathbf{D}_y\mathbf{u}(t_i) \tag{14-189a}$$

or

$$\hat{\mathbf{y}}_c(t_i^-) = \mathbf{C}\hat{\mathbf{x}}(t_i^-) + \mathbf{D}_y\mathbf{u}(t_i) \tag{14-189b}$$

using the outputs of a Kalman filter. To keep the system at a nonzero equilibrium condition such that steady state error is zero, the controller fed by the regulation error signal $[\mathbf{y}_d - \mathbf{y}_c(t_i)]$ must be able to deliver the appropriate *nonzero* steady state control when its *own input* is *zero*. This motivates the addition of *integral action* to form a *proportional-plus-integral* or *PI* controller, accomplished by introducing an additional set of dynamic variables, with defining difference equations, which is then augmented to the original states. These additional variables can correspond to either the *pseudointegral* or summation of the (negative) *regulation error*,

$$\begin{aligned}\mathbf{q}(t_i) &= \mathbf{q}(t_0) + \sum_{j=0}^{i-1} [\mathbf{y}_c(t_j) - \mathbf{y}_d] \\ &= \mathbf{q}(t_{i-1}) + [\mathbf{y}_c(t_{i-1}) - \mathbf{y}_d]\end{aligned} \tag{14-190}$$

or the *control differences* $\Delta\mathbf{u}(t_i)$, such that

$$\mathbf{u}(t_{i+1}) = \mathbf{u}(t_i) + \Delta\mathbf{u}(t_i) \tag{14-191}$$

where $\Delta\mathbf{u}(t_i)$ can be interpreted as a desired control rate times the sample period, $\Delta\mathbf{u}(t_i) \cong [\dot{\mathbf{u}}(t_i)\,\Delta t]$, and thus its alternate name of control "pseudorates."

For either augmented system model, a full-state feedback LQ optimal perturbation regulator can be designed. The LQ methodology provides a *systematic synthesis* procedure to generate a regulator with desired *stability properties ensured*, in which the appropriate *cross-feeds* in this multi-input/multi-output controller (whose evaluation may well not be very apparent from other design techniques) are dictated by system and cost descriptions, and in which *tradeoffs* between state and control amplitudes can be readily accomplished in the design process.

Before the PI controller synthesis [1, 2, 9, 20, 21, 32, 85, 86, 101, 133, 152, 169, 170] is detailed, two issues especially associated with PI controller design should be mentioned. First, *initial conditions* on controls and augmented states, as well as physical system states, must be considered carefully. Solving this initialization problem appropriately and *achieving good initial transient response* are one in the same important issue [21]. For our developments, we shall assume that, prior to time $t_0$, the input command was $\mathbf{y}_{\mathrm{d\,old}}$ and that the system had in fact reached the steady state conditions (see (14-156))

$$\mathbf{x}(t_j) = \mathbf{x}_{\mathrm{old}} = \mathbf{\Pi}_{12}\mathbf{y}_{\mathrm{d\,old}} \tag{14-192a}$$

$$\mathbf{u}(t_j) = \mathbf{u}_{\mathrm{old}} = \mathbf{\Pi}_{22}\mathbf{y}_{\mathrm{d\,old}} \tag{14-192b}$$

$$\mathbf{y}_{\mathrm{c}}(t_j) = \mathbf{y}_{\mathrm{d\,old}} \tag{14-192c}$$

for $j = -1, -2, \ldots$. At time $t_0$, a step change occurs in the command to $\mathbf{y}_{\mathrm{d}} \neq \mathbf{y}_{\mathrm{d\,old}}$ for $t_0, t_1, t_2, \ldots$, so that the desired steady state values to achieve from time $t_0$ forward are

$$\lim_{j\to\infty} \mathbf{x}(t_j) = \mathbf{x}_0 = \mathbf{\Pi}_{12}\mathbf{y}_{\mathrm{d}} \tag{14-193a}$$

$$\lim_{j\to\infty} \mathbf{u}(t_j) = \mathbf{u}_0 = \mathbf{\Pi}_{22}\mathbf{y}_{\mathrm{d}} \tag{14-193b}$$

$$\lim_{j\to\infty} \mathbf{y}_{\mathrm{c}}(t_j) = \mathbf{y}_{\mathrm{d}} \tag{14-193c}$$

as given by (14-155) and (14-156). Perturbation variables are defined with respect to these *new* steady state values as in (14-161), and we note specifically that

$$\begin{aligned}
\delta\mathbf{x}(t_{-1}) &= \mathbf{x}_{\mathrm{old}} - \mathbf{x}_0, & \delta\mathbf{x}(t_0) &= \mathbf{x}_{\mathrm{old}} - \mathbf{x}_0 \\
\delta\mathbf{u}(t_{-1}) &= \mathbf{u}_{\mathrm{old}} - \mathbf{u}_0, & \delta\mathbf{u}(t_0) &= \mathbf{u}(t_0) - \mathbf{u}_0 \\
\delta\mathbf{y}_{\mathrm{c}}(t_{-1}) &= \mathbf{y}_{\mathrm{d\,old}} - \mathbf{y}_{\mathrm{d}}, & \delta\mathbf{y}_{\mathrm{c}}(t_0) &= \mathbf{C}\mathbf{x}_{\mathrm{old}} + D_y\mathbf{u}(t_0) - \mathbf{y}_{\mathrm{d}}
\end{aligned} \tag{14-194}$$

Specifically, $\mathbf{y}_{\mathrm{d}}$ and $\mathbf{u}(t_0)$ can differ from their previous values, but $\mathbf{x}(t_0) = \mathbf{x}_{\mathrm{old}}$ since it cannot change instantaneously. This same conceptualization will be used in specifying appropriate initial conditions on the augmented state variables subsequently.

A second issue is that there are two possible implementation forms for discrete-time control laws, position and incremental [21]. In *position form*, the control "position" $\mathbf{u}(t_i)$ is specified in terms of the current position of the system state $\mathbf{x}(t_i)$, as in the steady state optimal LQ regulator

$$\mathbf{u}^*(t_i) = -\bar{\mathbf{G}}_{\mathrm{c}}^*\mathbf{x}(t_i) \tag{14-195}$$

In *incremental form*, by comparison, increments or changes in states and commands are used to generate appropriate increments in control relative to the previous control value; for example, subtracting $\mathbf{u}^*(t_{i-1})$ from $\mathbf{u}^*(t_i)$ in (14-195) yields the incremental form regulator as

$$\mathbf{u}^*(t_i) = \mathbf{u}^*(t_{i-1}) - \bar{\mathbf{G}}_c^*[\mathbf{x}(t_i) - \mathbf{x}(t_{i-1})] \tag{14-196}$$

The two forms are related to each other by a direct differencing manipulation, so they have the *same* basic input/output characteristics.

EXAMPLE 14.9 Consider the scalar state description

$$x(t_{i+1}) = \Phi x(t_i) + B_d u(t_i)$$

Applying the control law (14-195) yields a closed-loop system described by

$$x(t_{i+1}) = [\Phi - B_d \bar{G}_c^*] x(t_i)$$

with closed-loop system eigenvalue of $[\Phi - B_d \bar{G}_c^*]$. If the incremental law (14-196) is applied instead, the closed-loop system model is

$$x(t_{i+1}) = \Phi x(t_i) + B_d u^*(t_{i-1}) - B_d \bar{G}_c^*[x(t_i) - x(t_{i-1})]$$

or

$$\begin{bmatrix} x(t_{i+1}) \\ x(t_i) \end{bmatrix} = \begin{bmatrix} [\Phi - B_d \bar{G}_c^*] & B_d \bar{G}_c^* \\ 1 & 0 \end{bmatrix} \begin{bmatrix} x(t_i) \\ x(t_{i-1}) \end{bmatrix} + \begin{bmatrix} B_d \\ 0 \end{bmatrix} u^*(t_{i-1})$$

with closed-loop eigenvalues of $[\Phi - B_d \bar{G}_c^*]$ as before and 0, the latter corresponding to the one-sample pure delay that relates the lower partition of the augmented state to the upper partition of that augmented state at the previous sample time. ■

The position form is useful for gaining insights into the PI controller structure and operation that can enhance iterative design procedures. However, for actual implementation, an incremental form PI controller is generally preferable, for reasons to be discussed once the controllers are derived.

### *The PI Controller Based on Pseudointegral of Regulation Error* [1, 21]

Define a $p$-dimensional (same dimension as $\mathbf{y}_c(t_i)$ and $\mathbf{y}_d$) "pseudointegral" state $\mathbf{q}(t_i)$ as in (14-190), where

$$[\mathbf{y}_c(t_j) - \mathbf{y}_d] = [\mathbf{C}\mathbf{x}(t_j) - \mathbf{D}_y \mathbf{u}(t_j) - \mathbf{y}_d] \tag{14-197}$$

is the (negative) regulation error at time $t_j$. Since the system is assumed to be in equilibrium before $t_0$, $\mathbf{q}(t_0)$ is equal to the equilibrium value attained up to that time, $\mathbf{q}_{old}$, of value to be specified (done at the same time as $\mathbf{q}_0$ is established). Thus, this state satisfies

$$\mathbf{q}(t_{i+1}) = \mathbf{q}(t_i) + [\mathbf{y}_c(t_i) - \mathbf{y}_d] \tag{14-198}$$

The augmented system description then becomes

$$\begin{bmatrix} \mathbf{x}(t_{i+1}) \\ \mathbf{q}(t_{i+1}) \end{bmatrix} = \begin{bmatrix} \mathbf{\Phi} & \mathbf{0} \\ \mathbf{C} & \mathbf{I} \end{bmatrix} \begin{bmatrix} \mathbf{x}(t_i) \\ \mathbf{q}(t_i) \end{bmatrix} + \begin{bmatrix} \mathbf{B}_d \\ \mathbf{D}_y \end{bmatrix} \mathbf{u}(t_i) - \begin{bmatrix} \mathbf{0} \\ \mathbf{I} \end{bmatrix} \mathbf{y}_d \tag{14-199}$$

for $t_i \geq t_0$. As done previously, shift the defining coordinates and define perturbation variables as in (14-156), (14-161), and

$$\boldsymbol{\delta}\mathbf{q}(t_i) = \mathbf{q}(t_i) - \mathbf{q}_0 \tag{14-200}$$

where $\mathbf{q}_0$ is the new pseudointegral state steady state value, yet to be determined. This yields a zero setpoint regulation problem for the augmented perturbation system model

$$\begin{bmatrix} \boldsymbol{\delta}\mathbf{x}(t_{i+1}) \\ \boldsymbol{\delta}\mathbf{q}(t_{i+1}) \end{bmatrix} = \begin{bmatrix} \boldsymbol{\Phi} & \mathbf{0} \\ \mathbf{C} & \mathbf{I} \end{bmatrix} \begin{bmatrix} \boldsymbol{\delta}\mathbf{x}(t_i) \\ \boldsymbol{\delta}\mathbf{q}(t_i) \end{bmatrix} + \begin{bmatrix} \mathbf{B}_\mathrm{d} \\ \mathbf{D}_y \end{bmatrix} \boldsymbol{\delta}\mathbf{u}(t_i) \tag{14-201}$$

This can be combined with the quadratic cost

$$J = \sum_{i=0}^{N} \left\{ \frac{1}{2} \begin{bmatrix} \boldsymbol{\delta}\mathbf{x}(t_i) \\ \boldsymbol{\delta}\mathbf{q}(t_i) \\ \boldsymbol{\delta}\mathbf{u}(t_i) \end{bmatrix}^\mathrm{T} \begin{bmatrix} \mathbf{X}_{11} & \mathbf{X}_{12} & \mathbf{S}_1 \\ \mathbf{X}_{12}^\mathrm{T} & \mathbf{X}_{22} & \mathbf{S}_2 \\ \mathbf{S}_1^\mathrm{T} & \mathbf{S}_2^\mathrm{T} & \mathbf{U} \end{bmatrix} \begin{bmatrix} \boldsymbol{\delta}\mathbf{x}(t_i) \\ \boldsymbol{\delta}\mathbf{q}(t_i) \\ \boldsymbol{\delta}\mathbf{u}(t_i) \end{bmatrix} \right\} + \frac{1}{2} \begin{bmatrix} \boldsymbol{\delta}\mathbf{x}(t_{N+1}) \\ \boldsymbol{\delta}\mathbf{q}(t_{N+1}) \end{bmatrix}^\mathrm{T} \begin{bmatrix} \mathbf{X}_{\mathrm{f}11} & \mathbf{X}_{\mathrm{f}12} \\ \mathbf{X}_{\mathrm{f}12}^\mathrm{T} & \mathbf{X}_{\mathrm{f}22} \end{bmatrix} \begin{bmatrix} \boldsymbol{\delta}\mathbf{x}(t_{N+1}) \\ \boldsymbol{\delta}\mathbf{q}(t_{N+1}) \end{bmatrix} \tag{14-202}$$

to define the optimal regulator given by (14-6), with gains calculated via (14-14)–(14-17). Note that the desire to exert control to keep $\boldsymbol{\delta}\mathbf{q}(t_i)$ values small is reflected by explicitly including $\boldsymbol{\delta}\mathbf{q}(t_i)$ in this quadratic cost summation term. To generate a constant-gain control law, we can let $N \to \infty$ in (14-202), and solve for the steady state control law (14-77)–(14-82) under the assumptions associated with those equations. The solution is of the form

$$\boldsymbol{\delta}\mathbf{u}^*(t_i) = -[\bar{\mathbf{G}}_{\mathrm{c}1}^* \quad \bar{\mathbf{G}}_{\mathrm{c}2}^*] \begin{bmatrix} \boldsymbol{\delta}\mathbf{x}(t_i) \\ \boldsymbol{\delta}\mathbf{q}(t_i) \end{bmatrix} = -\bar{\mathbf{G}}_{\mathrm{c}1}^* \boldsymbol{\delta}\mathbf{x}(t_i) - \bar{\mathbf{G}}_{\mathrm{c}2}^* \boldsymbol{\delta}\mathbf{q}(t_i) \tag{14-203}$$

in which there is *no decoupling* of the algebraic Riccati equation of the form seen in Sections 14.6 and 14.8. Returning to the original coordinate system, (14-203) becomes

$$[\mathbf{u}^*(t_i) - \mathbf{u}_0] = -\bar{\mathbf{G}}_{\mathrm{c}1}^*[\mathbf{x}(t_i) - \mathbf{x}_0] - \bar{\mathbf{G}}_{\mathrm{c}2}^*[\mathbf{q}(t_i) - \mathbf{q}_0]$$

or

$$\mathbf{u}^*(t_i) = -\bar{\mathbf{G}}_{\mathrm{c}1}^*\mathbf{x}(t_i) - \bar{\mathbf{G}}_{\mathrm{c}2}^*\mathbf{q}(t_i) + [\mathbf{u}_0 + \bar{\mathbf{G}}_{\mathrm{c}1}^*\mathbf{x}_0 + \bar{\mathbf{G}}_{\mathrm{c}2}^*\mathbf{q}_0] \tag{14-204}$$

with $\mathbf{q}(t_i)$ calculated recursively by (14-197) and (14-198).

Now consider appropriately choosing $\mathbf{q}_0$ so as to yield best possible transient performance when (14-204) is applied. We seek the equilibrium value $\mathbf{q}_0$ that results in the lowest total cost to complete the process. From the LQ optimal perturbation controller design, we know that the total cost associated with using $\boldsymbol{\delta}\mathbf{u}^*(t_i)$ is

$$J_{\min} = \frac{1}{2} [\boldsymbol{\delta}\mathbf{x}^\mathrm{T}(t_0) \quad \boldsymbol{\delta}\mathbf{q}^\mathrm{T}(t_0)] \begin{bmatrix} \bar{\mathbf{K}}_{\mathrm{c}11} & \bar{\mathbf{K}}_{\mathrm{c}12} \\ \bar{\mathbf{K}}_{\mathrm{c}12}^\mathrm{T} & \bar{\mathbf{K}}_{\mathrm{c}22} \end{bmatrix} \begin{bmatrix} \boldsymbol{\delta}\mathbf{x}(t_0) \\ \boldsymbol{\delta}\mathbf{q}(t_0) \end{bmatrix} \tag{14-205}$$

The value of $\mathbf{q}_0$ that yields the lowest cost can be found by minimizing (14-205) with respect to $\boldsymbol{\delta}\mathbf{q}(t_0)$:

$$\mathbf{0} = \frac{\partial J_{\min}}{\partial[\boldsymbol{\delta}\mathbf{q}(t_0)]}^{\mathrm{T}} = \bar{\mathbf{K}}_{c22}\,\boldsymbol{\delta}\mathbf{q}(t_0) + \bar{\mathbf{K}}_{c12}^{\mathrm{T}}\,\boldsymbol{\delta}\mathbf{x}(t_0)$$

or, solving for $\boldsymbol{\delta}\mathbf{q}(t_0)$,

$$\begin{aligned}\boldsymbol{\delta}\mathbf{q}(t_0) = \mathbf{q}_{\text{old}} - \mathbf{q}_0 &= -\bar{\mathbf{K}}_{c22}^{-1}\bar{\mathbf{K}}_{c12}^{\mathrm{T}}[\mathbf{x}_{\text{old}} - \boldsymbol{\Pi}_{12}\mathbf{y}_{\text{d}}] \\ &= -\bar{\mathbf{K}}_{c22}^{-1}\bar{\mathbf{K}}_{c12}^{\mathrm{T}}\boldsymbol{\Pi}_{12}[\mathbf{y}_{\text{d old}} - \mathbf{y}_{\text{d}}]\end{aligned} \tag{14-206}$$

This can be viewed as the incremental verson of

$$\mathbf{q}_0 = -\bar{\mathbf{K}}_{c22}^{-1}\bar{\mathbf{K}}_{c12}^{\mathrm{T}}\boldsymbol{\Pi}_{12}\mathbf{y}_{\text{d}} \tag{14-207a}$$

$$\mathbf{q}_{\text{old}} = -\bar{\mathbf{K}}_{c22}^{-1}\bar{\mathbf{K}}_{c12}^{\mathrm{T}}\boldsymbol{\Pi}_{12}\mathbf{y}_{\text{d old}} \tag{14-207b}$$

which then establishes the desired steady state value of $\mathbf{q}(t_i)$ for $t_i \to \infty$ and for $t_i < t_0$. Now, substituting (14-207a) and (14-156) into (14-204) yields the final control law in position form as

$$\mathbf{u}^*(t_i) = -\bar{\mathbf{G}}_{c1}^*\mathbf{x}(t_i) - \bar{\mathbf{G}}_{c2}^*\mathbf{q}(t_i) + \mathbf{E}\mathbf{y}_{\text{d}}(t_i) \tag{14-208a}$$

$$\begin{aligned}\mathbf{q}(t_{i+1}) &= \mathbf{q}(t_i) + [\mathbf{y}_{\text{c}}(t_i) - \mathbf{y}_{\text{d}}(t_i)] \\ &= \mathbf{q}(t_i) + [\mathbf{C}\mathbf{x}(t_i) + \mathbf{D}_y\mathbf{u}(t_i) - \mathbf{y}_{\text{d}}(t_i)]\end{aligned} \tag{14-208b}$$

where $\bar{\mathbf{G}}_{c1}^*$ and $\bar{\mathbf{G}}_{c2}^*$ are from the LQ optimal perturbation regulator, and

$$\mathbf{E} = [\bar{\mathbf{G}}_{c1}^* - \bar{\mathbf{G}}_{c2}^*\bar{\mathbf{K}}_{c22}^{-1}\bar{\mathbf{K}}_{c12}^{\mathrm{T}}]\boldsymbol{\Pi}_{12} + \boldsymbol{\Pi}_{22} \tag{14-208c}$$

$$\mathbf{q}(t_0) = \mathbf{q}_{\text{old}} \tag{14-208d}$$

Notice that a time argument has been added to $\mathbf{y}_{\text{d}}$ to represent changes in setpoint, under the assumption that the system will essentially reach steady state conditions before $\mathbf{y}_{\text{d}}(t_i)$ changes again. The $t_i$ argument is clear in (14-208a), but one might question whether $\mathbf{y}_{\text{d}}(t_{i+1})$ should appear instead in (14-208b); the reason for being particularly careful about this will become more apparent in the next PI controller design. To resolve this, (14-208b) can be written for $i = -1$ to yield

$$\mathbf{q}(t_0) = \mathbf{q}(t_{-1}) + [\mathbf{y}_{\text{c}}(t_{-1}) - \mathbf{y}_{\text{d}}(t_{-1})] = \mathbf{q}_{\text{old}} + [\mathbf{y}_{\text{d old}} - \mathbf{y}_{\text{d}}(t_{-1})] = \mathbf{q}_{\text{old}}$$

as is appropriate; if $\mathbf{y}_{\text{d}}(t_i)$ were replaced by $\mathbf{y}_{\text{d}}(t_{i+1})$ in (14-208b), this equality would not hold. Thus, the *position form control law* is as given by (14-208), which is diagrammed in Fig. 14.11. Notice the sign changes in the feedforward portion of the loop to generate the error signal $[\mathbf{y}_{\text{d}}(t_i) - \mathbf{y}_{\text{c}}(t_i)]$ instead of $[\mathbf{y}_{\text{c}}(t_i) - \mathbf{y}_{\text{d}}(t_i)]$. The structure of this controller is seen to be composed of (1) *full-state feedback* through $\bar{\mathbf{G}}_{c1}^*$, (2) *command feedforward* through $\mathbf{E}$, and (3) *pseudointegration of the regulation error* in the forward path. As in past diagrams, Fig. 14.11 represents a "snapshot" of signals at time $t_i$. Specifically,

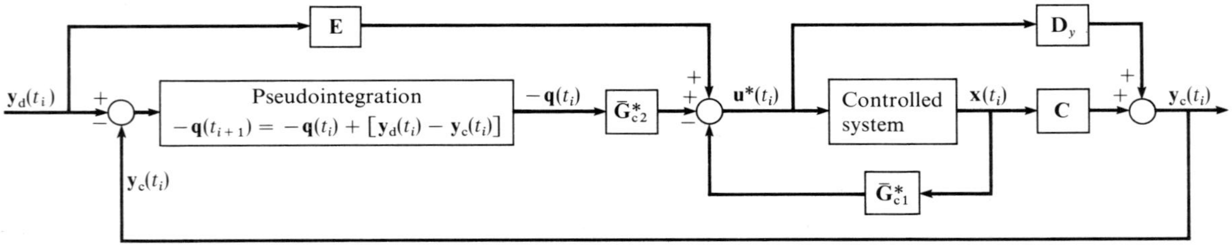


FIG. 14.11 Position form PI control law based on pseudointegral of regulation error.

$\mathbf{u}(t_i)$ depends on the pseudointegral (summation) of regulation errors through $[\mathbf{y}_d(t_{i-1}) - \mathbf{y}_c(t_{i-1})]$, and *not* through $[\mathbf{y}_d(t_i) - \mathbf{y}_c(t_i)]$, as expressed in this form (more will be said about this later).

By writing (14-208) for both $t_i$ and $t_{i-1}$, explicitly subtracting and rearranging, we get

$$\mathbf{u}^*(t_i) = \mathbf{u}^*(t_{i-1}) - \bar{\mathbf{G}}_{c1}^*[\mathbf{x}(t_i) - \mathbf{x}(t_{i-1})] - \bar{\mathbf{G}}_{c2}^*[\mathbf{q}(t_i) - \mathbf{q}(t_{i-1})] + \mathbf{E}[\mathbf{y}_d(t_i) - \mathbf{y}_d(t_{i-1})]$$

But, by using (14-208b), the *incremental form* of this PI law can be written conveniently as

$$\mathbf{u}^*(t_i) = \mathbf{u}^*(t_{i-1}) - \bar{\mathbf{G}}_{c1}^*[\mathbf{x}(t_i) - \mathbf{x}(t_{i-1})] + \bar{\mathbf{G}}_{c2}^*[\mathbf{y}_d(t_{i-1}) - \mathbf{y}_c(t_{i-1})] + \mathbf{E}[\mathbf{y}_d(t_i) - \mathbf{y}_d(t_{i-1})] \tag{14-209}$$

There is *no explicit pseudointegral state* in this form, so we do not have to be concerned about proper initialization of augmented states as we do in position forms. This is one reason why incremental forms are preferable; others will be discussed in Chapter 15 in conjunction with applying perturbation control laws to nonlinear systems.

It can be shown rigorously that either of these two forms has the desired type-1 property if $\det\{\bar{\mathbf{G}}_{c2}^*\} \neq 0$ [21, 155]. Heuristically, assume that the incremental form stabilizes the system so that *a* steady state is reached: from (14-209),

$$\mathbf{u}_{ss}^* = \mathbf{u}_{ss}^* - \bar{\mathbf{G}}_{c1}^*[\mathbf{x}_{ss} - \mathbf{x}_{ss}] + \bar{\mathbf{G}}_{c2}^*[\mathbf{y}_d - \mathbf{y}_{css}] + \mathbf{E}[\mathbf{y}_d - \mathbf{y}_d]$$

or

$$\bar{\mathbf{G}}_{c2}^*[\mathbf{y}_d - \mathbf{y}_{css}] = \mathbf{0} \tag{14-210}$$

Thus, if $\det\{\bar{\mathbf{G}}_{c2}^*\} \neq 0$, we can find $\bar{\mathbf{G}}_{c2}^{*-1}$ to solve this as

$$\mathbf{y}_d = \mathbf{y}_{css} \triangleq \lim_{i\to\infty} \mathbf{y}_c(t_i) \tag{14-211}$$

Moreover, consider the effect of *unmodeled* $\mathbf{d}$ on the original system, as in (14-188). Addition of $[\mathbf{d}^{\mathrm{T}} \,\vdots\, \mathbf{0}^{\mathrm{T}}]^{\mathrm{T}}$ to the right hand side of (14-199) does not affect the stability of the perturbation control law, so that *a* steady state value will be reached:

$$\lim_{i\to\infty} [\boldsymbol{\delta}\mathbf{x}(t_{i+1}) - \boldsymbol{\delta}\mathbf{x}(t_i)] = \mathbf{0} = \lim_{i\to\infty} [\mathbf{x}(t_{i+1}) - \mathbf{x}(t_i)] \tag{14-212a}$$

$$\lim_{i\to\infty} [\boldsymbol{\delta}\mathbf{q}(t_{i+1}) - \boldsymbol{\delta}\mathbf{q}(t_i)] = \mathbf{0} = \lim_{i\to\infty} [\mathbf{q}(t_{i+1}) - \mathbf{q}(t_i)] \tag{14-212b}$$

But, using (14-198), (14-212b) is equivalent to

$$\lim_{i\to\infty} [\mathbf{y}_c(t_i) - \mathbf{y}_d] = \mathbf{0} \tag{14-212c}$$

or, $\mathbf{y}_c(t_i)$ converges to $\mathbf{y}_d$ in the limit as $t_i \to \infty$, *despite* the effect of the unmodeled $\mathbf{d}$.

EXAMPLE 14.10 Recall the gimbaled platform problem introduced in Examples 14.2–14.4, with a scalar system model

$$\mathrm{x}(t_{i+1}) = \Phi \mathrm{x}(t_i) + B_{\mathrm{d}}\mathrm{u}(t_i) + \mathrm{w}_{\mathrm{d}}(t_i)$$
$$\mathrm{y}_{\mathrm{c}}(t_i) = \mathrm{x}(t_i) \qquad (C = 1, \quad D_y = 0)$$

with $\Phi = 0.82$ and $B_{\mathrm{d}} = 0.18$, but now we desire the system controlled output $y_{\mathrm{c}}(t_i)$ to track a (piecewise) constant $y_{\mathrm{d}}$ command initiated at $t_0$, in the face of *unmodeled* constant disturbances in the dynamics. Here, the nominal (i.e., without disturbance) augmented system model is, from (14-199)

$$\begin{bmatrix} x(t_{i+1}) \\ q(t_{i+1}) \end{bmatrix} = \begin{bmatrix} \Phi & 0 \\ 1 & 1 \end{bmatrix} \begin{bmatrix} x(t_i) \\ q(t_i) \end{bmatrix} + \begin{bmatrix} B_{\mathrm{d}} \\ 0 \end{bmatrix} u(t_i) - \begin{bmatrix} 0 \\ 1 \end{bmatrix} y_{\mathrm{d}}$$

From (14-155) and (14-156),

$$\begin{bmatrix} \Pi_{11} & \Pi_{12} \\ \Pi_{21} & \Pi_{22} \end{bmatrix} = \begin{bmatrix} (\Phi - 1) & B_{\mathrm{d}} \\ 1 & 0 \end{bmatrix}^{-1} = \begin{bmatrix} 0 & 1 \\ 1/B_{\mathrm{d}} & -(\Phi - 1)/B_{\mathrm{d}} \end{bmatrix}$$

and we can define perturbation states as in (14-161) and (14-200), with

$$x_0 = \Pi_{12} y_{\mathrm{d}} = y_{\mathrm{d}}, \qquad u_0 = \Pi_{22} y_{\mathrm{d}} = [(1 - \Phi)/B_{\mathrm{d}}] y_{\mathrm{d}}$$

and $q_0$ given by (14-207), i.e., $-(\bar{K}_{\mathrm{c}12}/\bar{K}_{\mathrm{c}22}) y_{\mathrm{d}}$, once the perturbation regulator is designed. The LQ synthesis via (14-77)–(14-82) yields

$$\delta u^*(t_i) = -\bar{G}_{\mathrm{c}1}^* \,\delta x(t_i) - \bar{G}_{\mathrm{c}2}^* \,\delta q(t_i)$$

where

$$[\bar{G}_{\mathrm{c}1}^* \ \vdots\ \bar{G}_{\mathrm{c}2}^*] = \frac{1}{U + B_{\mathrm{d}}{}^2 \bar{K}_{\mathrm{c}11}} [(B_{\mathrm{d}} \bar{K}_{\mathrm{c}11} \Phi + B_{\mathrm{d}} \bar{K}_{\mathrm{c}12} + S_1) \ \vdots\ (B_{\mathrm{d}} \bar{K}_{\mathrm{c}12} + S_2)]$$

in terms of the solution to the algebraic Riccati equation (14-77).

The position form control law is specified by (14-208),

$$u^*(t_i) = -\bar{G}_{\mathrm{c}1}^* x(t_i) - \bar{G}_{\mathrm{c}2}^* q(t_i) + E y_{\mathrm{d}}(t_i)$$
$$q(t_{i+1}) = q(t_i) + [x(t_i) - y_{\mathrm{d}}(t_i)]$$
$$E = [\bar{G}_{\mathrm{c}1}^* - \bar{G}_{\mathrm{c}2}^* \bar{K}_{\mathrm{c}12}/\bar{K}_{\mathrm{c}22}] + (1 - \Phi)/B_{\mathrm{d}}$$
$$q(t_0) = -(\bar{K}_{\mathrm{c}12}/\bar{K}_{\mathrm{c}22}) y_{\mathrm{d\,old}}$$

Applying this control to the original system produces a closed-loop deterministic system described by

$$\begin{bmatrix} x(t_{i+1}) \\ q(t_{i+1}) \end{bmatrix} = \begin{bmatrix} \Phi - B_{\mathrm{d}}\bar{G}_{\mathrm{c}1}^* & -B_{\mathrm{d}}\bar{G}_{\mathrm{c}2}^* \\ 1 & 1 \end{bmatrix} \begin{bmatrix} x(t_i) \\ q(t_i) \end{bmatrix} + \begin{bmatrix} E \\ -1 \end{bmatrix} y_{\mathrm{d}}(t_i)$$

$$y_{\mathrm{c}}(t_i) = [1 \quad 0] \begin{bmatrix} x(t_i) \\ q(t_i) \end{bmatrix}$$

with closed-loop characteristic equation

$$\lambda^2 - (\Phi - B_{\mathrm{d}}\bar{G}_{\mathrm{c}1}^* + 1)\lambda + (\Phi - B_{\mathrm{d}}\bar{G}_{\mathrm{c}1}^* + B_{\mathrm{d}}\bar{G}_{\mathrm{c}2}^*) = 0$$

The system is stable provided $|\lambda_i| < 1$ for $i = 1, 2$; if these are a complex conjugate pair, this condition becomes

$$(\Phi - B_{\mathrm{d}}\bar{G}_{\mathrm{c}1}^*)^2 - 2B_{\mathrm{d}}\bar{G}_{\mathrm{c}2}^* < 1$$

The incremental form control law is given by

$$u^*(t_i) = u^*(t_{i-1}) - \bar{G}^*_{c1}[x(t_i) - x(t_{i-1})] + \bar{G}^*_{c2}[y_d(t_{i-1}) - x(t_{i-1})] + E[y_d(t_i) - y_d(t_{i-1})]$$

Applying this to the original system yields the closed-loop system model

$$\begin{bmatrix} x(t_{i+1}) \\ x(t_i) \\ u(t_i) \end{bmatrix} = \begin{bmatrix} \Phi - B_d\bar{G}^*_{c1} & B_d(\bar{G}^*_{c1} - \bar{G}^*_{c2}) & B_d \\ 1 & 0 & 0 \\ -\bar{G}^*_{c1} & \bar{G}^*_{c1} - \bar{G}^*_{c2} & 1 \end{bmatrix} \begin{bmatrix} x(t_i) \\ x(t_{i-1}) \\ u(t_{i-1}) \end{bmatrix} + \begin{bmatrix} B_dE & B_d(\bar{G}^*_{c2} - E) \\ 0 & 0 \\ E & \bar{G}^*_{c2} - E \end{bmatrix} \begin{bmatrix} y_d(t_i) \\ y_d(t_{i-1}) \end{bmatrix}$$

with closed-loop characteristic equation

$$\lambda[\lambda^2 - (\Phi - B_d\bar{G}^*_{c1} + 1)\lambda + (\Phi - B_d\bar{G}^*_{c1} + B_d\bar{G}^*_{c2})] = 0$$

which has the same roots as in the position form case, plus the eigenvalue at zero that was seen to be characteristic of incremental forms in Example 14.9. ■

### *The PI Controller Based on Control Difference (Pseudorate)* [1, 21]

The $r$-dimensional control perturbation defined by (14-161b) is to be kept near zero in order to maintain the desired nonzero equilibrium condition. From this relation, one can write

$$\boldsymbol{\delta}\mathbf{u}(t_{i+1}) - \boldsymbol{\delta}\mathbf{u}(t_i) = [\mathbf{u}(t_{i+1}) - \mathbf{u}_0] - [\mathbf{u}(t_i) - \mathbf{u}_0]$$

or

$$\boldsymbol{\delta}\mathbf{u}(t_{i+1}) = \boldsymbol{\delta}\mathbf{u}(t_i) + [\mathbf{u}(t_{i+1}) - \mathbf{u}(t_i)] \tag{14-213}$$

This is the discrete-time equivalent of an integration with the term in brackets as an input. For instance, $[\mathbf{u}(t_{i+1}) - \mathbf{u}(t_i)]$ can be interpreted heuristically as $[\dot{\mathbf{u}}(t_i)]\,\Delta t$, so that (14-213) has the form of an Euler integration with "$\dot{\mathbf{u}}(t_i)$" as an input. If we define the *control difference* or *pseudorate* $\boldsymbol{\Delta}\mathbf{u}(t_i)$ as

$$\boldsymbol{\Delta}\mathbf{u}(t_i) \triangleq [\mathbf{u}(t_{i+1}) - \mathbf{u}(t_i)] \tag{14-214}$$

then (14-213) becomes

$$\boldsymbol{\delta}\mathbf{u}(t_{i+1}) = \boldsymbol{\delta}\mathbf{u}(t_i) + \boldsymbol{\Delta}\mathbf{u}(t_i) \tag{14-215}$$

This can be considered as an additional $r$-dimensional *state* equation to augment to the original system description, treating the *control pseudorate* as the *input* to a perturbation regulator problem. Note that, in steady state, we want $\mathbf{x}(t_i)$ and $\mathbf{u}(t_i)$ to approach the appropriate equilibrium values, i.e., $\boldsymbol{\delta}\mathbf{x}(t_i)$ and $\boldsymbol{\delta}\mathbf{u}(t_i)$ to stay small, and we also want $[\mathbf{u}(t_{i+1}) - \mathbf{u}(t_i)] \to \mathbf{0}$ as $t_i \to \infty$, so it makes sense to weight all of these quadratically in the cost definition.

For this alternate problem formulation, the perturbation state equations in augmented form are

$$\begin{bmatrix} \boldsymbol{\delta}\mathbf{x}(t_{i+1}) \\ \boldsymbol{\delta}\mathbf{u}(t_{i+1}) \end{bmatrix} = \begin{bmatrix} \boldsymbol{\Phi} & \mathbf{B}_d \\ \mathbf{0} & \mathbf{I} \end{bmatrix} \begin{bmatrix} \boldsymbol{\delta}\mathbf{x}(t_i) \\ \boldsymbol{\delta}\mathbf{u}(t_i) \end{bmatrix} + \begin{bmatrix} \mathbf{0} \\ \mathbf{I} \end{bmatrix} \boldsymbol{\Delta}\mathbf{u}(t_i) \tag{14-216}$$

and the cost to be minimized is

$$J = \sum_{i=-1}^{N} \left\{ \frac{1}{2} \begin{bmatrix} \boldsymbol{\delta}\mathbf{x}(t_i) \\ \boldsymbol{\delta}\mathbf{u}(t_i) \\ \boldsymbol{\Delta}\mathbf{u}(t_i) \end{bmatrix}^{\mathrm{T}} \begin{bmatrix} \mathbf{X}_{11} & \mathbf{X}_{12} & \mathbf{S}_1 \\ \mathbf{X}_{12}^{\mathrm{T}} & \mathbf{X}_{22} & \mathbf{S}_2 \\ \mathbf{S}_1^{\mathrm{T}} & \mathbf{S}_2^{\mathrm{T}} & \mathbf{U} \end{bmatrix} \begin{bmatrix} \boldsymbol{\delta}\mathbf{x}(t_i) \\ \boldsymbol{\delta}\mathbf{u}(t_i) \\ \boldsymbol{\Delta}\mathbf{u}(t_i) \end{bmatrix} \right\}$$
$$+ \frac{1}{2} \begin{bmatrix} \boldsymbol{\delta}\mathbf{x}(t_{N+1}) \\ \boldsymbol{\delta}\mathbf{u}(t_{N+1}) \end{bmatrix}^{\mathrm{T}} \begin{bmatrix} \mathbf{X}_{\mathrm{f}11} & \mathbf{0} \\ \mathbf{0} & \mathbf{0} \end{bmatrix} \begin{bmatrix} \boldsymbol{\delta}\mathbf{x}(t_{N+1}) \\ \boldsymbol{\delta}\mathbf{u}(t_{N+1}) \end{bmatrix} \tag{14-217}$$

As can be seen from (14-217), $\mathbf{U}$ specifies the weighting on control *differences* to prevent too high a rate of change of control commands, and $\mathbf{X}_{22}$ now specifies weights on control *magnitudes*. Since cost weighting on $\boldsymbol{\delta}\mathbf{u}(t_{N+1})$ does not make sense conceptually, $\mathbf{X}_{\mathrm{f}}$ has the form as indicated. Moreover, note that the summation term starts at $i = -1$ and not $i = 0$ as before. To understand why, recall (14-192)–(14-194): $\mathbf{u}_{\mathrm{old}} = \mathbf{u}(t_{-1}) = \mathbf{u}(t_{-2}) = \dots$, but $\mathbf{u}(t_0) \neq \mathbf{u}_{\mathrm{old}}$ in general, so that $\boldsymbol{\Delta}\mathbf{u}(t_i)$ given by (14-214) is zero for $i = -2, -3, \dots$, but

$$\boldsymbol{\Delta}\mathbf{u}(t_{-1}) = [\mathbf{u}(t_0) - \mathbf{u}(t_{-1})] = [\mathbf{u}(t_0) - \mathbf{u}_{\mathrm{old}}] \tag{14-218}$$

is not zero in general, and this change in control value should be weighted in the cost function *to provide desirable transient response* to a change in setpoint $\mathbf{y}_{\mathrm{d}}$—especially since this initial control difference is likely to have very large magnitude compared to all succeeding differences.

The optimal perturbation regulator is defined by (14-6) and (14-14)–(14-17). A constant-gain control law can be generated by letting $N \to \infty$ in (14-217), and solving for the steady state control law via (14-77)–(14-82). As in the previous PI controller case, there is no simplifying decoupling in the resulting Riccati equations. The steady state solution yields

$$\boldsymbol{\Delta}\mathbf{u}^*(t_i) = -\bar{\mathbf{G}}_{\mathrm{c}1}^*\, \boldsymbol{\delta}\mathbf{x}(t_i) - \bar{\mathbf{G}}_{\mathrm{c}2}^*\, \boldsymbol{\delta}\mathbf{u}(t_i) \tag{14-219}$$

Rather than leave this in terms of $\boldsymbol{\Delta}\mathbf{u}^*(t_i)$, we can combine it with (14-215) to yield the control law

$$\begin{aligned} \boldsymbol{\delta}\mathbf{u}^*(t_{i+1}) &= \boldsymbol{\delta}\mathbf{u}^*(t_i) - \bar{\mathbf{G}}_{\mathrm{c}1}^*\, \boldsymbol{\delta}\mathbf{x}(t_i) - \bar{\mathbf{G}}_{\mathrm{c}2}^*\, \boldsymbol{\delta}\mathbf{u}^*(t_i) \\ &= -\bar{\mathbf{G}}_{\mathrm{c}1}^*\, \boldsymbol{\delta}\mathbf{x}(t_i) + [\mathbf{I} - \bar{\mathbf{G}}_{\mathrm{c}2}^*]\, \boldsymbol{\delta}\mathbf{u}^*(t_i) \end{aligned} \tag{14-220}$$

Unfortunately, this control law does *not* yet have the type-1 property, and further manipulation is required to achieve this objective.

EXAMPLE 14.11 In the context of the gimbaled platform problem, the augmented perturbation state model is

$$\begin{bmatrix} \delta x(t_{i+1}) \\ \delta u(t_{i+1}) \end{bmatrix} = \begin{bmatrix} \Phi & B_{\mathrm{d}} \\ 0 & 1 \end{bmatrix} \begin{bmatrix} \delta x(t_i) \\ \delta u(t_i) \end{bmatrix} + \begin{bmatrix} 0 \\ 1 \end{bmatrix} \Delta u(t_i)$$

as defined about the equilibrium $x_0 = \Pi_{12} y_{\mathrm{d}} = y_{\mathrm{d}}$ and $u_0 = \Pi_{22} y_{\mathrm{d}} = [(1 - \Phi)/B_{\mathrm{d}}] y_{\mathrm{d}}$, with $\mathbf{\Pi}$ defined as in Example 14.10. The perturbation regulator is given by (14-219),

$$\Delta u^*(t_i) = -\bar{G}_{\mathrm{c}1}^*\, \delta x(t_i) - \bar{G}_{\mathrm{c}2}^*\, \delta u(t_i)$$

with

$$[\bar{G}_{c1}^* \,\vdots\, \bar{G}_{c2}^*] = \frac{1}{U + \bar{K}_{c22}} [(\bar{K}_{c12}\Phi + S_1) \,\vdots\, (\bar{K}_{c12}B_d + \bar{K}_{c22} + S_2)]$$

and $\bar{K}_{c11}$, $\bar{K}_{c12}$, and $\bar{K}_{c22}$ obtained from the solution to the steady state Riccati equation (14-77).

The control law (14-220) can be written as

$$\begin{aligned} u^*(t_{i+1}) &= u^*(t_i) - \bar{G}_{c1}^*[x(t_i) - \Pi_{12}y_d] - \bar{G}_{22}^*[u^*(t_i) - \Pi_{22}y_d] \\ &= [1 - \bar{G}_{c2}^*]u^*(t_i) - \bar{G}_{c1}^* x(t_i) + [\bar{G}_{c1}^*\Pi_{12} + \bar{G}_{c2}^*\Pi_{22}]y_d \end{aligned}$$

Consider applying this law to the original system with an unmodeled disturbance $d$ added, i.e.,

$$x(t_{i+1}) = \Phi x(t_i) + B_d u^*(t_i) + d$$

This yields a closed-loop system modeled as

$$\begin{bmatrix} x(t_{i+1}) \\ u(t_{i+1}) \end{bmatrix} = \begin{bmatrix} \Phi & B_d \\ -\bar{G}_{c1}^* & (1 - \bar{G}_{c2}^*) \end{bmatrix} \begin{bmatrix} x(t_i) \\ u(t_i) \end{bmatrix} + \begin{bmatrix} 0 \\ (\bar{G}_{c1}^*\Pi_{12} + \bar{G}_{c2}^*\Pi_{22} \end{bmatrix} y_d + \begin{bmatrix} 1 \\ 0 \end{bmatrix} d$$

Steady state conditions are achieved when

$$\begin{bmatrix} x(t_{i+1}) \\ u(t_{i+1}) \end{bmatrix} - \begin{bmatrix} x(t_i) \\ u(t_i) \end{bmatrix} = \begin{bmatrix} 0 \\ 0 \end{bmatrix}$$

or

$$\begin{bmatrix} (\Phi - 1) & B_d \\ -\bar{G}_{c1}^* & -\bar{G}_{c2}^* \end{bmatrix} \begin{bmatrix} x_{ss} \\ u_{ss} \end{bmatrix} + \begin{bmatrix} d \\ [\bar{G}_{c1}^* + \bar{G}_{c2}^*(1 - \Phi)/B_d]y_d \end{bmatrix} = \begin{bmatrix} 0 \\ 0 \end{bmatrix}$$

for which the solution is

$$\begin{aligned} \begin{bmatrix} x_{ss} \\ u_{ss} \end{bmatrix} &= \begin{bmatrix} (\Phi - 1) & B_d \\ -\bar{G}_{c1}^* & -\bar{G}_{c2}^* \end{bmatrix}^{-1} \begin{bmatrix} -d \\ -[\bar{G}_{c1}^* + \bar{G}_{c2}^*(1 - \Phi)/B_d]y_d \end{bmatrix} \\ &= \begin{bmatrix} y_d + \bar{G}_{c2}^*\{\bar{G}_{c1}^*B_d + \bar{G}_{c2}^*(1 - \Phi)\}^{-1}d \\ u_0 - \bar{G}_{c1}^*\{\bar{G}_{c1}^*B_d + \bar{G}_{c2}^*(1 - \Phi)\}^{-1}d \end{bmatrix} \end{aligned}$$

If $d$ were zero, $x_{ss} = y_d$ and $u_{ss} = u_0$ as desired. However, nonzero $d$ yields a steady state $x_{ss} = y_{css}$ that does *not* equal $y_d$, and thus the type-1 property is not attained. This controller lacks the type-1 property because of the appearance of nonzero $\bar{G}_{c2}^*$ in (14-220), destroying the *pure* numerical integration feature obtainable by accumulating (summing) past controls; note above that $x_{ss} = y_d$ if $\bar{G}_{c2}^*$ were zero. Nonzero $\bar{G}_{c2}^*$ provides *low-pass filtering* action though: taking the $z$-transform of the control law yields

$$zu(z) - [1 - \bar{G}_{c2}^*]u(z) = -\bar{G}_{c1}^* x(z) + [\bar{G}_{c1}^*\Pi_{12} + \bar{G}_{c2}^*\Pi_{22}]y_d(z)$$

so that the appropriate transfer functions are

$$\frac{u(z)}{x(z)} = \frac{-\bar{G}_{c1}^*}{z + [1 - \bar{G}_{c2}^*]}, \qquad \frac{u(z)}{y_d(z)} = \frac{[\bar{G}_{c1}^*\Pi_{12} + \bar{G}_{c2}^*\Pi_{22}]}{z + [1 - \bar{G}_{c2}^*]}$$

where the common denominator provides the low-pass filtering action, instead of pure integration as for $\bar{G}_{c2}^* = 0$. Such low-pass filtering is natural since we are penalizing high rates of change of control in the quadratic cost definition; it is also a useful characteristic since any system noises would also be low-pass filtered before impacting the control generation.

Finally, we note that the closed-loop system eigenvalues for this type-0 controller are solutions to

$$0 = [\lambda - \Phi][\lambda - (1 - \bar{G}_{c2}^*)] + B_d \bar{G}_{c1}^* = \lambda^2 - (\Phi - \bar{G}_{c2}^* + 1)\lambda + (\Phi - \Phi\bar{G}_{c2}^* + B_d \bar{G}_{c1}^*). \quad \blacksquare$$

Thus, we are motivated to manipulate (14-220) into a law with type-1 properties, if possible. Furthermore, we want to avoid leaving the control law in the form that merely attempts to drive deviations from *computed* steady state values (14-156) to zero, since $\mathbf{\Pi}_{12}$ and $\mathbf{\Pi}_{22}$ depend upon parameters of the system model, which may be uncertain: achieving those computed values in steady state do not ensure $\mathbf{y}_c(t_i)$ converging to $\mathbf{y}_d$ if there are modeling errors. Because of the stability afforded by the LQ regulator, $\mathbf{x}$, $\mathbf{u}$, and $\mathbf{y}_c$ *will* tend to *some* constant values, and to assure zero steady state error in response to a step, we want to incorporate a signal proportional to the (pseudo-)integral of the regulation error, $[\mathbf{y}_d - \mathbf{y}_c(t_i)]$, as in the previous PI controller.

It can be shown [1] that for the continuous-time analog of this problem, the result can be manipulated into the form

$$\mathbf{u}^*(t) = -\mathbf{K}_x\mathbf{x}(t) + \mathbf{K}_y[\mathbf{y}_d - \mathbf{y}_c(t)] + \mathbf{K}_\xi \int_{t_0}^{t} [\mathbf{y}_d - \mathbf{y}_c(\tau)]\, d\tau \qquad (14\text{-}221)$$

i.e., full-state feedback through a gain $\mathbf{K}_x$ and a PI controller in the forward path operating on the regulation error. From this analogy, it appears that a useful form to seek for the discrete-time case is

$$\mathbf{u}(t_i) = -\mathbf{K}_x\mathbf{x}(t_i) + \mathbf{K}_y[\mathbf{y}_d - \mathbf{y}_c(t_i)] + \mathbf{K}_\xi \sum_{j=0}^{i-1} [\mathbf{y}_d - \mathbf{y}_c(t_j)] \qquad (14\text{-}222)$$

Motivated by the structure of the previous discrete-time PI controller, (14-208) and (14-209), we will specifically seek a law of this form with $\mathbf{K}_y = \mathbf{0}$:

$$\mathbf{u}(t_i) = -\mathbf{K}_x\mathbf{x}(t_i) + \mathbf{K}_\xi \sum_{j=\{0 \text{ or } -1\}}^{i-1} [\mathbf{y}_d - \mathbf{y}_c(t_j)] \qquad (14\text{-}223\text{a})$$

in position form, where the lower limit might naturally be $-1$ instead of 0 due to the discussion surrounding (14-217) and (14-218) (to be discussed further, later), or

$$\mathbf{u}(t_i) = \mathbf{u}(t_{i-1}) - \mathbf{K}_x[\mathbf{x}(t_i) - \mathbf{x}(t_{i-1})] + \mathbf{K}_\xi[\mathbf{y}_d - \mathbf{y}_c(t_{i-1})] \quad (14\text{-}223\text{b})$$

in incremental form. Such an incremental law in terms of perturbation states becomes

$$\begin{aligned} \delta\mathbf{u}(t_{i+1}) &= \delta\mathbf{u}(t_i) - \mathbf{K}_x[\delta\mathbf{x}(t_{i+1}) - \delta\mathbf{x}(t_i)] + \mathbf{K}_\xi[-\delta\mathbf{y}_c(t_i)] \\ &= \delta\mathbf{u}(t_i) - \mathbf{K}_x[\delta\mathbf{x}(t_{i+1}) - \delta\mathbf{x}(t_i)] - \mathbf{K}_\xi[\mathbf{C}\,\delta\mathbf{x}(t_i) + \mathbf{D}_y\,\delta\mathbf{u}(t_i)] \end{aligned}$$

Noting that

$$\delta\mathbf{x}(t_{i+1}) - \delta\mathbf{x}(t_i) = [\mathbf{\Phi} - \mathbf{I}]\,\delta\mathbf{x}(t_i) + \mathbf{B}_d\,\delta\mathbf{u}(t_i)$$

this can be written as

$$\delta\mathbf{u}(t_{i+1}) = \delta\mathbf{u}(t_i) - [\mathbf{K}_x \quad \mathbf{K}_\xi]\begin{bmatrix} (\mathbf{\Phi} - \mathbf{I}) & \mathbf{B}_\mathrm{d} \\ \mathbf{C} & \mathbf{D}_y \end{bmatrix}\begin{bmatrix} \delta\mathbf{x}(t_i) \\ \delta\mathbf{u}(t_i) \end{bmatrix} \tag{14-224}$$

Now we compare this to the result obtained from the LQ regulator synthesis, namely (14-220), written as

$$\delta\mathbf{u}^*(t_{i+1}) = \delta\mathbf{u}^*(t_i) - [\bar{\mathbf{G}}_{\mathrm{c}1}^* \quad \bar{\mathbf{G}}_{\mathrm{c}2}^*]\begin{bmatrix} \delta\mathbf{x}(t_i) \\ \delta\mathbf{u}(t_i) \end{bmatrix} \tag{14-225}$$

Equations (14-224) and (14-225) are equivalent if we choose $\mathbf{K}_x$ and $\mathbf{K}_\xi$ to satisfy

$$[\mathbf{K}_x \quad \mathbf{K}_\xi]\begin{bmatrix} (\mathbf{\Phi} - \mathbf{I}) & \mathbf{B}_\mathrm{d} \\ \mathbf{C} & \mathbf{D}_y \end{bmatrix} = [\bar{\mathbf{G}}_{\mathrm{c}1}^* \quad \bar{\mathbf{G}}_{\mathrm{c}2}^*] \tag{14-226}$$

or

$$[\mathbf{K}_x \quad \mathbf{K}_\xi] = [\bar{\mathbf{G}}_{\mathrm{c}1}^* \quad \bar{\mathbf{G}}_{\mathrm{c}2}^*]\begin{bmatrix} (\mathbf{\Phi} - \mathbf{I}) & \mathbf{B}_\mathrm{d} \\ \mathbf{C} & \mathbf{D}_y \end{bmatrix}^{-1} = [\bar{\mathbf{G}}_{\mathrm{c}1}^* \quad \bar{\mathbf{G}}_{\mathrm{c}2}^*]\begin{bmatrix} \mathbf{\Pi}_{11} & \mathbf{\Pi}_{12} \\ \mathbf{\Pi}_{21} & \mathbf{\Pi}_{22} \end{bmatrix}$$

which yields the final result of

$$\mathbf{K}_x = \bar{\mathbf{G}}_{\mathrm{c}1}^*\mathbf{\Pi}_{11} + \bar{\mathbf{G}}_{\mathrm{c}2}^*\mathbf{\Pi}_{21} \tag{14-227a}$$

$$\mathbf{K}_\xi = \bar{\mathbf{G}}_{\mathrm{c}1}^*\mathbf{\Pi}_{12} + \bar{\mathbf{G}}_{\mathrm{c}2}^*\mathbf{\Pi}_{22} \tag{14-227b}$$

in terms of the previous regulator solution parameters and $\mathbf{\Pi}$, as the appropriate gains to use in (14-223).

Now assume that $\mathbf{y}_\mathrm{d}$ can undergo infrequent changes such that the system essentially reaches steady state again before the next change in $\mathbf{y}_\mathrm{d}$. Envision $\mathbf{y}_\mathrm{d}(t_i)$ changing from $\mathbf{y}_{\mathrm{d\,old}}$ to $\mathbf{y}_\mathrm{d}$ at time $t_0$, at which time we start to use the law (14-223b):

$$\mathbf{u}(t_0) = \mathbf{u}(t_{-1}) - \mathbf{K}_x[\mathbf{x}(t_0) - \mathbf{x}(t_{-1})] + \mathbf{K}_\xi[\mathbf{y}_\mathrm{d} - \mathbf{y}_\mathrm{c}(t_{-1})]$$

where $\mathbf{y}_\mathrm{d}$ can be replaced by $\mathbf{y}_\mathrm{d}(t_0) = \mathbf{y}_\mathrm{d}$ but *not* $\mathbf{y}_\mathrm{d}(t_{-1}) = \mathbf{y}_{\mathrm{d\,old}}$. Similar arguments at times before and after $t_0$ yield the final *incremental law* as

$$\mathbf{u}(t_i) = \mathbf{u}(t_{i-1}) - \mathbf{K}_x[\mathbf{x}(t_i) - \mathbf{x}(t_{i-1})] + \mathbf{K}_\xi[\mathbf{y}_\mathrm{d}(t_i) - \mathbf{y}_\mathrm{c}(t_{i-1})] \tag{14-228}$$

with gains evaluated as in (14-227); it is proper that the time indices of the last term do *not* match. The corresponding *position form law* is then

$$\mathbf{u}(t_i) = -\mathbf{K}_x\mathbf{x}(t_i) + \mathbf{K}_\xi \sum_{j=-1}^{i-1} [\mathbf{y}_\mathrm{d}(t_{j+1}) - \mathbf{y}_\mathrm{c}(t_j)] \tag{14-229}$$

where the lower limit oI $-1$ instead of 0 is appropriate not only due to the lower limit in the cost summation in (14-217), but also because this control law then *uses* $\mathbf{y}_\mathrm{d}(t_0)$, the value of the command from the instant it changes to the new

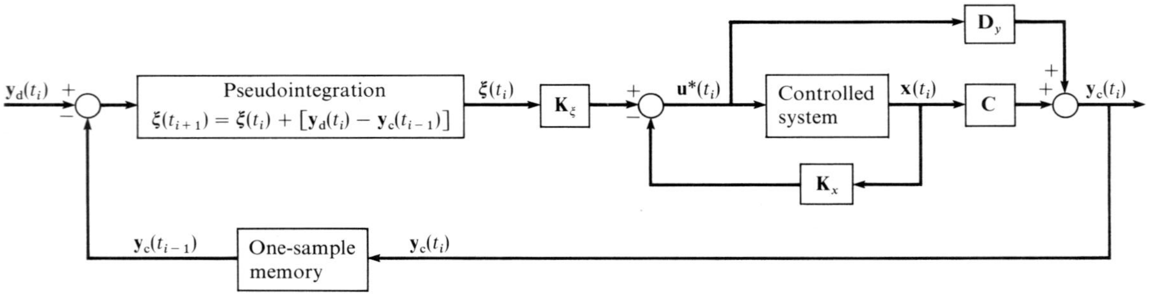


FIG. 14.12 Position form PI control law based on control differences (pseudorates).

desired value. This further motivates the decision to set $\mathbf{K}_y \equiv \mathbf{0}$ in (14-223), since such a term would become $\mathbf{K}_y[\mathbf{y}_d(t_{i+1}) - \mathbf{y}_c(t_i)]$ at time $t_i$, incorporating dependence on *future* commanded values. This position form control law can be expressed equivalently as

$$\mathbf{u}^*(t_i) = -\mathbf{K}_x\mathbf{x}(t_i) + \mathbf{K}_\xi\boldsymbol{\xi}(t_i) \tag{14-230a}$$

$$\boldsymbol{\xi}(t_{i+1}) = \boldsymbol{\xi}(t_i) + [\mathbf{y}_d(t_{i+1}) - \mathbf{y}_c(t_i)] \tag{14-230b}$$

with $\boldsymbol{\xi}(t_0) = [\mathbf{y}_d(t_0) - \mathbf{y}_c(t_{-1})]$ and gains given in (14-227); this is diagrammed in Fig. 14.12. It can be shown that this control law has the type-1 property, provided that $\det\{\mathbf{K}_\xi\} \neq 0$, using heuristic and rigorous arguments similar to those for the first PI controller form. The manipulation to achieve the type-1 characteristics, i.e., incorporation of the system dynamics model via (14-226), changed the input/output characteristics of the controller and removed the low-pass filtering operation from its structure.

EXAMPLE 14.12 Continuing from Example 14.11, we can generate a type-1 controller using the perturbation regulator result by writing

$$u^*(t_i) = -K_x x(t_i) + K_\xi \xi(t_i)$$
$$\xi(t_{i+1}) = \xi(t_i) + [y_d - x(t_i)]$$

where, from (14-227) and (14-155),

$$K_x = \bar{G}^*_{c1}\Pi_{11} + \bar{G}^*_{c2}\Pi_{21} = \bar{G}^*_{c2}/B_d$$
$$K_\xi = \bar{G}^*_{c1}\Pi_{12} + \bar{G}^*_{c2}\Pi_{22} = \bar{G}^*_{c1} + \bar{G}^*_{c2}(1 - \Phi)/B_d$$

Applying this control to the original system yields

$$\begin{bmatrix} x(t_{i+1}) \\ \xi(t_{i+1}) \end{bmatrix} = \begin{bmatrix} (\Phi - B_d K_x) & B_d K_\xi \\ -1 & 1 \end{bmatrix} \begin{bmatrix} x(t_i) \\ \xi(t_i) \end{bmatrix} + \begin{bmatrix} d \\ y_d \end{bmatrix}$$

The closed-loop eigenvalues are the solution to

$$0 = (\lambda - \Phi + B_d K_x)(\lambda - 1) + B_d K_\xi = \lambda^2 - (\Phi - \bar{G}^*_{c2} + 1)\lambda + (\Phi - \Phi\bar{G}^*_{c2} + B_d\bar{G}^*_{c1})$$

which are the *same* as for the previous type-0 controller (assuming no modeling errors in $\mathbf{\Pi}$). If the system is stabilized, some steady state is reached, so

$$0 = \lim_{i\to\infty} [\xi(t_{i+1}) - \xi(t_i)] = \lim_{i\to\infty} [y_d - x(t_i)]$$

or $x(t_i)$ converges to $y_d$ in steady state, even if there are unmodeled disturbances $d$ in the dynamics.

The type-1 controller in incremental form is

$$u^*(t_i) = u^*(t_{i-1}) - K_x[x(t_i) - x(t_{i-1})] + K_\xi[y_d - x(t_{i-1})]$$

and in steady state conditions, this yields

$$u_{ss} = u_{ss} - K_x[x_{ss} - x_{ss}] + K_\xi[y_d - x_{ss}]$$

and thus $x_{ss} = y_d$ provided $K_\xi \neq 0$. Putting this controller into the system yields a closed-loop configuration with eigenvalues as in the position form, plus an eigenvalue at zero, as discussed earlier. ■

*Generic Form of Digital PI Controller*

The two PI controllers just developed can be interrelated. In the latter development, we first obtained the incremental law (14-228) and then obtained the position form (14-229) as that law which would yield (14-228) by direct differencing. Another position form law that can yield (14-228) in this manner is

$$\mathbf{u}(t_i) = -\mathbf{K}_x\mathbf{x}(t_i) - \mathbf{K}_\xi\mathbf{q}(t_i) + \mathbf{K}_\xi\mathbf{y}_d(t_i) \tag{14-231a}$$

$$\mathbf{q}(t_{i+1}) = \mathbf{q}(t_i) + [\mathbf{y}_c(t_i) - \mathbf{y}_d(t_i)] \tag{14-231b}$$

where $\mathbf{q}(t_i)$ is identical to that given in (14-208) of the first PI controller. By comparing (14-208) and (14-231), a generic structural form becomes apparent:

$$\mathbf{u}^*(t_i) = -\mathscr{K}_x\mathbf{x}(t_i) + \mathscr{K}_1\boldsymbol{\varepsilon}(t_i) + \mathscr{K}_2\mathbf{y}_d(t_i) \tag{14-232a}$$

$$\boldsymbol{\varepsilon}(t_{i+1}) = \boldsymbol{\varepsilon}(t_i) + [\mathbf{y}_d(t_i) - \mathbf{y}_c(t_i)] \tag{14-232b}$$

where $\boldsymbol{\varepsilon}(t_i) = -\mathbf{q}(t_i)$ is the positive pseudointegral of the regulation error, and where the gains $(\mathscr{K}_x, \mathscr{K}_1, \mathscr{K}_2)$ are given by $(\bar{\mathbf{G}}_{c1}^*, \bar{\mathbf{G}}_{c2}^*, \mathbf{E})$ of (14-208) for the PI controller based on pseudointegrated regulation error and by $(\mathbf{K}_x, \mathbf{K}_\xi, \mathbf{K}_\xi)$ of (14-227) for the PI controller based on control differences. Note that $\mathscr{K}_1 = \mathscr{K}_2$ for the latter case. It can be shown rigorously that this generic controller has the type-1 property if $\det\{\mathscr{K}_1\} \neq 0$. The structure of (14-232) is a simple generalization of that depicted in Fig. 14.11.

To obtain another convenient means of expressing this generic law structure, add and subtract $\mathbf{y}_c(t_i)$ to the last term in (14-232a) to get

$$\mathbf{u}^*(t_i) = -\mathscr{K}_x\mathbf{x}(t_i) + \mathscr{K}_1\boldsymbol{\varepsilon}(t_i) + \mathscr{K}_2[\mathbf{y}_d(t_i) - \mathbf{y}_c(t_i) + \mathbf{C}\mathbf{x}(t_i) + \mathbf{D}_y\mathbf{u}(t_i)]$$

or

$$[\mathbf{I} - \mathscr{K}_2\mathbf{D}_y]\mathbf{u}^*(t_i) = -[\mathscr{K}_x - \mathscr{K}_2\mathbf{C}]\mathbf{x}(t_i) + \mathscr{K}_1\boldsymbol{\varepsilon}(t_i) + \mathscr{K}_2[\mathbf{y}_d(t_i) - \mathbf{y}_c(t_i)]$$

Assuming $[\mathbf{I} - \mathscr{K}_2\mathbf{D}_y]$ is nonsingular, this yields

$$\mathbf{u}^*(t_i) = [\mathbf{I} - \mathscr{K}_2\mathbf{D}_y]^{-1}\{-[\mathscr{K}_x - \mathscr{K}_2\mathbf{C}]\mathbf{x}(t_i) + \mathscr{K}_1\boldsymbol{\varepsilon}(t_i) + \mathscr{K}_2[\mathbf{y}_d(t_i) - \mathbf{y}_c(t_i)]\} \tag{14-233}$$

Thus, we have obtained the generic structure

$$\mathbf{u}^*(t_i) = -\mathscr{G}\mathbf{x}(t_i) + \mathscr{K}_p[\mathbf{y}_d(t_i) - \mathbf{y}_c(t_i)] + \mathscr{K}_i\boldsymbol{\varepsilon}(t_i) \tag{14-234a}$$

$$\boldsymbol{\varepsilon}(t_{i+1}) = \boldsymbol{\varepsilon}(t_i) + [\mathbf{y}_d(t_i) - \mathbf{y}_c(t_i)] \tag{14-234b}$$

where the *state feedback gain* $\mathscr{G}$ and *regulation error proportional gain* $\mathscr{K}_p$ *and integral gain* $\mathscr{K}_{\rm i}$ are given by

$$\begin{aligned}\mathscr{G} &= (\mathbf{I} - \mathbf{E}\mathbf{D}_y)^{-1}(\bar{\mathbf{G}}^*_{c1} - \mathbf{E}\mathbf{C})\\ \mathscr{K}_{\rm p} &= (\mathbf{I} - \mathbf{E}\mathbf{D}_y)^{-1}\bar{\mathbf{G}}^*_{c2}\\ \mathscr{K}_{\rm i} &= (\mathbf{I} - \mathbf{E}\mathbf{D}_y)^{-1}\mathbf{E}\end{aligned} \qquad (14\text{-}235)$$

for the PI controller based on augmenting pseudointegrated regulation error states, and by

$$\begin{aligned}\mathscr{G} &= (\mathbf{I} - \mathbf{K}_\xi\mathbf{D}_y)^{-1}(\mathbf{K}_x - \mathbf{K}_\xi\mathbf{C})\\ \mathscr{K}_{\rm p} &= (\mathbf{I} - \mathbf{K}_\xi\mathbf{D}_y)^{-1}\mathbf{K}_\xi\\ \mathscr{K}_{\rm i} &= \mathscr{K}_{\rm p}\end{aligned} \qquad (14\text{-}236)$$

for the PI controller based on augmenting control differences.

This structure is diagrammed in Fig. 14.13. Besides a full-state feedback, it encompasses a forward path of a proportional plus (pseudo-)integral compensator acting on the regulation error, where the "pseudointegral" is merely an implementation of (14-234b). Specifically evaluating $\mathscr{G}$, $\mathscr{K}_{\rm p}$, and $\mathscr{K}_{\rm i}$ associated with this structure provides meaningful insights for the design process. In the special case of no direct feedthrough in the controlled variable definition, i.e., $\mathbf{D}_y \equiv \mathbf{0}$, (14-234) and (14-235) simplify to $(\mathscr{G}, \mathscr{K}_{\rm p}, \mathscr{K}_{\rm i})$ given as $(\bar{\mathbf{G}}^*_{c1} - \mathbf{E}\mathbf{C}, \bar{\mathbf{G}}^*_{c2}, \mathbf{E})$ and $(\mathbf{K}_x - \mathbf{K}_\xi\mathbf{C}, \mathbf{K}_\xi, \mathbf{K}_\xi)$, respectively.

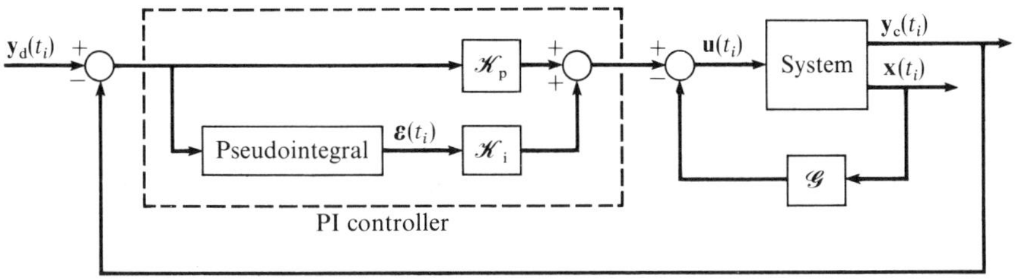


FIG. 14.13 Generic structure of controller.

The incremental form of this law is readily generated by direct differencing to yield

$$\begin{aligned}\mathbf{u}^*(t_i) = \mathbf{u}^*(t_{i-1}) &- \mathscr{G}[\mathbf{x}(t_i) - \mathbf{x}(t_{i-1})] + \mathscr{K}_{\rm p}[\mathbf{y}_d(t_i) - \mathbf{y}_c(t_i)]\\ &+ [\mathscr{K}_{\rm i} - \mathscr{K}_{\rm p}][\mathbf{y}_d(t_{i-1}) - \mathbf{y}_c(t_{i-1})]\end{aligned} \qquad (14\text{-}237)$$

For the PI controller based on augmenting control differences, the last term in (14-237) disappears. Note that there is no longer an explicit integrator state, with associated initial conditions, in (14-237). For applications in which the system is in fact nonlinear and linear techniques are being used to generate perturbation controls, i.e., for essentially *all* real applications of this synthesis, (14-237) provides simpler updating of nominals and initial conditions, and more direct compensation for an important phenomenon called windup (all of which will be discussed in Chapter 15), than (14-234), and thus is the preferred form.

EXAMPLE 14.13 The controller of Example 14.10 can be expressed as

$$u^*(t_i) = -\mathscr{G}x(t_i) + \mathscr{K}_{\mathrm{p}}[y_{\mathrm{d}} - x(t_i)] + \mathscr{K}_{\mathrm{i}}\varepsilon(t_i)$$

$$\varepsilon(t_{i+1}) = \varepsilon(t_i) + [y_{\mathrm{d}} - x(t_i)]$$

with

$$\mathscr{G} = \bar{G}^*_{\mathrm{c}1} - [\bar{G}^*_{\mathrm{c}1} - \bar{G}^*_{\mathrm{c}2}\bar{K}_{\mathrm{c}12}/\bar{K}_{\mathrm{c}22}] - (1-\Phi)/B_{\mathrm{d}}$$

$$\mathscr{K}_{\mathrm{p}} = \bar{G}^*_{\mathrm{c}2}$$

$$\mathscr{K}_{\mathrm{i}} = [\bar{G}^*_{\mathrm{c}1} - \bar{G}^*_{\mathrm{c}2}\bar{K}_{\mathrm{c}12}/\bar{K}_{\mathrm{c}22}] + (1-\Phi)/B_{\mathrm{d}}$$

where $\bar{G}^*_{\mathrm{c}1}$ is the gain on $\delta x(t_i)$ and $\bar{G}^*_{\mathrm{c}2}$ is the gain on $\delta q(t_i)$ in the perturbation controller given in (14-203). Similarly, the controller of Example 14.12 can be expressed in this generic form, with

$$\mathscr{G} = \bar{G}^*_{\mathrm{c}2}/B_{\mathrm{d}} - \bar{G}^*_{\mathrm{c}1} - \bar{G}^*_{\mathrm{c}2}(1-\Phi)/B_{\mathrm{d}} = \bar{G}^*_{\mathrm{c}2}\Phi/B_{\mathrm{d}} - \bar{G}^*_{\mathrm{c}1}$$

$$\mathscr{K}_{\mathrm{p}} = \bar{G}^*_{\mathrm{c}1} + \bar{G}^*_{\mathrm{c}2}(1-\Phi)/B_{\mathrm{d}}$$

$$\mathscr{K}_{\mathrm{i}} = \bar{G}^*_{\mathrm{c}1} + \bar{G}^*_{\mathrm{c}2}(1-\Phi)/B_{\mathrm{d}}$$

where $\bar{G}^*_{\mathrm{c}1}$ is the gain on $\delta x(t_i)$ and $\bar{G}^*_{\mathrm{c}2}$ is the gain on $\delta u(t_i)$ in the perturbation controller given by (14-219). ■

### *Extensions*

The PI controllers of this section have been derived by augmenting the original system state model with $p$ regulation error pseudointegral states or $r$ control difference variables (plus a manipulation using system dynamics), in order to achieve the type-1 property. If we augment the system states with *both* $p$ regulation error pseudointegrals and $r$ control differences, and solve for the optimal regulator of that augmented system, we can generate what is known as a *proportional plus integral plus filter*, or *PIF*, control law. In this form, the regulation pseudointegral states yield a type-1 PI structure, while the control differences yield a low-pass filter operation in series with the system input (*not* to be removed by a manipulation as invoked previously). Such a form allows specification of both tracking error desired characteristics and control rate constraints via separate terms in the cost function. This form is developed in detail in Problem 14.33.

On the other hand, by augmenting $Mp$ states corresponding to $M$ successive pseudointegrals of the regulation error, or $Mr$ states as $M$ successive differences (pseudoderivatives) of control, a type-$M$ system can be achieved: able to track constants, ramps, parabolas, . . . , $M$th order curves with zero steady state error, despite unmodeled disturbances of these same forms. A useful extension of these ideas leads to the command generator tracker developed in the next section.

The controllers of this section have assumed noise-free access to all states and controlled variables. This has been done only as the first step of the LQG synthesis process: a Kalman filter (or observer) can be used to generate $\hat{\mathbf{x}}(t_i^+)$ and $\hat{\mathbf{y}}_c(t_i^+)$ as in (14-189a), or suboptimally $\hat{\mathbf{x}}(t_i^-)$ and $\hat{\mathbf{y}}_c(t_i^-)$ as in (14-189b) for reduced computational delay time, to replace $\mathbf{x}(t_i)$ and $\mathbf{y}_c(t_i)$. This is accomplished at a loss to stability robustness, but robustness recovery techniques as discussed in Section 14.5 can be used to minimize this degradation, if required.

## 14.10 COMMAND GENERATOR TRACKING

A natural extension and generalization of the controllers of the previous sections is provided by a *command generator tracker* (*CGT*), also known as a model reference tracker [17, 18, 129]. Here we require a system to *respond to command inputs*, while *rejecting disturbances*, so that the states of the system maintain *desired trajectories* in real time. Both the desired trajectories (to be tracked) and the disturbances (to be rejected) are formulated as the outputs of linear system models: the first of these models is termed the *command generator* or model reference. This command generator can produce a time history of variables that the controlled system is to follow, such as a missile pitchover command sequence for ground launching, an approach sequence for V/STOL aircraft to transition from maneuvering flight to descent to hover, or an optimal evasive maneuver for a fighter aircraft. Or, the command generator could also specify a preferred system response model that the actual system should mimic, as for improved handling qualities of aircraft [26, 44].

In the open-loop conceptualization, the problem is to find the appropriate *feedforward* gains from the states of both the command generator and the disturbance linear model (shaping filter), to the system control inputs, that will yield "command generator tracking." Typically the objective is *asymptotic tracking*, in which we require the difference between the true system output and the command generator output to approach zero as time increases. "*Perfect tracking*" is a special case of this, in which the error must be zero for all time; it will provide useful insights into the synthesis of the desired controller.

If the original system is *unstable* or just *marginally stable*, or if *unmodeled disturbances* and *uncertainties* affect the system, we are motivated to add stabilizing control via *feedback*. Assuming full state knowledge, deterministic LQ regulator theory can be used to design regulators or compensators with PI structure to generate this feedback. In the case in which perfect full-state knowledge is not available, Kalman filters or observers can be used to generate state estimates for the feedback controller. Of course, eigenvalue/eigenvector placement procedures can be used in the design of the feedback controllers and/or observers, but we will develop the LQG synthesis methodology.

Consider a system described by the *linear time-invariant model*

$$\mathbf{x}(t_{i+1}) = \mathbf{\Phi}\mathbf{x}(t_i) + \mathbf{B}_d\mathbf{u}(t_i) + \mathbf{E}_x\mathbf{n}_d(t_i) + \mathbf{w}_d(t_i) \tag{14-238a}$$

$$\mathbf{y}_c(t_i) = \mathbf{C}\mathbf{x}(t_i) + \mathbf{D}_y\mathbf{u}(t_i) + \mathbf{E}_y\mathbf{n}_d(t_i) \tag{14-238b}$$

assuming initially that the whole state $\mathbf{x}(t_i)$ and controlled variable vector $\mathbf{y}_c(t_i)$ are perfectly accessible (to be replaced eventually by Kalman filter estimates) and that (14-238) is stabilizable and detectable. In these equations, $\mathbf{n}_d(\cdot,\cdot)$ is a disturbance, modeled by

$$\mathbf{n}_d(t_{i+1}) = \mathbf{\Phi}_n \mathbf{n}_d(t_i) + \mathbf{B}_{dn}\mathbf{n}_{cmd}(t_i) + \mathbf{G}_{dn}\mathbf{w}_{dn}(t_i) \tag{14-239}$$

with $\mathbf{n}_{cmd}(t_i)$ the commanded input to this disturbance shaping filter; for the simplest form of CGT law as developed here, we assume $\mathbf{n}_{cmd}(t_i) \equiv \mathbf{0}$; $\mathbf{w}_{dn}(\cdot,\cdot)$ is zero-mean white Gaussian discrete-time noise of covariance $\mathbf{Q}_{dn}$ that is independent of $\mathbf{w}_d(\cdot,\cdot)$. The controlled system modeled by (14-238) is to duplicate (as closely as possible) the output of a command generator model

$$\mathbf{x}_m(t_{i+1}) = \mathbf{\Phi}_m \mathbf{x}_m(t_i) + \mathbf{B}_{dm}\mathbf{u}_m \tag{14-240a}$$

$$\mathbf{y}_m(t_i) = \mathbf{C}_m \mathbf{x}_m(t_i) + \mathbf{D}_m\mathbf{u}_m \tag{14-240b}$$

where we note that, although $\mathbf{x}_m(t_i)$ and $\mathbf{u}_m$ need not be of the same dimension as $\mathbf{x}(t_i)$ and $\mathbf{u}(t_i)$, respectively, $\mathbf{y}_m(t_i)$ and $\mathbf{y}_c(t_i)$ are both of dimension $p$. Again, for the simplest CGT law generation, we assume that the command generator control input is a *constant*, $\mathbf{u}_m(t_i) \equiv \mathbf{u}_m$ for all $t_i \geq t_0$. In actual applications, the model reference control input is typically slowly varying relative to the chosen sample period, so $\mathbf{u}_m$ can be considered piecewise constant with long periods of essentially constant value. Usually, $\mathbf{u}_m$ is nonzero, and thus the command generator output is typically nonzero in steady state. The controller to induce the system to track this generated command should be able to track this nonzero steady state value (i.e., respond to a step input at least) with zero steady state error. Since we want a type-1 property in many cases, we often consider a PI controller design instead of a simple regulator in conjunction with a command generator, for eventual closed-loop implementation.

The command generator (14-240) in its simplest form could merely generate a *constant* desired output by letting $\mathbf{\Phi}_m = \mathbf{I}$, $\mathbf{B}_{dm} = \mathbf{0}$, $\mathbf{C}_m = \mathbf{0}$, and $\mathbf{D}_m = \mathbf{I}$ to yield

$$\mathbf{y}_m(t_i) = \mathbf{u}_m = \text{const} \tag{14-241a}$$

Or, a *constant rate-of-change* can be effected by letting $\mathbf{\Phi}_m = \mathbf{I}$, $\mathbf{B}_{dm} = \mathbf{I}\,\Delta t$, $\mathbf{C}_m = \mathbf{I}$, and $\mathbf{D}_m = \mathbf{0}$, giving:

$$\mathbf{x}_m(t_{i+1}) = \mathbf{x}_m(t_i) + \mathbf{u}_m\,\Delta t, \qquad \mathbf{y}_m(t_i) = \mathbf{x}_m(t_i) \tag{14-241b}$$

in which $\mathbf{u}_m$ is treated as the commanded constant value of $\dot{\mathbf{y}}_m$. Both of these special cases have been used in the design of digital flight control systems known as command augmentation systems (CAS) to regulate an aircraft about the pilot's command vector $\mathbf{u}_m$; for example, in lateral-directional dynamics, he can input either a constant commanded lateral velocity and yaw angle, or constant rates of change of these states (i.e., sideslip rate and turn rate) to undergo a turn. These and other special cases will be demonstrated in the examples.

Mathematically, the objective of the CGT is to force the error

$$\mathbf{e}(t_i) = \mathbf{y}_c(t_i) - \mathbf{y}_m(t_i)$$
$$= [\mathbf{C} \quad \mathbf{D}_y \quad \mathbf{E}_y] \begin{bmatrix} \mathbf{x}(t_i) \\ \mathbf{u}(t_i) \\ \mathbf{n}_d(t_i) \end{bmatrix} - [\mathbf{C}_m \quad \mathbf{D}_m] \begin{bmatrix} \mathbf{x}_m(t_i) \\ \mathbf{u}_m \end{bmatrix} \tag{14-242}$$

to zero. When this command error is zero (for the case of the zero-mean stochastic driving noises removed, or, when the mean error is zero in the stochastic case), the controlled system, or "plant," states and controls are said to be tracking the *ideal plant trajectory*. This ideal plant trajectory serves as a notational convenience and as an aid in constructing the CGT law itself. It is the set of time histories the plant states and controls must follow so that the true plant output perfectly matches the model output while modeled disturbances affect the plant: $\mathbf{x}_I(t_i)$ and $\mathbf{u}_I(t_i)$ for all $t_i$ that both satisfy the plant dynamics,

$$\mathbf{x}_I(t_{i+1}) = \mathbf{\Phi}\mathbf{x}_I(t_i) + \mathbf{B}_d\mathbf{u}_I(t_i) + \mathbf{E}_x\mathbf{n}_d(t_i) \tag{14-243}$$

and cause the command error to be zero,

$$[\mathbf{C} \quad \mathbf{D}_y \quad \mathbf{E}_y] \begin{bmatrix} \mathbf{x}_I(t_i) \\ \mathbf{u}_I(t_i) \\ \mathbf{n}_d(t_i) \end{bmatrix} = [\mathbf{C}_m \quad \mathbf{D}_m] \begin{bmatrix} \mathbf{x}_m(t_i) \\ \mathbf{u}_m \end{bmatrix} \tag{14-244}$$

A third requirement is also levied on the ideal plant response: that the ideal plant trajectory be a *linear* function of the model state, model control, and disturbance state (and disturbance control input for more complex forms where $\mathbf{n}_{cmd}(t_i)$ in (14-239) is not assumed identically equal to zero):

$$\begin{bmatrix} \mathbf{x}_I(t_i) \\ \mathbf{u}_I(t_i) \end{bmatrix} = \begin{bmatrix} \mathbf{A}_{11} & \mathbf{A}_{12} & \mathbf{A}_{13} \\ \mathbf{A}_{21} & \mathbf{A}_{22} & \mathbf{A}_{23} \end{bmatrix} \begin{bmatrix} \mathbf{x}_m(t_i) \\ \mathbf{u}_m \\ \mathbf{n}_d(t_i) \end{bmatrix} \tag{14-245}$$

where we recall that $\mathbf{u}_m(t_i) \equiv \mathbf{u}_m$ for all $t_i$ for this development. Equation (14-245) is an assumed, desirable form. To gain insight into the reasonableness of this form, consider the simple case of nonzero setpoint control while rejecting a *known* constant disturbance $\mathbf{d}_0$ (see Sections 14.7 and 14.8); i.e., (14-238)–(14-240), (14-243), and (14-244) with $\mathbf{E}_x$, $\mathbf{\Phi}_n$, $\mathbf{\Phi}_m$, and $\mathbf{D}_m$ all identity matrices and $\mathbf{w}_d$, $\mathbf{w}_{dn}$, $\mathbf{n}_{cmd}$, $\mathbf{E}_y$, $\mathbf{B}_{dm}$ and $\mathbf{C}_m$ all set to zero to yield

$$\mathbf{x}(t_{i+1}) = \mathbf{\Phi}\mathbf{x}(t_i) + \mathbf{B}_d\mathbf{u}(t_i) + \mathbf{d}_0, \qquad \mathbf{y}_c(t_i) = \mathbf{C}\mathbf{x}(t_i) + \mathbf{D}_y\mathbf{u}(t_i)$$
$$\mathbf{y}_m(t_i) = \mathbf{u}_m = \text{const} \quad (\mathbf{y}_d \text{ previously})$$

In equilibrium, $\mathbf{x}(t_{i+1}) = \mathbf{x}(t_i) = \mathbf{x}_I$, $\mathbf{u}(t_i) = \mathbf{u}_I$, and $\mathbf{y}_c(t_i) = \mathbf{y}_m(t_i) = \mathbf{u}_m$, or

$$\begin{bmatrix} (\mathbf{\Phi} - \mathbf{I}) & \mathbf{B}_d \\ \mathbf{C} & \mathbf{D}_y \end{bmatrix} \begin{bmatrix} \mathbf{x}_I \\ \mathbf{u}_I \end{bmatrix} = \begin{bmatrix} -\mathbf{d}_0 \\ \mathbf{u}_m \end{bmatrix}$$

so that we can solve for $\mathbf{x}_\text{I}$ and $\mathbf{u}_\text{I}$ as

$$\begin{bmatrix} \mathbf{x}_\text{I} \\ \mathbf{u}_\text{I} \end{bmatrix} = \begin{bmatrix} \mathbf{\Pi}_{11} & \mathbf{\Pi}_{12} \\ \mathbf{\Pi}_{21} & \mathbf{\Pi}_{22} \end{bmatrix} \begin{bmatrix} -\mathbf{d}_0 \\ \mathbf{u}_\text{m} \end{bmatrix}$$

which agrees with (14-156); this is of the desired form (14-245).

*Open-Loop CGT Law*

The solution of the CGT problem entails setting up and solving the matrix equations that the constant matrices $\mathbf{A}_{11}$ through $\mathbf{A}_{23}$ in (14-245) must satisfy. Using (14-238) and (14-243), we can write

$$\begin{bmatrix} \mathbf{x}_\text{I}(t_{i+1}) - \mathbf{x}_\text{I}(t_i) \\ \mathbf{y}_\text{I}(t_i) \end{bmatrix} = \begin{bmatrix} (\mathbf{\Phi} - \mathbf{I}) & \mathbf{B}_\text{d} \\ \mathbf{C} & \mathbf{D}_y \end{bmatrix} \begin{bmatrix} \mathbf{x}_\text{I}(t_i) \\ \mathbf{u}_\text{I}(t_i) \end{bmatrix} + \begin{bmatrix} \mathbf{E}_x \\ \mathbf{E}_y \end{bmatrix} \mathbf{n}_\text{d}(t_i)$$

$$= \begin{bmatrix} (\mathbf{\Phi} - \mathbf{I}) & \mathbf{B}_\text{d} \\ \mathbf{C} & \mathbf{D}_y \end{bmatrix} \begin{bmatrix} \mathbf{A}_{11} & \mathbf{A}_{12} & \mathbf{A}_{13} \\ \mathbf{A}_{21} & \mathbf{A}_{22} & \mathbf{A}_{23} \end{bmatrix} \begin{bmatrix} \mathbf{x}_\text{m}(t_i) \\ \mathbf{u}_\text{m} \\ \mathbf{n}_\text{d}(t_i) \end{bmatrix}$$

$$+ \begin{bmatrix} \mathbf{E}_x \\ \mathbf{E}_y \end{bmatrix} \mathbf{n}_\text{d}(t_i) \qquad \text{(14-246)}$$

where the last equality comes from incorporating (14-245). Now we seek a second expression for these quantities, to equate to (14-246). Writing $\mathbf{x}_\text{I}$ at both $t_{i+1}$ and $t_i$ from (14-245) and subtracting yields

$$[\mathbf{x}_\text{I}(t_{i+1}) - \mathbf{x}_\text{I}(t_i)]$$

$$= [\mathbf{A}_{11} \quad \mathbf{A}_{12} \quad \mathbf{A}_{13}] \begin{bmatrix} \mathbf{x}_\text{m}(t_{i+1}) - \mathbf{x}_\text{m}(t_i) \\ \mathbf{u}_\text{m} - \mathbf{u}_\text{m} \\ \mathbf{n}_\text{d}(t_{i+1}) - \mathbf{n}_\text{d}(t_i) \end{bmatrix}$$

$$= [\mathbf{A}_{11} \quad \mathbf{A}_{12} \quad \mathbf{A}_{13}] \begin{bmatrix} (\mathbf{\Phi}_\text{m} - \mathbf{I}) & \mathbf{B}_\text{dm} & \mathbf{0} \\ \mathbf{0} & \mathbf{0} & \mathbf{0} \\ \mathbf{0} & \mathbf{0} & (\mathbf{\Phi}_n - \mathbf{I}) \end{bmatrix} \begin{bmatrix} \mathbf{x}_\text{m}(t_i) \\ \mathbf{u}_\text{m} \\ \mathbf{n}_\text{d}(t_i) \end{bmatrix}$$

Furthermore, for ideal plant trajectories, (14-244) yields

$$\mathbf{y}_\text{I}(t_i) = [\mathbf{C}_\text{m} \quad \mathbf{D}_\text{m}] \begin{bmatrix} \mathbf{x}_\text{m}(t_i) \\ \mathbf{u}_\text{m}(t_i) \end{bmatrix}$$

Combining these two results yields

$$\begin{bmatrix} \mathbf{x}_\text{I}(t_{i+1}) - \mathbf{x}_\text{I}(t_i) \\ \mathbf{y}_\text{I}(t_i) \end{bmatrix} = \begin{bmatrix} \{\mathbf{A}_{11}(\mathbf{\Phi}_\text{m} - \mathbf{I})\} & \{\mathbf{A}_{11}\mathbf{B}_\text{dm}\} & \{\mathbf{A}_{13}(\mathbf{\Phi}_n - \mathbf{I})\} \\ \mathbf{C}_\text{m} & \mathbf{D}_\text{m} & \mathbf{0} \end{bmatrix} \begin{bmatrix} \mathbf{x}_\text{m}(t_i) \\ \mathbf{u}_\text{m} \\ \mathbf{n}_\text{d}(t_i) \end{bmatrix} \qquad \text{(14-247)}$$

Now, equating (14-246) and (14-247) yields, after rearrangement,

$$\left\{\begin{bmatrix} (\mathbf{\Phi}-\mathbf{I}) & \mathbf{B}_d \\ \mathbf{C} & \mathbf{D}_y \end{bmatrix}\begin{bmatrix} \mathbf{A}_{11} & \mathbf{A}_{12} & \mathbf{A}_{13} \\ \mathbf{A}_{21} & \mathbf{A}_{22} & \mathbf{A}_{23} \end{bmatrix} - \begin{bmatrix} \{\mathbf{A}_{11}(\mathbf{\Phi}_m-\mathbf{I})\} & \{\mathbf{A}_{11}\mathbf{B}_{dm}\} & \{\mathbf{A}_{13}(\mathbf{\Phi}_n-\mathbf{I})-\mathbf{E}_x\} \\ \mathbf{C}_m & \mathbf{D}_m & -\mathbf{E}_y \end{bmatrix}\right\}\begin{bmatrix} \mathbf{x}_m(t_i) \\ \mathbf{u}_m \\ \mathbf{n}_d(t_i) \end{bmatrix} = \mathbf{0} \quad (14\text{-}248)$$

Since $\mathbf{x}_m(t_i)$, $\mathbf{u}_m$, and $\mathbf{n}_d(t_i)$ are arbitrary and not necessarily zero, the matrix expression in the curly braces of (14-248) must be zero, yielding the equations for $\mathbf{A}_{11}$ through $\mathbf{A}_{23}$ as

$$\begin{bmatrix} \mathbf{A}_{11} & \mathbf{A}_{12} & \mathbf{A}_{13} \\ \mathbf{A}_{21} & \mathbf{A}_{22} & \mathbf{A}_{23} \end{bmatrix} = \begin{bmatrix} \mathbf{\Pi}_{11} & \mathbf{\Pi}_{12} \\ \mathbf{\Pi}_{21} & \mathbf{\Pi}_{22} \end{bmatrix}\begin{bmatrix} \{\mathbf{A}_{11}(\mathbf{\Phi}_m-\mathbf{I})\} & \{\mathbf{A}_{11}\mathbf{B}_{dm}\} & \{\mathbf{A}_{13}(\mathbf{\Phi}_n-\mathbf{I})-\mathbf{E}_x\} \\ \mathbf{C}_m & \mathbf{D}_m & -\mathbf{E}_y \end{bmatrix} \quad (14\text{-}249)$$

where $\mathbf{\Pi}$ is given by (14-155), (14-158), or (14-160), depending on the dimensionalities of $\mathbf{y}_c(t_i)$ versus $\mathbf{u}(t_i)$. Thus, in partitioned form, the equations to solve are

$$\mathbf{A}_{11} = \mathbf{\Pi}_{11}\mathbf{A}_{11}(\mathbf{\Phi}_m - \mathbf{I}) + \mathbf{\Pi}_{12}\mathbf{C}_m \quad (14\text{-}250a)$$

$$\mathbf{A}_{12} = \mathbf{\Pi}_{11}\mathbf{A}_{11}\mathbf{B}_{dm} + \mathbf{\Pi}_{12}\mathbf{D}_m \quad (14\text{-}250b)$$

$$\mathbf{A}_{13} = \mathbf{\Pi}_{11}\mathbf{A}_{13}(\mathbf{\Phi}_n - \mathbf{I}) - \mathbf{\Pi}_{11}\mathbf{E}_x - \mathbf{\Pi}_{12}\mathbf{E}_y \quad (14\text{-}250c)$$

$$\mathbf{A}_{21} = \mathbf{\Pi}_{21}\mathbf{A}_{11}(\mathbf{\Phi}_m - \mathbf{I}) + \mathbf{\Pi}_{22}\mathbf{C}_m \quad (14\text{-}250d)$$

$$\mathbf{A}_{22} = \mathbf{\Pi}_{21}\mathbf{A}_{11}\mathbf{B}_{dm} + \mathbf{\Pi}_{22}\mathbf{D}_m \quad (14\text{-}250e)$$

$$\mathbf{A}_{23} = \mathbf{\Pi}_{21}\mathbf{A}_{13}(\mathbf{\Phi}_n - \mathbf{I}) - \mathbf{\Pi}_{21}\mathbf{E}_x - \mathbf{\Pi}_{22}\mathbf{E}_y \quad (14\text{-}250f)$$

Of these, the nontrivial equations are (14-250a) and (14-250c) which, unless $\mathbf{\Phi}_m$ and/or $\mathbf{\Phi}_n$ equal $\mathbf{I}$, have the form $\mathbf{X} = \mathbf{AXB} + \mathbf{C}$ to be solved for $\mathbf{X}$. An available numerical algorithm [11, 74] can produce a unique solution provided that $\lambda_{\mathbf{A}i}\lambda_{\mathbf{B}j} \neq 1$ for all $i$ and $j$, where $\{\lambda_{\mathbf{A}i}\}$ are the eigenvalues of $\mathbf{A}$ and $\{\lambda_{\mathbf{B}j}\}$ are the eigenvalues of $\mathbf{B}$. Thus, a unique $\mathbf{A}_{11}$ can be found if

$$\lambda_{\mathbf{\Pi}_{11}i}\lambda_{(\mathbf{\Phi}_m - \mathbf{I})j} \neq 1 \qquad \text{for all } i \text{ and } j \quad (14\text{-}251a)$$

and a unique $\mathbf{A}_{13}$ can be found if

$$\lambda_{\mathbf{\Pi}_{11}i}\lambda_{(\mathbf{\Phi}_n - \mathbf{I})j} \neq 1 \qquad \text{for all } i \text{ and } j \quad (14\text{-}251b)$$

We note that $1/\lambda_{\mathbf{\Pi}_{11}i}$ are related to the transmission zeros of the plant (these are included in the set of inverses of eigenvalues of $(\mathbf{I} + \mathbf{\Pi}_{11})^{-1}\mathbf{\Pi}_{11}$) [17, 18, 33, 113].

Once the $\mathbf{A}_{ij}$'s are evaluated, the *open-loop command generator tracker law* can be written from the lower partition of (14-245) as

$$\mathbf{u}_{\mathrm{I}}(t_i) = \mathbf{A}_{21}\mathbf{x}_{\mathrm{m}}(t_i) + \mathbf{A}_{22}\mathbf{u}_{\mathrm{m}}(t_i) + \mathbf{A}_{23}\mathbf{n}_{\mathrm{d}}(t_i) \tag{14-252}$$

If desired, $\mathbf{x}_{\mathrm{I}}(t_i)$ can also be generated from (14-245). The structure of this control law is displayed in Fig. 14.14. This control law will cause the true system output to converge to tracking the command generator model output (14-240b), provided the true system is itself stable, there are no modeling errors, and there are no unmodeled disturbances affecting the system; feedback implementations are strongly motivated by these restrictions.

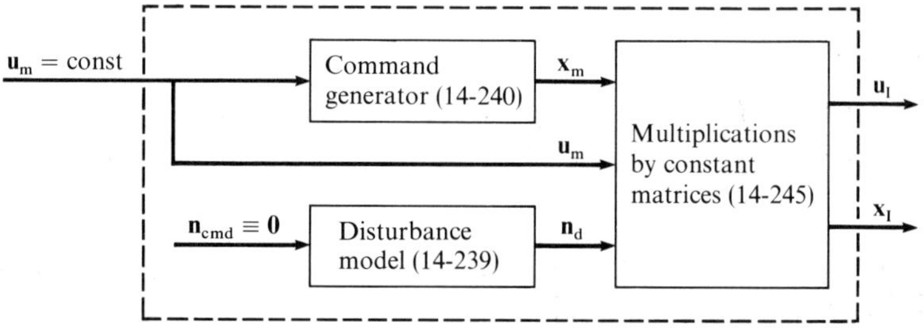


FIG. 14.14 Open-loop command generator tracker.

EXAMPLE 14.14 Again consider the gimbaled platform problem, and suppose we want to generate the appropriate open-loop CGT law to cause the gimbal angle to reach a nonzero setpoint despite a constant *modeled* disturbance. Thus, the system is

$$x(t_{i+1}) = \Phi x(t_i) + B_{\mathrm{d}}u(t_i) + n_{\mathrm{d}}, \qquad E_x = 1$$
$$y_{\mathrm{c}}(t_i) = x(t_i), \qquad C = 1, \quad D_y = E_y = 0$$

where the disturbance is modeled as

$$n_{\mathrm{d}}(t_{i+1}) = n_{\mathrm{d}}(t_i) = n_{\mathrm{d}} \qquad \Phi_n = 1, \quad n_{\mathrm{cmd}} \equiv 0, \quad w_{\mathrm{d}n} \equiv 0$$

and the system output is required to track the model output

$$x_{\mathrm{m}}(t_{i+1}) = x_{\mathrm{m}}(t_i), \qquad \Phi_{\mathrm{m}} = 1, \quad B_{\mathrm{dm}} = 0$$
$$y_{\mathrm{m}}(t_i) = u_{\mathrm{m}} = y_{\mathrm{m}} = \text{const} \qquad C_{\mathrm{m}} = 0, \quad D_{\mathrm{m}} = 1$$

This is a trivial command generator in that the output is simply a direct feedthrough of the input, and the state dynamics are arbitrary. From (14-155) and (14-249) we obtain

$$\begin{bmatrix} A_{11} & A_{12} & A_{13} \\ A_{21} & A_{22} & A_{23} \end{bmatrix} = \begin{bmatrix} 0 & 1 \\ 1/B_{\mathrm{d}} & (1-\Phi)/B_{\mathrm{d}} \end{bmatrix} \begin{bmatrix} \{A_{11}\cdot 0\} & \{A_{11}\cdot 0\} & \{(A_{13}\cdot 0) - 1\} \\ 0 & 1 & 0 \end{bmatrix}$$
$$= \begin{bmatrix} 0 & 1 & 0 \\ 0 & (1-\Phi)/B_{\mathrm{d}} & -1/B_{\mathrm{d}} \end{bmatrix}$$

Note that (14-251) is satisfied so a unique solution can be obtained, and "perfect tracking" is possible. The open-loop CGT law is then

$$\begin{aligned} u_1(t_i) &= A_{21}x_m(t_i) + A_{22}u_m + A_{23}n_d \\ &= [(1-\Phi)/B_d]y_m - [1/B_d]n_d \end{aligned}$$

Applying this to the system yields a result modeled by

$$\begin{aligned} x(t_{i+1}) &= \Phi x(t_i) + B_d\{[(1-\Phi)/B_d]y_m - [1/B_d]n_d\} + n_d \\ &= \Phi x(t_i) + y_m - \Phi y_m \end{aligned}$$

so that the error dynamics are expressible as

$$x(t_{i+1}) - y_m = \Phi[x(t_i) - y_m\}$$

which implies that $\{x(t_i) - y_m\} \to 0$, i.e., $x(t_i) \to y_m$, if $|\Phi| < 1$. ■

EXAMPLE 14.15 Consider the same example, but now assume we want the system to track a constant angular *rate* command despite the modeled disturbance. Thus, we have the same controlled system and disturbance models, but here the command generator becomes

$$\begin{aligned} x_m(t_{i+1}) &= x_m(t_i) + [\Delta t]\dot{y}_m, \qquad & \Phi_m = 1, \quad B_{dm} = \Delta t, \quad u_m = \dot{y}_m \\ y_m(t_i) &= x_m(t_i), & C_m = 1, \quad D_m = 0 \end{aligned}$$

so that (14-155) and (14-249) yield

$$\begin{aligned} \begin{bmatrix} A_{11} & A_{12} & A_{13} \\ A_{21} & A_{22} & A_{23} \end{bmatrix} &= \begin{bmatrix} 0 & 1 \\ 1/B_d & (1-\Phi)/B_d \end{bmatrix} \begin{bmatrix} \{A_{11}\cdot 0\} & \{A_{11}\,\Delta t\} & \{(A_{13}\cdot 0) - 1\} \\ 1 & 0 & 0 \end{bmatrix} \\ &= \begin{bmatrix} 1 & 0 & 0 \\ (1-\Phi)/B_d & (\Delta t)/B_d & -1/B_d \end{bmatrix} \end{aligned}$$

and the open-loop CGT law becomes

$$u_1(t_i) = [(1-\Phi)/B_d]x_m(t_i) + [(\Delta t)/B_d]\dot{y}_m - [1/B_d]n_d$$

Applying this to the system yields

$$\begin{aligned} x(t_{i+1}) &= \Phi x(t_i) + [(1-\Phi)x_m(t_i) + (\Delta t)\dot{y}_m - n_d] + n_d \\ &= \Phi\{x(t_i) - x_m(t_i)\} + x_m(t_i) + (\Delta t)\dot{y}_m \end{aligned}$$

so that

$$\{x(t_{i+1}) - x_m(t_{i+1})\} = \Phi\{x(t_i) - x_m(t_i)\}$$

and thus the plant must be stable, i.e., $|\Phi| < 1$, for the tracking error $\{x(t_i) - x_m(t_i)\}$ to decay to zero.

Notice that a CGT *can* be formulated so as to cause the controlled system to track a *constant rate command*—something many other controller synthesis techniques have trouble accomplishing. Here we let the dynamics of the command generator help us, as compared to the trivial case of the preceding example. ■

EXAMPLE 14.16 Consider the same example again, but now assume we want the gimbal to respond with the characteristics of some desired second order system, in the face of the modeled disturbance. The original system is modeled as a first order lag, so that it responds to a step input with an exponential rise. However, we would like the system to respond to a step input $u_m$ *as though it were a second order system*, exhibiting faster rise times and some degree of overshoot (impossible for a first order system) and rapid settling time.

Thus, the system and disturbance models remain unaltered, but the command generator model to be tracked is

$$\begin{bmatrix} x_{m1}(t_{i+1}) \\ x_{m2}(t_{i+1}) \end{bmatrix} = \begin{bmatrix} 0 & 1 \\ -a_0 & -a_1 \end{bmatrix} \begin{bmatrix} x_{m1}(t_i) \\ x_{m2}(t_i) \end{bmatrix} + \begin{bmatrix} 0 \\ 1 \end{bmatrix} u_m$$

$$y_m(t_i) = [1 \quad 0] \begin{bmatrix} x_{m1}(t_i) \\ x_{m2}(t_i) \end{bmatrix}$$

where the model eigenvalues are located at

$$\lambda = -(a_1/2) \pm \sqrt{a_1{}^2 - 4a_0}/2$$

yielding a complex pair of poles in the $z$ plane for $(a_1{}^2 - 4a_0) < 0$, and we also note that

$$\frac{y_m(t_{i+1})}{u_m(t_i)} = \frac{x_{m1}(t_{i+1})}{u_m(t_i)} = \frac{z}{z^2 + a_1 z + a_0}$$

or

$$y_m(t_{i+2}) = -a_1 y_m(t_{i+1}) - a_0 y_m(t_i) + u_m(t_i)$$

Thus, the $\mathbf{\Phi}_m$, $\mathbf{B}_{dm}$, and $\mathbf{C}_m$ are as given above and $D_m = 0$.

Now we solve (14-249) where, from (14-245), it can be seen that $\mathbf{A}_{11}$ and $\mathbf{A}_{21}$ must be 2-by-1, and all other $\mathbf{A}$ components are scalar. We obtain

$$\begin{bmatrix} \mathbf{A}_{11} & A_{12} & A_{13} \\ \mathbf{A}_{21} & A_{22} & A_{23} \end{bmatrix} = \begin{bmatrix} 0 & 1 \\ 1/B_d & (1-\Phi)/B_d \end{bmatrix} \begin{bmatrix} \left\{ \mathbf{A}_{11} \begin{bmatrix} -1 & 1 \\ -a_0 & -(a_1+1) \end{bmatrix} \right\} & \left\{ \mathbf{A}_{11} \begin{bmatrix} 0 \\ 1 \end{bmatrix} \right\} & \{(A_{13} \cdot 0) - 1\} \\ [1 \quad 0] & 0 & 0 \end{bmatrix}$$

$$= \begin{bmatrix} [\;1 & 0\;] & 0 & 0 \\ [(-\Phi/B_d) & (1/B_d)] & 0 & (-1/B_d) \end{bmatrix}$$

The open-loop CGT control is

$$\begin{aligned} u_1(t_i) &= \mathbf{A}_{21}\mathbf{x}_m(t_i) + A_{22}u_m + A_{23}n_d \\ &= [-\Phi/B_d]x_{m1}(t_i) + [1/B_d]x_{m2}(t_i) - [1/B_d]n_d \end{aligned}$$

Substituting this back into the system yields

$$\begin{aligned} x(t_{i+1}) &= \Phi x(t_i) + [-\Phi x_{m1}(t_i) + x_{m2}(t_i) - n_d] + n_d \\ &= \Phi\{x(t_i) - x_{m1}(t_i)\} + x_{m2}(t_i) \end{aligned}$$

so that the tracking error satisfies

$$x(t_{i+1}) - x_{m1}(t_{i+1}) = \Phi\{x(t_i) - x_{m1}(t_i)\}$$

and again this converges to zero provided $|\Phi| < 1$. ■

### *Closed-Loop Command Generator Tracking with Simple Regulator*

In general, the open-loop CGT law is not sufficient: the controlled system may be unstable or may exhibit undesirable performance that requires compensation, and the models upon which the open-loop law is based generally misrepresent the true system due to uncertain parameters and unmodeled disturbances. Thus, we seek a feedback control that will tend to drive perturbations from the ideal plant trajectory to zero. Assume that the open-loop CGT law $\mathbf{u}_I(t_i)$ has been evaluated for all $t_i$, such that the ideal state trajectory is $\mathbf{x}_I(t_i)$ for all $t_i$, as given by (14-245). Then define perturbation variables as

$$\delta\mathbf{x}(t_i) \triangleq \mathbf{x}(t_i) - \mathbf{x}_I(t_i) \tag{14-253a}$$

$$\delta\mathbf{u}(t_i) \triangleq \mathbf{u}(t_i) - \mathbf{u}_I(t_i) \tag{14-253b}$$

$$\delta\mathbf{y}_c(t_i) \triangleq \mathbf{y}_c(t_i) - \mathbf{y}_I(t_i) = \mathbf{y}_c(t_i) - \mathbf{y}_m(t_i) \tag{14-253c}$$

and we then desire to generate the LQ optimal regulator for the deterministic perturbation system model

$$\delta\mathbf{x}(t_{i+1}) = \mathbf{\Phi}\,\delta\mathbf{x}(t_i) + \mathbf{B}_d\,\delta\mathbf{u}(t_i) \tag{14-254}$$

which in steady state becomes

$$\delta\mathbf{u}(t_i) = -\bar{\mathbf{G}}_c{}^*\,\delta\mathbf{x}(t_i) \tag{14-255}$$

Rewriting this law in the original variables yields

$$\mathbf{u}(t_i) - \mathbf{u}_I(t_i) = -\bar{\mathbf{G}}_c{}^*[\mathbf{x}(t_i) - \mathbf{x}_I(t_i)]$$

or, finally,

$$\mathbf{u}(t_i) = \mathbf{u}_I(t_i) + \bar{\mathbf{G}}_c{}^*[\mathbf{x}_I(t_i) - \mathbf{x}(t_i)] \tag{14-256a}$$

This is diagrammed in Fig. 14.15a, using the open-loop CGT law from Fig. 14.14.

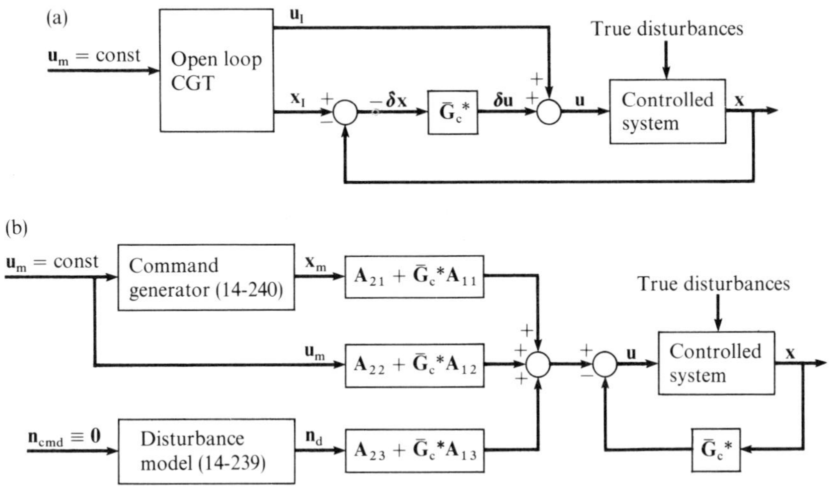


FIG. 14.15 Closed-loop command generator tracker with simple regulator.

Recalling (14-245), this can also be expressed as

$$\mathbf{u}(t_i) = -\bar{\mathbf{G}}_c{}^*\mathbf{x}(t_i) + [\mathbf{A}_{21} + \bar{\mathbf{G}}_c{}^*\mathbf{A}_{11}]\mathbf{x}_m(t_i) + [\mathbf{A}_{22} + \bar{\mathbf{G}}_c{}^*\mathbf{A}_{12}]\mathbf{u}_m + [\mathbf{A}_{23} + \bar{\mathbf{G}}_c{}^*\mathbf{A}_{13}]\mathbf{n}_d(t_i) \tag{14-256b}$$

as depicted in Fig. 14.15b. Note that the closed-loop system can be described by

$$\delta\mathbf{x}(t_{i+1}) = [\mathbf{\Phi} - \mathbf{B}_d\bar{\mathbf{G}}_c{}^*]\,\delta\mathbf{x}(t_i)$$
$$\{\mathbf{x}(t_{i+1}) - \mathbf{x}_I(t_{i+1})\} = [\mathbf{\Phi} - \mathbf{B}_d\bar{\mathbf{G}}_c{}^*]\{\mathbf{x}(t_i) - \mathbf{x}_I(t_i)\} \tag{14-257}$$

so that the perturbation error converges to zero asymptotically even if the original system were unstable, provided there are no modeling errors.

EXAMPLE 14.17 Consider the closed-loop CGT law for the problem of Example 14.14. From (14-252), the open-loop law was

$$u_I(t_i) = [0]x_m(t_i) + [(1 - \Phi)/B_d]u_m + [-1/B_d]n_d$$

where $u_m = y_m = y_d =$ const. and, from (14-245), we can also write

$$x_I(t_i) = [0]x_m(t_i) + [1]u_m + [0]n_d$$

Thus, the closed-loop CGT law is generated by (14-256a) as

$$u(t_i) = u_I(t_i) + \bar{\mathbf{G}}_c{}^*[x_I(t_i) - x(t_i)]$$
$$= \{[(1 - \Phi)/B_d]y_d - [1/B_d]n_d\} + \bar{G}_c{}^*[y_d - x(t_i)]$$

or, equivalently, from (14-256b), as

$$u(t_i) = -\bar{G}_c{}^*x(t_i) + [(1 - \Phi)/B_d + \bar{G}_c{}^*]y_d - [1/B_d]n_d$$

Applying this to the system model yields a close-loop

$$x(t_{i+1}) = \Phi x(t_i) + B_d u(t_i) + n_d$$
$$= [\Phi - B_d\bar{G}_c{}^*]x(t_i) + y_d - [\Phi - B_d\bar{G}_c{}^*]y_d$$

with error dynamics

$$\{x(t_{i+1}) - y_d\} = [\Phi - B_d\bar{G}_c{}^*]\{x(t_i) - y_d\}$$

as indicated in (14-257).

Similarly, in Example 14.15, $x_I(t_i) = x_m(t_i)$ by (14-245), and the closed-loop CGT law becomes

$$u(t_i) = \{[(1 - \Phi)/B_d]x_m(t_i) + [(\Delta t)/B_d]\dot{y}_m - [1/B_d]n_d\} + \bar{G}_c{}^*[x_m(t_i) - x(t_i)]$$

In Example 14.16, $x_I(t_i) = [1\ \ 0]\mathbf{x}_m(t_i) = x_{m1}(t_i)$, and the closed-loop CGT law becomes

$$u(t_i) = \{[-\Phi/B_d]x_{m1}(t_i) + [1/B_d]x_{m2}(t_i) - [1/B_d]n_d\} + \bar{G}_c{}^*[x_{m1}(t_i) - x(t_i)]$$

In both of these cases, (14-257) is again satisfied. ■

The full-state feedback perturbation regulator of (14-255) can also be expressed in the preferred incremental form as in (14-196), to yield an equivalent closed-loop CGT law of

$$\mathbf{u}(t_i) = \mathbf{u}(t_{i-1}) + [\mathbf{u}_I(t_i) - \mathbf{u}_I(t_{i-1})] + \bar{\mathbf{G}}_c^*[\mathbf{x}_I(t_i) - \mathbf{x}_I(t_{i-1})] \\ - \bar{\mathbf{G}}_c^*[\mathbf{x}(t_i) - \mathbf{x}(t_{i-1})] \quad (14\text{-}258)$$

Moreover, when only noise-corrupted measurements are available from the system, a Kalman filter or observer can be used to generate $\hat{\mathbf{x}}(t_i^+)$ (or, suboptimally, $\hat{\mathbf{x}}(t_i^-)$) to replace $\mathbf{x}(t_i)$ in either (14-256) or (14-258). If the disturbance to be rejected is not perfectly accessible, it too must be estimated, but, as already seen in Section 14.8, this does *not* yield two separate filters, one to produce $\hat{\mathbf{x}}(t_i^+)$ and the other outputting $\hat{\mathbf{n}}_d(t_i^+)$, but a single filter to estimate both.

*Initial Conditions: Practical Implementation Issue when $\mathbf{u}_m(t_i)$ Changes in Actual Application*

We have assumed that

$$\mathbf{u}_m(t_i) = \begin{cases} \mathbf{u}_m = \text{const} & i = 0, 1, 2, \ldots \\ \mathbf{u}_m^- \neq \mathbf{u}_m & i = \ldots, -2, -1 \end{cases} \quad (14\text{-}259)$$

i.e., that $\mathbf{u}_m$ is constant from $t_0$ forward. For practical implementation, any time that $\mathbf{u}_m$ changes (allowing infrequent step changes in value) will conceptually cause $i$ to be set to zero again. However, an inconsistency in the definition of the ideal trajectory arises at the time of step change in $\mathbf{u}_m$ (at $t_i = t_0$ conceptually). At $t_{-1}$, the ideal trajectory satisfies (14-245),

$$\begin{bmatrix} \mathbf{x}_I(t_{-1}) \\ \mathbf{u}_I(t_{-1}) \end{bmatrix} = \begin{bmatrix} \mathbf{A}_{11} & \mathbf{A}_{12} & \mathbf{A}_{13} \\ \mathbf{A}_{21} & \mathbf{A}_{22} & \mathbf{A}_{23} \end{bmatrix} \begin{bmatrix} \mathbf{x}_m(t_{-1}) \\ \mathbf{u}_m^- \\ \mathbf{n}_d(t_{-1}) \end{bmatrix} \quad (14\text{-}260a)$$

while at $t_0$ we obtain the corresponding result

$$\begin{bmatrix} \mathbf{x}_I(t_0) \\ \mathbf{u}_I(t_0) \end{bmatrix} = \begin{bmatrix} \mathbf{A}_{11} & \mathbf{A}_{12} & \mathbf{A}_{13} \\ \mathbf{A}_{21} & \mathbf{A}_{22} & \mathbf{A}_{23} \end{bmatrix} \begin{bmatrix} \mathbf{x}_m(t_0) \\ \mathbf{u}_m \\ \mathbf{n}_d(t_0) \end{bmatrix} \quad (14\text{-}260b)$$

But the ideal trajectory transient is also to satisfy (14-243),

$$\mathbf{x}_I(t_0) = \mathbf{\Phi}\mathbf{x}_I(t_{-1}) + \mathbf{B}_d\mathbf{u}_I(t_{-1}) + \mathbf{E}_x\mathbf{n}_d(t_{-1}) \quad (14\text{-}261)$$

Combining (14-261) and (14-260a) does *not* yield the $\mathbf{x}_I(t_0)$ of (14-260b); they would agree if $\mathbf{u}_m(t_{-1}) = \mathbf{u}_m^-$ had equalled $\mathbf{u}_m(t_0) = \mathbf{u}_m$ (and note there is no discrepancy at future samples in which $\mathbf{u}_m$ does not change). To avoid this inconsistency, whenever $\mathbf{u}_m(t_i)$ changes, we go back to time $t_{i-1}$ and *restart* all variables under our control, assuming that $\mathbf{u}_m(t_{i-1}) = \mathbf{u}_m$ instead of its previous value $\mathbf{u}_m^-$. Of course, $\mathbf{x}(t_{i-1})$ and $\mathbf{u}(t_{i-1})$ cannot be changed in causal systems, but $\mathbf{x}_{\mathrm{m}}(t_{i-1})$, $\mathbf{u}_{\mathrm{m}}(t_{i-1})$, $\mathbf{x}_{\mathrm{I}}(t_{i-1})$, and $\mathbf{u}_{\mathrm{I}}(t_{i-1})$ can be altered. In actual implementation, this simply results in the command generator model equation (14-240a) becoming

$$\mathbf{x}_{\mathrm{m}}(t_i) = \mathbf{\Phi}_{\mathrm{m}}\mathbf{x}_{\mathrm{m}}(t_{i-1}) + \mathbf{B}_{\mathrm{dm}}\mathbf{u}_{\mathrm{m}}(t_i) \tag{14-262}$$

where $\mathbf{u}_{\mathrm{m}}(t_i)$ has replaced $\mathbf{u}_{\mathrm{m}}(t_{i-1})$. Thus, the $\mathbf{x}_{\mathrm{m}}(t_i)$ and $\mathbf{u}_{\mathrm{m}}$ that appear in the feedback control law (and PI law to come) *cannot* be computed completely until time $t_i$ when $\mathbf{u}_{\mathrm{m}}(t_i)$ becomes available, precluding precomputation of these terms in the background between $t_{i-1}$ and $t_i$ to speed up control computations, except for problems in which you know the entire $\mathbf{u}_{\mathrm{m}}$ time history a priori.

*Closed-Loop Command Generator Tracking with PI Controller*

It is useful to consider a PI closed-loop law in conjunction with command generator tracking because (1) it *forces* the difference between actual system output and command generator model output to zero in steady state, even in the face of some modeling errors, (2) it can accommodate *unmodeled* constant disturbances as well as reject the modeled ones, and (3) it will also turn out to include the model control's feedforward contribution to the actual system control signal. In this section, the PI/CGT controller based on *control difference weighting* will be developed, in both position and preferred incremental forms, to cause the actual system controlled variable outputs $\mathbf{y}_{\mathrm{c}}(t_i)$ to track the model outputs $\mathbf{y}_{\mathrm{m}}(t_i)$. The PI controller based on regulation error weighting and the generic PI controller are pursued in Problem 14.37.

Assume that the open-loop CGT law $\mathbf{u}_{\mathrm{I}}(t_i)$ and the associated ideal state trajectory $\mathbf{x}_{\mathrm{I}}(t_i)$ have been evaluated for all $t_i$. For this PI law formulation, it is also necessary to specify the ideal trajectory of control differences defined by (14-214); using (14-245) we get

$$\begin{aligned}\Delta\mathbf{u}_{\mathrm{I}}(t_i) &= \mathbf{u}_{\mathrm{I}}(t_{i+1}) - \mathbf{u}_{\mathrm{I}}(t_i)\\ &= \mathbf{A}_{21}[\mathbf{x}_{\mathrm{m}}(t_{i+1}) - \mathbf{x}_{\mathrm{m}}(t_i)] + \mathbf{A}_{23}[\mathbf{n}_{\mathrm{d}}(t_{i+1}) - \mathbf{n}_{\mathrm{d}}(t_i)]\end{aligned} \tag{14-263}$$

where the term $\mathbf{A}_{22}[\mathbf{u}_{\mathrm{m}}(t_{i+1}) - \mathbf{u}_{\mathrm{m}}(t_i)]$ is zero because $\mathbf{u}_{\mathrm{m}}$ is constant from $t_0$ forward and it is zeroed at $t_i = t_{-1}$ by the restart procedure of (14-262) for proper initialization. Letting the perturbation variables be defined as in (14-253) and

$$\delta\,\Delta\mathbf{u}(t_i) = \Delta\mathbf{u}(t_i) - \Delta\mathbf{u}_{\mathrm{I}}(t_i) \tag{14-264}$$

we form the deterministic augmented perturbation system state equation by direct subtraction of deterministic system and ideal trajectory relations, to achieve (14-216) with $\Delta\mathbf{u}(t_i)$ replaced by $\delta\,\Delta\mathbf{u}(t_i)$. The objective is to transfer the controlled system from perfectly tracking the model output $\mathbf{y}_{\mathrm{m}}$ corresponding to input $\mathbf{u}_{\mathrm{m}}^-$, to perfectly tracking the model output for $\mathbf{u}_{\mathrm{m}}$, while minimizing a quadratic cost as in (14-217) in the limit as $N \to \infty$ (to achieve steady state *constant* gains), again with $\Delta\mathbf{u}(t_i)$ replaced by $\delta\,\Delta\mathbf{u}(t_i)$. The LQ optimal perturbation regulator solution yields

$$\delta\,\Delta\mathbf{u}^*(t_i) = -\bar{\mathbf{G}}_{c1}^*\,\delta\mathbf{x}(t_i) - \bar{\mathbf{G}}_{c2}^*\,\delta\mathbf{u}(t_i) \tag{14-265}$$

Combining this with $\delta\mathbf{u}(t_{i+1}) = \delta\mathbf{u}(t_i) + \delta\,\Delta\mathbf{u}(t_i)$ yields a *type*-0 controller with certain *low-pass filtering* characteristics, as given by (14-220), which can be combined with perturbation variable definitions (14-253) and (14-264), and ideal trajectory definition (14-245), to yield

$$\begin{aligned}\mathbf{u}(t_i) = \mathbf{u}(t_{i-1}) &+ \mathbf{A}_{21}[\mathbf{x}_m(t_i) - \mathbf{x}_m(t_{i-1})] + \mathbf{A}_{23}[\mathbf{n}_d(t_i) - \mathbf{n}_d(t_{i-1})]\\ &- \bar{\mathbf{G}}_{c1}^*[\mathbf{x}(t_{i-1}) - \mathbf{A}_{11}\mathbf{x}_m(t_{i-1}) - \mathbf{A}_{12}\mathbf{u}_m(t_i) - \mathbf{A}_{13}\mathbf{n}_d(t_{i-1})]\\ &- \bar{\mathbf{G}}_{c2}^*[\mathbf{u}(t_{i-1}) - \mathbf{A}_{21}\mathbf{x}_m(t_{i-1}) - \mathbf{A}_{22}\mathbf{u}_m(t_i) - \mathbf{A}_{23}\mathbf{n}_d(t_{i-1})]\end{aligned} \tag{14-266}$$

in incremental form, with $\mathbf{x}_m(t_i)$ computed according to (14-262).

To achieve the more desired type-1 performance, (14-226) is solved to yield (14-227), with which we can write (14-223b) as

$$\delta\mathbf{u}(t_i) = \delta\mathbf{u}(t_{i-1}) - [\mathbf{K}_x \quad \mathbf{K}_\xi]\begin{bmatrix}\delta\mathbf{x}(t_i) - \delta\mathbf{x}(t_{i-1})\\ \delta\mathbf{y}_c(t_{i-1})\end{bmatrix}$$

Using (14-253), this can be written as

$$\begin{aligned}\mathbf{u}(t_i) = \mathbf{u}_I(t_i) + \mathbf{u}(t_{i-1}) - \mathbf{u}_I(t_{i-1}) - \mathbf{K}_x[\mathbf{x}(t_i) - \mathbf{x}(t_{i-1})]\\ + \mathbf{K}_x[\mathbf{x}_I(t_i) - \mathbf{x}_I(t_{i-1})] + \mathbf{K}_\xi[\mathbf{y}_m(t_{i-1}) - \mathbf{y}_c(t_{i-1})]\end{aligned}$$

Note that, unlike the case in Section 14.9, the time arguments of $\mathbf{y}_m$ and $\mathbf{y}_c$ agree here. Within this expression, we can substitute

$$[\mathbf{K}_x \quad \mathbf{I}]\begin{bmatrix}\mathbf{x}_I(t_i) - \mathbf{x}_I(t_{i-1})\\ \mathbf{u}_I(t_i) - \mathbf{u}_I(t_{i-1})\end{bmatrix} = [\mathbf{K}_x \quad \mathbf{I}]\begin{bmatrix}\mathbf{A}_{11} & \mathbf{A}_{12} & \mathbf{A}_{13}\\ \mathbf{A}_{21} & \mathbf{A}_{22} & \mathbf{A}_{23}\end{bmatrix}\begin{bmatrix}\mathbf{x}_m(t_i) - \mathbf{x}_m(t_{i-1})\\ \mathbf{0}\\ \mathbf{n}_d(t_i) - \mathbf{n}_d(t_{i-1})\end{bmatrix}$$

to yield the final *incremental form PI/CGT law* as

$$\begin{aligned}\mathbf{u}(t_i) = \mathbf{u}(t_{i-1}) &- \mathbf{K}_x[\mathbf{x}(t_i) - \mathbf{x}(t_{i-1})]\\ &+ \mathbf{K}_\xi\left\{[\mathbf{C}_m \quad \mathbf{D}_m]\begin{bmatrix}\mathbf{x}_m(t_{i-1})\\ \mathbf{u}_m(t_i)\end{bmatrix} - [\mathbf{C} \quad \mathbf{D}_y]\begin{bmatrix}\mathbf{x}(t_{i-1})\\ \mathbf{u}(t_{i-1})\end{bmatrix}\right\}\\ &+ [\mathbf{K}_x\mathbf{A}_{11} + \mathbf{A}_{21}][\mathbf{x}_m(t_i) - \mathbf{x}_m(t_{i-1})]\\ &+ [\mathbf{K}_x\mathbf{A}_{13} + \mathbf{A}_{23}][\mathbf{n}_d(t_i) - \mathbf{n}_d(t_{i-1})]\end{aligned} \tag{14-267}$$

Note the time argument $t_i$ on $\mathbf{u}_m$, due to the restart procedure as in (14-262); this is the direct feedthrough of $\mathbf{u}_m(t_i)$ to $\mathbf{u}(t_i)$ mentioned earlier. The corresponding, less preferred, position form law can be expressed as

$$\begin{aligned}\mathbf{u}(t_i) = &-\mathbf{K}_x\mathbf{x}(t_i) + [\mathbf{K}_x\mathbf{A}_{11} + \mathbf{A}_{21}]\mathbf{x}_m(t_i)\\ &+ [\mathbf{K}_x\mathbf{A}_{13} + \mathbf{A}_{23}]\mathbf{n}_d(t_i) + \mathbf{K}_\xi\boldsymbol{\xi}(t_i)\end{aligned} \tag{14-268a}$$

where

$$\begin{aligned}\xi(t_i) &= \xi(t_{i-1}) + \{\mathbf{y}_m(t_{i-1}) - \mathbf{y}_c(t_{i-1})\}\\ &= \xi(t_{i-1}) + \left\{[\mathbf{C}_m \quad \mathbf{D}_m]\begin{bmatrix}\mathbf{x}_m(t_{i-1})\\ \mathbf{u}_m(t_i)\end{bmatrix} - [\mathbf{C} \quad \mathbf{D}_y]\begin{bmatrix}\mathbf{x}(t_{i-1})\\ \mathbf{u}(t_{i-1})\end{bmatrix}\right\}\end{aligned} \tag{14-268b}$$

Note the structural similarity between this and (14-256b), but with the deletion of the term involving $\mathbf{u}_m$ and the addition of the $\mathbf{K}_\xi\xi(t_i)$ term that incorporates the pseudointegral of the tracking error. Since (14-268a) is also expressible as

$$\mathbf{u}(t_i) = \mathbf{u}_I(t_i) + \mathbf{K}_x[\mathbf{x}_I(t_i) - \mathbf{x}(t_i)] + \mathbf{K}_\xi\xi(t_i) - [\mathbf{K}_x\mathbf{A}_{12} + \mathbf{A}_{22}]\mathbf{u}_m(t_i)$$

it can be seen (by letting $\mathbf{u} = \mathbf{u}_I$ and $\mathbf{x} = \mathbf{x}_I$ above) that the ideal integrator state trajectory $\xi_I$ is a constant that satisfies

$$\mathbf{K}_\xi\xi_I = [\mathbf{K}_x\mathbf{A}_{12} + \mathbf{A}_{22}]\mathbf{u}_m \tag{14-269a}$$

and thus the appropriate integrator initial condition is

$$\mathbf{K}_\xi\xi(t_0) = [\mathbf{K}_x\mathbf{A}_{12} + \mathbf{A}_{22}]\mathbf{u}_m^- \tag{14-269b}$$

assuming that perfect tracking was obtained prior to time $t_0$. An alternate position law that also yields (14-267) by direct subtraction, since $\mathbf{u}_m(t_i) - \mathbf{u}_m(t_{i-1}) = \mathbf{0}$, is

$$\mathbf{u}(t_i) = \mathbf{u}_I(t_i) + \mathbf{K}_x[\mathbf{x}_I(t_i) - \mathbf{x}(t_i)] + \mathbf{K}_\xi\xi(t_i) \tag{14-270}$$

which is also combined with (14-268b); however, by the same line of reasoning, the appropriate initial condition is

$$\mathbf{K}_\xi\xi(t_0) = \mathbf{K}_\xi\xi_I = \mathbf{0} \tag{14-271}$$

The structure of this law is even more similar to (14-256b), and is in the form of the open-loop CGT law control $\mathbf{u}_I(t_i)$, plus a proportional gain times state deviations from the ideal state trajectory, plus an integral gain times the tracking error pseudointegral to provide the desired type-1 characteristics.

EXAMPLE 14.18 Consider the PI/CGT law for the problem of Example 14.14. The open-loop CGT $u_I$ and $x_I$ are as evaluated in that example and Example 14.17:

$$u_I(t_i) = [(1 - \Phi)/B_d]u_m + [-1/B_d]n_d, \qquad x_I(t_i) = u_m = y_m = y_d$$

Equations (14-270) and (14-268b) then yield the PI/CGT law as

$$\begin{aligned}u(t_i) &= \{[(1 - \Phi)u_m - n_d]/B_d\} + K_x[y_d - x(t_i)] + K_\xi\xi(t_i)\\ \xi(t_{i+1}) &= \xi(t_i) + [y_d - x(t_i)]\end{aligned}$$

with an initial condition of $\xi(t_0) = \xi_I = 0$. This is in the form of the open-loop CGT law plus the output of a PI controller operating on the tracking error $[y_d - x(t_i)]$; such a simplification of the general structure is due to the fact that $y_c = x$ in this scalar state problem. Applying this controller to the original system model yields a closed-loop configuration described by

$$\begin{aligned}x(t_{i+1}) &= \Phi x(t_i) + B_d u(t_i) + n_d\\ &= [\Phi - B_dK_x]x(t_i) + y_d - [\Phi - B_dK_x]y_d + B_dK_\xi\xi(t_i)\end{aligned}$$

with error dynamics described via

$$\begin{bmatrix}[x(t_{i+1}) - y_d] \\ \xi(t_{i+1})\end{bmatrix} = \begin{bmatrix}(\Phi - B_d K_x) & B_d K_\xi \\ -1 & 1\end{bmatrix}\begin{bmatrix}[x(t_i) - y_d] \\ \xi(t_i)\end{bmatrix}$$

Thus, if the closed-loop eigenvalues, i.e., the roots of

$$0 = [\lambda - (\Phi - B_d K_x)][\lambda - 1] + B_d K_\xi$$

both have magnitude less than unity, the tracking error decays to zero and $x(t_i)$ converges to $y_d = u_m$ in steady state, and $\xi(t_i)$ converges to the ideal value $\xi_I = 0$ in steady state. Note that this agrees with the result of Example 14.12, as expected, since the command generator is a trivial direct feedthrough in this case. As shown in that previous example, $x(t_i)$ converges to $y_d$ in steady state even if there are *unmodeled constant* disturbances in the dynamics as well as disturbances which the controller is specifically designed to reject. This law can also be expressed according to (14-268) and (14-269) if desired.

The preferable incremental form is as given by (14-267),

$$\begin{aligned} u(t_i) &= u(t_{i-1}) - K_x[x(t_i) - x(t_{i-1})] + K_\xi[y_d - x(t_{i-1})] \\ &\quad + [K_x \cdot 0 + 0][x_m(t_i) - x_m(t_{i-1})] + [K_x \cdot 0 - (1/B_d)][n_d - n_d] \\ &= u(t_{i-1}) - K_x[x(t_i) - x(t_{i-1})] + K_\xi[y_d - x(t_{i-1})] \end{aligned}$$

The eigenvalues of the closed-loop system with this controller applied are the same as with the position form law applied, with an additional eigenvalue at zero, as discussed previously.

Note that the type-0 controller would be given via (14-266) as

$$u(t_i) = u(t_{i-1}) - \bar{G}^*_{c1}[x(t_{i-1}) - y_d] - \bar{G}^*_{c2}[u(t_{i-1}) + \{(\Phi - 1)y_d + n_d\}/B_d]$$

which agrees with Example 14.11. ■

EXAMPLE 14.19 Consider the PI/CGT law for the problems in which the command generator is *not* a trivial feedthrough. First recall Example 14.15. With the open-loop CGT $u_I$ evaluated as in that example and $x_I = x_m$ by (14-245), Eqs. (14-270) and (14-268b) yield

$$\begin{aligned} u(t_i) &= \{[(1 - \Phi)/B_d]x_m(t_i) + [(\Delta t)/B_d]\dot{y}_m - [1/B_d]n_d\} \\ &\quad + K_x[x_m(t_i) - x(t_i)] + K_\xi \xi(t_i) \\ \xi(t_{i+1}) &= \xi(t_i) + [x_m(t_i) - x(t_i)]; \qquad \xi(t_0) = \xi_I = 0 \end{aligned}$$

Applying this to the original system dynamics yields

$$\begin{aligned} x(t_{i+1}) &= \Phi x(t_i) + B_d u(t_i) + n_d \\ &= [\Phi - B_d K_x][x(t_i) - x_m(t_i)] + x_m(t_i) + \Delta t\, \dot{y}_m + B_d K_\xi \xi(t_i) \end{aligned}$$

and, since $x_m(t_{i+1}) = x_m(t_i) + \Delta t\, \dot{y}_m$, the error dynamics satisfy

$$\begin{bmatrix}[x(t_{i+1}) - x_m(t_{i+1})] \\ \xi(t_{i+1})\end{bmatrix} = \begin{bmatrix}(\Phi - B_d K_x) & B_d K_\xi \\ -1 & 1\end{bmatrix}\begin{bmatrix}[x(t_i) - x_m(t_i)] \\ \xi(t_i)\end{bmatrix}$$

which are the same as in the previous example.

The better incremental form (14-267) is

$$u(t_i) = u(t_{i-1}) - K_x[x(t_i) - x(t_{i-1})] + K_\xi[x_m(t_{i-1}) - x(t_{i-1})] + [K_x + (1 - \Phi)/B_d][\Delta t\, \dot{y}_m] + 0$$

and the closed-loop system eigenvalues are again as for the position form, plus an additional eigenvalue at zero. When the controlled system is perfectly tracking the command generator model, this incremental form reduces to

$$\begin{aligned} u(t_i) &= u(t_{i-1}) - K_x[\Delta t\, \dot{y}_m] + K_\xi[0] + [K_x + (1-\Phi)/B_d][\Delta t\, \dot{y}_m] \\ &= u(t_{i-1}) + [(1-\Phi)/B_d][\Delta t\, \dot{y}_m] \end{aligned}$$

and this equals $u_I(t_i)$; its magnitude grows unbounded for nonzero desired $\dot{y}_m$. Here the PI tracker *cannot* perfectly track the command if there are modeling errors: "perfect tracking" of the true system would require $u(t_i)$ to be generated according to this equation, using true $\Phi$ and $B_d$. If different design values of $\Phi$ and $B_d$ are used instead, the computed $u(t_i)$ does not equal the control to yield that perfect tracking. In this case, augmenting a *single* set of integrators to the original problem description does *not* compensate for inevitable modeling errors (as is often claimed in the literature); *two* augmented integrators would accommodate such errors here.

The PI/CGT law for the problem of Example 14.16 is similarly found to be

$$\begin{aligned} u(t_i) &= u_I(t_i) + K_x[x_{m1}(t_i) - x(t_i)] + K_\xi \xi(t_i) \\ \xi(t_{i+1}) &= \xi(t_i) + [x_{m1}(t_i) - x(t_i)] \end{aligned}$$

in position form, again with the same error dynamics as in the preceding two cases. In incremental form, this becomes

$$\begin{aligned} u(t_i) = u(t_{i-1}) &- K_x[x(t_i) - x(t_{i-1})] + K_\xi[x_{m1}(t_{i-1}) - x(t_{i-1})] \\ &+ [K_x - (\Phi/B_d)][x_{m1}(t_i) - x_{m1}(t_{i-1})] \\ &+ [1/B_d][x_{m2}(t_i) - x_{m2}(t_{i-1})] \end{aligned}$$ ■

As in the preceding cases, Kalman filters or observers can be used to generate $\hat{\mathbf{x}}(t_i^+)$, or possibly $\hat{\mathbf{x}}(t_i^-)$, to replace $\mathbf{x}(t_i)$ for cases in which only noise-corrupted measurements are available at each $t_i$ instead of perfect knowledge of the entire state. When $\mathbf{n}_d(t_i)$ also requires estimation, one must use a single filter that does not decompose into two independent filters. Open-loop CGT's and closed-loop CGT's based on either regulators or PI controllers can also be developed for more general model inputs of a specified structure, i.e., for nonconstant $\mathbf{u}_m$ in (14-240) and nonzero $\mathbf{n}_{cmd}$ in (14-239) [17, 18].

## 14.11 PERFORMANCE EVALUATION OF LINEAR SAMPLED-DATA CONTROLLERS

We want to develop a systematic approach to the design of linear stochastic controllers for realistic applications in which computer limitations, such as computational speed and memory size, are important constraints on the design. We will exploit the LQG optimal controller synthesis capability (and other viable methods) to generate a number of prospective controllers, as based upon models of different complexity and "tuning" conditions. It is then necessary to compare their performance capabilities and loading requirements in a trade-off analysis, to yield the best controller for actual implementation. Thus, an essential aspect of controller design is the availability of a performance analysis algorithm to evaluate *any* form of sampled-data linear controller in a *realistic* environment. (This section is an extension of results in [189].)

Similar to the filter performance analysis discussed in Chapter 6 of Volume 1 [115], a "truth model" is first developed to portray the real system to be controlled, and the effects of the external environment upon the system, as well as can be modeled (regardless of computational loading). Then the various proposed controllers are inserted into this simulation of the "real world," and their performance evaluated. This is depicted in Fig. 14.16. The truth model is generally described by the $n_t$-dimensional Itô state equation

$$\mathbf{dx}_t(t) = \mathbf{f}_t[\mathbf{x}_t(t), \mathbf{u}(t), t]\, dt + \mathbf{G}_t[\mathbf{x}_t(t), t]\, \mathbf{d}\boldsymbol{\beta}_t(t) \tag{14-272}$$

with $\boldsymbol{\beta}_t(\cdot,\cdot)$ as Brownian motion of diffusion $\mathbf{Q}_t(t)$ for all time $t$ (with hypothetical derivative $\mathbf{w}_t(\cdot,\cdot)$ as zero-mean white Gaussian noise of strength $\mathbf{Q}_t(t)$). Sampled-data measurements are available to the controller as the $m$-vector

$$\mathbf{z}_t(t_i) = \mathbf{h}_t[\mathbf{x}_t(t_i), t_i] + \mathbf{v}_t(t_i) \tag{14-273}$$

with $\mathbf{v}_t(\cdot,\cdot)$ as discrete-time zero-mean white Gaussian noise of covariance $\mathbf{R}_t(t_i)$ for all $t_i$. Note that Fig. 14.16 also admits a desired controlled variable command $\mathbf{y}_d(t_i)$ as an input to the controller, as seen in Fig. 13.3. For the most useful form of performance evaluation algorithm, one would desire the ability to specify such a continuous-time description of the system to be controlled and of its environment, and a discrete-time description of the controller (to be implemented as software in an online digital computer eventually) with a sample frequency that is a variable design parameter.

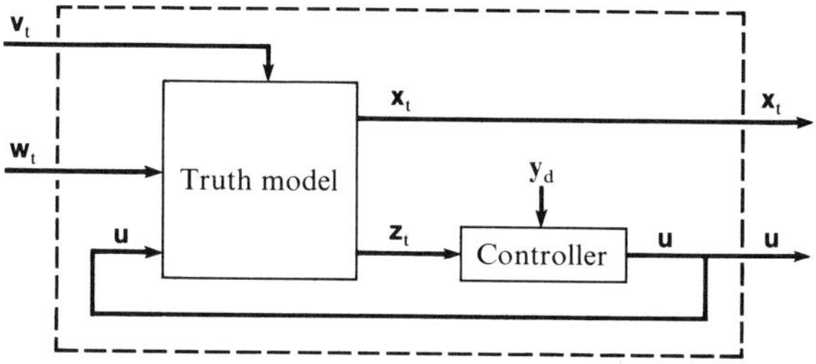


FIG. 14.16 Performance evaluation of linear sampled-data controller.

We want to evaluate specific *linear* control laws that use measurements up to and (possibly) including the $i$th measurement, to compute the control $\mathbf{u}(t_i)$ to be applied as a constant value over the ensuing sample period. Let the control algorithm entail an $n_c$th order linear difference equation for a controller state vector $\mathbf{x}_c(t_i)$ and an associated output relation to generate $\mathbf{u}(t_i)$:

$$\mathbf{u}(t_i) = \mathbf{G}_{cx}(t_i)\mathbf{x}_c(t_i) + \mathbf{G}_{cz}(t_i)\mathbf{z}_i + \mathbf{G}_{cy}(t_i)\mathbf{y}_d(t_i) \tag{14-274a}$$

$$\mathbf{x}_c(t_{i+1}) = \boldsymbol{\Phi}_c(t_{i+1}, t_i)\mathbf{x}_c(t_i) + \mathbf{B}_{cz}(t_i)\mathbf{z}_i + \mathbf{B}_{cy}(t_i)\mathbf{y}_d(t_i) \tag{14-274b}$$

where $\mathbf{z}_i$ is the realization of the measurement at $t_i$, $\mathbf{z}(t_i, \omega_k) = \mathbf{z}_i$. It is usually assumed that the controller state is zero at $t_0$: $\mathbf{x}_c(t_0) = \mathbf{0}$. Note that computation of $\mathbf{u}(t_i)$ precedes computation of the controller updated state in (14-274), in order to *minimize the destabilizing computational delay time between taking a measurement and producing the appropriate control signal*; (14-274b) can then be computed "in the background" to set up computations for the next sample time (in fact, the first term of (14-274a) can also be precomputed for further reduction of this delay). Computation time can be further reduced by exploiting the state structure of (14-274): transforming to canonical form or phase variable form (the latter being possible even for *time-varying* systems by Eqs. (10-74)–(10-84) of Volume 2) yields many ones and zeros in defining matrices and thus few required additions and multiplications.

EXAMPLE 14.20 To show that (14-274) is a useful generic result, first consider the simple LQ regulator that assumes perfect measurements of state are available: $\mathbf{z}(t_i) = \mathbf{x}(t_i)$. Then (14-6) is equivalent to (14-274a) with $\mathbf{G}_{cx}(t_i) \equiv \mathbf{0}$, $\mathbf{G}_{cz}(t_i) = -\mathbf{G}_c^*(t_i)$, and $\mathbf{y}_d(t_i) \equiv \mathbf{0}$. It is *not* a dynamic compensator, so there is no internal computer state ($n_c = 0$), and (14-274b) is not maintained.

The LQG optimal regulator (14-7) can be expressed in this form by letting the controller state be $\mathbf{x}_c(t_i) \triangleq \hat{\mathbf{x}}(t_i^-)$ and writing

$$\mathbf{u}^*(t_i) = -\mathbf{G}_c^*(t_i)[\hat{\mathbf{x}}(t_i^-) + \mathbf{K}(t_i)\{\mathbf{z}_i - \mathbf{H}(t_i)\hat{\mathbf{x}}(t_i^-)\}]$$

which, along with (14-118), yields (14-274) with $\mathbf{y}_d(t_i) \equiv \mathbf{0}$ and

$$\begin{aligned}
\mathbf{G}_{cx}(t_i) &= -\mathbf{G}_c^*(t_i)[\mathbf{I} - \mathbf{K}(t_i)\mathbf{H}(t_i)] \\
\mathbf{G}_{cz}(t_i) &= -\mathbf{G}_c^*(t_i)\mathbf{K}(t_i) \\
\mathbf{\Phi}_c(t_{i+1}, t_i) &= [\mathbf{\Phi}(t_{i+1}, t_i) - \mathbf{B}_d(t_i)\mathbf{G}_c^*(t_i)][\mathbf{I} - \mathbf{K}(t_i)\mathbf{H}(t_i)] \\
\mathbf{B}_{cz}(t_i) &= [\mathbf{\Phi}(t_{i+1}, t_i) - \mathbf{B}_d(t_i)\mathbf{G}_c^*(t_i)]\mathbf{K}(t_i)
\end{aligned}$$

On the other hand, the suboptimal law $\mathbf{u}(t_i) = -\mathbf{G}_c^*(t_i)\hat{\mathbf{x}}(t_i^-)$ as suggested in (14-95) can be expressed as in (14-274) with

$$\begin{aligned}
\mathbf{G}_{cx}(t_i) &= -\mathbf{G}_c^*(t_i) \\
\mathbf{G}_{cz}(t_i) &= \mathbf{0} \\
\mathbf{\Phi}_c(t_{i+1}, t_i) &= \mathbf{\Phi}(t_{i+1}, t_i)[\mathbf{I} - \mathbf{K}(t_i)\mathbf{H}(t_i)] - \mathbf{B}_d(t_i)\mathbf{G}_c^*(t_i) \\
\mathbf{B}_{cz}(t_i) &= \mathbf{\Phi}(t_{i+1}, t_i)\mathbf{K}(t_i)
\end{aligned}$$

and, since $\mathbf{G}_{cz}(t_i) \equiv \mathbf{0}$, the control $\mathbf{u}(t_i)$ can be computed *completely* before time $t_i$, and $\mathbf{u}(t_i)$ can be applied with *no* computational delay time.

The generic PI controller that assumes perfect state knowledge was given by (14-232). Letting the controller state be $\boldsymbol{\varepsilon}(t_i)$, this is equivalent to (14-274) with

$$\begin{aligned}
&\mathbf{G}_{cx} = \mathcal{K}_1, && \mathbf{G}_{cz} = -\mathcal{K}_x, && \mathbf{G}_{cy} = \mathcal{K}_2 \\
&\mathbf{\Phi}_c = (\mathbf{I} - \mathbf{D}_y\mathcal{K}_1), && \mathbf{B}_{cz} = -(\mathbf{C} - \mathbf{D}_y\mathcal{K}_x), && \mathbf{B}_{cy} = (\mathbf{I} - \mathbf{D}_y\mathcal{K}_2)
\end{aligned}$$

The PI controller based on noisy measurement inputs can be expressed as in (14-274) by using the augmented controller state $\mathbf{x}_c(t_i) \triangleq [\hat{\mathbf{x}}(t_i^-)^T \,\vdots\, \boldsymbol{\varepsilon}(t_i)^T]^T$.

The CGT/regulator law (14-256) can be expressed in this manner by letting $\mathbf{x}_c(t_i) \triangleq [\mathbf{x}_m(t_i)^T \,\vdots\, \mathbf{n}_d(t_i)^T]^T$ and $\mathbf{y}_d \triangleq \mathbf{u}_m$, and combining (14-239) with $\mathbf{w}_{dn} \equiv \mathbf{0}$, (14-240), and (14-256b) to yield (14-274), with

$$\mathbf{G}_{cx} = [\mathbf{A}_{21} + \bar{\mathbf{G}}_c{}^*\mathbf{A}_{11} \vdots \mathbf{A}_{23} + \bar{\mathbf{G}}_c{}^*\mathbf{A}_{13}]$$
$$\mathbf{G}_{cz} = -\bar{\mathbf{G}}_c{}^*$$
$$\mathbf{G}_{cy} = [\mathbf{A}_{22} + \bar{\mathbf{G}}_c{}^*\mathbf{A}_{12}]$$

$$\mathbf{\Phi}_c = \begin{bmatrix} \mathbf{\Phi}_m & \mathbf{0} \\ \mathbf{0} & \mathbf{\Phi}_n \end{bmatrix}, \qquad \mathbf{B}_{cz} = \begin{bmatrix} \mathbf{0} \\ \mathbf{0} \end{bmatrix}, \qquad \mathbf{B}_{cy} = \begin{bmatrix} \mathbf{B}_{dm} \\ \mathbf{0} \end{bmatrix}$$

If a filter is inserted to generate a state estimate from noisy measurements, $\mathbf{x}_c(t_i)$ becomes $[\hat{\mathbf{x}}(t_i^-)^T \vdots \mathbf{x}_m(t_i)^T \vdots \mathbf{n}_d(t_i)^T]^T$; if a CGT/PI controller is to be evaluated, (14-274) is again valid, with $\boldsymbol{\varepsilon}(t_i)$ augmented to the controller states just discussed. ■

Returning to Fig. 14.16, so far we have described the truth model and controller blocks. For performance evaluation purposes, we want to portray the characteristics of both the truth model state process $\mathbf{x}_t(\cdot,\cdot)$ and the commanded controls $\mathbf{u}(\cdot,\cdot)$. It is the *truth model state*, depicting *actual* system characteristics, and not the state of the model that the controller assumes to be an adequate description (especially in reduced-order controller designs), that is of importance. For instance, if the controller is designed on the basis of ignoring certain system modes, it is critical to evaluate the effect of that controller on total system behavior, including the modes neglected during the design process. Moreover, the *continuous-time* process $\mathbf{x}_t(\cdot,\cdot)$ should be investigated, not just the discrete-time process as seen only at the controller sample times: desirable behavior of this discrete-time process is not sufficient to ensure proper control of the continuous-time system, due to such problems as aliasing (misrepresentation of frequency content of signals) and excessive delays caused by choosing an inappropriately long sample period. Sample frequency is a *critical design parameter* [23, 121, 140, 179, 189]. A sampled-data controller cannot control a system as "tightly" as one which continuously absorbs information about the process and continuously modifies the control signals (assuming all required calculations can be performed in real time). If the truth model is in fact linear, then the optimal LQG sampled-data controller, in which both the Kalman filter and deterministic controller gains are evaluated on the basis of the truth model itself, is optimal *only* in the sense that no other controller *at that same sample frequency* can outperform it. When the capacity of the control computer is limited, a simplified controller operating at a high sample frequency may well outperform the "optimal" controller operating at a necessarily lower sample rate. Nevertheless, the LQG synthesis capability will be of substantial assistance in the design of such simplified controllers as well, such as in generating the LQG design based on a purposely reduced-order and simplified model, and possibly considering only steady state form for actual implementation.

It is also desired to characterize the properties of $\mathbf{u}(\cdot,\cdot)$. One is interested in the level of commanded controls, as to verify that they do not exceed certain physical limits, such as control actuator saturation, gimbal stops, maximum admissible $g$-loading of structures, etc. If the truth model is linear and all processes are Gaussian, one might constrain the $3\sigma$ (three times the standard deviation) values about the mean not to violate these constraints.

Thus, it is desired to generate the mean and rms time histories of the truth model states and commanded controls. It is also possibly of interest to evaluate a scalar expected cost involved in using a particular controller design, such as given by (14-20) using $\mathbf{x}_t(\cdot,\cdot)$ and $\mathbf{u}(\cdot,\cdot)$. However, the single scalar number $J_c$ for each controller is *not* the primary performance indicator: the designer must ensure adequate system *time responses* in *individual channels* of interest, and this is not necessarily accomplished merely by minimizing $J_c$ of a specified functional form. Rather, the time histories may indicate constraint violations by a controller with lowest $J_c$, and this would indicate the need for an iterative redefinition of the cost weighting matrices in (14-20) so that the lowest $J_c$ really does reflect the desired controller performance.

Figure 14.16 can form the basis of a *Monte Carlo analysis* to evaluate the statistical properties of $\mathbf{x}_t(\cdot,\cdot)$ and $\mathbf{u}(\cdot,\cdot)$. Numerous samples of $\boldsymbol{\beta}_t(\cdot,\cdot)$ (or $\mathbf{w}_t(\cdot,\cdot)$) and $\mathbf{v}_t(\cdot,\cdot)$ are used to drive (14-272)–(14-274) to generate individual samples of $\mathbf{x}_t(\cdot,\cdot)$ and $\mathbf{u}(\cdot,\cdot)$ by direct simulation, and then sample statistics can be computed. By restricting attention to cases in which the truth model is itself linear, the mean and rms time histories and the overall mean quadratic cost (14-20) can be calculated *explicitly*, *without* requiring a Monte Carlo analysis of many simulation runs. Therefore, let us assume that the truth model is adequately represented by

$$\begin{aligned} \mathbf{dx}_t(t) &= \mathbf{F}_t(t)\mathbf{x}_t(t)\,dt + \mathbf{B}_t(t)\mathbf{u}(t)\,dt + \mathbf{G}_t(t)\,\mathbf{d}\boldsymbol{\beta}_t(t) \\ \dot{\mathbf{x}}_t(t) &= \mathbf{F}_t(t)\mathbf{x}_t(t) + \mathbf{B}_t(t)\mathbf{u}(t) + \mathbf{G}_t(t)\mathbf{w}_t(t) \end{aligned} \tag{14-275a}$$

$$\mathbf{z}_t(t_i) = \mathbf{H}_t(t_i)\mathbf{x}_t(t_i) + \mathbf{v}_t(t_i) \tag{14-275b}$$

as a special case of (14-272) and (14-273). The initial condition is known with only some uncertainty in general, so $\mathbf{x}_t(t_0)$ is modeled as Gaussian, with mean $\bar{\mathbf{x}}_{t0}$ and covariance $\mathbf{P}_{t0}$. Also, it is assumed that the measurement samples will be made at a single predetermined (by the designer) sampling frequency; applications in which multiple sampling rates are appropriate can be handled by letting the basic sample period correspond to the largest interval common to all sample periods (for example, 50 Hz inner loops of an aircraft flight control system can be combined with 10 Hz guidance and navigation loops by operating with a sample period of 0.02 sec: the outer loops change value every fifth short period interval). As before, we assume $\mathbf{u}(t) = \mathbf{u}(t_i)$ for all $t \in [t_i, t_{i+1})$.

Based on (14-274)–(14-275), we now wish to evaluate the first two moments of the Gaussian augmented output vector for Fig. 14.16,

$$\mathbf{y}_a(\cdot,\cdot) \triangleq \begin{bmatrix} \mathbf{x}_t(\cdot,\cdot) \\ \mathbf{u}(\cdot,\cdot) \end{bmatrix} \tag{14-276}$$

All quantities of primary interest in performance analysis are assumed to be components, or linear combinations of components, of $\mathbf{y}_a(\cdot,\cdot)$. Let $q_k$ be a scalar quantity of interest, expressed as

$$\mathrm{q}_k(t) = \mathbf{q}_k^{\mathrm{T}}\mathbf{y}_a(t) \tag{14-277a}$$

Then its statistical description includes

$$\mathrm{mean}\{\mathrm{q}_k(t)\} = \mathbf{q}_k^{\mathrm{T}}\mathbf{m}_{ya}(t) \tag{14-277b}$$

$$\mathrm{rms}\{\mathrm{q}_k(t)\} = [\mathbf{q}_k^{\mathrm{T}}\boldsymbol{\Psi}_{ya}(t)\mathbf{q}_k]^{1/2} \tag{14-277c}$$

$$1\sigma\{\mathrm{q}_k(t)\} = [\mathbf{q}_k^{\mathrm{T}}\mathbf{P}_{ya}(t)\mathbf{q}_k]^{1/2} \tag{14-277d}$$

in terms of $\mathbf{m}_{ya}(t) \triangleq E\{\mathbf{y}_a(t)\}$, $\boldsymbol{\Psi}_{ya}(t) \triangleq E\{\mathbf{y}_a(t)\mathbf{y}_a^{\mathrm{T}}(t)\}$, and $\mathbf{P}_{ya}(t) \triangleq E\{[\mathbf{y}_a(t) - \mathbf{m}_{ya}(t)][\mathbf{y}_a(t) - \mathbf{m}_{ya}(t)]^{\mathrm{T}}\}$ for all $t$.

EXAMPLE 14.21 In the design of an automatic landing system, two truth model states might be $x_{t1}$ and $x_{t2}$ as aircraft position error in the north and east directions, respectively, and one control might be aileron deflection angle $u_1$. Assume that two quantities of direct interest are $q_1$ = cross-track error in approaching a runway at heading angle $\psi$ from north and $q_2$ = aileron angle itself. Then

$$\begin{aligned} q_1(t) &= x_{t2}(t)\cos\psi - x_{t1}(t)\sin\psi \\ &= [-\sin\psi \,\vdots\, \cos\psi \,\vdots\, 0 \,\vdots\, \cdots \,\vdots\, 0]\mathbf{y}_a(t) = \mathbf{q}_1^{\mathrm{T}}\mathbf{y}_a(t) \\ q_2(t) &= [\underbrace{0 \,\vdots\, \cdots \,\vdots\, 0}_{n \text{ zeros}} \,\vdots\, 1 \,\vdots\, 0 \,\vdots\, \cdots \,\vdots\, 0]\mathbf{y}_a(t) = \mathbf{q}_2^{\mathrm{T}}\mathbf{y}_a(t) \end{aligned}$$

with statistics as in (14-277). ■

We further assume that the total cost of a given control operation can be adequately described by (14-20b) with $\mathbf{x}$ replaced by $\mathbf{x}_t$. Once the quantities of interest are identified as $q_1$ to $q_K$, a positive penalty weighting $w_k(t)$ is assigned to each, and the rate of increase of the cost can be defined as

$$\frac{dJ_c}{dt} = E\left\{\frac{1}{2}\sum_{k=1}^{K} w_k(t)\mathrm{q}_k^2(t)\right\} \tag{14-278a}$$

from which the composite matrix $\mathbf{W}_{ya}(t)$ in (14-20b) can be generated as

$$\begin{aligned} \frac{dJ_c}{dt} &= E\left\{\frac{1}{2}\sum_{k=1}^{K} w_k(t)[\mathbf{q}_k^{\mathrm{T}}\mathbf{y}_a(t)]^2\right\} \\ &\triangleq E\{\tfrac{1}{2}\mathbf{y}_a^{\mathrm{T}}(t)\mathbf{W}_{ya}(t)\mathbf{y}_a(t)\} = \tfrac{1}{2}\mathrm{tr}\{\mathbf{W}_{ya}(t)\boldsymbol{\Psi}_{ya}(t)\} \end{aligned} \tag{14-278b}$$

from which it can be seen that

$$\mathbf{W}_{ya}(t) = \sum_{k=1}^{K} w_k(t)\mathbf{q}_k\mathbf{q}_k^{\mathrm{T}} \tag{14-278c}$$

$\mathbf{X}_{\mathrm{ft}}$ in (14-20b) is evaluated similarly. Again, this performance indicator can be generated from knowledge of moments of $\mathbf{y}_{\mathrm{a}}(\cdot,\cdot)$.

To evaluate the statistical properties of $\mathbf{y}_{\mathrm{a}}(\cdot,\cdot)$, we first characterize the augmented internal state process associated with Fig. 14.16, i.e., the Gaussian process composed of both truth model and controller states,

$$\mathbf{x}_{\mathrm{a}}(\cdot,\cdot) = \begin{bmatrix} \mathbf{x}_{\mathrm{t}}(\cdot,\cdot) \\ \mathbf{x}_{\mathrm{c}}(\cdot,\cdot) \end{bmatrix} \tag{14-279}$$

First we generate an equivalent discrete-time system model to evaluate statistics at the sample times, and then establish the appropriate differential equations to interject results between sample times; actual performance evaluation software is often implemented in this manner to provide efficient computation of discrete-time process results, with an *option* for the more complete continuous-time results when desired.

Given (14-275), (14-276), and (possibly) (14-20b), we can generate the equivalent discrete-time truth model (14-1) and (possibly) cost (14-24)–(14-27) by performing the integrations indicated in (14-28). Once this is accomplished, (14-274a) and (14-275b) are used to eliminate $\mathbf{u}(t_i)$ and $\mathbf{z}_t(t_i)$ from the truth model state equation (14-1) with $\mathbf{G}_{\mathrm{dt}}(t_i) \equiv \mathbf{I}$ and controller state equation (14-274b) to yield

$$\begin{aligned} \mathbf{x}_{\mathrm{t}}(t_{i+1}) = {} & [\mathbf{\Phi}_{\mathrm{t}}(t_{i+1},t_i) + \mathbf{B}_{\mathrm{dt}}(t_i)\mathbf{G}_{\mathrm{cz}}(t_i)\mathbf{H}_{\mathrm{t}}(t_i)]\mathbf{x}_{\mathrm{t}}(t_i) \\ & + \mathbf{B}_{\mathrm{dt}}(t_i)\mathbf{G}_{\mathrm{cx}}(t_i)\mathbf{x}_{\mathrm{c}}(t_i) + \mathbf{B}_{\mathrm{dt}}(t_i)\mathbf{G}_{\mathrm{cy}}(t_i)\mathbf{y}_{\mathrm{d}}(t_i) \\ & + \mathbf{I}\mathbf{w}_{\mathrm{dt}}(t_i) + \mathbf{B}_{\mathrm{dt}}(t_i)\mathbf{G}_{\mathrm{cz}}(t_i)\mathbf{v}_{\mathrm{t}}(t_i) \end{aligned} \tag{14-280a}$$

$$\begin{aligned} \mathbf{x}_{\mathrm{c}}(t_{i+1}) = {} & \mathbf{\Phi}_{\mathrm{c}}(t_{i+1},t_i)\mathbf{x}_{\mathrm{c}}(t_i) + \mathbf{B}_{\mathrm{cz}}(t_i)\mathbf{H}_{\mathrm{t}}(t_i)\mathbf{x}_{\mathrm{t}}(t_i) \\ & + \mathbf{B}_{\mathrm{cy}}(t_i)\mathbf{y}_{\mathrm{d}}(t_i) + \mathbf{B}_{\mathrm{cz}}(t_i)\mathbf{v}_{\mathrm{t}}(t_i) \end{aligned} \tag{14-280b}$$

Defining the augmented driving noise vector $\mathbf{w}_{\mathrm{da}}(\cdot,\cdot)$ as the zero-mean, white, Gaussian discrete-time noise

$$\mathbf{w}_{\mathrm{da}}(\cdot,\cdot) \triangleq \begin{bmatrix} \mathbf{w}_{\mathrm{dt}}(\cdot,\cdot) \\ \mathbf{v}_{\mathrm{t}}(\cdot,\cdot) \end{bmatrix} \tag{14-281a}$$

with covariance for all $t_i$ as

$$\mathbf{Q}_{\mathrm{da}}(t_i) = \begin{bmatrix} \mathbf{Q}_{\mathrm{dt}}(t_i) & \mathbf{0} \\ \mathbf{0} & \mathbf{R}_{\mathrm{t}}(t_i) \end{bmatrix} \tag{14-281b}$$

(with correlated $\mathbf{w}_{\mathrm{dt}}(t_i)$ and $\mathbf{v}_{\mathrm{t}}(t_i)$ readily handled by nonzero off-diagonal blocks), Eq. (14-280) can be written conveniently as

$$\mathbf{x}_{\mathrm{a}}(t_{i+1}) = \mathbf{\Phi}_{\mathrm{a}}(t_{i+1},t_i)\mathbf{x}_{\mathrm{a}}(t_i) + \mathbf{B}_{\mathrm{da}}(t_i)\mathbf{y}_{\mathrm{d}}(t_i) + \mathbf{G}_{\mathrm{da}}(t_i)\mathbf{w}_{\mathrm{da}}(t_i) \tag{14-282}$$

where (dropping time arguments for compactness),

$$\mathbf{\Phi}_{\mathrm{a}} \triangleq \begin{bmatrix} \mathbf{\Phi}_{\mathrm{t}} + \mathbf{B}_{\mathrm{dt}}\mathbf{G}_{\mathrm{c}z}\mathbf{H}_{\mathrm{t}} & \mathbf{B}_{\mathrm{dt}}\mathbf{G}_{\mathrm{c}x} \\ \mathbf{B}_{\mathrm{c}z}\mathbf{H}_{\mathrm{t}} & \mathbf{\Phi}_{\mathrm{c}} \end{bmatrix}$$

$$\mathbf{B}_{\mathrm{da}} \triangleq \begin{bmatrix} \mathbf{B}_{\mathrm{dt}}\mathbf{G}_{\mathrm{c}y} \\ \mathbf{B}_{\mathrm{c}y} \end{bmatrix} \tag{14-283}$$

$$\mathbf{G}_{\mathrm{da}} \triangleq \begin{bmatrix} \mathbf{I} & \mathbf{B}_{\mathrm{dt}}\mathbf{G}_{\mathrm{c}z} \\ \mathbf{0} & \mathbf{B}_{\mathrm{c}z} \end{bmatrix}$$

Because the initial controller state is assumed to be zero, the initial $\mathbf{x}_{\mathrm{a}}(t_0)$ is Gaussian with

$$E\{\mathbf{x}_{\mathrm{a}}(t_0)\} = \begin{bmatrix} \bar{\mathbf{x}}_{\mathrm{t}0} \\ \mathbf{0} \end{bmatrix} \triangleq \bar{\mathbf{x}}_{\mathrm{a}0} \tag{14-284a}$$

$$E\{[\mathbf{x}_{\mathrm{a}}(t_0) - \bar{\mathbf{x}}_{\mathrm{a}0}][\mathbf{x}_{\mathrm{a}}(t_0) - \bar{\mathbf{x}}_{\mathrm{a}0}]^{\mathrm{T}}\} = \begin{bmatrix} \mathbf{P}_{\mathrm{t}0} & \mathbf{0} \\ \mathbf{0} & \mathbf{0} \end{bmatrix} \tag{14-284b}$$

A known nonzero $\mathbf{x}_{\mathrm{c}}(t_0)$ would simply change the lower component in (14-284a). Also, if desired, the cost equation (14-24) and (14-26) can be written as

$$\begin{aligned} J = E\Big\{&\frac{1}{2}\,\mathbf{x}_{\mathrm{a}}^{\mathrm{T}}(t_{N+1})\mathbf{X}_{\mathrm{fa}}\mathbf{x}_{\mathrm{a}}(t_{N+1}) \\ &+ \frac{1}{2}\sum_{i=0}^{N} [\mathbf{x}_{\mathrm{a}}^{\mathrm{T}}(t_i)\mathbf{X}_{\mathrm{a}}(t_i)\mathbf{x}_{\mathrm{a}}(t_i) + \mathbf{y}_{\mathrm{d}}^{\mathrm{T}}(t_i)\mathbf{Y}_{\mathrm{d}}(t_i)\mathbf{y}_{\mathrm{d}}(t_i) \\ &+ 2\mathbf{x}_{\mathrm{a}}^{\mathrm{T}}(t_i)\mathbf{S}_{\mathrm{a}}(t_i)\mathbf{y}_{\mathrm{d}}(t_i) + \mathbf{w}_{\mathrm{da}}^{\mathrm{T}}(t_i)\mathbf{W}_{\mathrm{a}}(t_i)\mathbf{w}_{\mathrm{da}}(t_i) + L_{\mathrm{r}}(t_i)]\Big\} \end{aligned} \tag{14-285}$$

where (again dropping time arguments)

$$\mathbf{X}_{\mathrm{fa}} \triangleq \begin{bmatrix} \mathbf{X}_{\mathrm{ft}} & \mathbf{0} \\ \mathbf{0} & \mathbf{0} \end{bmatrix}$$

$$\mathbf{X}_{\mathrm{a}} \triangleq \begin{bmatrix} \mathbf{X}_{\mathrm{t}} + \mathbf{H}_{\mathrm{t}}^{\mathrm{T}}\mathbf{G}_{\mathrm{c}z}^{\mathrm{T}}\mathbf{U}_{\mathrm{t}}\mathbf{G}_{\mathrm{c}z}\mathbf{H}_{\mathrm{t}} + \mathbf{S}_{\mathrm{t}}\mathbf{G}_{\mathrm{c}z}\mathbf{H}_{\mathrm{t}} + \mathbf{H}_{\mathrm{t}}^{\mathrm{T}}\mathbf{G}_{\mathrm{c}z}^{\mathrm{T}}\mathbf{S}_{\mathrm{t}}^{\mathrm{T}} & \mathbf{H}_{\mathrm{t}}^{\mathrm{T}}\mathbf{G}_{\mathrm{c}z}^{\mathrm{T}}\mathbf{U}_{\mathrm{t}}\mathbf{G}_{\mathrm{c}x} + \mathbf{S}_{\mathrm{t}}\mathbf{G}_{\mathrm{c}x} \\ (\mathbf{H}_{\mathrm{t}}^{\mathrm{T}}\mathbf{G}_{\mathrm{c}z}^{\mathrm{T}}\mathbf{U}_{\mathrm{t}}\mathbf{G}_{\mathrm{c}x} + \mathbf{S}_{\mathrm{t}}\mathbf{G}_{\mathrm{c}x})^{\mathrm{T}} & \mathbf{G}_{\mathrm{c}x}^{\mathrm{T}}\mathbf{U}_{\mathrm{t}}\mathbf{G}_{\mathrm{c}x} \end{bmatrix}$$

$$\mathbf{Y}_{\mathrm{d}} \triangleq \mathbf{G}_{\mathrm{c}y}^{\mathrm{T}}\mathbf{U}_{\mathrm{t}}\mathbf{G}_{\mathrm{c}y}$$

$$\mathbf{S}_{\mathrm{a}} \triangleq [\mathbf{H}_{\mathrm{t}}^{\mathrm{T}}\mathbf{G}_{\mathrm{c}z}^{\mathrm{T}}\mathbf{U}_{\mathrm{t}}\mathbf{G}_{\mathrm{c}y} + \mathbf{S}_{\mathrm{t}}\mathbf{G}_{\mathrm{c}y} \quad \mathbf{G}_{\mathrm{c}x}^{\mathrm{T}}\mathbf{U}_{\mathrm{t}}\mathbf{G}_{\mathrm{c}y}]$$

$$\mathbf{W}_{\mathrm{a}} \triangleq \begin{bmatrix} \mathbf{0} & \mathbf{0} \\ \mathbf{0} & \mathbf{G}_{\mathrm{c}z}^{\mathrm{T}}\mathbf{U}_{\mathrm{t}}\mathbf{G}_{\mathrm{c}z} \end{bmatrix} \tag{14-286}$$

From (14-282), it can be seen that the mean $\mathbf{m}_{x\mathrm{a}}$ and covariance $\mathbf{P}_{x\mathrm{a}}$ of $\mathbf{x}_{\mathrm{a}}(\cdot,\cdot)$

can be propagated by

$$\mathbf{m}_{xa}(t_{i+1}) = \mathbf{\Phi}_a(t_{i+1}, t_i)\mathbf{m}_{xa}(t_i) + \mathbf{B}_{da}(t_i)\mathbf{y}_d(t_i) \tag{14-287}$$

$$\mathbf{P}_{xa}(t_{i+1}) = \mathbf{\Phi}_a(t_{i+1}, t_i)\mathbf{P}_{xa}(t_i)\mathbf{\Phi}_a^T(t_{i+1}, t_i) + \mathbf{G}_{da}(t_i)\mathbf{Q}_{da}(t_i)\mathbf{G}_{da}^T(t_i) \tag{14-288}$$

using the augmented matrices defined in (14-281)–(14-284). The correlation matrix for $\mathbf{x}_a(\cdot,\cdot)$ can then be evaluated as

$$\mathbf{\Psi}_{xa}(t_i) \triangleq E\{\mathbf{x}_a(t_i)\mathbf{x}_a^T(t_i)\} = \mathbf{P}_{xa}(t_i) + \mathbf{m}_{xa}(t_i)\mathbf{m}_{xa}^T(t_i) \tag{14-289}$$

In terms of these propagated quantities, the cost (14-285) can be expressed as

$$\begin{aligned} J = {} & \tfrac{1}{2}\operatorname{tr}[\mathbf{X}_{fa}\,\mathbf{\Psi}_{xa}(t_{N+1})] \\ & + \frac{1}{2}\sum_{i=0}^{N} \{\operatorname{tr}[\mathbf{X}_a(t_i)\mathbf{\Psi}_{xa}(t_i)] + \mathbf{y}_d^T(t_i)\mathbf{Y}_d(t_i)\mathbf{y}_d(t_i) \\ & + 2\mathbf{m}_{xa}^T(t_i)\mathbf{S}_a(t_i)\mathbf{y}_d(t_i) + \operatorname{tr}[\mathbf{W}_a(t_i)\mathbf{Q}_{da}(t_i)] + J_r(t_i)\} \end{aligned} \tag{14-290}$$

using (14-286) and (14-26). An efficient performance evaluation algorithm would sequentially compute $\mathbf{m}_{xa}(t_i)$, $\mathbf{P}_{xa}(t_i)$ and $\mathbf{\Psi}_{xa}(t_i)$, for $i = 0, 1, \ldots, N$, and if desired, compute one summation term of (14-290) for each $i$ and add it to a running sum; at the final time $t_{N+1}$, the complete time histories of $\mathbf{x}_a(\cdot,\cdot)$ moments are available, and $\frac{1}{2}\operatorname{tr}[\mathbf{X}_{fa}\mathbf{\Psi}_{xa}(t_{N+1})]$ can be added to the running sum to provide the corresponding $J$ value.

As discussed before, the time histories of the moments of $\mathbf{y}_a(\cdot,\cdot)$ are needed above and beyond the scalar $J$ value for a complete performance analysis. Since $\mathbf{y}_a(\cdot,\cdot)$ can be related to $\mathbf{x}_a(\cdot,\cdot)$ by

$$\begin{aligned} \begin{bmatrix} \mathbf{x}_t(t_i) \\ \mathbf{u}(t_i) \end{bmatrix} = {} & \begin{bmatrix} \mathbf{I} & \mathbf{0} \\ \mathbf{G}_{cz}(t_i)\mathbf{H}_t(t_i) & \mathbf{G}_{cx}(t_i) \end{bmatrix} \begin{bmatrix} \mathbf{x}_t(t_i) \\ \mathbf{x}_c(t_i) \end{bmatrix} \\ & + \begin{bmatrix} \mathbf{0} \\ \mathbf{G}_{cy}(t_i) \end{bmatrix} \mathbf{y}_d(t_i) + \begin{bmatrix} \mathbf{0} \\ \mathbf{G}_{cz}(t_i) \end{bmatrix} \mathbf{v}_t(t_i) \end{aligned} \tag{14-291}$$

the desired statistics can be generated as

$$\mathbf{m}_{ya}(t_i) = \begin{bmatrix} \mathbf{I} & \mathbf{0} \\ \mathbf{G}_{cz}(t_i)\mathbf{H}_t(t_i) & \mathbf{G}_{cx}(t_i) \end{bmatrix} \mathbf{m}_{xa}(t_i) + \begin{bmatrix} \mathbf{0} \\ \mathbf{G}_{cy}(t_i) \end{bmatrix} \mathbf{y}_d(t_i) \tag{14-292a}$$

$$\begin{aligned} \mathbf{P}_{ya}(t_i) = {} & \begin{bmatrix} \mathbf{I} & \mathbf{0} \\ \mathbf{G}_{cz}(t_i)\mathbf{H}_t(t_i) & \mathbf{G}_{cx}(t_i) \end{bmatrix} \mathbf{P}_{xa}(t_i) \begin{bmatrix} \mathbf{I} & \mathbf{H}_t^T(t_i)\mathbf{G}_{cz}^T(t_i) \\ \mathbf{0} & \mathbf{G}_{cx}^T(t_i) \end{bmatrix} \\ & + \begin{bmatrix} \mathbf{0} \\ \mathbf{G}_{cz}(t_i) \end{bmatrix} \mathbf{R}_t(t_i)[\mathbf{0} \quad \mathbf{G}_{cz}^T(t_i)] \end{aligned} \tag{14-292b}$$

$$\mathbf{\Psi}_{ya}(t_i) = \mathbf{P}_{ya}(t_i) + \mathbf{m}_{ya}(t_i)\mathbf{m}_{ya}^T(t_i) \tag{14-292c}$$

using the state statistics already evaluated in (14-287)–(14-289).

Finally, to evaluate the desired moments between sample instants, we can write the differential equation for $\mathbf{y}_a(\cdot,\cdot)$ that is valid for all $t \in [t_i, t_{i+1})$:

$$\mathbf{dy}_a(t) \triangleq \begin{bmatrix} \mathbf{dx}_t(t) \\ \mathbf{du}(t) \end{bmatrix} = \begin{bmatrix} \mathbf{F}_t(t) & \mathbf{B}_t(t) \\ \mathbf{0} & \mathbf{0} \end{bmatrix} \begin{bmatrix} \mathbf{x}_t(t) \\ \mathbf{u}(t) \end{bmatrix} dt + \begin{bmatrix} \mathbf{G}_t(t) \\ \mathbf{0} \end{bmatrix} \mathbf{d\beta}_t(t) \tag{14-293}$$

integrated forward to $t_{i+1}$, whereupon $\mathbf{u}(t)$ undergoes a step change. Thus, starting from the initial conditions given by (14-292) at $t_i$, we can propagate

$$\dot{\mathbf{m}}_{ya}(t) = \begin{bmatrix} \mathbf{F}_t(t) & \mathbf{B}_t(t) \\ \mathbf{0} & \mathbf{0} \end{bmatrix} \mathbf{m}_{ya}(t) \tag{14-294a}$$

$$\dot{\mathbf{P}}_{ya}(t) = \begin{bmatrix} \mathbf{F}_t(t) & \mathbf{B}_t(t) \\ \mathbf{0} & \mathbf{0} \end{bmatrix} \mathbf{P}_{ya}(t) + \mathbf{P}_{ya}(t) \begin{bmatrix} \mathbf{F}_t^{\mathrm{T}}(t) & \mathbf{0} \\ \mathbf{B}_t^{\mathrm{T}}(t) & \mathbf{0} \end{bmatrix} + \begin{bmatrix} \mathbf{G}_t(t)\mathbf{Q}_t(t)\mathbf{G}_t^{\mathrm{T}}(t) & \mathbf{0} \\ \mathbf{0} & \mathbf{0} \end{bmatrix} \tag{14-294b}$$

and calculate $\mathbf{\Psi}_{ya}(t)$ as $[\mathbf{P}_{ya}(t) + \mathbf{m}_{ya}(t)\mathbf{m}_{ya}^{\mathrm{T}}(t)]$, for all $t \in [t_i, t_{i+1})$. The values of $\mathbf{m}_{ya}(t_{i+1})$, $\mathbf{P}_{ya}(t_{i+1})$, and $\mathbf{\Psi}_{ya}(t_{i+1})$ from the discrete-time recursion of (14-287)–(14-289) and (14-292) then provide the appropriate initial conditions for the succeeding sample period. Use of such a performance evaluation algorithm to provide useful outputs through (14-277) (and possibly (14-286) and (14-290)) will be demonstrated in the next section. Note that it is capable of evaluating the performance of *any* linear sampled-data controller, and not just designs based on LQG synthesis.

## 14.12 SYSTEMATIC DESIGN PROCEDURE

One primary use of stochastic linear optimal control theory is to provide a *synthesis* capability for *stabilizing* control laws [83, 168, 189]. If the "truth model" is adequately represented as linear, LQG synthesis also provides a *benchmark* or standard of performance against which *practical designs* can be measured. If there were no computer constraints, the *benchmark* optimal LQG stochastic controller *based upon the truth model* would be the best controller design for any given sample frequency, "best" in the sense that it will minimize the expected value of the given quadratic penalty function (averaged over the ensemble of operations of the process). By synthesizing and evaluating this benchmark controller for various sample frequencies, the designer can determine if it is at all possible to meet his conflicting design objectives: this benchmark controller exhibits the *theoretical limits of performance* imposed on any design by imperfect information about the system state, driving disturbances and uncertainties, constraints on control levels, and the natural dynamics of the system. Typically, this requires an iteration on choice of cost weighting matrices to provide a benchmark with all desired performance characteristics.

EXAMPLE 14.22 Consider an air-to-air missile intercept problem [189] described by Fig. 14.17. An air-to-air missile is fired at a target that is performing evasive maneuvers, such that the launch direction is along the initial line-of-sight; the problem is simplified by considering only the pitch plane. Onboard the missile is an inertially stabilized tracking radar that provides noisy measurements of the target angular location $\varepsilon(t)$ with respect to the *initial* line-of-sight. We want to use LQG synthesis to design the guidance system to accept these noisy measurements every $\Delta t$ sec (initially chosen to be 0.8 sec) and determine the appropriate steering command of missile lateral acceleration to intercept the target. It is assumed that the closed-loop autopilot responds rapidly enough to these commands, that the missile achieves the commanded acceleration instantaneously. As a further simplification, we assume that the closing velocity between the missile and target is essentially constant (3000 ft/sec) and that the total time interval of interest is known a priori to be 12 sec (in actual implementation, "time-to-go" to intercept would be continually estimated onboard). The guidance system is to be designed so as to minimize rms terminal miss distance, while not ever commanding lateral accelerations that might jeopardize aerodynamic stability or structural integrity of the missile (i.e., never exceeding 1200 ft/sec$^2$ lateral acceleration command).

One possible choice of truth model states for a linearized problem description would be the *relative* lateral distance $y_{TM}(t)$ between the target and missile as measured normal to the initial line-of-sight, the *relative* lateral velocity $v_{TM}(t)$, and the *total* lateral acceleration $a_T(t)$ of the target; missile lateral acceleration $a_M(t)$ is the control variable. Other choices, such as other combinations of relative and total variables or a line-of-sight coordinate system instead of an inertial set of axes, are also possible. In terms of these states, the truth model state equations become

$$\dot{\mathsf{y}}_{TM}(t) = \mathsf{v}_{TM}(t)$$

$$\dot{\mathsf{v}}_{TM}(t) = \mathsf{a}_T(t) - \mathsf{a}_M(t)$$

$$\dot{\mathsf{a}}_T(t) = -[1/T]\mathsf{a}_T(t) + \mathsf{w}_t(t)$$

where the uncertain target evasive acceleration has been modeled as a stationary first order Gauss–Markov process with mean zero and autocorrelation kernel

$$E\{\mathsf{a}(t)\mathsf{a}(t+\tau)\} = \sigma_a^2 \exp\{-|\tau|/T\}$$

To generate this autocorrelation, white noise $\mathsf{w}_t(\cdot,\cdot)$ driving the truth model has $Q_t = 2\sigma_a^2/T$. The mean square acceleration $\sigma_a^2$ and correlation time $T$ are treated as design parameters: a bomber would be characterized by low $\sigma_a^2$ and large $T$ while an air superiority fighter would be modeled with larger $\sigma_a^2$ and shorter $T$; here we let $\sigma_a = 1\ g$ and $T = 2.5$ sec. The truth model initial conditions are $\bar{\mathbf{x}}_{t0} = \mathbf{0}$ and $\mathbf{P}_{t0}$ given by a diagonal matrix with diagonal elements of 0 ft$^2$ (since the initial line-of-sight establishes the coordinate system for measuring $y_{TM}$), $[200 \text{ ft/sec}]^2$ (a reasonable value), and $[32.2 \text{ ft/sec}^2]^2$ (the steady state variance of $\mathsf{a}_T$ above), respectively.

The measured line-of-sight angle can be represented as

$$\begin{aligned} \mathsf{z}_t(t_i) &= \varepsilon_t(t_i) + \mathsf{v}_t(t_i) \\ &= \tan^{-1}\{\mathsf{y}_{TM}(t_i)/(\text{horizontal range})\} + \mathsf{v}_t(t_i) \end{aligned}$$

but, for small angles the arctangent can be replaced by the angle itself, and range $\cong$ (constant closing velocity) $\times$ (time-to-go), to yield an approximate expression,

$$\mathsf{z}_t(t_i) = \frac{1}{(3000)(12 - t_i)} \mathsf{y}_{TM}(t_i) + \mathsf{v}_t(t_i)$$

where $\mathsf{v}(\cdot,\cdot)$ is zero-mean white Gaussian discrete-time noise with mean squared value of

$$E\{\mathsf{v}_t(t_i)^2\} = R_t(t_i) = \frac{\{R_1/[(3000)(12 - t_i)]^2\} + R_2}{\Delta t}$$

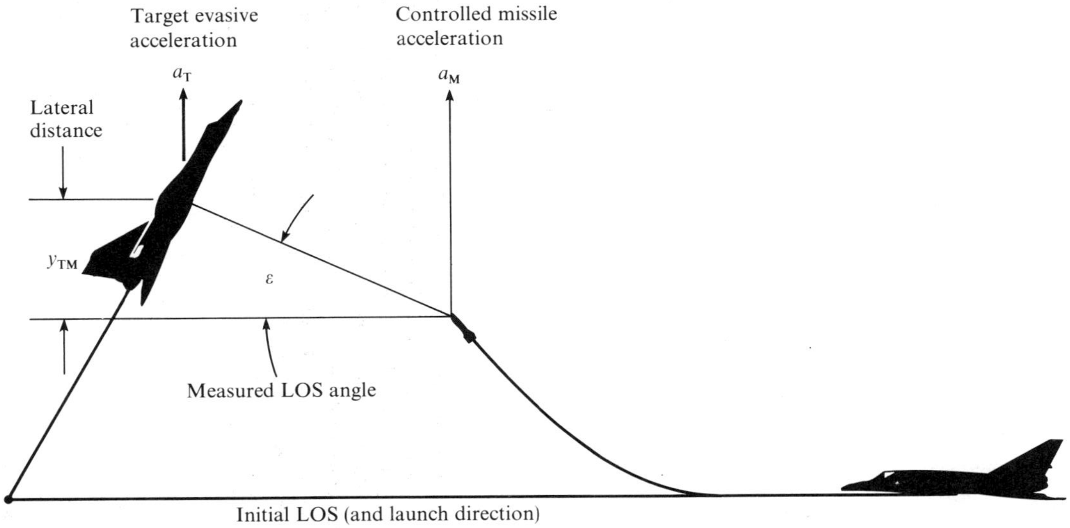


FIG. 14.17 Air-to-air missile problem geometry. From Widnall [189] with permission of The MIT Press.

where the first term grows larger as the range decreases (because of growing target size and thus increased dispersion of radar returns from different points on the target) and the second term $R_2$ is related to the fundamental angular resolution of the radar; $R_t(t_i)$ increases with decreasing $\Delta t$ since this results in less internal processing time for the radar between sample times and thus larger measurement errors. For this example, $R_1 = 15\ \text{ft}^2\ \text{sec}$ and $R_2 = 1.5 \times 10^{-5}\ \text{rad}^2\ \text{sec}$.

Thus, the basic truth model is

$$\begin{bmatrix} \dot{y}_{TM}(t) \\ \dot{v}_{TM}(t) \\ \dot{a}_T(t) \end{bmatrix} = \begin{bmatrix} 0 & 1 & 0 \\ 0 & 0 & 1 \\ 0 & 0 & -0.4 \end{bmatrix} \begin{bmatrix} y_{TM}(t) \\ v_{TM}(t) \\ a_T(t) \end{bmatrix} + \begin{bmatrix} 0 \\ -1 \\ 0 \end{bmatrix} a_M(t) + \begin{bmatrix} 0 \\ 0 \\ 1 \end{bmatrix} w_t(t)$$

$$z_t(t_i) = \begin{bmatrix} \dfrac{1}{3000(12 - t_i)} & 0 & 0 \end{bmatrix} \begin{bmatrix} y_{TM}(t_i) \\ v_{TM}(t_i) \\ a_T(t_i) \end{bmatrix} + v_t(t_i)$$

Although the state dynamics model is time invariant with stationary input, the measurements involve both a time-varying $\mathbf{H}_t(t_i)$ and nonstationary $v_t(\cdot,\cdot)$.

To accomplish the stated objectives, let us design a guidance algorithm that will minimize the average quadratic cost

$$J_c = E\left\{\frac{1}{2} y_{TM}(t_f)^2 + \frac{1}{2}\int_{t_0}^{t_f} W_{uu} a_M(t)^2\, dt\right\}$$

which is of the form given by (14-20) with

$$\mathbf{X}_{ft} = \begin{bmatrix} 1 & 0 & 0 \\ 0 & 0 & 0 \\ 0 & 0 & 0 \end{bmatrix}, \qquad \mathbf{W}_{x_t x_t} = \begin{bmatrix} 0 & 0 & 0 \\ 0 & 0 & 0 \\ 0 & 0 & 0 \end{bmatrix}, \qquad \mathbf{W}_{x_t u} = \begin{bmatrix} 0 \\ 0 \\ 0 \end{bmatrix}, \qquad \mathbf{W}_{uu} = W_{uu}$$

This cost includes a quadratic penalty on the terminal rms miss distance plus a weighted integral of mean squared missile lateral acceleration. The constant $W_{uu}$ was selected iteratively, starting with a very small value ($10^{-9}$; *not* zero so that the optimization problem does not yield a trivial solution of applying *no* control until the *last* sample period). If at any time during the 12-sec operating period, the rms missile acceleration exceeded 400 ft/sec$^2$ (i.e., $3\sigma \geq 1200$ ft/sec$^2$), the design of the benchmark controller was deemed unacceptable; the value of the constant $W_{uu}$ was then iteratively adjusted until the peak rms acceleration equalled 400 ft/sec$^2$.

The LQG synthesis was used to generate a 3-state Kalman filter and deterministic optimal controller 1-by-3 gain matrix in cascade, which was transformed by (10-74)–(10-84) of Volume 2 into the equivalent

$$u(t_i) = [1 \quad 0 \quad 0]\mathbf{x}_c(t_i) + G_{cz}(t_i) z_i$$

$$\begin{bmatrix} x_{c1}(t_{i+1}) \\ x_{c2}(t_{i+1}) \\ x_{c3}(t_{i+1}) \end{bmatrix} = \begin{bmatrix} 0 & 1 & 0 \\ 0 & 0 & 1 \\ \phi_{c1}(t_i) & \phi_{c2}(t_i) & \phi_{c3}(t_i) \end{bmatrix} \begin{bmatrix} x_{c1}(t_i) \\ x_{c2}(t_i) \\ x_{c3}(t_i) \end{bmatrix} + \begin{bmatrix} B_{cz1}(t_i) \\ B_{cz2}(t_i) \\ B_{cz3}(t_i) \end{bmatrix} z_i$$

which is of the form given generically in (14-274). The matrix elements above are all precomputable and can be stored; they are *not* well approximated as essentially constant over the desired interval. Note the minimal delay time due to computations in this form: when $z_i$ becomes available, a single multiplication by $G_{cz}(t_i)$ and addition to $x_{c1}(t_i)$ yields the commanded acceleration for application; all other computations are done "in the background" before the next sample time.

The performance of this benchmark design was evaluated by the algorithm developed in the previous section. Figure 14.18 plots the time history of rms lateral displacement $y_{TM}$ and rms missile

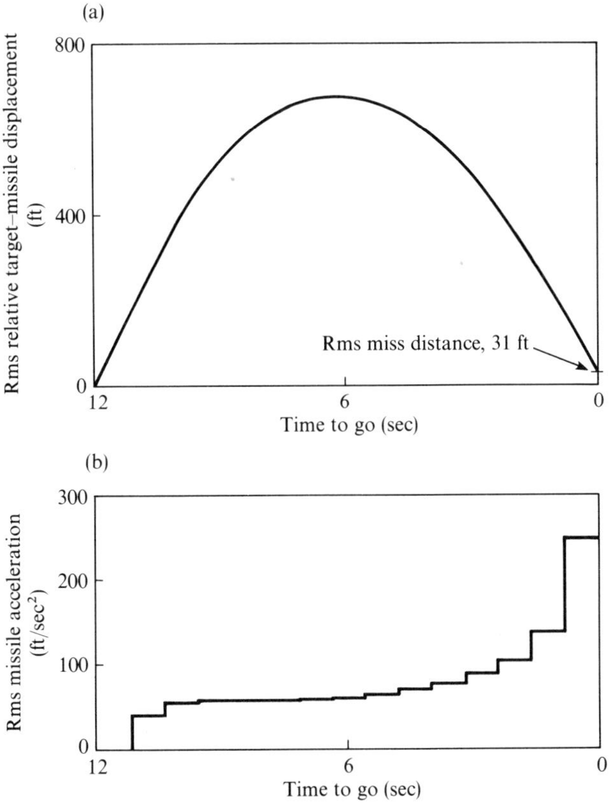


FIG. 14.18 Performance of the missile using benchmark LQG-based guidance system. (a) Time history of rms lateral displacement $y_{TM}$. (b) Rms missile acceleration $a_M$. From Widnall [189] with permission of The MIT Press.

acceleration $a_M$. The rms miss distance is 31.3 ft, and the peak rms missile lateral acceleration is 249 ft/sec$^2$—since this is below the tolerance level of 400 ft/sec$^2$, it is not necessary to repeat the design procedure with larger $W_{uu}$ value (in fact, changing $W_{uu}$ by an order of magnitude did not change the rms miss distance to three significant figures).

However, the result is not really very desirable. The missile acceleration profile increases sharply near the final time, and the physical result is a high-$g$ turn into a pursuit of the target; the target can often pull evasive high-$g$ turns to outmaneuver the missile. One would prefer a flatter rms missile acceleration history, higher in the beginning and not rising so much at the end (causing the missile to "lead" the target more). This can be accomplished by making $W_{uu}$ a *time-varying* value that *increases* in time, placing heavier penalty on rms accelerations at the end of the interval than at the beginning; LQG synthesis and performance analysis could be iterated until desired time history characteristics are achieved. ■

Since the sample rate is a major design factor [121], we seek a systematic means of choosing it [189]. Although the benchmark controller might be computationally infeasible for very short sample periods, the theoretical performance bound it represents can be useful in setting the appropriate sample rate for an eventual implementable design. The performance of the benchmark at various sampling frequencies indicates how much benefit can be fundamentally derived from decreasing the sample period. This sets the *appropriate range of sample periods*, as well as a *performance standard* to which proposed, implementable designs can be compared.

EXAMPLE 14.23 The LQG synthesis and performance evaluation of the previous example were repeated at sample periods of 0.1, 0.2, 0.4, 0.6, 0.8, and 1.0 sec. For the three largest sample periods, the negligibly small $W_{uu}$ of $10^{-9}$ provided acceptable results, but larger $W_{uu}$ values were required, and found iteratively, to yield designs whose peak rms acceleration levels were just at the limit of 400 ft/sec$^2$ for shorter sample periods.

Figure 14.19 plots the resulting rms terminal miss distance versus sample period $\Delta t$, for the missile under benchmark LQG design control. From this figure, one can see that reducing the sample period from the original 0.8 sec allows the *benchmark* to perform considerably better. A point of diminishing returns is reached at 0.2 sec in that further reduction in sample period yields little performance benefit. An appropriate choice would be between 0.2 and 0.6 sec. Assume that the benchmark cannot be implemented at sample periods less than 0.8 sec because of computational loading. Nevertheless, this theoretical bound of performance has provided insight into a *good range of sample periods* for implementable designs. *Within that range*, we can now seek controllers of reduced complexity (and thus able to be implemented in that desirable range), and strive to generate designs that *approach* the *bound* given in the figure. By having this bound at his disposal, the designer can avoid both a premature termination of design efforts at short sample periods, and a lengthy effort to achieve a substantial performance improvement by decreasing the sample period when such improvement is actually unobtainable. ■

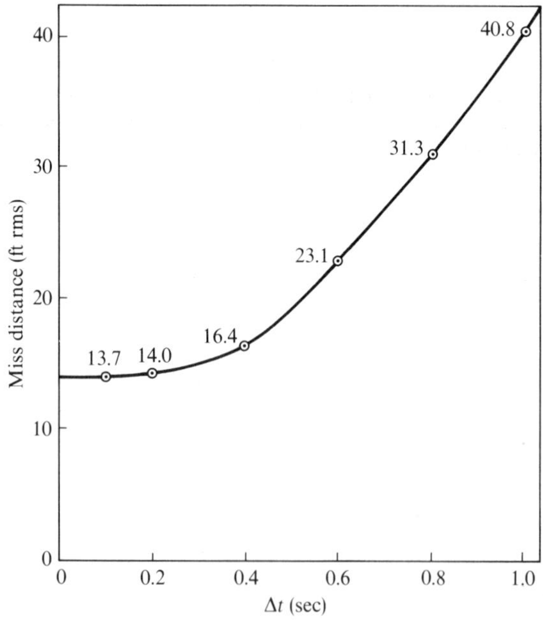


FIG. 14.19 Rms miss distance for missile under benchmark control for different sample periods. From Widnall [189] with permission of The MIT Press.

We seek a simplified controller design while still exploiting the LQG synthesis capability. Three ways to accomplish this are (1) *transformation of the controller state space description* to a phase variable or canonical form to reduce the number of arithmetic operations required, and similar efficient means of implementing a given design, (2) *generation of simpler, lower-dimensioned models* upon which to base an LQG design, and (3) *approximating the controller structure* generated by LQG methods. The first of these has already been discussed, and does not affect performance. The other two methods will now be pursued, and the adequacy of any resulting designs can be evaluated through the performance analysis algorithm of Section 14.11 and Fig. 14.16.

First consider generation of low-order, simplified models that retain enough of the essential aspects of dynamics and measuring devices to be an "adequate" model for LQG controller synthesis. *State order reduction* [52, 56, 95, 106, 107, 158, 162–165, 183] can be accomplished with good understanding of the physics of a problem: (1) high frequency modes with sufficient natural damping can be ignored as not requiring active attention from the controller; (2) higher order modes which typically have lower amplitudes or energy associated with them (such as high-order bending modes or fuel slosh modes in an aerospace vehicle) can be neglected, retaining only the dominant modes; (3) other noncritical modes might also be discarded, especially if they are only weakly observable or controllable (consider eigenvalues of the appropriate Gramians; singular value decomposition is of use here as well [80, 123]); (4) a modal cost analysis [162–165] can be performed on the benchmark controller, and the modes that contribute only insignificantly to the total cost to complete the process can be discarded; (5) complicated truth model shaping filters for generating noises of appropriate characteristics can be approximated by lower order shaping filters that capture the fundamental characteristics. Of these techniques, the fourth perhaps warrants further explanation. Once a cost $J_c$ is established of the form (14-20) such that the benchmark LQG controller achieves desired performance characteristics, the analog to the *error budget* of Section 6.8 of Volume 1 [115] can be conducted. The contributions to $J_c$ by individual system modes are separately evaluated, and modes are deleted that cause a negligible proportion of the total $J_c$.

Simplification of system model matrices for a given dimension state model is also possible. *Dominated terms in a single matrix element can be neglected*, as in replacing $\mathbf{\Phi}$ by $[\mathbf{I} + \mathbf{F}\,\Delta t]$ in a time-invariant model, ignoring higher order terms in each element. Moreover, entire *weak coupling terms can be removed*, producing matrix elements of zero and thus fewer required multiplications. Sometimes such removal allows *decoupling the controller states* to generate separate, smaller controllers. In numerous applications, the terms that can be ignored comprise the time-varying nature of the model description, or at least the most rapid variations, so that a *time-invariant model* (or at least one that admits quasi-static coefficients) can be used as the basis of a significantly simplified controller. Furthermore, a given problem can often be decomposed via singular perturbations [7, 24, 46, 81, 82, 141, 153, 157, 178, 187, 202] into a number of simpler nested problems, with "inner loops" operating at fast sample rates to control "high frequency" dynamics states (as designed assuming slower states are unspecified constants to be rejected) and "outer loops" operating with longer sample periods to control slower dynamics (designed assuming that "faster" states have reached steady state conditions).

Such simplifications typically require *tuning* of the Kalman filter in the LQG design, to account for neglected states or matrix elements, as discussed in Chapter 6 of Volume 1. Fictitious noise, or pseudonoise, is added to the reduced-order model to account for the additional uncertainty caused by the model deletions, and $\mathbf{Q}_d$ and $\mathbf{R}$ (and possibly $\mathbf{P}_0$) are established iteratively to yield good state estimation performance from the reduced-order, simplified filter; robustness recovery desires may also impact on this tuning process (recall (14-127)–(14-133)). A dual "tuning process" on the weighting matrices defining the cost for the reduced-order LQ regulator, readjusting the penalties on the remaining states to achieve the desired performance characteristics from the real (truth model) system when driven by this controller output, may be warranted as well.

EXAMPLE 14.24 Reconsider the air-to-air missile problem of the previous two examples. One could attempt to replace the time-correlated target acceleration process model with a white noise having the same low-frequency power spectral density value, thereby eliminating one state variable (reducing the acceleration shaping filter from one to zero states). Thus, the model upon which to base the controller is

$$\begin{bmatrix} \dot{\mathrm{y}}_{\mathrm{TM}}(t) \\ \dot{\mathrm{v}}_{\mathrm{TM}}(t) \end{bmatrix} = \begin{bmatrix} 0 & 1 \\ 0 & 0 \end{bmatrix} \begin{bmatrix} \mathrm{y}_{\mathrm{TM}}(t) \\ \mathrm{v}_{\mathrm{TM}}(t) \end{bmatrix} + \begin{bmatrix} 0 \\ -1 \end{bmatrix} a_{\mathrm{M}}(t) + \begin{bmatrix} 0 \\ 1 \end{bmatrix} \mathrm{a}_{\mathrm{T}}(t)$$

where $\mathrm{a}_{\mathrm{T}}(\cdot,\cdot)$ is now stationary zero-mean white Gaussian noise with $E\{\mathrm{a}_{\mathrm{T}}(t)\mathrm{a}_{\mathrm{T}}(t+\tau)\} = Q\,\delta(\tau)$ and thus a power spectral density value of $Q$ over all $\omega$. $Q$ is equated to the low frequency PSD value of the original $\mathrm{a}_{\mathrm{T}}(\cdot,\cdot)$ process, i.e.,

$$\Psi_{a_T a_T}(\omega)\Big|_{\omega=0} = \frac{2\sigma_a^{\,2}/T}{\omega^2 + (1/T)^2}\Big|_{\omega=0} = 2\sigma_a^{\,2} T$$

as shown in Fig. 14.20. The measurement model, $R(t_i)$, and $J_c$ evaluations can remain as in Example 14.22 (with matrices appropriately decreased in dimension), and $\mathbf{P}_0$ can be equated to the upper left 2-by-2 block of the original $\mathbf{P}_0$.

For the sake of this example, we shall assume that the LQG design based on this reduced-order model can be implemented with a sample period of 0.6 sec instead of the 0.8 sec as for the benchmark. There is *no* assurance that the resulting LQG controller, which is optimal for the second order simplified problem, can operate effectively in the realistic environment as depicted by the truth model. Thus, it is important to use the performance evaluation algorithm of Section 14.11 to investigate its true capabilities. When this was accomplished, the result was an rms miss distance of 24.8 ft, with a peak rms missile acceleration of 345 ft/sec$^2$ (satisfactorily below 400 ft/sec$^2$).

Compare this to the performance bound of Fig. 14.19. First of all, this second order controller operating at a period of 0.6 sec outperforms the benchmark LQG design operating at the necessarily longer 0.8 sec period (31.3 ft rms miss distance). Second, at the sample period of 0.6 sec, little can be

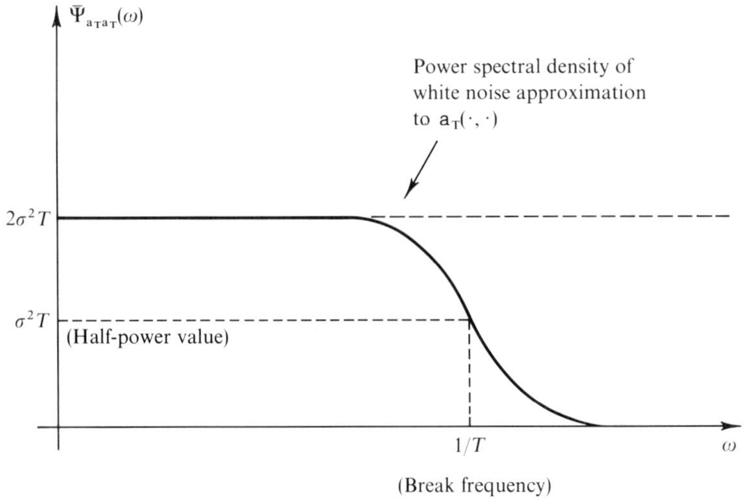


FIG. 14.20 White noise approximation to exponentially time-correlated noise. From Maybeck [115].

done to improve performance significantly, since the 24.8 ft rms miss distance this implementable controller attains is very close to the theoretical bound of 23.1 ft at that sample period. If performance improvement is still required, it is best accomplished by further decrease of sample period, as by further model simplification (though there are no guarantees of *implementable* designs at that sample period achieving performance close to that of the theoretical bound).

Note that filter tuning could be achieved in this case by setting $Q$ greater than $2\sigma_a{}^2 T$, to account for the fact that the reduced-order model is a less adequate portrayal of the real world. Such pseudonoise addition can also enhance robustness of the LQG controller. ■

To this point we have considered simplifying system models and then using LQG synthesis to generate a potential design. Once such a design is achieved, it is also useful to consider approximating the controller structure itself. The controller gains in (14-274) can be precomputed and stored rather than calculated online. Such a *precomputed gain history* can often be *approximated closely by curve-fitted simple functions*, such as piecewise constant, piecewise linear, weighted exponential, or polynomial functions. This removes the majority of the computational burden inherent in an LQG design.

If the system model is time invariant and the dynamics driving noise and measurement noise are stationary, then for computation times sufficiently after the initial time, the Kalman filter gains essentially reach their constant steady state values (independent of $\mathbf{P}_0$). In addition, if the cost weighting matrices $\mathbf{X}$, $\mathbf{U}$, and $\mathbf{S}$ are not time varying, then for sample times sufficiently before the terminal time, the optimal deterministic controller gains are constant (independent of $\mathbf{X}_f$). Thus, for an intermediate interval of time, the optimal controller of (14-274) is a constant-gain time-invariant algorithm. In certain applications in which this intermediate period is a substantial fraction of the entire interval of interest, one might generate a practical controller design by using a constant-gain steady state LQG controller over the entire operating period (once these constant values had been established by standard LQG synthesis). The performance degradation of ignoring the initial and final controller transients could readily be evaluated by the performance evaluation algorithm developed previously, and if it is acceptable, the constant-gain approximation implemented as the actual real-time controller.

In all of these methods, the LQG synthesis capability has been exploited to generate potential designs in a systematic fashion. The performance potential of each must be evaluated in a realistic environment, and a tradeoff conducted between performance capabilities and computer burden, to generate the final algorithm for implementation. Reliable, efficient, and numerically precise software has been developed [11, 43, 49, 62, 72–74, 79, 80, 83, 93, 94, 98, 99, 122, 132, 138, 142, 168, 177, 189] to aid this iterative design procedure. Considerable practical experience indicates that this is a very fruitful design methodology [3, 5, 8, 22, 25, 34, 76–78, 117, 150, 151, 156, 168–170, 198].

## 14.13 THE LQG CONTROLLER FOR CONTINUOUS-TIME MEASUREMENTS

In this chapter, attention has been concentrated on the sampled-data measurement case, as is appropriate for digital controller implementation. However, it is also appropriate to consider *continuous-time measurements*: especially for the case of *time-invariant* system models driven by *stationary* noises, with a quadratic cost defined in terms of *constant* weighting matrices, the *steady state* LQG controller is a *time-invariant, constant-gain* algorithm that can be readily implemented in analog form [1, 3–6, 42, 47, 62, 64, 71, 75, 92, 104, 118, 148, 198–200].

Consider a continuous-time system described by (14-19), from which continuous-time measurements are available in the form of

$$\mathbf{dy}(t) = \mathbf{H}(t)\mathbf{x}(t)\,dt + \mathbf{d\beta}_{\mathrm{m}}(t) \tag{14-295}$$

where $\boldsymbol{\beta}_{\mathrm{m}}(\cdot,\cdot)$ is Brownian motion typically assumed independent of $\boldsymbol{\beta}(\cdot,\cdot)$ in (14-19a) and of diffusion $\mathbf{R}_{\mathrm{c}}(t)$ for all $t \in [t_0, t_{\mathrm{f}}]$:

$$E\{\boldsymbol{\beta}_{\mathrm{m}}(t)\} = \mathbf{0} \tag{14-296a}$$

$$E\{\mathbf{d\beta}_{\mathrm{m}}(t)\,\mathbf{d\beta}_{\mathrm{m}}^{\mathrm{T}}(t)\} = \mathbf{R}_{\mathrm{c}}(t)\,dt \tag{14-296b}$$

$$E\{\mathbf{d\beta}_{\mathrm{m}}(t)\,\mathbf{d\beta}^{\mathrm{T}}(t)\} = \mathbf{0} \tag{14-296c}$$

Heuristically, (14-295) can be divided through by $dt$ to yield

$$\mathbf{z}(t) = \mathbf{H}(t)\mathbf{x}(t) + \mathbf{v}_{\mathrm{c}}(t) \tag{14-297}$$

where $\mathbf{v}_c(\cdot,\cdot)$ is the white Gaussian noise formed as the hypothetical derivative of $\boldsymbol{\beta}_m(\cdot,\cdot)$, with

$$E\{\mathbf{v}_c(t)\} = \mathbf{0} \tag{14-298a}$$

$$E\{\mathbf{v}_c(t)\mathbf{v}_c^T(t+\tau)\} = \mathbf{R}_c(t)\,\delta(t) \tag{14-298b}$$

$$E\{\mathbf{v}_c(t)\mathbf{w}^T(t+\tau)\} = \mathbf{0} \tag{14-298c}$$

The subscript c on $\mathbf{R}_c(t)$ and $\mathbf{v}_c(\cdot,\cdot)$ denotes continuous-time, to distinguish these terms from $\mathbf{R}(t_i)$ and $\mathbf{v}(\cdot,\cdot)$ associated with the discrete-time measurement case. Now it is desired to generate the optimal stochastic controller, optimal in the sense that it minimizes the quadratic cost given by (14-20). As seen in Chapter 5 of Volume 1 [115], one means of deriving this result is to consider the sampled-data algorithm based on the discrete-time measurements

$$\mathbf{z}(t_i) = \mathbf{H}(t_i)\mathbf{x}(t_i) + \mathbf{v}(t_i) \tag{14-299}$$

where $\mathbf{v}(\cdot,\cdot)$ is a discrete-time, zero-mean, white Gaussian noise with

$$E\{\mathbf{v}(t_i)\mathbf{v}^T(t_i)\} = \mathbf{R}(t_i) = \mathbf{R}_c(t_i)/\Delta t \tag{14-300}$$

in the limit as the sample period $\Delta t \to 0$.

When this is accomplished, the resulting *continuous-measurement LQG optimal stochastic controller* has the same certainty equivalence property as seen in the discrete-measurement case [3, 4, 12, 197]. That is to say, the controller algorithm is the *cascade of the appropriate Kalman filter and the associated deterministic LQ optimal regulator gains* (under the continuous-time analogy of sufficient conditions discussed in Section 14.4).

The Kalman filter for this problem was developed in Section 5.11 of Volume 1. It produces the conditional mean estimate of $\mathbf{x}(t)$,

$$\hat{\mathbf{x}}(t) = E\{\mathbf{x}(t) \,|\, \mathbf{z}(\tau) = \mathbf{z}(\tau);\ t_0 \le \tau \le t\} \tag{14-301}$$

according to

$$\dot{\hat{\mathbf{x}}}(t) = \mathbf{F}(t)\hat{\mathbf{x}}(t) + \mathbf{B}(t)\mathbf{u}(t) + \mathbf{K}(t)[\mathbf{z}(t) - \mathbf{H}(t)\hat{\mathbf{x}}(t)] \tag{14-302a}$$

$$\hat{\mathbf{x}}(t_0) = \hat{\mathbf{x}}_0 \tag{14-302b}$$

where the Kalman filter gain $\mathbf{K}(t)$ is precomputable as

$$\mathbf{K}(t) = \mathbf{P}(t)\mathbf{H}^T(t)\mathbf{R}_c^{-1}(t) \tag{14-303}$$

and $\mathbf{P}(t)$ is the associated (conditional and unconditional) error covariance found as the solution to the forward Riccati equation

$$\begin{aligned}\dot{\mathbf{P}}(t) = {} & \mathbf{F}(t)\mathbf{P}(t) + \mathbf{P}(t)\mathbf{F}^T(t) + \mathbf{G}(t)\mathbf{Q}(t)\mathbf{G}^T(t) \\ & - \mathbf{P}(t)\mathbf{H}^T(t)\mathbf{R}_c^{-1}(t)\mathbf{H}(t)\mathbf{P}(t)\end{aligned} \tag{14-304a}$$

$$\mathbf{P}(t_0) = \mathbf{P}_0 \tag{14-304b}$$

As in the discrete-time Kalman filter, (14-302a) has the structure of an internal dynamics model with an additive correction term of a gain times the measurement residual, the difference between the observed values $\mathbf{z}(t)$ and the best estimate of what they would be based on previous information and an internal model of measuring devices, i.e., $\mathbf{H}(t)\hat{\mathbf{x}}(t)$. In (14-304a), the first two terms indicate the homogeneous system effects (often stabilizing), the third term is the covariance increasing effect due to continual input of dynamics driving noise $\mathbf{w}(\cdot,\cdot)$, and the fourth term is the error covariance decreasing effect of incorporating the continuous-measurement information. Sometimes (14-304b) is written with the fourth term expressed equivalently as $[\mathbf{K}(t)\mathbf{H}(t)\mathbf{P}(t)]$ or $[\mathbf{K}(t)\mathbf{R}_c(t)\mathbf{K}^T(t)]$; it was shown in Volume 1 that when the last term is written as $[\mathbf{K}(t)\mathbf{R}_c(t)\mathbf{K}^T(t)]$, (14-304) generalizes to the evaluation of estimation error covariance associated with a state estimator (of unaltered state dimension) that generates its gain $\mathbf{K}(t)$ in some manner other than by (14-303). The filter can be appropriately "tuned" by choice of $\mathbf{Q}(\cdot)$ and $\mathbf{R}_c(\cdot)$ time histories, and initial condition $\mathbf{P}_0$, in direct analogy to the sampled-data measurement case. Offline precomputation of $\mathbf{P}(t)$ and $\mathbf{K}(t)$ is especially motivated here due to the severe numerical difficulties often encountered in integrating Riccati differential equations [57, 72–74, 94, 98, 99, 139, 195]. Note that, unlike the discrete-time case, it is *essential* that $\mathbf{R}_c(t)$ be positive definite for all $t$, since its inverse is required directly in the algorithm [57, 139, 195].

Applying the same limiting procedure to the deterministic LQ optimal regulator (14-6) and (14-14)–(14-17) with $\mathbf{W}_{xu}(t) \equiv \mathbf{0}$ for all $t$ in (14-20) yields the result

$$\mathbf{u}^*(t) = -\mathbf{G}_c^*(t)\mathbf{x}(t) \tag{14-305}$$

with the optimal controller gain $\mathbf{G}_c^*(t)$ given by

$$\mathbf{G}_c^*(t) = \mathbf{W}_{uu}^{-1}(t)\mathbf{B}^T(t)\mathbf{K}_c(t) \tag{14-306}$$

obtained by solving the backward Riccati differential equation for

$$d\mathbf{K}_c(t)/d(-t) = -\dot{\mathbf{K}}_c(t),$$

$$-\dot{\mathbf{K}}_c(t) = \mathbf{F}^T(t)\mathbf{K}_c(t) + \mathbf{K}_c(t)\mathbf{F}(t) + \mathbf{W}_{xx}(t) - \mathbf{K}_c(t)\mathbf{B}(t)\mathbf{W}_{uu}^{-1}(t)\mathbf{B}^T(t)\mathbf{K}_c(t) \tag{14-307a}$$

$$\mathbf{K}_c(t_f) = \mathbf{X}_f \tag{14-307b}$$

offline before real-time implementation. The minimum cost to go from initial time $t_0$ to final time $t_f$ using this control can be evaluated as $[\frac{1}{2}\mathbf{x}^T(t_0)\mathbf{K}_c(t_0)\mathbf{x}(t_0)]$. It can readily be seen that, again there is a *duality* relationship between this optimal regulator and the previous Kalman filter, and thus (1) the Riccati differential equation (14-307) has similar inherent numerical problems as in (14-304), (2) the software developed for filter design can also be used for effective regulator design, (3) "tuning" of the controller can be accomplished by adjusting the $\mathbf{W}_{xx}(\cdot)$ and $\mathbf{W}_{uu}(\cdot)$ time histories and terminal condition $\mathbf{X}_f$, and (4) it is essential for $\mathbf{W}_{uu}(t)$ to be positive definite for all $t$ since its inverse is used explicitly. Moreover, an expression equivalent to (14-307a),

$$-\dot{\mathbf{K}}_c(t) = [\mathbf{F}(t) - \mathbf{B}(t)\mathbf{G}_c(t)]^T\mathbf{K}_c(t) + \mathbf{K}_c(t)[\mathbf{F}(t) - \mathbf{B}(t)\mathbf{G}_c(t)] + \mathbf{W}_{xx}(t) + \mathbf{G}_c^T(t)\mathbf{W}_{uu}(t)\mathbf{G}_c(t) \tag{14-308}$$

also generalizes to yield proper cost evaluations as $[\frac{1}{2}\mathbf{x}^T(t_0)\mathbf{K}_c(t_0)\mathbf{x}(t_0)]$ for controllers with gains chosen by means other than the optimal evaluation (14-306).

The Kalman filter can be generalized to the case of *correlated* measurement and dynamics driving noises, and analogously the regulator can be generalized to the case of $\mathbf{W}_{xu}(t) \neq \mathbf{0}$ in (14-20). However, (14-302)–(14-308) are the most common forms used in actual practice.

The structure of the continuous-measurement LQG optimal stochastic controller is portrayed in Fig. 14.21. The Kalman filter accepts the incomplete, noise-corrupted, real-time measurements, and uses its internal dynamics model and measurement model, as well as gain evaluated on the basis of these models and statistical descriptions of uncertainties, to create the optimal state estimate $\hat{\mathbf{x}}(t)$. This is premultiplied by the optimal gains for the deterministic LQ regulator, to produce the optimal control $\mathbf{u}^*(t)$ to feed back into the system under control.

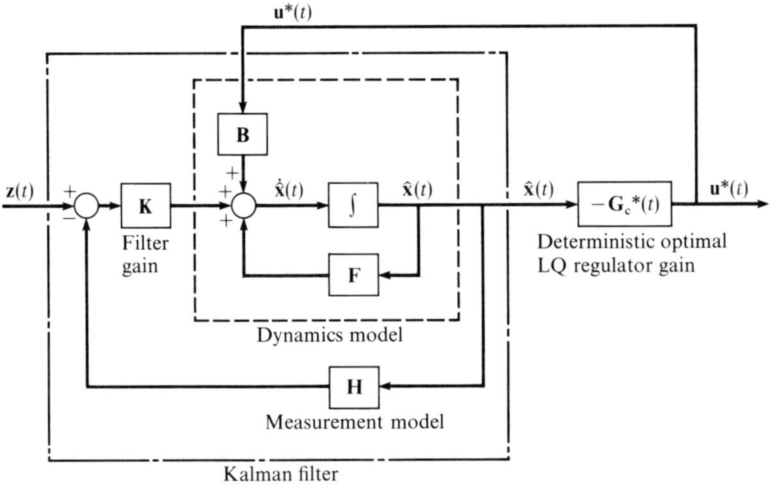


FIG. 14.21 Continuous-measurement LQG optimal stochastic controller.

Under the sufficient conditions as discussed in Section 14.4, this is a stabilizing controller, though one with robustness properties impaired due to the appearance of the Kalman filter when perfect state knowledge is not available (see Section 14.5). The optimal cost to complete the process in the stochastic case with noise-corrupted measurements is expressible as

$$\mathscr{C}^*[\hat{\mathbf{x}}_0, t_0] = \tfrac{1}{2}[\hat{\mathbf{x}}_0{}^{\mathrm{T}}\mathbf{K}_{\mathrm{c}}(t_0)\hat{\mathbf{x}}_0 + g(t_0)] \tag{14-309}$$

by exploiting sufficient statistics as in the discrete-time measurement case, where $\hat{\mathbf{x}}_0$ is the mean of the a priori state density, $\mathbf{K}_{\mathrm{c}}(t_0)$ is found by solving (14-307) backward in time, and the scalar $g(t_0)$ is found as the solution to the backward differential equation

$$dg(t)/d(-t) = \mathrm{tr}\{\mathbf{W}_{xx}(t)\mathbf{P}(t) + [\mathbf{K}(t)\mathbf{H}(t)\mathbf{P}(t)]\mathbf{K}_{\mathrm{c}}(t)\} \tag{14-310a}$$

$$g(t_{\mathrm{f}}) = \mathrm{tr}\{\mathbf{X}_{\mathrm{f}}\mathbf{P}(t_{\mathrm{f}})\} \tag{14-310b}$$

which is the direct analog of (13-20) and (13-21).

In actual applications, the *constant-gain time-invariant* approximation to the LQG stochastic regulator is often sought because of its ease of implementation. If the system model is time invariant (constant **F**, **B**, **G**, and **H**) and noise inputs are stationary (constant **Q** and $\mathbf{R}_{\mathrm{c}}$), then one can ignore the initial filter transient and implement the steady state filter

$$\dot{\hat{\mathbf{x}}}(t) = \mathbf{F}\hat{\mathbf{x}}(t) + \mathbf{B}\mathbf{u}(t) + \bar{\mathbf{K}}[\mathbf{z}(t) - \mathbf{H}\hat{\mathbf{x}}(t)] \tag{14-311a}$$

$$= [\mathbf{F} - \bar{\mathbf{K}}\mathbf{H}]\hat{\mathbf{x}}(t) + \mathbf{B}\mathbf{u}(t) + \bar{\mathbf{K}}\mathbf{z}(t) \tag{14-311b}$$

for all time, where $\bar{\mathbf{K}} = \bar{\mathbf{P}}\mathbf{H}^{\mathrm{T}}\mathbf{R}_{\mathrm{c}}^{-1}$ and $\bar{\mathbf{P}}$ is the solution to the steady state (i.e., algebraic) Riccati equation

$$\dot{\bar{\mathbf{P}}} = \mathbf{0} = \mathbf{F}\bar{\mathbf{P}} + \bar{\mathbf{P}}\mathbf{F}^{\mathrm{T}} + \mathbf{G}\mathbf{Q}\mathbf{G}^{\mathrm{T}} - \bar{\mathbf{P}}\mathbf{H}^{\mathrm{T}}\mathbf{R}_{\mathrm{c}}^{-1}\mathbf{H}\bar{\mathbf{P}} \tag{14-312}$$

Similarly, if the system model is time invariant and the cost weighting matrices $\mathbf{W}_{xx}$ and $\mathbf{W}_{uu}$ are constant, the steady state constant-gain controller corresponding to the infinite horizon problem ($[t_{\mathrm{f}} - t_0] \to \infty$) can be used for all time, ignoring the terminal transient:

$$\mathbf{u}(t) = -\bar{\mathbf{G}}_{\mathrm{c}}{}^*\hat{\mathbf{x}}(t) = -[\mathbf{W}_{uu}^{-1}\mathbf{B}^{\mathrm{T}}\bar{\mathbf{K}}_{\mathrm{c}}]\hat{\mathbf{x}}(t) \tag{14-313}$$

where $\bar{\mathbf{K}}_{\mathrm{c}}$ is the solution to the steady state (algebraic) Riccati equation

$$-\dot{\bar{\mathbf{K}}}_{\mathrm{c}} = \mathbf{0} = \mathbf{F}^{\mathrm{T}}\bar{\mathbf{K}}_{\mathrm{c}} + \bar{\mathbf{K}}_{\mathrm{c}}\mathbf{F} + \mathbf{W}_{xx} - \bar{\mathbf{K}}_{\mathrm{c}}\mathbf{B}\mathbf{W}_{uu}^{-1}\mathbf{B}^{\mathrm{T}}\bar{\mathbf{K}}_{\mathrm{c}} \tag{14-314}$$

EXAMPLE 14.25 Again return to the gimbaled platform problem introduced in Example 14.2. Letting $\mathrm{x}(t)$ be the gimbal angle of the platform, we have

$$\dot{\mathrm{x}}(t) = [-1/T]\mathrm{x}(t) + [1/T]u(t) + \mathrm{w}(t)$$

where $\mathrm{w}(\cdot,\cdot)$ is zero-mean white Gaussian noise of constant strength $Q$. Now assume that *continuous-time* measurements are available in the form

$$\mathrm{z}(t) = \mathrm{x}(t) + \mathrm{v}_{\mathrm{c}}(t)$$

with $\mathrm{v}_{\mathrm{c}}(\cdot,\cdot)$ being zero-mean white Gaussian noise of constant strength $R_{\mathrm{c}}$. Assume the a priori state information is in the form of a Gaussian density for $\mathrm{x}(t_0)$, with mean $\hat{x}_0$ and variance $P_0$.

Further assume that we want to generate the control law that minimizes the cost

$$J_c = E\left\{\frac{1}{2} X_f x^2(t_f) + \frac{1}{2}\int_{t_f}^{t} [W_{xx}x^2(t) + W_{uu}u^2(t)]\, dt\right\}$$

The deterministic optimal LQ regulator is found through (14-305)–(14-307) to be

$$u^*(t) = -G_c^*(t)x(t)$$

where $G_c^*(t) = K_c(t)/[W_{uu}T]$, and $K_c(t)$ is the precomputable solution to the backward Riccati equation

$$-\dot{K}_c(t) = -2K_c(t)/T + W_{xx} - K_c^2(t)/[W_{uu}T^2]$$
$$K_c(t_f) = X_f$$

Since $x(t)$ is not known perfectly, it is replaced (via certainty equivalence) by the $\hat{x}(t)$ produced by the Kalman filter given by (14-302)–(14-304):

$$\dot{\hat{x}}(t) = [-1/T]\hat{x}(t) + [1/T]u(t) + K(t)[z(t) - \hat{x}(t)]$$

where the Kalman filter gain is $K(t) = P(t)/R_c$, where $P(t)$ is the precomputable solution to the forward Riccati equation

$$\dot{P}(t) = -2P(t)/T + Q - P^2(t)/R_c, \qquad P(t_0) = P_0$$

For implementation, one would consider the steady state constant-gain LQG controller,

$$u^*(t) = -\bar{G}_c^*\hat{x}(t)$$

where $\bar{G}_c^*$ is given as $\bar{K}_c/[W_{uu}T]$ in terms of the positive solution to $-\dot{\bar{K}}_c(t) = 0$, or

$$\bar{G}_c^* = \sqrt{1 + (W_{xx}/W_{uu})} - 1$$

and $\hat{x}(t)$ is generated by the steady state filter

$$\dot{\hat{x}}(t) = [-1/T]\hat{x}(t) + [1/T]u(t) + \bar{K}[z(t) - \hat{x}(t)]$$

where $\bar{K}$ is given as $\bar{P}/R_c$ in terms of the positive solution to $\dot{\bar{P}}(t) = 0$, or

$$\bar{K} = \sqrt{(1/T^2) + (Q/R_c)} - (1/T)$$

Note that these steady state gains are dependent upon the "tuning" ratios $(W_{xx}/W_{uu})$ and $(Q/R_c)$ for the regulator and filter designs, respectively, and are independent of $X_f$ and $P_0$. ■

As in the discrete-time measurement case, the importance of this LQG controller is the *synthesis capability* associated with it. In direct analogy to Sections 14.6–14.10, it can be used to formulate more than just simple stochastic regulators: trackers [92], nonzero setpoint controllers [92], controllers with specific disturbance rejection properties [58–61], PI controllers [1], and command generator trackers [17, 19, 180] can all be readily synthesized. A systematic design approach would then exploit this synthesis capability to propose simplified, implementable controllers, and evaluate the performance capabilities of each tuned algorithm in a realistic environment, as already accomplished for sampled-data controllers in Sections 14.11 and 14.12.

## 14.14 SUMMARY

This chapter developed the *design synthesis* and *performance analysis* capabilities for linear sampled-data (and also continuous-measurement) controllers associated with the LQG stochastic optimal control problem. Under the assumptions of *linear* systems driven by white *Gaussian* noises being adequate, the controller to minimize a *quadratic* cost function is the cascade of the Kalman filter for the given problem with the gain matrix of the associated deterministic optimal control problem.

First, Section 14.2 extended the simple LQG problem introduced in Chapter 13, to admit a physically motivated, generalized quadratic cost for sampled-data control applications. For a system described by state dynamics (14-1) (and controlled variables (14-2) or (14-29)), the full-state feedback regulator that minimizes the generalized quadratic cost (14-13) is given by

$$\mathbf{u}^*(t_i) = -\mathbf{G}_c^*(t_i)\mathbf{x}(t_i) \tag{14-315}$$

with $\mathbf{G}_c^*(t_i)$ evaluated by (14-14)–(14-17), a precomputed backward Riccati difference equation solution. This is the appropriate context for sampled-data controller design, starting from a continuous-time specification of dynamics and cost as in (14-18)–(14-20), and producing an equivalent discrete-time formulation via (14-28); see also (14-30). The LQG stochastic optimal controller for the case of only incomplete, noise-corrupted measurements being available, uses the gain matrix so generated to produce the optimal control

$$\mathbf{u}^*(t_i) = -\mathbf{G}_c^*(t_i)\hat{\mathbf{x}}(t_i^+) \tag{14-316}$$

where $\hat{\mathbf{x}}(t_i^+)$ is produced by the Kalman filter designed initially without concern for controller characteristics (i.e., without regard for the control cost function to be minimized). To minimize the computational delay time between sampling the measuring devices and producing the control to feed back, one might also consider the suboptimal law

$$\mathbf{u}(t_i) = -\mathbf{G}_c^*(t_i)\hat{\mathbf{x}}(t_i^-) \tag{14-317}$$

where the delay reduction is gained at the expense of using a less precise estimate of the current state $\mathbf{x}(t_i)$.

Since closed-loop system stability is a primary concern, Section 14.3 established appropriate stability concepts, both zero-input and bounded input/bounded output forms applicable to systems represented by (1) stochastic and deterministic, (2) nonlinear and linear, and (3) time-varying and time-invariant models, along with some sufficient conditions for such stability types. Section 14.4 then developed the stability characteristics of LQG controllers and the Riccati equation solutions associated with these controllers. If $\mathbf{X}_f \geq \mathbf{0}$, $\mathbf{X}(t_i) \geq \mathbf{0}$, $\mathbf{U}(t_i) > \mathbf{0}$, and $\mathbf{S}(t_i)$ is chosen so that the composite matrix in the summation of (14-13b) is positive semidefinite for all $t_i$; if all defining matrices are bounded for all $t_i$; if the system model composed of (14-70) and (14-71) is uniformly completely controllable and reconstructible; then (1) a bounded positive definite *steady state* $\bar{\mathbf{K}}_c(t_i)$ exists that is approached by *all* backward Riccati difference equation solutions, (2) the corresponding steady state deterministic optimal LQ regulator (14-73) *minimizes* the cost to complete the process as $[\frac{1}{2}\mathbf{x}_0{}^T\bar{\mathbf{K}}_c(t_0)\mathbf{x}_0]$, and (3) the closed-loop system representation (14-75) is *exponentially stable* (asymptotic stability in the time-invariant system model case is assured by that model being stabilizable and detectable). A *dual* relationship yielded the stability results for the Kalman filter, and the intersection of these sufficient conditions produced the sufficient conditions for a stabilizing LQG controller. These results were also discussed in the context of feedback controllers with regulator and observer gains set through pole placement methods. In Section 14.5, these results were shown to be very *robust* to variations of the true system from assumed design conditions *if full-state feedback is available*, but with (potentially greatly) reduced robustness caused by the introduction of the Kalman filter to generate estimates of the state if the state itself is not perfectly and totally accessible. Robustness can be evaluated in terms of properties of the loop gain and return difference transformation of (14-98) and (14-99). Preservation of stability in the face of additive or multiplicative alterations in the original system is guaranteed by (14-104) or its sufficient condition (14-106) in the general case, by (14-113) in terms of a singular value decomposition for the more restrictive time-invariant discrete-time case in which both the original system and alteration are representable by $z$-transform operations, and by (14-123) for the corresponding continuous-time case in which original system and alterations are expressible as Laplace transform operations. *Robustness guarantees* for the full-state feedback controller were given in the discussion surrounding (14-119) and (14-120), and these minimal guaranteed properties may be substantially exceeded by specific individual designs. *No* such minimum guarantees are available with a Kalman filter or observer in the loop, and the robustness of any such design must be evaluated. Robustness recovery procedures were also discussed as a means of enhancing LQG stochastic controller robustness: specific forms of retuning the filter or observer were seen to provide such enhancement (thus filter design is *not* totally independent of controller characteristics, when robustness is considered).

Sections 14.6–14.10 extended LQG synthesis beyond simple regulator design. Target *trackers* described by (14-144), (14-146) and (14-147), and *controllers with specific disturbance rejection properties*, given by (14-175)–(14-177), were generated by solving an LQ regulator problem for a system description augmented with states describing variables to be tracked or rejected. One major component of both designs was a *system state feedback* through the gains of the regulator that *ignores* these additional variables. For the LQG controller in the case of only noise-corrupted measurements being available, the Kalman filter often decomposes in a similar manner for the tracker, but not for the disturbance rejection controller.

Nonzero setpoint controllers and constant disturbance rejection were seen to require consideration as special cases, and Section 14.7 initiated development of *perturbation variables* (14-161) appropriately defined about some nonzero desired nominal state, control, and output values, (14-156). Here $\mathbf{\Pi}$ is defined by (14-155), (14-158), or (14-160) depending on the dimensionality of the controls $\mathbf{u}(t_i)$ versus that of the controlled variables $\mathbf{y}_c(t_i)$. Although (14-166) represents a potential nonzero setpoint design, it is generally better to achieve the *type-1 property*, i.e., the ability to reach a desired setpoint with zero steady state error (mean error in the stochastic case) despite *unmodeled* constant disturbances.

This objective is accomplished by the PI controllers of Section 14.9. They can be designed via LQG synthesis by augmenting the original system states with the pseudointegral of the regulation error (14-190) and generating the augmented perturbation regulator, to yield the controller given in position form (14-208) or the preferable incremental form (14-209). Alternately, the controls can be written in terms of control differences or pseudorates as in (14-191), augmenting these relations to the original system states and using the augmented perturbation LQG regulator to define the final PI design, (14-230) with the aid of (14-227), or, in incremental form, (14-228). These can be interrelated by the generic form of PI controller, either (14-232) or (14-234)–(14-236) in position form for design insights: having the structure of a full-state feedback around the system, to which is added a forward path of a proportional plus (pseudo-) integral compensator acting on the regulation error, with gains dictated by the LQG synthesis. Proper initialization of the augmented state variables was shown to be tantamount to designing for desirable transient characteristics of the closed-loop system. For actual implementation, the incremental form is given by (14-237).

The command generator trackers of Section 14.10 are a culmination of these controller designs, causing a system modeled by (14-238) to track the output of a command generator model (14-240) while rejecting specific disturbances modeled by a disturbance shaping filter model (14-239), assuming the command generator input to be essentially constant and the disturbance model to have no nonzero-mean inputs. The open-loop CGT generates the ideal state and control trajectories via (14-245), (14-249), and (14-250), using $\mathbf{\Pi}$ as defined by (14-155), (14-158), or (14-160). A CGT/perturbation regulator can be produced as in (14-256), but a CGT/PI law with type-1 tracking capability is superior and is given by (14-267) in implementable incremental form, or by (14-270) in position form for design insights. For actual applications in which the command generator input can undergo infrequent changes, the restart modification of (14-262) enhances the transient characteristics of the command generator model and thus of the overall closed-loop system.

Once LQG synthesis and possibly other techniques have been used to generate a number of prospective designs of general form (14-274), their performance must be evaluated in a realistic environment, as described by (14-272), (14-273), and Fig. 14.16. The desired performance measure, i.e., the statistical characteristics of the truth model states and controls, can be evaluated by a Monte Carlo analysis of (14-277); if the truth model is adequately represented as linear, this can be replaced by direct computation of statistics as in (14-281)–(14-294). The "benchmark" LQG design based on the truth model can be used as a theoretical performance bound and as an aid in selecting an appropriate controller sample period. Prospective implementable designs can be derived from (1) LQG synthesis performed on reduced-order and simplified models, (2) approximation of the LQG design itself (as using a steady state constant-gain controller for all time), and (3) transformation of the controller state space description to an efficient form and specific ordering of the algorithm to minimize both the delay between measurement input and control generation, and the total computational burden per iteration period.

Finally, Section 14.13 presented the corresponding results for the case of continuous-time measurements, (14-295)–(14-298). Again, the optimal LQG stochastic controller is the cascade of a Kalman filter (14-302)–(14-304), with the gains of the corresponding deterministic optimal controller (14-305)–(14-308). In practice, this is usually constrained to an implementation of the steady state constant-gain controller (14-311)–(14-314) for all time, in applications described by time-invariant system models, constant cost weighting matrices, and stationary noises.

The next chapter extends stochastic controller design to applications in which the LQG assumptions are not a totally adequate portrayal of the real-world problem.

## REFERENCES

1. Anderson, B. D. O., and Moore, J. B., "Linear Optimal Control." Prentice Hall, Englewood Cliffs, New Jersey, 1971.
2. Athans, M., On the design of P-I-D controllers using optimal linear regulator theory, *Automatica* **7**, 643–647 (1971).
3. Athans, M. (ed.), Special issue on linear-quadratic-Gaussian problem, *IEEE Trans. Automatic Control* **AC-16** (6) (1971).
4. Athans, M., The role and use of the Linear-Quadratic-Gaussian problem in control system design, *IEEE Trans. Automat. Control* **AC-16** (6), 529–552 (1971).
5. Athans, M., The discrete time Linear-Quadratic-Gaussian stochastic control problem, *Ann. Econ. Social Measurement* **1** (4), 449–491 (1972).
6. Athans, M., and Falb, P. L., "Optimal Control: An Introduction to the Theory and its Applications." McGraw-Hill, New York, 1966.
7. Balas, M. J., Observer stabilization of singularly perturbed systems, *AIAA J. Guidance and Control* **1** (1), 93–95 (1978).

8. Balas, M. J., Feedback control of flexible systems, *IEEE Trans. Automat. Control* **AC-23** (4), 673–679 (1978).
9. Balchen, J. G., Endresen, T., Fjeld, M., and Olsen, T. O., Multivariable PID estimation and control in systems with biased disturbances, *Automatica* **9** 295–307 (1973).
10. Barnett, S., Insensitivity of optimal linear discrete-time regulators, *Internat. J. Control* **21** (5), 843–848 (1975).
11. Barraud, A. Y., A numerical algorithm to solve $A^TXA\text{-}X = Q$, *IEEE Trans. Automat. Control* **AC-22** (5), 883–884 (1977).
12. Bar-Shalom, Y., and Tse, E., Dual effect, certainty equivalence, and separation in stochastic control, *IEEE Trans. Automat. Control* **AC-19** (5), 494–500 (1974).
13. Belletrutti, J. J., and MacFarlane, A. G. J., Characteristic loci techniques in multivariable-control-system design *Proc. IEE* **118**, 1291–1296 (1971).
14. Bertsekas, D., and Shreve, S., "Stochastic Optimal Control: The Discrete Time Case." Academic Press, New York, 1979.
15. Brasch, F. M., Jr., and Pearson, J. B., Pole placement using dynamic compensators, *IEEE Trans. Automat. Control* **AC-15** (1), 34–43 (1970).
16. Brockett, R. W., "Finite Dimensional Dynamical Systems." Wiley, New York, 1970.
17. Broussard, J. R., Command Generator Tracking, Rep. #TIM-612-3. The Analytic Sciences Corp., Reading, Massachusetts (March 1978).
18. Broussard, J. R., Command Generator Tracking—The Discrete Time Case, Rep. #TIM-612-2. The Analytic Sciences Corp., Reading, Massachusetts (March 1978).
19. Broussard, J. R., and Berry, P. W., Command Generator Tracking—The Continuous Time Case, Rep. #TIM-612-1. The Analytic Sciences Corp., Reading, Massachusetts (February 1978).
20. Broussard, J. R., Berry, P. W., and Stengel, R. F., Modern Digital Flight Control System Design for VTOL Aircraft, NASA Contractor Rep. 159019. The Analytic Sciences Corp., Reading, Massachusetts (March 1979).
21. Broussard, J. R., and Safonov, M. G., Design of Generalized Discrete Proportional-Integral Controllers by Linear-Optimal Control Theory, Rep. #TIM-804-1. The Analytic Sciences Corp., Reading, Massachusetts (October 1976).
22. Bryson, A. E., Jr., and Ho, Y., "Applied Optimal Control." Ginn (Blaisdell), Waltham, Massachusetts, 1969.
23. Cadzow, J. A., and Martens, H. R., "Discrete-Time and Computer Control Systems." Prentice-Hall, Englewood Cliffs, New Jersey, 1970.
24. Calise, A. J., A new boundary layer matching procedure for singly perturbed systems, *IEEE Trans. Automat. Control* **AC-23** (3), 434–438 (1978).
25. Cassidy, J. F., Jr., Athans, M., and Lee, W.-H., On the design of electronic automative engine controls using linear quadratic control theory, *IEEE Trans. Automat. Control* **AC-25** 901–912 (1980).
26. Chalk, C. R. *et al.*, Background Information and Users Guide for MIL-F-8785B(ASG), "Military Specification—Flying Qualities of Piloted Airplanes, Air Force Flight Dynamics Lab., AFFDL-TR-69-72. Wright-Patterson AFB, Ohio (August 1969).
27. Churchill, R. V., "Introduction to Complex Variables and Applications." McGraw-Hill, New York, 1948.
28. Cook, P. A., Circle criteria for stability in Hilbert space, *SIAM J. Control* **13** (3), 593–610 (1975).
29. Corsetti, C. D., Analysis and Synthesis of Regulators and Observers for a Class of Linear, Discrete-Time Stochastic Systems Via Eigenvalue and Generalized Eigenvector Assignment Techniques. Ph.D. dissertation, Air Force Institute of Technology, Wright-Patterson AFB, Ohio (1982).
30. Cruz, J. B., and Perkins, W. R., A new approach to the sensitivity problem in multivariable feedback system design, *IEEE Trans. Automat. Control* **AC-9** (3), 216–223 (1964).

31. D'Appolito, J. A., and Hutchinson, C. E., Low sensitivity filters for state estimation in the presence of large parameter uncertainties, *IEEE Trans. Automat. Control* **AC-14** (3), 310–312 (1969).
32. Davison, E. J., Comments on optimal control of the linear regulator with constant disturbances, *IEEE Trans. Automat. Control* **AC-15** (2), 264 (1970).
33. Davison, E. J., and Wang, S. H., Properties and calculation of transmission zeroes of linear multivariable systems, *Automatica* **10**, 643–658 (1974).
34. DeHoff, R. L., and Hall, W. E., Jr., Multivariable quadratic synthesis of an advanced turbofan engine controller, *AIAA J. Guidance and Control* **1** (2), 136–142 (1978).
35. Desoer, C. A., "Notes for a Second Course on Linear Systems." Van Nostrand-Reinhold, Princeton, New Jersey, 1970.
36. Desoer, C. A., and Vidyasagar, M., "Feedback Systems: Input—Output Properties." Academic Press, New York, 1975.
37. Deyst, J. J., Jr., Optimal Control in the Presence of Measurement Uncertainties. MIT Exp. Astro. Lab. Rep. TE-17, Cambridge, Massachusetts (January 1967).
38. Dorato, P., and Levis, A. H., Optimal linear regulators: The discrete-time case, *IEEE Trans. on Automat. Control* **AC-16** (6), 613–620 (1971).
39. Doyle, J. C., Guaranteed margins for LQG regulators, *IEEE Trans. on Automat. Control* **AC-23** (4), 756–757 (1978).
40. Doyle, J. C., Robustness of multiloop linear feedback systems, *Proc. 1978 IEEE Conf. Decision and Control, San Diego, California* pp. 12–18 (January 1979).
41. Doyle, J. C., and Stein, G., Robustness with observers, *IEEE Trans. on Automat. Control* **AC-24** (4), 607–611 (1979).
42. Florentin, J. J., Optimal control of continuous-time, Markov, stochastic systems, *J. Electron. Control* **10** (6), 473–488 (1961).
43. Golub, G. H., Nash, S., and VanLoan, C., A Hessenberg-Schur method for the problem $AX + XB = C$, *IEEE Trans. Automat. Control* **AC-24** (6), 909–913 (1979).
44. Govindaraj, K. S., and Rynaski, E. G., Design criteria for optimal flight control systems, *AIAA J. Guidance and Control* **3** (4), 376–378 (1980).
45. Gunckel, T. L., and Franklin, G. F., A general solution for linear sampled data control, *Trans. ASME, J. Basic Eng.* **85D** 197–201 (1963).
46. Haddad, A. H., and Kokotovic, P. V., Stochastic control of linear singularly perturbed systems, *IEEE Trans. Automat. Control* **AC-22** (5), 815–821 (1976).
47. Hager, W. W., and Horowitz, L. L., Convergence and stability properties of the discrete Riccati operator equation and the associated optimal control and filtering problems, *SIAM J. Control and Opt.* **14** (2), 295–312 (1976).
48. Hahn, W., "Theory and Application of Liapunov's Direct Method." Prentice-Hall, Englewood Cliffs, New Jersey, 1963.
49. Hamilton, E. L., Chitwood, G., and Reeves, R. M., The General Covariance Analysis Program (GCAP), An Efficient Implementation of the Covariance Analysis Equations. Air Force Avionics Lab., Wright-Patterson AFB, Ohio (June 1976).
50. Harvey, C. H., and Stein, G., Quadratic weights for asymptotic regulator properties, *IEEE Trans. Automat. Control* **AC-23** (3), 378–387 (1978).
51. Hill, D. J., and Moylan, P. J., Connections between finite gain and asymptotic stability, *IEEE Trans. Automat. Control* **AC-25** (5), 931–936 (1980).
52. Hirzinger, G., and Kreisselmeier, G., On optimal approximation of high-order linear systems by low-order models, *Internat. J. Control* **22** (23), 399–408 (1975).
53. Ho, Y. C., and Lee, R. C. K., A Bayesian approach to problems in stochastic estimation and control, *IEEE Trans. Automat. Control* **AC-9** (5), 333–339 (1964).
54. Houpis, C. H., and Constantinides, C. T., Relationship between conventional-control theory figures of merit and quadratic performance index in optimal control theory for single-input/single-output system, *Proc. IEE* **120** (1), 138–142 (1973).

55. Hsu, J. C., and Meyer, A. U., "Modern Control Principles and Applications." McGraw-Hill, New York, 1968.
56. Iwens, R. P., Challenges in stable and robust control systems design for large space structures, *Proc. IEEE Conf. Decision and Control, Albuquerque, New Mexico* pp. 999–1002 (December 1980).
57. Jacobson, D., Totally singular quadratic minimization, *IEEE Trans. Automat. Control* **AC-16** (6), 651–658 (1971).
58. Johnson, C. D., Optimal control of the linear regulator with constant disturbances, *IEEE Trans. Automat. Control* **AC-13** (4), 416–421 (1968).
59. Johnson, C. D., Further study of the linear regulator with disturbances satisfying a linear differential equation, *IEEE Trans. Automat. Control* **AC-15** (2), 222–228 (1970).
60. Johnson, C. D., Accommodation of external disturbances in linear regulator and servo-mechanism problems, *IEEE Trans. Automat. Control* **AC-16** (6), 635–644 (1971).
61. Johnson, C. D., Theory of disturbance-accommodating controllers, "Control and Dynamic Systems: Advances in Theory and Applications" (C. T. Leondes, ed.), Vol. 12, pp. 387–490. Academic Press, New York, 1976.
62. Jonckheere, E. A., and Silverman, L. M., Spectral theory of the linear-quadratic optimal control problem: A new algorithm for spectral computations, *IEEE Trans. Automat. Control* **AC-25** (5), 880–888 (1980).
63. Joseph, P. D., and Tou, J. T., On linear control theory, *AIEE Trans.* **80**, 193–196 (1961).
64. Kailath, T., and Ljung, L., Asymptotic behavior of constant-coefficient Riccati differential equations, *IEEE Trans. Automat. Control* **AC-21** (3), 385–388 (1976).
65. Kalman, R. E., Contributions to the theory of optimal control, *Bol. Soc. Mat. Mexicana* **5**, 102–119 (1960).
66. Kalman, R. E., A new approach to linear filtering and prediction problems, *Trans. ASME (J. Basic Eng.)* **82**, 34–45 (1960).
67. Kalman, R. E., Mathematical description of dynamic systems, *SIAM J. Control, Ser. A* **1** (2), 152–192 (1963).
68. Kalman, R. E., When is a linear control system optimal? *Trans. ASME, Ser. D, J. Basic Eng.* **86**, 51–60 (1964).
69. Kalman, R. E., and Bertram, J. E., Control system analysis and design via the "second method" of Lyapunov, *Trans. ASME (J. Basic Eng.), Ser. D* **82**, 371–400 (1960).
70. Kalman, R. E., and Koepcke, R. W., Optimal synthesis of linear sampling control systems using generalized performance indices, *Trans. the ASME* **80**, 1800–1826 (1958).
71. Kleinman, D. L., On the Linear Regulator Problem and the Matrix Riccati Equation, Tech. Rep. ESL-R-271. MIT Electronic Systems Lab., Cambridge, Massachusetts (June 1966).
72. Kleinman, D. L., On an iterative technique for Riccati equation computations, *IEEE Trans. Automatic Control* **AC-13** (2), 114–115 (1968).
73. Kleinman, D. L., Stabilizing a discrete, constant, linear system with application to iterative methods for solving the Riccati equation, *IEEE Trans. on Automat. Control* **AC-19** (3), 252–254 (1974).
74. Kleinman, D. L., A Description of Computer Programs Useful in Linear Systems Studies, Tech. Rep. TR-75-4. Univ. of Connecticut, Storrs, Connecticut (October 1975).
75. Kleinman, D. L., and Athans, M., The Discrete Minimum Principle with Application to the Linear Regulator Problem, Tech. Rep. ESL-R-260. Electronic Systems Lab., MIT, Cambridge, Massachusetts (June 1966).
76. Kleinman, D. L., Baron, S., and Levison, W. H., An optimal control model of human response, Part I: Theory and validation, Part II: Prediction of human performance in a complex task, *Automatica* **6**, 357–383 (1970).
77. Kleinman, D. L., Baron, S., and Levison, W., A control theoretic approach to manned-vehicle systems analysis, *IEEE Trans. Automat. Control* **AC-16** (6), 824–832 (1971).

78. Kleinman, D. L., and Perkins, T. R., Modeling human performance in a time-varying anti aircraft tracking loop, *IEEE Trans. Automat. Control* **AC-19** (4), 297–306 (1974).
79. Kleinman, D. L., and Rao, P. K., Extensions to the Bartels-Stewart algorithm for linear matrix equations, *IEEE Trans. Automat. Control* **AC-23** (1), 85–87 (1978).
80. Klema, V. C., and Laub, A. J., The singular value decomposition: Its computation and some applications, *IEEE Trans. Automat. Control* **AC-25** (2), 164–176 (1980).
81. Kokotovic, P. V., O'Malley, R. E. Jr., and Sannuti, P., Singular perturbations and order reduction in control theory—An overview, *Automatica* **12**, 123–132 (1976).
82. Kokotovic, P. V., and Yackel, R. A., Singular perturbation of linear regulators: Basic theorems, *IEEE Trans. Automat. Control* **AC-17** (1), 29–37 (1972).
83. Konar, A. F., Mahesh, J. K., and Kizilos, B., Digital Flight Control Systems for Tactical Fighters. Honeywell, Inc., Air Force Flight Dynamics Lab., AFFDL-TR-73-119 and AFFDL-TR-74-69, Wright-Patterson AFB, Ohio, (December 1973 and July 1974).
84. Kozin, F., A survey of stability of stochastic systems, *Proc. 1968 Joint Automat. Control Conf. Stochastic Problems in Control, Ann Arbor, Michigan*, pp. 39–86 (June 1968); also *Automatica* **5**, 95–112 (1969).
85. Kriendler, E., On the linear optimal servo problem, *Internat. J. Control* **9** (4), 465–472 (1969).
86. Kriendler, E., and Rothschild, D., Model following in linear-quadratic optimization, *AIAA J.* **14** (7), 835–842 (1976).
87. Kushner, H. J., Stability of stochastic dynamical systems, "Control and Dynamic Systems: Advances in Theory and Applications" (C. T. Leondes, ed.), Vol. 4, pp. 73–102. Academic Press, New York, 1965.
88. Kushner, H. J., "Stochastic Stability and Control." Academic Press, New York, 1967.
89. Kushner, H. J., "Introduction to Stochastic Control." Holt, New York, 1971.
90. Kwakernaak, H., Optimal low-sensitivity linear feedback systems, *Automatica* **5**, 279–285 (1969).
91. Kwakernaak, H., and Sivan, R., The maximally achievable accuracy of linear optimal regulators and linear optimal filters, *IEEE Trans. Automat. Control* **AC-17** (1), 79–86 (1972).
92. Kwakernaak, H., and Sivan, R., "Linear Optimal Control Systems." Wiley (Interscience), New York, 1972.
93. Lainiotis, D. G., Discrete Riccati equation solutions: Partitioned algorithms, *IEEE Trans. Automat. Control* **AC-20**, 555–556 (1975).
94. Lainiotis, D. G., Partitioned Riccati solutions and integration-free doubling algorithms, *IEEE Trans. Automat. Control* **AC-21**, 677–689 (1976).
95. Larson, V., and Likins, P. W., Optimal estimation and control of elastic spacecraft, *in* "Control and Dynamic Systems, Advances in Theory and Applications" (C. T. Leondes, ed.), Vol. 13. Academic Press, New York, 1977.
96. LaSalle, J., and Lefschetz, S. "Stability by Lyapunov's Second Method." Academic Press, New York, 1961.
97. Laub, A. J., An inequality and some computations related to the robust stability of linear dynamic systems, *IEEE Trans. on Automat. Control* **AC-24** (2), 318–320 (1979).
98. Laub, A. J., Linear Multivariable Control: Numerical Considerations, Tech. Rep. ESL-P-833. MIT Electronic Systems Lab., Cambridge, Massachusetts (July 1978).
99. Laub, A. J., A Schur method for solving algebraic Riccati equations, *IEEE Trans. on Automatic Control* **AC-24** (6), 913–921 (1979); also Tech. Rep. LIDS-R-859. MIT. Lab. for Information and Decision Systems, Cambridge, Massachusetts (October 1978).
100. Lee, R. C. K., "Optimal Estimation, Identification and Control." MIT Press, Cambridge, Massachusetts, 1964.
101. Lee, W., and Athans, M., The Discrete-Time Compensated Kalman Filter, Rep. ESL-P-791. MIT Electronic Systems Lab., Cambridge, Massachusetts (December 1977).
102. Leondes, C. T., and Pearson, J. O., Suboptimal estimation of systems with large parameter uncertainties, *Automatica* **10**, 413–424 (1974).

103. Letov, A. M., "Stability in Nonlinear Control Systems." Princeton Univ. Press, Princeton, New Jersey, 1961.
104. Levine, W. S., Johnson, T. L., and Athans, M., Optimal limited state-variable feedback controllers for linear systems, *IEEE Trans. Automat. Control* **AC-16** (6), 785–793 (1971).
105. Levis, A. H., Schlueter, R. A., and Athans, M., On the behavior of optimal sampled data regulators, *Internat. J. Control*, **13** (2), 343–361 (1971).
106. Likins, P. W., Ohkami, Y., and Wong, C., Appendage modal coordinate truncation in hybrid coordinate dynamic analysis, *J. Spacecraft* **13** (10), 611–617 (1976).
107. Lin, J. G., Three steps to alleviate control and observation spillover problems of large space structures, *Proc. IEEE Conf. Decision and Control, Albuquerque, New Mexico*, pp. 438–444 (December 1980).
108. Luenberger, D. G., Observing the state of a linear system, *IEEE Trans. Military Electron.* **MIL-8**, 74–80 (1964).
109. Luenberger, D. G., Observers for multivariable systems, *IEEE Trans. Automat. Control* **AC-11** (2), 190–197 (1966).
110. MacFarlane, A. G. J., Return-difference and return-ratio matrices and their use in the analysis and design of multivariable feedback control systems, *Proc. IEE* **117** (10), 2037–2049 (1970).
111. MacFarlane, A. G. J., A survey of some recent results in linear multivariable feedback theory, *Automatica* **8**, 455–492 (1972).
112. MacFarlane, A. G. J., and Belletrutti, J. J., The characteristic locus design method, *Automatica* **9**, 575–588 (1973).
113. MacFarlane, A. G. J., and Karcanias, N., Poles and zeroes of linear multivariable systems: A survey of the algebraic, geometric and complex-variable theory, *Internat. J. Control* **24** (1), 33–74 (1976).
114. MacFarlane, A. G. J., and Postlethwaite, I., The generalized Nyquist stability criterion and multivariable root loci, *Internat. J. Control* **25**, 81–127 (1977).
115. Maybeck, P. S., "Stochastic Models, Estimation and Control," Vol. 1. Academic Press, New York, 1979.
116. McMorran, P. D., Extension of the inverse Nyquist method, *Electron. Lett.* **6**, 800–801 (1970).
117. McMuldrock, C. G., VTOL Controls for Shipboard Landing, M.S. thesis, Report LIDS-TH-928. MIT Lab. for Information and Decision Systems, Cambridge, Massachusetts (August 1979).
118. Meditch, J. S., "Stochastic Optimal Linear Estimation and Control." McGraw-Hill, New York, 1969.
119. Meier, L., Larson, R. E., and Tether, A. J., Dynamic programming for stochastic control of discrete systems, *IEEE Trans. Automat. Control* **AC-16** (6), 767–775 (1971).
120. Meirovitch, L., and Öz, H., Observer modal control of dual-spin flexible spacecraft, *AIAA J. Guidance and Control* **2** (2), 101–110 (1979).
121. Melzer, S. M., and Kuo, B. C., Sampling period sensitivity of the optimal sampled-data linear regulator, *Automatica* **7**, 367–370 (1971).
122. Moler, C., and Van Loan, C., Nineteen dubious ways to compute the exponential of a matrix, *SIAM Rev.* **20** (4), 801–836 (1978).
123. Moore, B. C., Principal component analysis in linear systems: Controllability, observability, and model reduction, *IEEE Trans. Automat. Control* **AC-26** (1), 17–32 (1981).
124. Morris, J. M., The Kalman filter: A robust estimator for some classes of linear quadratic problems, *IEEE Trans. Informat. Theory* **IT-22** (5), 526–534 (1976).
125. Newton, G. C., Jr., Gould, L. A., and Kaiser, J. F., "Analytical Design of Linear Feedback Controls." Wiley, New York, 1964.
126. Novak, L. M., Optimal minimal-order observers for discrete-time systems—A unified theory, *Automatica* **8**, 379–387 (1972).

127. Novak, L. M., Discrete-time optimal stochastic observers, "Control and Dynamic Systems: Advances in Theory and Applications" (C. T. Leondes, ed.), Vol. 12, pp. 259–312. Academic Press, New York, 1976.
128. Nuzman, D. W., and Sandell, N. R., Jr., An inequality arising in robustness analysis of multivariable systems, *IEEE Trans. Automat. Control* **AC-24** (3), 492–493 (1979).
129. O'Brien, M. J., and Broussard, J. R., Feedforward control to track the output of a forced model, *Proc. 1978 IEEE Conf. Decision and Control*, San Diego, California, pp. 1149–1155 (January 1979).
130. Ogata, K., "State Space Analysis of Control Systems." Prentice-Hall, Englewood Cliffs, New Jersey, 1967.
131. O'Reilly, J., and Newmann, M. M., Minimal order optimal estimators for discrete-time linear stochastic systems, "Recent Theoretical Developments in Control" (M. J. Gregson, ed.). Academic Press, New York, 1978.
132. Pappas, T., Laub, A. J., and Sandell, N. R., Jr., On the numerical solution of the discrete-time algebraic Riccati equation, *IEEE Trans. Automat. Control* **AC-25** (4), 631–641 (1980).
133. Parker, K. T., Design of proportional-integral-derivative controllers by the use of optimal-linear-regulator theory, *Proc. IEE* **119** (7) (1972).
134. Patel, R. V., Toda, M., and Sridhar, B., Robustness of linear quadratic state feedback designs, *Proc. Joint Automat. Control Conf., San Francisco, California*, pp. 1668–1673, (June, 1977).
135. Perkins, W. R., and Cruz, J. B., The parameter variation problem in state feedback control systems, *Trans. ASME, J. Basic Eng., Ser. D* **87**, 120–124 (1965).
136. Perkins, W. R., and Cruz, J. B., Feedback properties of linear regulators, *IEEE Trans. Automat. Control* **AC-16** (6), 659–664 (1971).
137. Potter, J. E., A Guidance-Navigation Separation Theorem, Rep. RE-11. MIT Experimental Astronomy Lab., Cambridge, Massachusetts (1964); also *Proc. AIAA/ION Astrodynam. Guidance and Control Conf., Los Angeles, California*, AIAA Paper No. 64-653 (1964).
138. Potter, J. E., Matrix quadratic solutions, *SIAM J. Appl. Math.* **14**, 496–501 (1966).
139. Powers, W. F., Cheng, B.-D., and Edge, E. R., Singular optimal control computation, *AIAA J. Guidance and Control* **1** (1), 83–89 (1978).
140. Ragazzini, J. K., and Franklin, G. F., "Sampled-Data Control Systems." McGraw-Hill, New York, 1958.
141. Rauch, H. E., Order reduction in estimation with singular perturbation, *Symp. Nonlinear Estimat. its Appl.* pp. 231–241 (September 1973).
142. Repperger, D. W., A square root of a matrix approach to obtain the solution to a steady state matrix Riccati equation, *IEEE Trans. Automat. Control* **AC-21**, 786–787 (1976).
143. Roseau, M., "Vibrations Non linéaires et Théorie de la Stabilité." Springer-Verlag, Berlin and New York, 1966.
144. Rosenbrock, H. H., Design of multivariable control systems using inverse Nyquist array, *Proc. IEE* **116**, 1929–1936 (1969).
145. Rosenbrock, H. H., "State Space and Multivariable Theory." Wiley (Interscience), New York, 1970.
146. Rosenbrock, H. H., Multivariable circle theorems, *in* "Recent Mathematical Developments in Control" (D. J. Ball, ed.). Academic Press, New York, 1973.
147. Rynaski, E. G., and Whitbeck, R. F., The Theory and Application of Linear Optimal Control, Tech. Rep. AFFDL-TR-65-28. Air Force Flight Dynamics Lab., Wright-Patterson AFB, Ohio (October 1965).
148. Sage, A. P., "Optimum Systems Control." Prentice-Hall, Englewood Cliffs, New Jersey, 1968.
149. Safonov, M. G., Robustness and Stability Aspects of Stochastic Multivariable Feedback System Design, Ph.D. dissertation and Rep. ESL-R-763. MIT, Cambridge, Massachusetts (September 1977).

150. Safonov, M. G., and Athans, M., Gain and phase margin of multiloop LQG regulators, *IEEE Trans. Automat. Control* **AC-22** (2), 173–179 (1977).

151. Safonov, M. G., and Athans, M., Robustness and computational aspects of nonlinear stochastic estimators and regulators, *IEEE Trans. Automat. Control* **AC-23** (4), 717–725 (1978).

152. Sandell, N. R., Jr., Optimal Linear Tracking Systems, Rep. No. ESL-R-456. Electronics Systems Lab., MIT, Cambridge, Massachusetts (September 1971).

153. Sandell, N. R., Jr., Robust stability of linear dynamic systems with application to singular perturbation theory, Recent Developments in the Robustness Theory of Multivariable Systems (N. R. Sandell, ed.), Tech. Rep. LIDS-R-954. MIT Lab. for Information and Decision Systems, Cambridge, Massachusetts and Rep. ONR-CR215-271-1F. Office of Naval Research, Arlington, Virginia (August 1979); also *Automatica* (July 1979).

154. Sandell, N. R., Jr. (ed.), Recent Developments in the Robustness Theory of Multivariable Systems, Rep. ONR-CR215-271-1F of the Office of Naval Research, and Rep. LIDS-R-954 of the MIT. Lab. for Information and Decision Systems, Cambridge, Massachusetts (August 1979).

155. Sandell, N. R., Jr., and Athans, M., On type-L multivariable linear systems, *Automatica* **9** (1), 131–136 (1973).

156. Schmidt, D. K., Optimal flight control synthesis via pilot modelling, *AIAA J. Guidance and Control* **2** (4), 308–312 (1979).

157. Sesak, J. R., and Higgins, T. J., Sensitivity-constrained linear optimal control via forced singular perturbation model reduction, *Proc. 1978 IEEE Conf. Decision and Control, San Diego, California* pp. 1145–1148 (January 1979).

158. Sesak, J. R., Likins, P., and Coradetti, T., Flexible spacecraft control by model error sensitivity suppression, *J. Astronaut. Sci.* **27** (2), 131–156 (1979).

159. Shaked, U., The asymptotic behavior of the root-loci of multivariable optimal regulators, *IEEE Trans. Automat. Control* **AC-23** (3), 425–430 (1978).

160. Silverman, L. M., Discrete Riccati equations: Alternative algorithms, asymptotic properties, and system theory interpretations, "Control and Dynamic Systems: Advances in Theory and Applications" (C. T. Leondes, ed.), Vol. 12, pp. 313–386. Academic Press, New York, 1976.

161. Simon, H. A., Dynamic programming under uncertainty with a quadratic criterion function, *Econometrica* **24**, 74–81 (1956).

162. Skelton, R. E., On cost sensitivity controller design methods for uncertain dynamic systems, *J. Astronaut. Sci.* **27** (2), 181–205 (1979).

163. Skelton, R. E., Hughes, P. C., and Hablani, H., Order reduction for models of space structures using modal cost analysis, *J. Guidance and Control* (1982), special issue on Large Space Structures (to appear).

164. Skelton, R. E., and Likins, P. W., Techniques of modeling and model error compensation in linear regulator problems *in* "Advances in Control and Dynamic Systems" (C. T. Leondes, ed.), Vol. 14. Academic Press, New York, 1978.

165. Skelton, R. E., and Yedavalli, R. K., Modal cost analysis of flexible space structures with uncertain modal data, *Proc. IEEE Conf. Decision and Control, Albuquerque, New Mexico* pp. 792–794 (1980).

166. Sorenson, H. W., Controllability and observability of linear, stochastic, time-discrete control systems, "Control and Dynamic Systems: Advances in Theory and Applications" (C. T. Leondes, ed.) Vol. 6, pp. 95–158. Academic Press, New York, 1968.

167. Stein, G., Generalized quadratic weights for asymptotic regulator properties, *IEEE Trans. Automat. Control* **AC-24** (4), 559–566 (1979).

168. Stein, G., and Doyle, J. C., Multivariable feedback design: Concepts for a classical/modern synthesis, *IEEE Trans. Automat. Control* **AC-26** (1), 4–16 (1981).

169. Stengel, R. F., Broussard, J. R., and Berry, P. W., Digital controllers for VTOL aircraft, *IEEE Trans. Aerospace and Electron. Syst.* **AES-14** (1), 54–63 (1978).
170. Stengel, R. F., Broussard, J. R., and Berry, P. W., Digital flight control design for a tandem-rotor helicopter, *Automatica* **14**, 301–312 (1978).
171. Stewart, G. W., "Introduction to Matrix Computations." Academic Press, New York, 1973.
172. Strang, G., "Linear Algebra and its Applications." Academic Press, New York, 1976.
173. Streibel, C., Sufficient statistics in the optimum control of stochastic systems, *J. Math. Anal. Appl.* **12**, 576–592 (1965).
174. Sundareshan, M. K., and Vidyasagar, M., $L_2$-stability of large-scale dynamical systems—Criteria via positive operator theory, *IEEE Trans. Automat. Control* **AC-22** (3), 396–399 (1977).
175. Takahashi, Y., Rabins, M. J., and Auslander, D. M., "Control and Dynamic Systems." Addison-Wesley, Reading, Massachusetts, 1970.
176. Theil, H., A note on certainty equivalence in dynamic planning, *Econometrica* **25**, 346–349 (1957).
177. Thornton, C. L., and Jacobson, R. A., Linear stochastic control using $UDU^T$ matrix factorization, *AIAA J. Guidance and Control* **1** (4), 232–236 (1978).
178. Teneketzis, D., and Sandell, N. R., Jr., Linear regulator design for stochastic systems by a multiple time scales method, *IEEE Trans. Automat. Control* **AC-22** (4), 615–621 (1977).
179. Tou, J. S., "Optimum Design of Digital Systems." Academic Press, New York, 1963.
180. Trankle, T. L., and Bryson, A. E., Jr., Control logic to track outputs of a command generator, *AIAA J. Guidance and Control* **1** (2), 130–135 (1978).
181. Tse, E., Observer-estimators for discrete-time systems, *IEEE Trans. Automat. Control* **AC-18** (1), 10–16 (1973).
182. Tse, E., and Athans, M., Optimal minimal-order observer estimators for discrete linear time-varying systems, *IEEE Trans. Automat. Control* **AC-15** (4), 416–426 (1970).
183. Tse, E. C. Y., Medanic, J. V., and Perkins, W. R., Generalized Hessenberg transformations for reduced order modeling of large scale systems, *Internat. J. Control* **27** (4), 493–512 (1978).
184. Uttam, B. J., On the stability of time-varying systems using an observer for feedback control, *Internat. J. Control* **14** (6), 1081–1087 (1971).
185. Uttam, B. J., and O'Halloran, W. F., Jr., On observers and reduced order optimal filters for linear stochastic systems, *Proc. Joint Automat. Control Conf.* (August 1972).
186. Uttam, B. J., and O'Halloran, W. F., Jr., On the computation of optimal stochastic observer gains, *IEEE Trans. Automat. Control* **AC-20** (1), 145–146 (1975).
187. Washburn, R. B., Mehra, R. K., and Sajan, S., Application of singular perturbation technique (SPT) and continuation methods for on-line aircraft trajectory optimization, *Proc. 1978 IEEE Conf. Decision and Control, San Diego, California*, 983–990 (January 1979).
188. Whitbeck, R. F., Wiener-Hopf Approaches to Regulator, Filter/Observer, and Optimal Coupler Problems, Tech. Rep. ONR-CR215-260-1. Office of Naval Research, Systems Technology, Inc., Hawthorne, California (January 1980).
189. Widnall, W. S., "Applications of Optimal Control Theory to Computer Controller Design." MIT Press, Cambridge, Massachusetts, 1968.
190. Willems, J. C., "The Analysis of Feedback Systems." MIT Press, Cambridge, Massachusetts, 1970.
191. Willems, J. C., Mechanisms for the stability and instability of feedback systems, *Proc. IEEE* **64** (1), 24–35 (1976).
192. Willems, J. L., "Stability Theory of Dynamical Systems." Wiley, New York, 1970.
193. Willsky, A. S., "Digital Signal Processing and Control and Estimation Theory: Points of Tangency, Areas of Intersection, and Parallel Directions." MIT Press, Cambridge, Massachusetts, 1979.

194. Witsenhausen, H. S., Separation of estimation and control for discrete time systems, *Proc. IEEE* **59** (9), 1557–1566 (1971).
195. Womble, M. E., The Linear Quadratic Gaussian Problem with Ill-Conditioned Riccati Matrices. Ph.D. dissertation, MIT and Rep. T-570, C. S. Draper Lab., Cambridge, Massachusetts (May 1972).
196. Wong, P. K., and Athans, M., Closed-loop structural stability for linear-quadratic optimal systems, *IEEE Trans. Automat. Control* **AC-22** (1), 94–99 (1977).
197. Wonham, W. M., On the separation theorem of stochastic control, *J. SIAM Control* **6** (2), 312–326 (1968).
198. Wonham, W. M., Optimal stochastic control, *Proc. Joint Automatic Control Conf., Stochastic Problems in Control, Ann Arbor, Michigan* pp. 107–120 (June 1968).
199. Wonham, W. M., On a matrix Riccati equation of stochastic control, *J. SIAM Control* **6** (4), 681–697 (1968).
200. Wonham, W. M., "Linear Multivariable Control: A Geometric Approach." Springer-Verlag, Berlin and New York, 1974.
201. Woodhead, M. A., and Porter, B., Optimal modal control, *Measurement and Control* **6**, 301–303 (1973).
202. Yackel, R. A., and Kokotovic, P. V., A boundary layer method for the matrix Riccati equations, *IEEE Trans. Automat. Control* **AC-18** (1), 17–24 (1973).
203. Yoshikawa, T., and Kobayashi, H., Comments on "optimal minimal order observer-estimators for discrete linear time-varying systems," *IEEE Trans. Automat. Control* **AC-17** (2), 272–273 (1972).

## PROBLEMS

**14.1** Show that (14-12) is algebraically equivalent to (14-9), as claimed in the text.

**14.2** (a) Derive (14-14)–(14-17) by the same procedure as used in Chapter 13 to develop the controller for the case of the cross-weighting matrix $\mathbf{S}(t_i) \equiv \mathbf{0}$.

(b) Show that this result can also be obtained [92] by defining an associated system equation

$$\mathbf{x}(t_{i+1}) = \mathbf{\Phi}'(t_{i+1}, t_i)\mathbf{x}(t_i) + \mathbf{B}_d(t_i)\mathbf{u}'(t_i)$$

where

$$\mathbf{\Phi}'(t_{i+1}, t_i) = \mathbf{\Phi}(t_{i+1}, t_i) - \mathbf{B}_d(t_i)\mathbf{U}^{-1}(t_i)\mathbf{S}^T(t_i)$$
$$\mathbf{u}'(t_i) = \mathbf{u}(t_i) + \mathbf{U}^{-1}(t_i)\mathbf{S}^T(t_i)\mathbf{x}(t_i)$$

and then finding the control $\mathbf{u}'^*(t_i)$ to minimize the quadratic cost (*without* cross-weighting terms)

$$J' = \frac{1}{2}\mathbf{x}^T(t_{N+1})\mathbf{X}_f\mathbf{x}(t_{N+1}) + \sum_{i=0}^{N} \frac{1}{2}\{\mathbf{x}^T(t_i)\mathbf{X}'(t_i)\mathbf{x}(t_i) + \mathbf{u}'^T(t_i)\mathbf{U}(t_i)\mathbf{u}'(t_i)\}$$

where

$$\mathbf{X}'(t_i) = \mathbf{X}(t_i) - \mathbf{S}(t_i)\mathbf{U}^{-1}(t_i)\mathbf{S}^T(t_i)$$

One can solve the equivalent problem for the $\mathbf{K}_c'$ and $\mathbf{G}_c^{*\prime}$ time histories by the results of Chapter 13, summarized in (14-1)–(14-10). Then, since

$$\begin{aligned}\mathbf{u}(t_i) &= \mathbf{u}'(t_i) - \mathbf{U}^{-1}(t_i)\mathbf{S}^T(t_i)\mathbf{x}(t_i)\\ &= -\mathbf{G}_c^{*\prime}(t_i)\mathbf{x}(t_i) - \mathbf{U}^{-1}(t_i)\mathbf{S}^T(t_i)\mathbf{x}(t_i)\end{aligned}$$

we can then solve for $\mathbf{G}_c^*(t_i)$ according to

$$\mathbf{G}_c^*(t_i) = \mathbf{G}_c^{*\prime}(t_i) + \mathbf{U}^{-1}(t_i)\mathbf{S}^T(t_i)$$

Note that the closed-loop characteristics agree, i.e.,

$$[\mathbf{\Phi}(t_{i+1}, t_i) - \mathbf{B}_d(t_i)\mathbf{G}_c{}^*(t_i)] = [\mathbf{\Phi}'(t_{i+1}, t_i) - \mathbf{B}_d(t_i)\mathbf{G}_c^{*\prime}(t_i)]$$

(c) Consider the LQG stochastic optimal control assuming perfect knowledge of the state at $t_i$. Derive the extension of the $g(t_i)$ relation of Example 13.7 to the case of $\mathbf{S}(t_i) \neq \mathbf{0}$.

(d) Consider the LQG stochastic optimal control assuming only incomplete noise-corrupted measurements available at each $t_i$. Derive the extension of the $g(t_i)$ relation of Example 13.12 to the case of $\mathbf{S}(t_i) \neq \mathbf{0}$.

**14.3** Consider the steady state constant-gain exponentially-weighted LQ optimal state regulator, in which the constant gain $\bar{\mathbf{G}}_c{}^*$ is to be chosen so as to minimize the cost

$$J = \lim_{N\to\infty} \frac{1}{N} \sum_{i=0}^{N} \frac{1}{2} \alpha^{2i} \begin{bmatrix} \mathbf{x}(t_i) \\ \mathbf{u}(t_i) \end{bmatrix}^{\mathrm{T}} \begin{bmatrix} \mathbf{X} & \mathbf{S} \\ \mathbf{S}^{\mathrm{T}} & \mathbf{U} \end{bmatrix} \begin{bmatrix} \mathbf{x}(t_i) \\ \mathbf{u}(t_i) \end{bmatrix}$$

where the composite matrix in the summation is assumed positive definite and $\alpha \geq 1$ is an exponential weighting constant. Use (14-13)–(14-17) to show that the optimal control is $\mathbf{u}^*(t_i) = -\bar{\mathbf{G}}_c{}^*\mathbf{x}(t_i)$, where

$$\bar{\mathbf{G}}_c{}^* = [\mathbf{U} + \mathbf{B}_d{}^{\mathrm{T}}\bar{\mathbf{K}}_c\mathbf{B}_d]^{-1}[\mathbf{B}_d{}^{\mathrm{T}}\bar{\mathbf{K}}_c\mathbf{\Phi} + \mathbf{S}^{\mathrm{T}}]$$

and $\bar{\mathbf{K}}_c$ is the unique symmetric positive definite solution to the steady state (algebraic) discrete-time Riccati equation

$$\bar{\mathbf{K}}_c = \alpha^2\{\mathbf{X} + \mathbf{\Phi}^{\mathrm{T}}\bar{\mathbf{K}}_c\mathbf{\Phi} - [\mathbf{B}_d{}^{\mathrm{T}}\bar{\mathbf{K}}_c\mathbf{\Phi} + \mathbf{S}^{\mathrm{T}}]^{\mathrm{T}}[\mathbf{U} + \mathbf{B}_d{}^{\mathrm{T}}\bar{\mathbf{K}}_c\mathbf{B}_d]^{-1}[\mathbf{B}_d{}^{\mathrm{T}}\bar{\mathbf{K}}_c\mathbf{\Phi} + \mathbf{S}^{\mathrm{T}}]\}$$

**14.4** Derive (14-28), the differential equations to be integrated to specify the equivalent discrete-time problem corresponding to (14-18)–(14-20), using (14-23), (14-25), and (14-27).

**14.5** (a) Consider Example 14.2. Show that $\bar{K}_c$ and $\bar{G}_c{}^*$ depend on the ratio $(X/U)$ and are independent of $X_f$. Obtain the steady state constant filter gain $K_{ss}$. Obtain the controller form given at the end of the example. Use this controller to provide feedback to the original system, and explicitly evaluate the closed-loop system eigenvalues.

(b) Repeat part (a) for Example 14.3, with any necessary generalizations.

(c) Repeat part (a) for Example 14.4, with any necessary generalizations.

**14.6** Consider the LQG stochastic controller. Under what conditions can a constant-gain steady state controller be reached? Show that the steady state values of $\mathbf{P}(t_i^-)$, $\mathbf{P}(t_i^+)$, and $\mathbf{K}(t_i)$ of the Kalman filter are independent of $\mathbf{P}_0$, and that the steady state $\mathbf{K}_c(t_i)$ and $\mathbf{G}_c{}^*(t_i)$ are independent of $\mathbf{X}_f$. Express the controller in the standard form (14-274) and explicitly evaluate the constant matrices in this form.

**14.7** (a) Consider the LQG optimal stochastic controller with cost function given by (14-13). Again let (14-1) and (14-31) describe the system model, but now let $\mathbf{w}_d(\cdot,\cdot)$ and $\mathbf{v}(\cdot,\cdot)$ be correlated according to

$$E\{\mathbf{w}_d(t_i)\mathbf{v}^{\mathrm{T}}(t_j)\} = \mathbf{C}(t_i)\,\delta_{ij}$$

(b) Repeat part (a), but with correlation specified by

$$E\{\mathbf{w}_d(t_{i-1})\mathbf{v}^{\mathrm{T}}(t_j)\} = \mathbf{C}(t_j)\,\delta_{ij}$$

(c) For a measurement model of (14-32), show that (14-33) is the only modification necessary to the LQG optimal stochastic controller.

**14.8** (a) Prove that, if (14-70) is completely controllable, then the solution $\mathbf{K}_c(t_i)$ to the backward Riccati equation (14-14)–(14-16) with $\mathbf{K}_c(t_{N+1}) = \mathbf{0}$ converges to a positive semidefinite bounded steady state solution sequence $\bar{\mathbf{K}}_c(t_i)$ for all $t_i$ as $t_i \to -\infty$, and this $\bar{\mathbf{K}}_c(t_i)$ sequence for all $t_i$ is itself a solution to the backward equation. Use the method of proof outlined in the text between (14-70) and (14-71).

(b) Show that (14-73) minimizes the limiting cost (14-74).

(c) Show that the closed-loop system representation (14-75) is exponentially stable.

(d) Demonstrate the corresponding results for time-invariant system models, namely, that the claims corresponding to (14-76)–(14-81) are valid.

**14.9** Consider the LQ optimal state regulator problem, and define the costate, or adjoint vector, associated with the state of this problem as

$$\mathbf{p}(t_i) = \mathbf{K}_c(t_{i+1})\mathbf{x}(t_{i+1})$$

(a) Using (14-14)–(14-17), show that $\mathbf{p}(t_i)$ satisfies

$$\mathbf{p}(t_{i-1}) = \mathbf{\Phi}^T(t_{i+1}, t_i)\mathbf{p}(t_i) + [\mathbf{X}(t_i) + \mathbf{S}(t_i)\mathbf{G}_c^*(t_i)]\mathbf{x}(t_i)$$

(b) Now write out $[-\mathbf{U}^{-1}(t_i)\mathbf{B}_d^T(t_i)\mathbf{p}(t_i)]$ using the defining relation for $\mathbf{p}(t_i)$, and substitute in the dynamics equation for $\mathbf{x}(t_{i+1})$. Use (14-14) premultiplied by $[\mathbf{U}(t_i) + \mathbf{B}_d^T(t_i)\mathbf{K}_c(t_{i+1})\mathbf{B}_d(t_i)]$ to derive

$$-\mathbf{U}^{-1}(t_i)\mathbf{B}_d^T(t_i)\mathbf{p}(t_i) = \mathbf{U}^{-1}(t_i)\mathbf{S}^T(t_i)\mathbf{x}(t_i) + \mathbf{u}^*(t_i)$$

(c) Combine the results of (a) and (b) to develop the two-point boundary value problem

$$\begin{aligned}
\mathbf{x}(t_{i+1}) &= [\mathbf{\Phi}(t_{i+1}, t_i) - \mathbf{B}_d(t_i)\mathbf{U}^{-1}(t_i)\mathbf{S}^T(t_i)]\mathbf{x}(t_i) - \mathbf{B}_d(t_i)\mathbf{U}^{-1}(t_i)\mathbf{B}_d^T(t_i)\mathbf{p}(t_i)\\
\mathbf{p}(t_{i-1}) &= [\mathbf{X}(t_i) + \mathbf{S}(t_i)\mathbf{G}_c^*(t_i)]\mathbf{x}(t_i) + \mathbf{\Phi}^T(t_{i+1}, t_i)\mathbf{p}(t_i)\\
\mathbf{x}(t_0) &= \mathbf{x}_0\\
\mathbf{p}(t_{N+1}) &= \mathbf{K}_c(t_{N+1})\mathbf{x}(t_{N+1})
\end{aligned}$$

(d) For the case of time-invariant system models and constant cost weighting matrices, take the $z$ transform of these equations and show that the result can be expreseed as

$$\begin{bmatrix}\mathbf{x}(z)\\ \mathbf{p}(z)\end{bmatrix} = \begin{bmatrix} z\mathbf{I} - [\mathbf{\Phi} - \mathbf{B}_d\mathbf{U}^{-1}\mathbf{S}^T] & \mathbf{B}_d\mathbf{U}^{-1}\mathbf{B}_d^T\\ -[\mathbf{X} + \mathbf{S}\bar{\mathbf{G}}_c^*] & z^{-1}\mathbf{I} - \mathbf{\Phi}^T\end{bmatrix}\begin{bmatrix} z\mathbf{x}_0\\ -[\mathbf{X} + \mathbf{S}\bar{\mathbf{G}}_c^*]\mathbf{x}_0 - \mathbf{\Phi}^T\mathbf{p}_0\end{bmatrix}$$

and use this to demonstrate the validity of (14-82).

**14.10** (a) Consider a scalar system model of the form (14-1). Develop the state feedback control law that provides output deadbeat response as described below (14-87). Can such a controller always be generated with an LQ deterministic optimal controller by appropriate choice of $X$, $U$, and $S$? What if $S \equiv 0$? Show the closed-loop system response to a nonzero initial state.

(b) Consider a scalar measurement as in (14-31) or (14-32). Develop the observer (14-94) with deadbeat response. Can such an estimator always be generated with a Kalman filter by appropriate choice of $Q_d$, $R$, and $E\{\mathrm{w}_d(t_i)\mathrm{v}(t_i)\}$? What if $E\{\mathrm{w}_d(t_i)\mathrm{v}(t_i)\} \equiv 0$?

(c) Using the results of (a) and (b), develop an output feedback state deadbeat control system. Show the closed-loop system response to a nonzero initial state.

(d) Repeat parts (a) to (c) for a general second order, single-input/single-output system model. Consider the cases of real and complex conjugate poles, and with and without a zero in the corresponding $z$-transform transfer function. If output deadbeat response is achieved, does this guarantee state deadbeat response?

**14.11** Recall the duality relationships of (14-89) and Table 14.1. Show that the LQ optimal deterministic state regulator for the case of $\mathbf{S}(t_i) \neq \mathbf{0}$ is related in a dual manner to the Kalman filter developed for the case of $E\{\mathbf{w}_d(t_i)\mathbf{v}^T(t_i)\} \neq \mathbf{0}$. (See Section 5.9 of Volume 1.)

**14.12** (a) Demonstrate the validity of (14-115).

(b) Show that (14-116) is also correct.

**14.13** (a) Develop the results corresponding to the block diagram in Fig. 14.4 for the control law $\mathbf{u}(t_i) = -\bar{\mathbf{G}}_c{}^*\mathbf{x}(t_i)$.

(b) Repeat for $\mathbf{u}(t_i) = -\bar{\mathbf{G}}_c{}^*\hat{\mathbf{x}}(t_i{}^+)$, as portrayed in Fig. 14.5.

(c) Repeat for $\mathbf{u}(t_i) = -\bar{\mathbf{G}}_c{}^*\hat{\mathbf{x}}(t_i{}^-)$, as given by Fig. 14.6.

(d) Repeat part (b) for the case of $\mathbf{D}_z \neq \mathbf{0}$ in the measurement model, (14-32).

(e) Repeat part (c) for the case of $\mathbf{D}_z \neq \mathbf{0}$.

**14.14** Using the results of Section 14.13, develop the result used in Section 14.5, (14-124).

**14.15** (a) Show that, if (14-127) were satisfied, then the return difference matrix of the control law $\mathbf{u}(t_i) = -\bar{\mathbf{G}}_c{}^*\hat{\mathbf{x}}(t_i{}^-)$ would duplicate the return difference matrix of the control law $\mathbf{u}(t_i) = -\bar{\mathbf{G}}_c{}^*\mathbf{x}(t_i)$ that assumes perfect access to all states at time $t_i$.

(b) Show that if (14-128) were satisfied, then (14-127) would be satisfied in the limit as $q \to \infty$.

(c) Show the analogous results for continuous-time measurements, (14-132) and (14-133), are valid.

(d) Show that these conditions can be achieved in the continuous-time measurement case by adding pseudonoise of strength $[q^2\mathbf{BVB}^T]$ to the $\mathbf{Gw}(t)$ dynamics driving noise term used in the Kalman filter formulation for this problem, i.e., adding pseudonoise where $\mathbf{u}(t)$ enters the dynamics.

**14.16** Consider the equivalent discrete-time model for the cascade of two first order lags, $(s+2)^{-1}$ and $(s+5)^{-1}$, driven by a scalar input and by two independent zero-mean white Gaussian discrete-time noises of unit variance each, given approximately by

$$\begin{bmatrix} x_1(t_{i+1}) \\ x_2(t_{i+1}) \end{bmatrix} = \begin{bmatrix} 0.8 & 0 \\ 0.1 & 0.6 \end{bmatrix} \begin{bmatrix} x_1(t_i) \\ x_2(t_i) \end{bmatrix} + \begin{bmatrix} 0.2 \\ 0 \end{bmatrix} u(t_i) + \begin{bmatrix} w_{d1}(t_i) \\ w_{d2}(t_i) \end{bmatrix}$$

for a sample period of 0.1 sec.

(a) Develop the LQG optimal controller that assumes perfect knowledge of $\mathbf{x}(t_i)$ at each $t_i$, optimal in the sense that it minimizes

$$J = E\left\{\frac{1}{2}x_1{}^2(t_{N+1}) + \frac{1}{2}x_2{}^2(t_{N+1}) + \sum_{i=0}^{N}\frac{1}{2}[x_1{}^2(t_i) + x_2{}^2(t_i) + Uu^2(t_i)]\right\}$$

Evaluate for cases of $U = 0.1$, 1, and 10.

(b) Obtain the constant-gain steady state controllers corresponding to results in (a).

(c) Develop the covariance performance analysis of the generated controllers acting upon a "truth model" of the same second order system model as that used for controller design. Show how steady state rms state regulation errors degrade when the $\Phi_{11}$ element in the truth model changes to 0.9 and 0.7 while the controller assumes it stays at 0.8. Repeat for the truth model $\Phi_{22}$ element changing from 0.6 to 0.7 and 0.5.

(d) Now develop the LQG controller based on noise-corrupted measurements

$$z(t_i) = x_2(t_i) + v(t_i)$$

with $v(\cdot,\cdot)$ of variance $R$. Evaluate for $R = 0.1$, 1, and 10.

(e) Obtain the corresponding constant-gain steady state controllers.

(f) Repeat part (c) for these controllers.

(g) Now robustify these designs by replacing the Kalman filters with full-order observers as in (14-94), with gains chosen as in (14-127)–(14-131), with $q = 1, \sqrt{10}$, 10, and 100. Repeat parts (e) and (f).

(h) Compare this result to the result of tuning the original Kalman filter by adding pseudonoise of variance $q^2\mathbf{G}_d\mathbf{Q}_d\mathbf{G}_d{}^T$ or $q^2\mathbf{B}_d V\mathbf{B}_d{}^T$ $(V = 1)$ for $q^2 = 1$, 10, 100, and $10^4$. Repeat parts (e) and (f).

**14.17** Display the robustness characteristics of the closed-loop systems based on the controllers of the previous problem by plotting

$$\sigma_{\min}[\mathbf{I} + \mathbf{G}_L(\exp\{2\pi j\omega/\omega_s\}] \quad \text{and} \quad \sigma_{\min}[\mathbf{I} + \mathbf{G}_L^{-1}\exp(2\pi j\omega/\omega_s)\}]$$

for $(\omega/\omega_s) \in [0, 1]$ (at enough discrete values to see the characteristics), as in (14-113). Do this for controllers of

(a) part (b).

(b) part (e).

(c) part (g).

(d) part (h).

**14.18** Consider the design of the optimal feedback controller in Fig. 14.P1. The physical system is an integrator driven by both control $u(t)$ and noise modeled as zero-mean white and Gaussian, with

$$E\{\mathrm{w}(t)\mathrm{w}(t+\tau)\} = 2\delta(\tau)$$

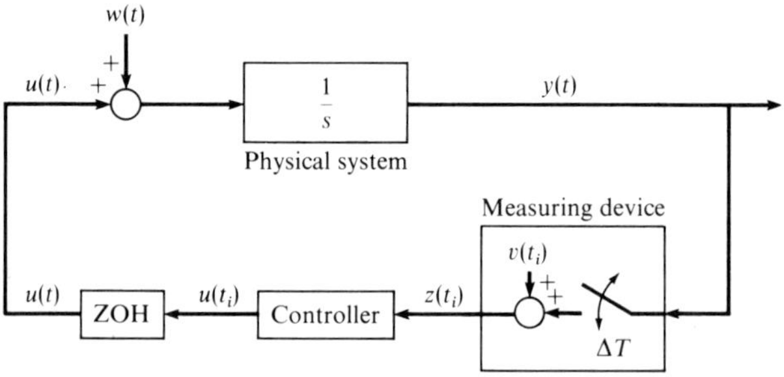


FIG. 14.P1

The system output $y(t)$ is sampled every 0.1 sec ($\Delta T = 0.1$ sec) by a measuring device whose output is corrupted by discrete-time zero-mean white Gaussian noise with

$$E\{\mathrm{v}(t_i)\mathrm{v}(t_j)\} = \begin{cases} 1 & \text{if} \quad t_i = t_j \\ 0 & \text{if} \quad t_i \neq t_j \end{cases}$$

This measured output is passed to the digital controller to be designed. The discrete-time output of the controller is passed through a zero-order hold (ZOH), which maintains that value over the next sample period, so that

$$u(t) = u(t_i) \qquad \text{for} \quad t_i \le t < t_{i+1}$$

It is desired to design the controller so that it will perform well over a time interval of 5 min, with the measure of performance being a small value of

$$\frac{1}{2} E\left\{ \sum_{i=0}^{2999} [\mathrm{y}(t_i)^2 + 0.1\mathrm{u}(t_i)^2] + \mathrm{y}(t_f = 5 \text{ min})^2 \right\}$$

(a) Define the equivalent discrete-time model for the physical system.

(b) Completely define the controller algorithm, including all recursions necessary for final implementation. Which ones can be precomputed and stored?

(c) Obtain the steady state controller. Is it time invariant? Would it be reasonable to use this steady state controller for the entire time interval of interest? What are the benefits and disadvantages? If the steady state controller is used, is the closed-loop system stable?

(d) Instead of regulating $y(t_i)$ to zero for all $t_i$, it is now desired to regulate it to some nonzero constant value, which can be arbitrarily set by a human operator. Letting the desired value be $d$, generate the controller to minimize

$$\frac{1}{2}E\left\{\sum_{i=0}^{2999}([\mathrm{y}(t_i)-d]^2+0.1\mathrm{u}(t_i)^2)+[\mathrm{y}(t_f)-d]^2\right\}$$

and relate its structure to the previous controllers. What are the shortcomings of this design?

(e) Now it is desired instead to force $y(t_i)$ to follow the position behavior of a target whose position is described as a first order Gauss–Markov process with mean zero and

$$E\{\mathrm{p}(t)\mathrm{p}(t+\tau)\}=\sigma^2\exp\{-|\tau|/T\}$$

Generate the controller to minimize

$$\frac{1}{2}E\left\{\sum_{i=0}^{2999}([\mathrm{y}(t_i)-\mathrm{p}(t_i)]^2+0.1\mathrm{u}(t_i)^2)+[\mathrm{y}(t_f)-\mathrm{p}(t_f)]^2\right\}$$

assuming $p(t_i)$ can be measured perfectly, and relate its structure to the previous controllers. How would you design a steady state controller? Are you assured of a stable design?

(f) How does the controller structure in (e) change if only noise-corrupted measurements of target position are available as

$$\mathrm{p}_{\mathrm{meas}}(t_i)=\mathrm{p}(t_i)+\mathrm{v}_{\mathrm{p}}(t_i)$$

where $\mathrm{v}_{\mathrm{p}}(\cdot,\cdot)$ is zero-mean white Gaussian discrete-time noise of variance $R_{\mathrm{p}}$?

(g) Can the same methodology as in (e) and (f) be used if the target's position were again to be tracked by $y(t_i)$, but its *acceleration* were described as first order Gauss–Markov of zero mean and

$$E\{\mathrm{a}(t)\mathrm{a}(t+\tau)\}=\sigma_{\mathrm{a}}{}^2\exp\{-|\tau|/T_{\mathrm{a}}\}?$$

Explain your answer fully.

(h) Return to part (d) and generate the PI controller based on regulation error pseudointegral weighting to cause $y(t_i)$ to track a piecewise constant desired signal $d$, expressing the result in both position and incremental form. Show that unmodeled constant disturbances in the dynamics do not affect steady state tracking errors.

(i) Repeat part (h), using the PI controller based on control difference weighting.

(j) Repeat part (h) using the generic PI controller.

(k) Design the open-loop CGT law designed to cause $y(t_i)$ to track the desired $d$ with the input/output response of a first order lag, $(s+1)^{-1}$, while simultaneously rejecting first order Gauss–Markov $\mathrm{n}_{\mathrm{d}}(\cdot,\cdot)$ in (14-238a) as well as $\mathrm{w}_{\mathrm{d}}(\cdot,\cdot)$ as considered previously, where

$$E\{\mathrm{n}_{\mathrm{d}}(t_i)\mathrm{n}_{\mathrm{d}}(t_j)\}=\exp\{-|t_i-t_j|/2\}$$

(l) Design the CGT/regulator corresponding to part (k).

(m) Design the CGT/PI controller corresponding to part (k).

**14.19** Consider a simple spacecraft attitude control problem in which only planar rigid body dynamics are modeled. Thus, if $\theta$ is the angle that the rigid-body centerline of the spacecraft makes with some inertial coordinate frame, the fundamental system characteristics are described by

$$I\ddot{\theta}(t)=\mathrm{u}(t)+\mathrm{w}(t)$$

where $I$ is the moment of inertia, $u(t)$ is a feedback control torque (as applied by an inertia wheel or attitude control vernier rocket firings), and $w(\cdot,\cdot)$ is zero-mean white Gaussian noise of strength $Q$ meant to model interfering torques and to account for model inadequacy. Assume the spacecraft has an inertial navigation system, from which it can obtain sampled-data measurements of angular orientation every $\Delta t$ sec, as

$$z(t_i) = \theta(t_i) + v(t_i)$$

where $v(\cdot,\cdot)$ is discrete-time zero-mean white Gaussian noise independent of $w(\cdot,\cdot)$, and of variance $R$, describing measurement corruptions. Let $\theta(t_0)$ and $\dot{\theta}(t_0)$ be modeled as independent Gaussian random variables, each independent of $w(\cdot,\cdot)$ and $v(\cdot,\cdot)$, and of means $\bar{\theta}_0$ and 0, respectively, and variances $\sigma_{\theta 0}^2$ and $\sigma_{\dot{\theta} 0}^2$, respectively.

Assume that you wish to develop the sampled-data controller to minimize the cost

$$J_c = E\left\{\frac{1}{2}\,\theta^2(t_{N+1}) + \frac{1}{2}X_{f\dot{\theta}}\dot{\theta}^2(t_{N+1}) + \int_{t_0}^{t_{N+1}} \frac{1}{2}\,[\theta^2(t) + W_{\dot{\theta}}\dot{\theta}^2(t) + W_u u^2(t)]\,dt\right\}$$

(a) Establish the equivalent discrete-time problem formulation via (14-23), (14-25), and (14-27). Write the differential equations, (14-28), that could also be used to obtain this system model.

(b)–(m) Repeat parts (b)–(m) of Problem 14.18, using obvious extensions to the above case to cause $\theta$ to track desired $d$, and letting $n_d(\cdot,\cdot)$ in part (k) enter the dynamics as $I\ddot{\theta}(t) = u(t) + n_d(t) + w(t)$.

**14.20** Consider a system described by the linear second order transfer function model

$$G(s) = \frac{\tau s + 1}{s^2 + 2\zeta\omega_n s + \omega_n{}^2}$$

(noting the special cases of $\tau = 0$, $\zeta = 0$, $\omega_n = 0$, and combinations of these). Let this system be driven by scalar input $[u(t) + w(t)]$, where $u(t)$ is applied control that is to be generated in feedback form, and where $w(\cdot,\cdot)$ is zero-mean white Gaussian noise of strength $Q$, meant to model both disturbances and the uncertainty that the proposed model is a completely adequate depiction. Assume that the system is known to start from rest. Denote the output variable as $y(t)$, and assume that sampled-data noise-corrupted measurements of it are available every $\Delta t$ sec as

$$z(t_i) = y(t_i) + v(t_i)$$

where $v(\cdot,\cdot)$ is zero-mean white Gaussian discrete-time noise of variance $R$. Assume that you wish to develop the sampled-data controller to minimize the cost

$$J_c = E\left\{\frac{1}{2}\,y^2(t_{N+1}) + \int_{t_0}^{t_{N+1}} \frac{1}{2}\,[y^2(t) + W_{uu}u^2(t) + 2W_{yu}x(t)u(t)]\,dt\right\}$$

(a) Establish the equivalent discrete-time problem formulation via (14-23), (14-25), and (14-27). Also write out the corresponding (14-28).

(b)–(m) Repeat parts (b)–(m) of Problem 14.18, using obvious extensions to the above cost to cause the scalar output $y$ to track desired $d$, and letting $n_d(\cdot,\cdot)$ in part (k) enter the dynamics in the same manner as $[u(t) + w(t)]$.

**14.21** A control problem can be depicted as in Fig. 14.P2. The input $y_{com}(t_i)$ is the discrete-time prescheduled commanded time history for the system output $y(t)$ to follow; it can be viewed here as the output of a deterministic stable time-invariant first order linear system with nonzero initial conditions and no inputs. Furthermore, $n(t)$ is a time-correlated Gaussian zero-mean noise disturbance (continuous-time white Gaussian noise really is a mathematical fiction!). To be more specific, let the system characteristic time $T$ be 0.1 sec. and let the autocorrelation of $n(\cdot,\cdot)$ be

$$E\{n(t)n(t + \tau)\} = \exp[-100|\tau|]$$

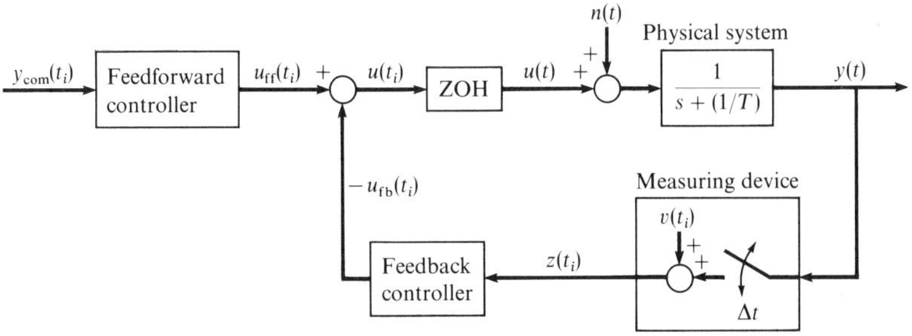


FIG. 14.P2

The system output $y(t)$ is sampled every 0.01 sec by the noise-corrupted measuring device, where $\mathsf{v}(\cdot,\cdot)$ is white Gaussian discrete-time noise, independent of $\mathsf{w}(\cdot,\cdot)$, with statistics

$$E\{\mathsf{v}(t_i)\} = 0, \qquad E\{\mathsf{v}(t_i)^2\} = 1$$

These measurements are then available to a digital feedback controller algorithm used to generate $u_{\text{fb}}(t_i)$, the discrete-time feedback control, for all $t_i$. This can then be combined with the feedforward control $u_{\text{ff}}(t_i)$ to generate the total control $u(t_i)$. A zero-order hold (ZOH) then maintains that value over the next sample period, so that the control $u(t)$ applied to the physical system is $u(t) = u(t_i)$ for $t_i \leq t < t_{i+1}$.

It is desired to design the digital controller so that the output $y(t)$ follows the commands $y_{\text{com}}(t_i)$ as closely as possible over a one-minute period while not expending too much control effort. Therefore, a cost function of the form

$$J = \frac{1}{2} E\left\{ \sum_{i=0}^{6000} [\mathsf{y}(t_i) - y_{\text{com}}(t_i)]^2 + \sum_{i=0}^{5999} U\mathsf{u}(t_i)^2 \right\}$$

has been postulated for the problem, with $U$ initially set at 0.1.

(a) It is claimed that the noise $\mathsf{n}(\cdot,\cdot)$ can be replaced with a white Gaussian noise model without serious inadequacy of the resulting model. Why? What is the strength of the white Gaussian noise to use for this replacement?

(b) What is the equivalent discrete-time model of the physical system assuming the white noise replacement in (a) is valid? Give complete details. (Note: It may be of use to know that $\exp(-0.1) \cong 0.9$; $\int_0^{0.01} \exp(-20t)\,dt \cong 0.009$.)

(c) It is claimed that, to design only the feedback controller (and not the feedforward controller), the specific form of the time history of $y_{\text{com}}$ values is not required. Why? Give the structure of the total controller.

Develop the optimal stochastic feedback controller based upon the model from (b) and the cost specified above. Express the result in the standard form

$$u(t_i) = G_{cx}(t_i)x_c(t_i) + G_{cz}(t_i)z_i$$
$$x_c(t_{i+1}) = \Phi_c(t_{i+1}, t_i)x_c(t_i) + B_{cz}(t_i)z_i$$

Why is this standard form of importance?

(d) An engineer claims that the design will be simpler, and almost equivalent in performance, if the model for the physical system is replaced by a simple integrator. Is this a reasonable model? Look at the steady state controller for $1/T$ equal to 10 versus 0, and compare results. What conditions on $1/T$ would make his proposal reasonable?

(e) Explicitly evaluate all quantities required in performance analysis of this steady state controller based on an integrator model of the physical plant. Let the truth model be defined using the white noise approximation of parts (a) and (b). Do not perform any complicated computations; just define the components necessary for the analysis and tell how they would be obtained if they are not evaluated explicitly.

(f) It is claimed that a more generalized cost function might be appropriate for this sampled-data controller design than the cost $J$ defined above. Why? How does inclusion of such generalization affect the evaluation of controller gains?

(g) It is claimed that designing the controller with a time-correlated model for $\mathrm{n}(\cdot,\cdot)$ instead of a white model will not affect the design of the feedback controller designed in preceding parts. Draw a block diagram of the controller structure, and tell why this is, or is not, a valid claim. Do you need measurements of additional quantities? Do filter and deterministic controller structures both behave in a manner consistent with this claim? Are approximations required?

**14.22** Consider generalizations of the trackers described in Section 14.6.

(a) Show that a reference variable model

$$\mathbf{y}_r(t_i) = \mathbf{C}_r(t_i)\mathbf{x}_r(t_i) + \mathbf{D}_r(t_i)\mathbf{u}(t_i)$$

instead of (14-134b) can be handled by $\mathbf{S}(t_i) \neq \mathbf{0}$ in the cost definition. Does this make sense conceptually?

(b) Show that $\mathbf{w}_{dr}(\cdot,\cdot)$ of (14-134) and (14-135) *can* be correlated with $\mathbf{w}_d(\cdot,\cdot)$ of the basic system dynamics, with *no* effect on the full-state feedback controller. Show that this generalization *does* affect the structure of the controller that inputs noisy measurements instead of perfect knowledge of all states.

(c) Show that a reference variable model

$$\mathbf{y}_r(t_i) = \mathbf{C}_r(t_i)\mathbf{x}_r(t_i) + \mathbf{D}_{rw}(t_i)\mathbf{w}_{dr}(t_i)$$

can be used to replace (14-134b) in the tracker development, and display both the full-state feedback case and the noisy measurement feedback case.

(d) Show the effect of replacing (14-148a) with (14-32).

(e) Show the effect of letting $\mathbf{v}(\cdot,\cdot)$ and $\mathbf{v}_r(\cdot,\cdot)$ be correlated in (14-148c).

(f) Consider the special case of noisy system measurements (14-148a) being available, but in which reference variables $\mathbf{y}_r(t_i)$ are known perfectly instead of through (14-148b).

**14.23** Repeat Examples 14.5 and 14.6 for the case of tracking a prespecified deterministic input, by letting $\mathrm{w}_r(\cdot,\cdot) \equiv 0$ and initial $x_r(t_0)$ be a deterministic nonzero value. This provides some basic insights into command generator tracking. In Example 14.6, is it advisable to set $\sigma_r^2$ to zero as done in Example 14.5? Discuss completely.

**14.24** (a) To understand right inverses better, consider the simple example of $\mathbf{Ax} = \mathbf{b}$ given by

$$\begin{bmatrix} 1 & 1 \end{bmatrix} \begin{bmatrix} x_1 \\ x_2 \end{bmatrix} = 1$$

Evaluate the Euclidean length and generalized norm length of $\mathbf{x}$ for the cases of

$$\mathbf{W} = \begin{bmatrix} 1 & 0 \\ 0 & 1 \end{bmatrix}, \qquad \mathbf{W}' = \begin{bmatrix} 2 & 0 \\ 0 & 1 \end{bmatrix}$$

$$\mathbf{W}'' = \begin{bmatrix} 1 & \frac{1}{2} \\ \frac{1}{2} & 1 \end{bmatrix}, \qquad \mathbf{W}''' = \begin{bmatrix} 2 & \frac{1}{2} \\ \frac{1}{2} & 1 \end{bmatrix}$$

Also evaluate $\sqrt{\mathbf{W}}^{\mathrm{T}}\mathbf{x}$, $\mathbf{A}^{\mathrm{RMW}}$, and $\mathbf{x} = \mathbf{A}^{\mathrm{RMW}}\mathbf{b}$ for each case. Show that, by letting $W_{ii}$ be larger, $x_i$ will be smaller.

(b) To understand left inverses better, compare (14-159) to the development of (3-145)–(3-151) in Volume 1.

**14.25** Another, algebraically equivalent, means of expressing the nonzero setpoint controller (14-166b), can be developed [92]. Let (14-166b) be written structurally as

$$\mathbf{u}^*(t_i) = -\bar{\mathbf{G}}_c{}^*\mathbf{x}(t_i) + \mathbf{u}_0{}'$$

in the case of steady state constant gain $\bar{\mathbf{G}}_c{}^*$ being used, where $\mathbf{u}_0{}'$ is a constant to be determined. Consider the case of $\mathbf{D}_y \equiv \mathbf{0}$ in (14-152). Substitute the $\mathbf{u}^*(t_i)$ above into the dynamics to obtain steady state conditions

$$[(\mathbf{\Phi} - \mathbf{B}_d\bar{\mathbf{G}}_c{}^*) - \mathbf{I}]\mathbf{x}_0 + \mathbf{B}_d\mathbf{u}_0{}' = \mathbf{0}$$

How are you sure a steady state condition exists? Why is the matrix in brackets invertible? Show that this result combined with $\mathbf{y}_d = \mathbf{C}\mathbf{x}_0$ yields, assuming the dimension of $\mathbf{y}_c(t_i)$ equals that of $\mathbf{u}(t_i)$,

$$\mathbf{u}_0{}' = \{\mathbf{C}[\mathbf{I} - (\mathbf{\Phi} - \mathbf{B}_d\bar{\mathbf{G}}_c{}^*)]^{-1}\mathbf{B}_d\}^{-1}\mathbf{y}_d$$

if the outer inverse exists. Note that the *closed-loop* transfer function from $\mathbf{u}_0{}'$ to $\mathbf{y}_c$ (see Fig. 14.9) is

$$\mathbf{T}_{\mathrm{CL}}(z) = \mathbf{C}[z\mathbf{I} - (\mathbf{\Phi} - \mathbf{B}_d\bar{\mathbf{G}}_c{}^*)]^{-1}\mathbf{B}_d$$

so that $\mathbf{u}_0{}' = \mathbf{T}_{\mathrm{CL}}(z = 1)^{-1}\mathbf{y}_d$. It can be shown [92] that $\mathbf{T}_{\mathrm{CL}}(z = 1)$ has an inverse if and only if the *open-loop* transfer function matrix $\mathbf{T}_{\mathrm{OL}}(z)$ associated with the original system,

$$\mathbf{T}_{\mathrm{OL}}(z) = \mathbf{C}[z\mathbf{I} - \mathbf{\Phi}]^{-1}\mathbf{B}_d$$

has a nonzero numerator polynomial with no zeros at $z = 1$. Express the final result in a block diagram as in Fig. 14.9b, followed by a multiplication by $\mathbf{C}$ to generate $\mathbf{y}_c(t)$. Use this block diagram and the final value theorem for $z$-transforms,

$$\lim_{i\to\infty} \mathbf{y}(t_i) = \lim_{z\to 1} \{[z - 1]\mathbf{y}(z)\}$$

to show that steady state $\mathbf{y}_c(t_i)$ is in fact $\mathbf{y}_d$. (Note that the $z$-transform of a constant $\mathbf{y}_d$ for all time is given by $z\,\mathbf{y}_d/[z - 1]$.)

**14.26** Develop the stochastic optimal regulator that specifically rejects a deterministic disturbance as in (14-170) but with $\mathbf{Q}_{dn} \equiv \mathbf{0}$ (with nonzero initial condition).

**14.27** (a) Develop in explicit detail the controller for Example 14.8 that inputs noise-corrupted measurements

$$\mathbf{z}_1(t_i) = \mathbf{H}(t_i)\mathbf{x}(t_i) + \mathbf{v}(t_i), \qquad \mathbf{z}_2(t_i) = \mathbf{H}_n(t_i)\mathbf{x}_n(t_i) + \mathbf{v}_n(t_i)$$

(b) Repeat when only $\mathbf{z}_1(t_i)$ measurements are available.

(c) Repeat for the case of measurements

$$\mathbf{z}_3(t_i) = \mathbf{H}(t_i)\mathbf{x}(t_i) + \mathbf{H}_n(t_i)\mathbf{x}_n(t_i) + \mathbf{v}(t_i)$$

being available only.

**14.28** Consider the use of a control of the form

$$\mathbf{u}(t_i) = -\bar{\mathbf{G}}_c{}^*\mathbf{x}(t_i) + \mathbf{u}_0$$

to suppress the effects of the constant disturbance $\mathbf{d}$ in

$$\mathbf{x}(t_{i+1}) = \mathbf{\Phi}\mathbf{x}(t_i) + \mathbf{B}_d\mathbf{u}(t_i) + \mathbf{d}$$

and regulate $\mathbf{y}_c(t_i) = \mathbf{C}\mathbf{x}(t_i)$ to zero. Let the dimension of $\mathbf{y}_c(t_i)$ equal the dimension of $\mathbf{u}(t_i)$. Find the conditions of equilibrium steady state (how are you assured such a steady state condition exists?) and show that they can be expressed as

$$\lim_{i\to\infty} \mathbf{y}_c(t_i) = \mathbf{C}[\mathbf{I} - (\mathbf{\Phi} - \mathbf{B}_d\bar{\mathbf{G}}_c{}^*)]^{-1}\mathbf{B}_d\mathbf{u}_0 + \mathbf{C}[\mathbf{I} - (\mathbf{\Phi} - \mathbf{B}_d\bar{\mathbf{G}}_c{}^*)]^{-1}\mathbf{d}$$

Solve for the $\mathbf{u}_0$ that yields steady state $\mathbf{y}_c$ of zero. Using the results of Problem 14.25, give the necessary and sufficient conditions for the existence of $\mathbf{u}_0$. Combine this with the results of Section 14.4 to state the necessary and sufficient conditions for a law of this proposed form to be asymptotically stabilizing, allowing $\bar{\mathbf{G}}_c{}^*$ to be generalized to arbitrary constant $\mathbf{G}_c$. Similarly state the conditions under which the poles of the closed-loop system can be arbitrarily placed with such a law. Note that the augmentation of states in the PI controllers allows such a constant $\mathbf{u}_0$ to be developed automatically to counteract the effect of $\mathbf{d}$, but that it does so without requiring perfect system modeling (never really achievable) to drive steady state $\mathbf{y}_c$ to zero.

**14.29** The PI controllers of Section 14.9 were designed for systems modeled by (14-188), where the disturbance $\mathbf{d}$ was assumed unknown and unmodeled in the design synthesis. Show that *if you do know* $\mathbf{d}$ a priori (a restrictive case), that a different equilibrium solution would be appropriate, namely

$$\mathbf{x}_0 = \mathbf{\Pi}_{12}\mathbf{y}_d - \mathbf{\Pi}_{11}\mathbf{d}, \qquad \mathbf{u}_0 = \mathbf{\Pi}_{22}\mathbf{y}_d - \mathbf{\Pi}_{21}\mathbf{d}$$

**14.30** (a) Develop the control law (14-203) in explicit detail, demonstrating that the algebraic Riccati equation for the augmented system does not decouple.

(b) Similarly develop (14-219) in detail, showing lack of decoupling of the Riccati equation as well as its exact structural form.

**14.31** (a) In Example 14.10, demonstrate the validity of the displayed closed-loop system descriptions for the position and incremental form laws. Also show the validity of the condition for closed-loop stability.

(b) In Example 14.11, show that the given closed-loop system description is valid.

(c) In Example 14.12, show the closed-loop system description for the position form law is correct. Develop the same description for the incremental form law, and show the eigenvalues of this system description to be the same as for the case of the position form law, plus an additional eigenvalue at zero.

(d) Extend these results to Example 14.13.

**14.32** Show that the position form law (14-231) satisfies the incremental law (14-228).

**14.33** Develop the proportional plus integral plus filter (PIF) control law described at the end of Section 14.9.

(a) First write the description of the system in terms of the augmented state vector $[\mathbf{x}^T(t_i) \vdots \mathbf{u}^T(t_i) \vdots \mathbf{q}^T(t_i)]^T$.

(b) Shift the coordinate origin and develop the perturbation state equations for the augmented system description.

(c) Specifically define the cost function to be minimized (allowing all possible cross-weightings, and choosing the lower limit on the summation carefully) to yield a solution of the form

$$\Delta\mathbf{u}^*(t_i) = -[\bar{\mathbf{G}}_{c1}^* \quad \bar{\mathbf{G}}_{c2}^* \quad \bar{\mathbf{G}}_{c3}^*]\begin{bmatrix}\mathbf{x}(t_i)\\ \mathbf{u}(t_i)\\ \mathbf{q}(t_i)\end{bmatrix}$$

and evaluate all gains explicitly.

(d) Show that the best choice of nominal $\mathbf{q}_0$, best in the sense of yielding minimal total cost to complete the process, is

$$\mathbf{q}_0 = -\bar{\mathbf{K}}_{c33}^{-1}[\bar{\mathbf{K}}_{c23}^{T}\mathbf{\Pi}_{22} + \bar{\mathbf{K}}_{c13}^{T}\mathbf{\Pi}_{12}]\mathbf{y}_d$$

using $\mathbf{\Pi}$ as in Sections 14.7 and 14.9, and partitioning of $\mathbf{K}_c$ in obvious correspondence to those in Section 14.9.

(e) Show that the position form PIF law is

$$\mathbf{u}(t_i) = \mathbf{u}(t_{i-1}) - \bar{\mathbf{G}}_{c1}^{*}\mathbf{x}(t_{i-1}) - \bar{\mathbf{G}}_{c2}^{*}\mathbf{u}(t_{i-1}) - \bar{\mathbf{G}}_{c3}^{*}\mathbf{q}(t_{i-1}) + \mathbf{E}'\mathbf{y}_d(t_i)$$
$$\mathbf{q}(t_i) = \mathbf{q}(t_{i-1}) + [\mathbf{C}\mathbf{x}(t_{i-1}) + \mathbf{D}_y\mathbf{u}(t_{i-1}) - \mathbf{y}_d(t_i)]$$
$$\mathbf{E}' = \bar{\mathbf{G}}_{c1}^{*}\mathbf{\Pi}_{12} + \bar{\mathbf{G}}_{c2}^{*}\mathbf{\Pi}_{22} - \bar{\mathbf{G}}_{c3}^{*}\bar{\mathbf{K}}_{c33}^{-1}[\bar{\mathbf{K}}_{c13}^{T}\mathbf{\Pi}_{12} + \bar{\mathbf{K}}_{c23}^{T}\mathbf{\Pi}_{22}]$$

and specifically argue that the time indices on $\mathbf{y}_d$ in the first two equations are correct. Draw a block diagram of the structure of this law.

(f) Show that the preferred incremental form PIF law is

$$\mathbf{u}(t_i) = \mathbf{u}(t_{i-1}) + \Delta\mathbf{u}(t_{i-1})$$
$$\Delta\mathbf{u}(t_i) = [\mathbf{I} - \bar{\mathbf{G}}_{c2}^{*}]\,\Delta\mathbf{u}(t_{i-2}) - \bar{\mathbf{G}}_{c1}^{*}[\mathbf{x}(t_{i-1}) - \mathbf{x}(t_{i-2})]$$
$$+ \bar{\mathbf{G}}_{c3}^{*}[\mathbf{y}_d(t_{i-1}) - \mathbf{y}_c(t_{i-2})] + \mathbf{E}'[\mathbf{y}_d(t_i) - \mathbf{y}_d(t_{i-1})]$$

(g) Show (heuristically) that the PIF has the type-1 property provided that $\det\{\bar{\mathbf{G}}_{c3}^{*}\} \neq 0$ and $\det\{\bar{\mathbf{G}}_{c2}^{*}\} \neq 0$ (the latter is not a necessary condition, but the two together provide sufficient conditions).

(h) Demonstrate the ability of the PIF law to reject constant unmodeled disturbances in the dynamics.

(i) Develop the PIF control law for the same problem as investigated in Examples 14.10–14.13 of Section 14.9. Specify the position form law, show that steady state $y_c(t_i) = y_d$ despite unmodeled constant disturbances in dynamics, and display the characteristic equation to yield the closed-loop system eigenvalues. Repeat for the incremental form law.

(j) Relate the laws of parts (e) and (f) to the generic form controller gain evaluations, as done in Section 14.9.

**14.34** (a) Develop in explicit detail the PI controllers as in Section 14.9, but for the case in which perfect knowledge of $\mathbf{x}(t_i)$ and $\mathbf{y}_c(t_i)$ is replaced by the generation of $\hat{\mathbf{x}}(t_i^+)$ and $\hat{\mathbf{y}}_c(t_i^+)$ by means of a Kalman filter, recalling (14-189). Draw the appropriate block diagrams for position and incremental form laws.

(b) Develop the suboptimal law in which $\hat{\mathbf{x}}(t_i^-)$ and $\hat{\mathbf{y}}_c(t_i^-)$ replace $\mathbf{x}(t_i)$ and $\mathbf{y}_c(t_i)$, respectively. Demonstrate the savings in computational delay, compared to the controllers of part (a).

(c) Show the problems associated with developing Kalman filters corresponding to the augmented system descriptions: consider the observability of regulator error pseudointegral states, and the controllability of augmented control states from the points of entry of the dynamics driving noise. Since these are artificial states for design purposes, the ability to generate a stable filter for the augmented state description is actually of secondary importance, in view of parts (a) and (b).

**14.35** (a) Consider Example 14.17. Generate the closed-loop system descriptions for Examples 14.14–14.16, and show that the results claimed in Example 14.17 are valid, specifically that (14-257) is satisfied in all cases.

(b) Rework Example 14.17 for the incremental form regulator law (14-258), and compare results.

(c) Show that the stochastic CGT/regulator law can be generated from (14-256) or (14-258), and by replacing $\mathbf{x}(t_i)$ by $\hat{\mathbf{x}}(t_i^+)$. Why might you choose to use $\hat{\mathbf{x}}(t_i^-)$ instead? Show this form.

(d) Rework Example 14.17 with the controllers derived in part (c).

**14.36** Show that the entire restart implementation discussed in Section 14.10 to avoid inconsistency in definition, and poor transient performance in practice, simply entails implementation of (14-262) instead of (14-240a).

**14.37** (a) Section 14.10 developed the CGT/PI controller based on control difference weighting. Develop the CGT/PI controller instead based upon regulation error pseudointegral weighting.

(b) Develop the generic form of CGT/PI controller.

(c) Develop the CGT/PIF controller (see Problem 14.33).

**14.38** (a) Consider Example 14.18. Develop the error dynamics equation displayed for the CGT/PI law in position form. Develop the same type of description for the incremental form law, and explicitly compare closed-loop eigenvalues to those of the position form.

(b) Repeat part (a) for Example 14.19 (both applications).

**14.39** At the end of Section 14.5, it was mentioned that purposeful limiting of controller bandwidth is often accomplished during or after controller design synthesis, in order to address the inadequacy of assumed linear models beyond a specific frequency range (a robustness issue). This problem explores direct incorporation of such a desirable characteristic into the design of a CGT/regulator or CGT/PI control law.

(a) Reconsider Examples 14.16, 14.17, and 14.19, in which the original system, modeled as a first order lag with lag coefficient $(1/T) = 2$ rad/sec (see Example 14.2), was to be caused to behave like a second order system model. Now assume we wish it to respond as though it were a first order lag with more rapid response, i.e., with a model lag coefficient of 3 rad/sec. But, because we question adequacy of the original system description at higher frequencies, we want additional high frequency rolloff; therefore, the command generator is the cascade of two first order lags, with lag coefficients of 3 and 5 rad/sec, respectively. Develop the equivalent discrete-time command generator, the open-loop CGT law, CGT/regulator, and CGT/PI controller.

(b) As an alternative to part (a), let the command generator model be the equivalent discrete-time model for a single first order lag with lag coefficient of 3 rad/sec. Now incorporate the desire for additional rolloff into the disturbance model, whose outputs are to be rejected. Add to the original disturbance model, the equivalent discrete-time model for the output of a lead/lag filter, $[(s + 5)/(s + 50)]$, driven by white Gaussian noise. Demonstrate the effect of $\mathbf{E}_x = \mathbf{0}$ versus $\mathbf{E}_y = \mathbf{0}$ in (14-238). Compare these designs to that of part (a), characterizing open-loop and closed-loop properties and complexity.

(c) Repeat part (a) for the case of $x(t_i)$ and $n_d(t_i)$ being replaced by $\hat{x}(t_i^+)$ and $\hat{n}_d(t_i^+)$ from a Kalman filter based on measurements of the form

$$\mathsf{z}(t_i) = \mathsf{x}(t_i) + \mathsf{v}(t_i)$$

with $\mathsf{v}(\cdot,\cdot)$ discrete-time zero-mean white Gaussian noise of variance $R$.

(d) Repeat part (b), but this time use $\hat{x}(t_i^+)$ and $\hat{\mathbf{n}}_d(t_i^+)$ from the appropriate Kalman filter based on the measurement model of part (c). Compare these results to those of part (c).

**14.40** In the text it was suggested that a control of the form $\mathbf{u}(t_i) = -\mathbf{G}_c^*(t_i)\hat{\mathbf{x}}(t_i^-)$ be used instead of the LQG optimal control $\mathbf{u}^*(t_i) = -\mathbf{G}_c^*(t_i)\hat{\mathbf{x}}(t_i^+)$ due to the computational delay in the latter form; this alteration is in fact often made in practice. Section 14.11 also discussed appropriate ordering of computations and choice of state variables to minimize the delay between sample time $t_i$ and the generation of $\mathbf{u}(t_i)$ itself.

Assume that the calculations required to implement the LQG optimal control, even in the most efficient form, take $\Delta$ sec, where $\Delta$ is not negligible compared to the sample period, $[t_{i+1} - t_i] =$ const for all $i$. Thus, a measurement, $\mathbf{z}(t_i, \omega_k) = \mathbf{z}_i$ is taken at time $t_i$, but the LQG control (to be held constant over the next sample period) cannot be applied until $t_i + \Delta$, as seen in Fig. 14.P3. If the standard LQG control $\mathbf{u}^*(t_i)$ were not actually applied until time $(t_i + \Delta)$, the induced lag could yield significantly degraded performance, even to the extent of causing destabilization.

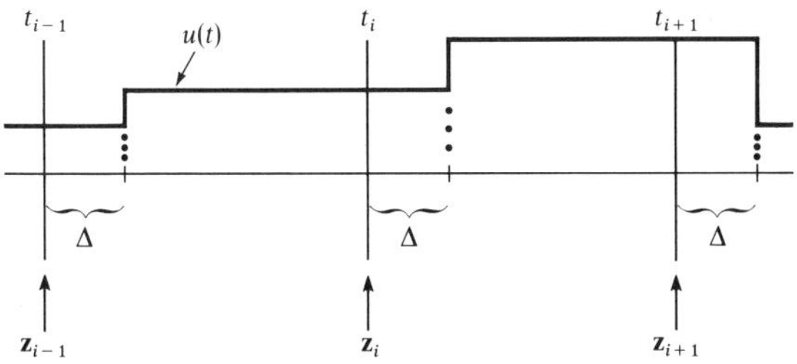


FIG. 14.P3

Assume Δ can be evaluated by analysis of the online controller algorithm, accounting for time for additions, multiplications, overhead, etc., in the computer.

(a) Develop the details of implementing instead a control law

$$\mathbf{u}(t_i + \Delta) = -\mathbf{G}_c^*(t_i + \Delta)\hat{\mathbf{x}}(t_i + \Delta^-)$$

where $\hat{\mathbf{x}}(t_i + \Delta^-)$ is the optimal prediction of the $\mathbf{x}(t_i + \Delta)$ based on the measurement history through $\mathbf{z}(t_i, \omega_k) = \mathbf{z}_i$. How would you obtain $\mathbf{G}_c^*(t_i + \Delta)$? Express the result straightforwardly as a measurement update at $t_i$, propagation to $(t_i + \Delta)$, calculation and application of $\mathbf{u}(t_i + \Delta)$, and propagation to next sample time $t_{i+1}$.

(b) Now we want to minimize the number of operations between $t_i$ and $(t_i + \Delta)$, and then allow all other required calculations to "set up" for the measurement sample at $t_{i+1}$ to be performed "in the background" between $(t_i + \Delta)$ and $t_{i+1}$. Is $\mathbf{P}(t_i + \Delta^-)$ needed explicitly? Show an equivalent, efficient control algorithm in the form of (14-274), letting $\mathbf{x}_c(t_i + \Delta) = E\{\mathbf{x}(t_i + \Delta) | \mathbf{Z}(t_{i-1}) = \mathbf{Z}_{i-1}\}$.

(c) As an alternate to (b), show that the computations in (a) can simply be rearranged, recognizing which calculations can be performed before $\mathbf{z}_i$ becomes available at time $t_i$. First show that $\hat{\mathbf{x}}(t_i + \Delta^-)$ can be written in the form

$$\hat{\mathbf{x}}(t_i + \Delta^-) = \hat{\mathbf{x}}_h(t_i + \Delta^-) + \mathbf{\Phi}(t_i + \Delta, t_i)\mathbf{K}(t_i)\mathbf{z}_i$$

where the first "homogeneous" term can be evaluated without knowledge of $\mathbf{z}_i$. Explicitly evaluate $\hat{\mathbf{x}}_h(t_i + \Delta^-)$. Now, assuming that $\mathbf{u}^*(t_{i-1} + \Delta)$ has just been evaluated and applied, and that $\hat{\mathbf{x}}_h(t_{i-1} + \Delta^-)$ is available from the previous iteration period, show that the setup in the background between $(t_{i-1} + \Delta)$ and $t_i$ can be expressed by explicit equations for $\hat{\mathbf{x}}(t_i^-)$, $\mathbf{P}(t_{i-1}^+)$, $\mathbf{P}(t_i^-)$, $\mathbf{K}(t_i)$, $\hat{\mathbf{x}}_h(t_i + \Delta^-)$, $\mathbf{u}_h^*(t_i + \Delta) = -\mathbf{G}_c^*(t_i + \Delta)\mathbf{x}_h(t_i + \Delta^-)$, and $\mathbf{G}_{cz}(t_i + \Delta)$. At sample time $t_i$, $\mathbf{z}_i$ becomes available, and $\mathbf{u}^*(t_i + \Delta)$ is calculated as

$$\mathbf{u}^*(t_i + \Delta) = \mathbf{u}_h^*(t_i + \Delta) + \mathbf{G}_{cz}(t_i + \Delta)\mathbf{z}_i$$

(d) Show that if part (b) is accomplished, but with $\mathbf{x}_c(t_i + \Delta) = E\{\mathbf{x}(t_i + \Delta) | \mathbf{Z}(t_i) = \mathbf{Z}_i\}$, the Kalman filter must be based on a measurement model

$$\mathbf{z}(t_i) = \mathbf{H}'(t_i)\mathbf{x}(t_{i-1} + \Delta) + \mathbf{D}_z'(t_i)\mathbf{u}(t_{i-1} + \Delta) + \mathbf{v}'(t_i)$$

where $\mathbf{v}'(\cdot,\cdot)$ is correlated with dynamics driving noise $\mathbf{w}(\cdot,\cdot)$. Generate the resulting controller in detail, and compare the complexity to that of previous controllers.

**14.41** Consider Example 14.22. Explicitly generate the LQG control law. Apply the transformation of (10-74)–(10-84) to express the controller in the form of (14-274), as done in that example.

**14.42** Derive the results of Example 14.20 for controllers in the generic form of (14-274):

(a) LQG full-state feedback regulator.

(b) LQG optimal control $\mathbf{u}^*(t_i) = -\mathbf{G}_c^*(t_i)\hat{\mathbf{x}}(t_i^+)$.

(c) suboptimal law $\mathbf{u}(t_i) = -\mathbf{G}_c^*(t_i)\hat{\mathbf{x}}(t_i^-)$.

(d) generic PI controller with perfect knowledge of full state; special cases of PI controller based on regulation error pseudointegral weighting and PI controller based on control difference weighting.

(e) generic PI controller with $\hat{\mathbf{x}}(t_i^+)$ replacing $\mathbf{x}(t_i)$; special cases as in (d).

(f) generic PI controller with $\hat{\mathbf{x}}(t_i^-)$ replacing $\mathbf{x}(t_i)$; special cases as in (d).

(g) CGT/regulator based on full state knowledge at each $t_i$.

(h) CGT/regulator based on $\hat{\mathbf{x}}(t_i^+)$ from filter.

(i) CGT/regulator based on $\hat{\mathbf{x}}(t_i^-)$ from filter.

(j) CGT/PI controller based on knowledge of $\mathbf{x}(t_i)$.

(k) CGT/PI controller based on $\hat{\mathbf{x}}(t_i^+)$ from filter.

(l) CGT/PI controller based on $\hat{\mathbf{x}}(t_i^-)$ from filter.

**14.43** Develop the controller performance analysis algorithm as in Section 14.11, but for the generalized case of $\mathbf{w}_{dt}(\cdot,\cdot)$ and $\mathbf{v}_t(\cdot,\cdot)$ being correlated (see (14-281)).

**14.44** Develop an efficient software implementation of the performance evaluation algorithm of Section 14.11. Generate both a Monte Carlo tool that admits a nonlinear truth model, and a mean and covariance analysis package based on a linear truth model assumption.

**14.45** A control problem can be depicted as in the Fig. 14.P4.

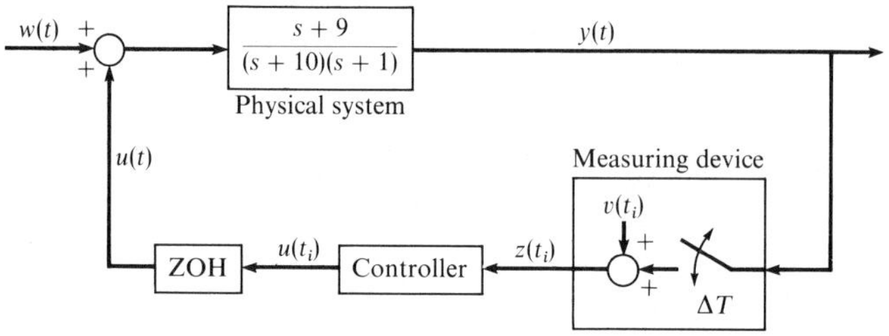


FIG. 14.P4

The system to be controlled can be described through a linear time-invariant model, with a frequency-domain transfer function of

$$T(s) = \frac{y(s)}{u(s) + w(s)} = \frac{s + 9}{(s + 10)(s + 1)}$$

It is driven by white Gaussian noise, $\mathrm{w}(\cdot,\cdot)$, with statistics

$$E\{\mathrm{w}(t)\} = 0, \qquad E\{\mathrm{w}(t)\mathrm{w}(t + \tau)\} = 2.5\,\delta(\tau)$$

A measuring device samples the system output $y(t)$ every 0.5 sec ($\Delta T = 0.5$ sec), and provides measurements at this sample rate of the form

$$\mathrm{z}(t_i) = \mathrm{y}(t_i) + \mathrm{v}(t_i)$$

where $\mathrm{v}(\cdot,\cdot)$ is discrete-time white Gaussian noise, independent of $\mathrm{w}(\cdot,\cdot)$ for all time, with statistics

$$E\{\mathrm{v}(t_i)\} = 0, \qquad E\{\mathrm{v}^2(t_i)\} = 1$$

These measurements are then available to a digital controller algorithm, to be designed, which then generates $u(t_i)$ based on knowledge of the time history of measurement values. A zero-order hold (ZOH) then maintains that value over the next sample interval, so that

$$u(t) = u(t_i) \qquad \text{for} \quad t_i \le t < t_{i+1}$$

It is desired to build a digital controller that will perform well over a time interval of 10 min, with the measure of performance being a small value of

$$J = E\left\{\frac{1}{2}\sum_{i=0}^{1200} \mathrm{y}(t_i)^2\right\}$$

The system is known to start from rest.

To generate a simple controller design, it has been suggested to build an optimal LQG controller based upon a system model of

$$T'(s) = \frac{y(s)}{u(s) + w'(s)} = \frac{1}{s+1}$$

where $\mathrm{w}'(\cdot,\cdot)$ is white Gaussian noise with statistics

$$E\{\mathrm{w}'(t)\} = 0, \qquad E\{\mathrm{w}'(t)\mathrm{w}'(t+\tau)\} = 3\,\delta(\tau)$$

(a) Why might that particular model be proposed as a good approximation?

(b) Develop the optimal stochastic controller based upon this model, and express it in the standard form of

$$u(t_i) = G_{cx}(t_i)x_c(t_i) + G_{cz}(t_i)z_i$$
$$x_c(t_{i+1}) = \Phi_c(t_{i+1}, t_i)x_c(t_i) + B_{cz}(t_i)z_i$$

Note: It may be of use to know that the inverse Laplace transform of $1/(s+1)$ is $e^{-t}$ and that $e^{-0.5} \cong 0.61$, $\int_0^{0.5} e^{-2\tau}\,d\tau \cong 0.316$.

(c) Have you arrived at a time-invariant controller? If not, would a time-invariant approximation be reasonable? Why? What would this design be? Give details and complete evaluation.

(d) Explicitly evaluate all quantities required in a performance analysis of this controller: specify (in as complete detail as possible) the algorithm to evaluate the mean quadratic cost of operating time-invariant controller from part (c) with the actual physical system. How would you determine RMS values of $\mathrm{y}$, $\dot{\mathrm{y}}$, and $\mathrm{u}$?

Note: Part of the solution entails developing an equivalent discrete-time model. First generate the standard controllable state space description of the system modeled in the frequency domain by $T(s)$. Describe the computations to generate the equivalent discrete-time model from this form. When such computations are carried out, the result is

$$\begin{bmatrix} \mathrm{x}_1(t_i+0.5) \\ \mathrm{x}_2(t_i+0.5) \end{bmatrix} = \begin{bmatrix} 0.677 & 0.067 \\ -0.670 & -0.060 \end{bmatrix}\begin{bmatrix} \mathrm{x}_1(t_i) \\ \mathrm{x}_2(t_i) \end{bmatrix} + \begin{bmatrix} 0.032 \\ 0.067 \end{bmatrix}\mathrm{u}(t_i) + \begin{bmatrix} \mathrm{w}_{d1}(t_i) \\ \mathrm{w}_{d2}(t_i) \end{bmatrix}$$

$$\mathrm{z}(t_i) = \begin{bmatrix} 9 & 1 \end{bmatrix}\begin{bmatrix} \mathrm{x}_1(t_i) \\ \mathrm{x}_2(t_i) \end{bmatrix} + \mathrm{v}(t_i)$$

with $\mathbf{w}_d(\cdot,\cdot)$ a zero-mean white Gaussian sequence with covariance

$$\mathbf{Q}_d(t_i) = \begin{bmatrix} 0.0057 & 0.0056 \\ 0.0056 & 0.1082 \end{bmatrix}$$

**14.46** Consider the system configuration in Fig. 14.P5, where the control $u(t)$ is a constant input for all time, $u(t) = 2$. The three independent noise sources are all zero-mean, Gaussian, with autocorrelation kernels

$$E\{w_1(t)w_1(t+\tau)\} = 1\,\delta(\tau)$$
$$E\{w_2(t)w_2(t+\tau)\} = 3\,\delta(\tau)$$
$$E\{n(t)n(t+\tau)\} = 4e^{-2|\tau|}$$

$y(t)$ is known to start at rest.

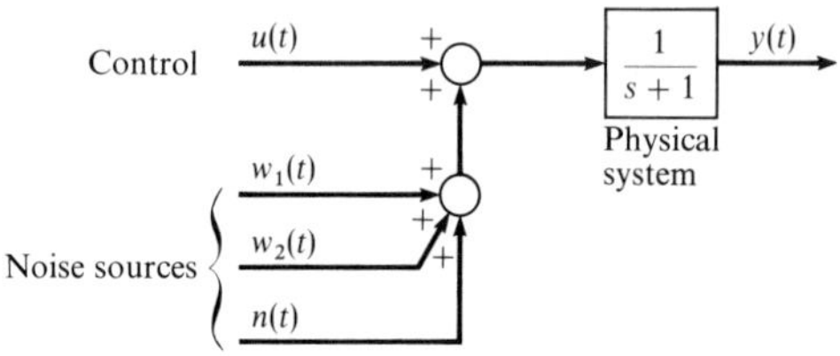


FIG. 14.P5

(a) Show that an equivalent system model is given by Fig. 14.P6, and evaluate $A$. Show *all reasoning* that this is in fact equivalent.

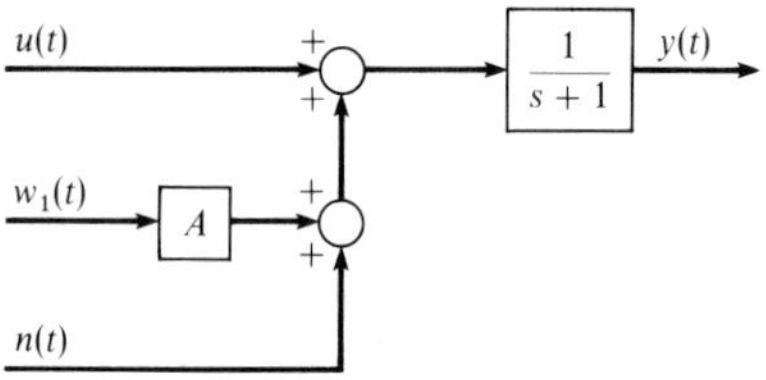


FIG. 14.P6

(b) Completely describe the shaping filter to generate $n(\cdot,\cdot)$.

(c) Evaluate the *correlation kernel* of the output $y(\cdot,\cdot)$: $E\{y(t)y(t+\tau)\}$.

(d) Suppose you had discrete-time, noise-corrupted measurements of $y$:

$$z(t_i) = y(t_i) + v(t_i)$$

where $v(\cdot,\cdot)$ is discrete-time white Gaussian noise, independent of the state or y processes, with statistics (for all sample times $t_i$ and $t_j$):

$$E\{v(t_i)\} = 1$$

$$E\{v(t_i)v(t_j)\} = \begin{cases} 2 & \text{for } t_i = t_j \\ 1 & \text{for } t_i \neq t_j \end{cases}$$

Evaluate the *mean*, *variance kernel*, and *correlation kernel* for z.

(e) An engineer has suggested a simpler model to use as an adequate approximation. He suggests replacing $n(t)$ by a white Gaussian zero-mean noise, and then combining this with the $w_1(t)$ and $w_2(t)$ in the original system diagram. Why might this be a reasonable approximation? What should be the strength of the white Gaussian noise to replace $n(t)$? (Suggestion: Consider the power spectral density description for $\mathsf{n}(t)$.) Write out the overall *system model* for $\mathsf{y}(t)$ and $\mathsf{z}(t_i)$ based on this approximation.

**14.47** One means of reducing controller dimensionality and complexity is to exploit *weak coupling* through *nonsingular perturbations* to yield approximate decoupled system representations, with independent controllers for each of the decoupled subsystems. For instance, consider

$$\begin{bmatrix} \dot{\mathbf{x}}_1(t) \\ \dot{\mathbf{x}}_2(t) \end{bmatrix} = \begin{bmatrix} \mathbf{F}_{11} & \varepsilon\mathbf{F}_{12} \\ \varepsilon\mathbf{F}_{21} & \mathbf{F}_{22} \end{bmatrix} \begin{bmatrix} \mathbf{x}_1(t) \\ \mathbf{x}_2(t) \end{bmatrix} + \begin{bmatrix} \mathbf{B}_{11} & \mathbf{0} \\ \mathbf{0} & \mathbf{B}_{22} \end{bmatrix} \begin{bmatrix} \mathbf{u}_1(t) \\ \mathbf{u}_2(t) \end{bmatrix}$$

where $\varepsilon$ is a small positive parameter. ("Nonsingular perturbations" refers to the fact that $\varepsilon$ only appears on the right hand side of the equation.) If $\varepsilon$ is approximated as zero, the system above decouples completely.

(a) Generate the sampled-data LQ deterministic optimal full-state feedback controller for each subsystem. Develop the steady state constant-gain approximation.

(b) Consider the corresponding sampled-data LQG optimal full-state feedback controller. Specify any additional requirements for the controller design to decouple into two separate controllers. Again attain a constant-gain approximate controller.

(c) Consider the sampled-data LQG optimal stochastic controller based on measurements

$$\mathbf{z}_1(t_i) = \mathbf{H}_1\mathbf{x}_1(t_i) + \mathbf{v}_1(t_i)$$
$$\mathbf{z}_2(t_i) = \mathbf{H}_2\mathbf{x}_2(t_i) + \mathbf{v}_2(t_i)$$

Specify any additional requirements for the controller to decouple into two separate LQG controllers. Can measurements of a more general form be considered? Develop the steady state constant-gain approximate controller design as well.

(d) Set up the performance analysis of the controller designed in part (c), using the methodology of Section 14.11. Let $\varepsilon = 0$ for the controller designs, but let $\varepsilon \neq 0$ in the truth model. Notice the *decentralized* structure of the controller: two separate, independent controllers are interacting with a single physical system as represented by the truth model.

(e) Repeat parts (a)–(d) for the continuous-time controller case. Specifically consider the steady state constant-gain controllers as important special cases.

(f) Carry out the above parts in detail for the case of all variables being scalar.

**14.48** Another means of reducing complexity of controller designs is to view a given system as an interconnection of subsystems with different natural time scales, i.e., to decompose it into two (or more) intercoupled parts, one involving high-speed dynamics and the other having slower dynamics. It is then appropriate to build high sample-rate "inner-loop" or "lower-level" controllers to exert influence on the high-speed dynamic states (treating low-speed coupling terms as essentially constant), and lower sample-rate "outer-loop" or "upper-level" controllers to act on slower variables (treating high-speed coupling terms as having reached steady state conditions). To do this, consider *singular perturbations* (i.e., perturbations to the left hand side of the differential equation, instead of the right hand side as in the previous case; see references listed in Section 14.12):

$$\begin{bmatrix} \dot{\mathbf{x}}_1(t_i) \\ \varepsilon\dot{\mathbf{x}}_2(t) \end{bmatrix} = \begin{bmatrix} \mathbf{F}_{11} & \mathbf{F}_{12} \\ \mathbf{F}_{21} & \mathbf{F}_{22} \end{bmatrix} \begin{bmatrix} \mathbf{x}_1(t) \\ \mathbf{x}_2(t) \end{bmatrix}$$

where $\mathbf{x}_1(t)$ denotes the slow system states and $\mathbf{x}_2(t)$, the fast system states. If we set $\varepsilon = 0$, we obtain

$$\dot{\mathbf{x}}_1(t) = (\mathbf{F}_{11} - \mathbf{F}_{12}\mathbf{F}_{22}^{-1}\mathbf{F}_{21})\mathbf{x}_1(t)$$
$$\mathbf{x}_2(t) = -\mathbf{F}_{22}^{-1}\mathbf{F}_{21}\mathbf{x}_1(t)$$

where the second equation is an algebraic relation for "steady state" $\mathbf{x}_2(t)$ in terms of $\mathbf{x}_1(t)$, rather than an additional differential equation. Thus, the "slow" eigenvalues of the original system are well approximated by those of $(\mathbf{F}_{11} - \mathbf{F}_{12}\mathbf{F}_{22}^{-1}\mathbf{F}_{21})$, and the "fast" system poles by the eigenvalues of $\mathbf{F}_{22}/\varepsilon$ (note that we can write $\dot{\mathbf{x}}_2(t) = (\mathbf{F}_{22}/\varepsilon)\mathbf{x}_2(t) + (\mathbf{F}_{21}/\varepsilon)\mathbf{x}_1(t))$.

(a) Show the corresponding results if the original system is also driven by terms

$$\begin{bmatrix} \mathbf{B}_1 \\ \mathbf{B}_2 \end{bmatrix} \mathbf{u}(t) \quad \text{and/or} \quad \begin{bmatrix} \mathbf{G}_1 \\ \mathbf{G}_2 \end{bmatrix} \mathbf{w}(t)$$

with $\mathbf{w}(\cdot,\cdot)$ zero-mean white Gaussian noise of strength $\mathbf{Q}$. Interpret the algebraic equation for $\mathbf{x}_2(t)$ carefully. Because of the difficulty that arises, we can choose to incorporate the approximation corresponding to the $\dot{\mathbf{x}}_1(t)$ equations for $\varepsilon = 0$, but maintain the $\dot{\mathbf{x}}_2(t)$ equation for the fast states.

(b) Divide the lower partition, $\varepsilon\dot{\mathbf{x}}_2(t)$, through by $\varepsilon$ and develop the full-state feedback LQ deterministic optimal state regulator (continuous-time case) as

$$\mathbf{u}^*(t) = -\bar{\mathbf{G}}_c{}^*\mathbf{x}(t) = -\bar{\mathbf{G}}_{c1}^*\mathbf{x}_1(t) - \bar{\mathbf{G}}_{c2}^*\mathbf{x}_2(t)$$

in steady state, designed to minimize

$$J_c = \lim_{t_f \to \infty} \frac{1}{t_f} \left\{ \frac{1}{2} \int_0^{t_f} \mathbf{x}^T(t)\mathbf{W}_{xx}\mathbf{x}(t) + \mathbf{u}^T(t)\mathbf{W}_{uu}\mathbf{u}(t)\, dt \right\}$$

Do the same for the reduced-order system model based on $\varepsilon = 0$, and compare.

(c) Now consider the state estimator based on

$$\begin{aligned}
\dot{\mathbf{x}}_1(t) &= \mathbf{F}_{11}\mathbf{x}_1(t) + \mathbf{F}_{12}\mathbf{x}_2(t) + \mathbf{B}_1\mathbf{u}(t) + \mathbf{w}_1(t) \\
\varepsilon\dot{\mathbf{x}}_2(t) &= \mathbf{F}_{21}\mathbf{x}_1(t) + \mathbf{F}_{22}\mathbf{x}_2(t) + \mathbf{B}_2\mathbf{u}(t) + \mathbf{w}_2(t) \\
\mathbf{z}(t) &= \mathbf{H}_1\mathbf{x}_1(t) + \mathbf{H}_2\mathbf{x}_2(t) + \mathbf{v}_c(t)
\end{aligned}$$

with $\mathbf{u}$ prespecified and $\mathbf{w}_1$, $\mathbf{w}_2$, and $\mathbf{v}_c$ independent of each other. First divide $\varepsilon\dot{\mathbf{x}}_2(t)$ through by $\varepsilon$ and generate the full-order filter. Set $\varepsilon = 0$ to obtain a relation with which to eliminate $\hat{\mathbf{x}}_2$ from the $\hat{\mathbf{x}}_1$ equation.

Then consider the equations above in the limit as $\varepsilon \to 0$ and develop the reduced-order estimator for $\mathbf{x}_1$. Relative to this "slow filter", the oscillations of $\mathbf{x}_2$ about its "steady state value" look like white noise and cannot be estimated. Now construct the "fast filter", in which $\mathbf{x}_1$ can be treated essentially as a bias, with current estimated value of $\hat{\mathbf{x}}_1$ from the slow filter, of the form

$$\begin{aligned}
\varepsilon\dot{\hat{\mathbf{y}}}_2(t) &= \mathbf{F}_{22}\hat{\mathbf{y}}_2(t) + \mathbf{K}_2[\mathbf{H}_1\hat{\mathbf{x}}_1(t) + \mathbf{H}_2\hat{\mathbf{y}}_2(t) - \mathbf{z}(t)] \\
\hat{\mathbf{x}}_2(t) &= -\mathbf{F}_{22}^{-1}[\mathbf{F}_{21}\hat{\mathbf{x}}_1(t) + \mathbf{B}_2\mathbf{u}(t)] + \hat{\mathbf{y}}_2(t)
\end{aligned}$$

Note the hierarchial structure.

(d) Now consider the LQG regulator based on the models of part (c), but where $\mathbf{u}(t) = -\bar{\mathbf{G}}_c{}^*\hat{\mathbf{x}}(t)$ with $\bar{\mathbf{G}}_c{}^*$ designed to minimize the expected value of the cost in (b). Set up the full-scale equations, let $\varepsilon \to 0$ to eliminate $\hat{\mathbf{x}}_2$ from the slow filter relations, and express the final result. Observe the structure of a slow filter passing a state estimate to a fast filter (but not vice versa), with both filters having $\mathbf{z}(t)$ as input and with both state estimates multiplied by gains to generate the control to apply to the system: a *hierarchial structure*. Note that the eigenvalues of $(\mathbf{F} - \mathbf{B}\bar{\mathbf{G}}_c{}^*)$ and $(\mathbf{F} - \mathbf{K}\mathbf{H})$, as well as those of $\mathbf{F}$, must be separated: feedback gains in the filter and controller can disrupt such separation. Considerable physical insight is required in actual applications to establish such separations.

(e) Develop the analogous results to parts (a)–(d) for the sampled-data case.

(f) Show how such a multirate controller can be evaluated as in Section 14.11 if the longer sample period is an integer number of shorter sample periods.

**14.49** Consider the continuous-time measurement LQG controllers of Section 14.13.

(a) Develop the extension to the case of $\mathbf{w}(\cdot,\cdot)$ of (14-19) and $\mathbf{v}_c(\cdot,\cdot)$ of (14-297) being nonzero-mean.

(b) Develop the extension to the case of $\mathbf{w}(\cdot,\cdot)$ of (14-19) and $\mathbf{v}_c(\cdot,\cdot)$ of (14-297) being correlated with one another.

(c) Derive (14-305)–(14-308) by the limiting procedure described in the text.

(d) Derive the results corresponding to those of part (c), but with $\mathbf{W}_{xu}(t) \neq \mathbf{0}$.

(e) Show that (14-308) generalizes to yield proper cost evaluations as $[\frac{1}{2}\mathbf{x}^T(t_0)\mathbf{K}_c(t_0)\mathbf{x}(t_0)]$ for controllers with gains $\mathbf{G}_c$ chosen by means other than the optimal evaluation (14-306), whereas (14-307) does not.

(f) Derive (14-310).

(g) Derive the optimal stochastic controller based on dynamics model (14-19) and *both* discrete-time measurements as (14-31) *and* continuous-time measurements as (14-297).

**14.50** Consider a system modeled as the cascade of two first order lags driven by a scalar input and by two independent unit-strength zero-mean white Gaussian noises (comparable to the model of Problems 14.16 and 14.17):

$$\begin{bmatrix} \dot{x}_1(t) \\ \dot{x}_2(t) \end{bmatrix} = \begin{bmatrix} -2 & 0 \\ 1 & -5 \end{bmatrix}\begin{bmatrix} x_1(t) \\ x_2(t) \end{bmatrix} + \begin{bmatrix} 1 \\ 0 \end{bmatrix} u(t) + \begin{bmatrix} w_1(t) \\ w_2(t) \end{bmatrix}$$

(a) Develop the LQG optimal controller that assumes continuous-time perfect knowledge of the entire state, optimal in the sense that it minimizes

$$J_c = E\left\{\frac{1}{2}x_1^2(t_{N+1}) + \frac{1}{2}x_2^2(t_{N+1}) + \int_{t_0}^{t_f}\frac{1}{2}[x_1^2(t) + x_2^2(t) + W_{uu}u^2(t)]\,dt\right\}$$

Evaluate for cases of $W_{uu} = 0.1$, 1, and 10.

(b) Obtain the constant-gain steady state controllers corresponding to results in (a).

(c) Display robustness characteristics of the closed-loop systems based on these controllers by plotting $\sigma_{\min}[\mathbf{I} + \mathbf{G}_L(j\omega)]$ and $\sigma_{\min}[\mathbf{I} + \mathbf{G}_L^{-1}(j\omega)]$ for $\omega \in [0, 20 \text{ rad/sec}]$ (at enough discrete values to see the characteristics), as in (14-123).

(d) Develop the covariance performance analysis of the generated controllers acting upon a "truth model" of the same second order system model as that used for controller design. Show how steady state rms state regulation errors degrade when the pole at $(-2)$ in the truth model changes to $(-1)$ and $(-3)$ while the controller assumes it stays at $(-2)$. Repeat for the second truth model pole changing from $(-5)$ to $(-3)$ or $(-7)$.

(e) Now develop the LQG optimal controller based on noise-corrupted measurements

$$z(t) = x_2(t) + v_c(t)$$

with $v_c(\cdot,\cdot)$ of strength $R_c$. Evaluate for $R_c = 0.1$, 1, and 10.

(f) Obtain the corresponding constant-gain steady state controllers.

(g) Display the robustness characteristics of the closed-loop systems based on these controllers, as in (c).

(h) Repeat part (d) for these controllers.

(i) Now robustify these designs by adding pseudonoise of strength $(\mathbf{q}^2\mathbf{B}V\mathbf{B}^T)$ to the model upon which the filter is based, for $V = 1$ and $q^2 = 1$, 10, 100 (see the discussion of (14-132) and (14-133)). Repeat parts (f)–(h).

(j) Compare this result to the result of tuning the filter with pseudonoise of different structural characteristics: instead add noise of strength $(q^2\mathbf{GQG}^T)$ for $q^2 = 1$, 10, and 100, and demonstrate that the robustness recovery of part (i) is superior.

**14.51** Develop the following LQG optimal stochastic controllers based upon continuous-time perfect knowledge of the entire state vector:

(a) trackers analogous to those of Section 14.6,

(b) nonzero setpoint controllers analogous to those of Section 14.7,

(c) controllers with specific disturbance rejection properties, analogous to those of Section 14.8,

(d) PI controllers based on regulation error integral weighting, based on control rate weighting, and of generic form, analogous to those of Section 14.9,

(e) command generator trackers analogous to those of Section 14.10 for the cases of open-loop, CGT/regulator, and CGT/PI laws.

**14.52** Repeat the preceding problem, parts (a)–(e), with continuous-time incomplete noise-corrupted measurements as (14-297) being available instead of perfect state knowledge.

CHAPTER **15**

# Nonlinear stochastic controllers

## 15.1 INTRODUCTION

This chapter extends stochastic controller design to the class of problems in which model nonlinearities are significant. Section 15.2 formulates the problem and characterizes potential solutions. Sections 15.3–15.5 then consider linear perturbation control, assumed certainty equivalence design, and an approximate dual control as viable means of synthesizing suboptimal controllers, as an alternative to attempted solution of full-scale dynamic programming. The important special case of adaptive control to address uncertain parameters in the underlying system models is developed in Section 15.6. Finally, Section 15.7 extracts from these results a philosophy for the design of stochastic controllers, and Section 15.8 concludes with an overall summary and indication of developing technologies and future trends.

## 15.2 BASIC PROBLEM FORMULATION AND CONTROLLER CHARACTERISTICS

Consider controlling a system modeled by the Itô stochastic differential equation

$$\mathbf{dx}(t) = \mathbf{f}[\mathbf{x}(t), \mathbf{u}(t), t]\,dt + \mathbf{G}[\mathbf{x}(t), t]\,\mathbf{d}\boldsymbol{\beta}(t) \tag{15-1}$$

based upon sampled-data measurements of the form

$$\mathbf{z}(t_i) = \mathbf{h}[\mathbf{x}(t_i), t_i] + \mathbf{v}(t_i) \tag{15-2}$$

where the controller to be designed is to apply $r$ control inputs to the system so as to cause $p$ controlled variables of the form

$$\mathbf{y}_c(t) = \mathbf{c}[\mathbf{x}(t), \mathbf{u}(t), t] \tag{15-3}$$

to exhibit desirable behavior over an interval of time corresponding to $(N+1)$ sample periods. This problem context was introduced and discussed to some degree in Section 13.1. The same basic context can be used to develop controllers for systems described by the stochastic difference equation (13-47) from one sample time to the next, given measurements as in (15-2) and controlled variables as in (15-3) evaluated at the sample times $t_i$.

In order to achieve desired performance, we defined a cost function as in (13-96),

$$J = E\left\{\sum_{i=0}^{N} L[\mathbf{x}(t_i), \mathbf{u}(t_i), t_i] + L_{\mathrm{f}}[\mathbf{x}(t_{N+1})]\right\} \tag{15-4}$$

composed of the expected value of the sum of loss functions $L$ associated with each sample period plus a loss function $L_{\mathrm{f}}$ associated with state values at the final time. As observed previously, the loss functions might also be functions of the dynamics driving noise in some problem contexts. We then sought the controller function $\mathbf{u}[\cdot,\cdot]$, as a function of available information and current time (see (13-52), (13-73), and (13-89)), that minimized this cost, in anticipation that this controller function will cause the system to behave "properly." In actual practice, this might entail an iterative redefinition of the loss functions in order to provide for all aspects of performance as desired. Sections 13.5 and 13.6 developed the stochastic dynamic programming approach to solving such a general stochastic optimal control problem, and the control function generated by (13-90)–(13-92) is the optimal $\mathbf{u}^*[\cdot,\cdot]$ to be applied. This control is of *closed-loop* form, and it is important in the general case to distinguish its characteristics from those of *feedback* controls as well as *open-loop* control [27, 28, 30, 46]. Under the LQG assumptions of the previous chapter, there is no distinction between closed-loop and feedback laws: the reason for this and the significance of the difference in the more general problem setting will now be developed.

To state the differences between the various classes of controllers, it is convenient to introduce some notation to describe the elements that define the control problem. Let the knowledge of the dynamics model structure, i.e., $\mathbf{f}[\cdot,\cdot,t]$ and $\mathbf{G}[\cdot,t]$ for all $t$ in (15-1), be denoted as

$$\{\mathbf{f},\mathbf{G}\} \triangleq \{\mathbf{f}[\cdot,\cdot,t], \mathbf{G}[\cdot,t] \text{ for all } t \in [t_0, t_{N+1}]\} \tag{15-5}$$

and the knowledge of the measurement structure (15-2) through the $j$th measurement as

$$\{\mathbf{h}\}^j \triangleq \{\mathbf{h}[\cdot,t_k], k = 0, 1, \ldots, j\} \tag{15-6}$$

where we allow for the possibility of the first measurement being made at time $t_0$ instead of waiting until $t_1$ (as discussed previously). Let the joint probability density characterizing the uncertainties associated with the dynamics, i.e.,

$\mathbf{x}(t_0)$ and $\boldsymbol{\beta}(\cdot,\cdot)$ for all $t \in [t_0, t_{N+1}]$, be denoted as

$$p^{\mathrm{d}} \triangleq f_{\mathbf{x}(t_0),\,\boldsymbol{\beta}(\cdot)}(\boldsymbol{\xi}, \boldsymbol{\eta}) \tag{15-7a}$$

(This is really an abuse of notation: To specify the properties of $\boldsymbol{\beta}(\cdot,\cdot)$ above, the joint density for the infinite number of random vectors $\boldsymbol{\beta}(t,\cdot)$ for all $t \in [t_0, t_{N+1}]$ can be described as Gaussian, with mean of zero, and covariance matrix depicted by $s$-by-$s$ blocks of $\int_{t_0}^{t} \mathbf{Q}(\tau)\,d\tau$ along the diagonal and also repeated to the right and downward from the block diagonal position since $E\{\boldsymbol{\beta}(t)\boldsymbol{\beta}^T(t+\tau)\} = E\{\boldsymbol{\beta}(t)\boldsymbol{\beta}^T(t)\}$ for all $\tau \geq 0$; see Section 4.5 in Volume 1.) Furthermore, let the joint probability of these dynamics characteristics and the measurement corruption noise through the $j$th measurement be denoted as

$$p^{j} \triangleq f_{\mathbf{x}(t_0),\,\boldsymbol{\beta}(\cdot),\,\mathbf{v}(t_0),\,\mathbf{v}(t_1),\,\ldots,\,\mathbf{v}(t_j)}(\boldsymbol{\xi}, \boldsymbol{\eta}, \boldsymbol{\rho}_0, \boldsymbol{\rho}_1, \ldots, \boldsymbol{\rho}_j) \tag{15-7b}$$

The actual data that is potentially available to the controller at time $t_i$ is as given in (13-73),

$$\mathbf{I}_i \triangleq \{\mathbf{z}_0^{\mathrm{T}}, [\mathbf{u}^{\mathrm{T}}(t_0), \mathbf{z}_1^{\mathrm{T}}], \ldots, [\mathbf{u}^{\mathrm{T}}(t_{i-1}), \mathbf{z}_i^{\mathrm{T}}]\}^{\mathrm{T}} \tag{15-8}$$

starting from $\mathbf{I}_0 = \mathbf{z}_0$: knowledge of the time history of measurement values and controls that have already been applied in the past, viewed as a specific realization of random vector $\mathbf{I}(t_i,\cdot)$ of dimension $\{i(m+r)+m\}$. Thus, the controller is causal—it cannot be a function of realizations of the information vector components corresponding to future time (more rigorously, $\mathbf{u}(t_i)$ is measurable with respect to the $\sigma$-algebra generated by $\mathbf{I}(t_i)$, i.e., by $\mathbf{Z}(t_i)$ corresponding to some fixed function $\mathbf{u}[\cdot, t_j]$ or known time history $\mathbf{u}[t_j]$ for all $t_j < t_i$, where random variables are defined as identity mappings on the real Euclidean sample space $\Omega$). Finally, we denote knowledge of the loss functions defining the cost (15-4) as

$$\{L\} \triangleq \{L[\cdot,\cdot,t_i] \text{ for all } i \in [0, N],\ L_{\mathrm{f}}[\cdot]\} \tag{15-9}$$

In terms of this notation, we can say that an *open-loop* control law has the form

$$\mathbf{u}^{\mathrm{OL}}(t_i) = \mathbf{u}^{\mathrm{OL}}[t_i; \{\mathbf{f}, \mathbf{G}\}, p^{\mathrm{d}}, \{L\}] \tag{15-10}$$

It is a function of time only and the entire time history of control *values* can be determined before real-time usage, evaluated on the basis of the structure and statistical description of the dynamics model, and the definition of the cost to be minimized. It does not use any real-time measurement data from the system. On the other hand, both feedback and closed-loop laws allow access to the real-time data, but they differ as to the availability of information provided about the measurements to be taken in the future.

A *feedback* control law is defined to be a law that is continually provided real-time measurement data, as fed back from the actual system under control. However, no knowledge is provided to it about the existence, form, and precision of future measurements. Thus, it has the form

$$\mathbf{u}^{\mathrm{F}}(t_i) = \mathbf{u}[\mathbf{I}_i, t_i; \{\mathbf{f}, \mathbf{G}\}, \{\mathbf{h}\}^i, p^i, \{L\}] \tag{15-11}$$

It is a function of both the information $\mathbf{I}_i$ and $t_i$, and it is evaluated on the basis of modeling elements available to the open-loop law plus the description of the structure and statistical characteristics of the time history of measurements through time $t_i$.

In contrast, a *closed-loop* law directly exploits the knowledge that the loop will remain closed throughout the future interval to the final time that control is to be applied, $t_N$. It adds to the information provided to a feedback controller, explicit knowledge of the types of measurements to be taken and the statistical characteristics of the noise corrupting them:

$$\mathbf{u}^{\mathrm{CL}}(t_i) = \mathbf{u}[\mathbf{I}_i, t_i; \{\mathbf{f}, \mathbf{G}\}, \{\mathbf{h}\}^N, p^N, \{L\}] \tag{15-12}$$

It anticipates that measurements of the indicated form will be taken in the future, and it uses this in determining the control to apply at the current time $t_i$. Such statistical anticipation allows prior assessment of the impact of future measurements on attaining the control objectives, and this can yield a significantly different optimal (or appropriate) control than that achieved by a feedback controller ignorant of these future observations [2, 7, 27, 28, 30, 53, 82, 120, 152].

There is also a class of controllers in between feedback and closed-loop [28, 41], known as *M-measurement feedback*, or limited-future closed-loop. Here the controller is allowed to anticipate only the next $M$ measurements into the future, rather than all. It has the form

$$\mathbf{u}^{M\mathrm{F}}(t_i) = \mathbf{u}^{M\mathrm{F}}[\mathbf{I}_i, t_i; \{\mathbf{f}, \mathbf{G}\}, \{\mathbf{h}\}^{i+M}, p^{i+M}, \{L\}] \tag{15-13}$$

and it is intended to provide some of the characteristics of a closed-loop controller, but at less computational expense, or to be used in applications where prespecification of the entire future program of measurements cannot be accomplished with absolute assurance. In some applications, it makes sense to redefine the cost iteratively to include loss functions only out to time $t_{i+M}$ as well.

Although the *dynamic programming* algorithm can be used to synthesize optimal closed-loop controllers, it can also be altered to generate control laws in the other classes just described as well. An *optimal open-loop* law can be produced by (13-90)–(13-92) by replacing conditional expectations by unconditional ones, i.e., by letting $\mathbf{S}(t_k) \equiv \mathbf{0}$ for all $k$. One form of feedback law, known as *open-loop optimal feedback*, can be produced by letting $\mathbf{S}(t_k) \equiv \mathbf{0}$ for all $k > i$ when evaluating $\mathbf{u}(t_i)$: this controller specifically assumes that although the current measurement history can be used in control computation, *no observations will be made in the future*, and that an optimal *open-loop* policy will be adopted at all future instants until the final time [2, 21, 26, 28, 46, 47,

51, 80, 133]. This control is applied only over the single sample period from $t_i$ to $t_{i+1}$; when the next measurement does arrive at $t_{i+1}$, it is in fact used to compute the control $\mathbf{u}(t_{i+1})$, but again assuming that it will be optimal open-loop control from that time forward. Specific forms of $M$-measurement feedback laws can be generated analogously by setting $\mathbf{S}(t_k) \equiv \mathbf{0}$ for all $k > (i + M)$ when evaluating $\mathbf{u}(t_i)$, or by iteratively redefining the objective of a finite-horizon problem so as to use (13-90)–(13-92) directly but with the terminal condition (13-91) replaced by

$$\mathscr{C}^*[\mathbf{S}_{i+M}, t_{i+M}] = \bar{L}'[\mathbf{S}_{i+M}, \mathbf{u}\{\mathbf{S}_{i+M}, t_{i+M}\} = \mathbf{0}, t_{i+M}] \qquad (15\text{-}14)$$

for all $t_{i+M} < t_{N+1}$.

However, the use of dynamic programming as a computational algorithm has been shown to be inherently burdensome. There are other useful *synthesis* approaches that can yield *suboptimal* control laws much more readily than direct application of dynamic programming, and these will be the subject of succeeding sections. Before this is done, it is useful to characterize some properties of general stochastic controllers that may or *may not* be shared with the linear controllers of the preceding two chapters [28, 148].

As mentioned in Section 13.3, the stochastic optimal control problem may, but only under special conditions, have the *certainty equivalence* property, in which the stochastic closed-loop optimal control $\mathbf{u}^{\mathrm{SCLO}}(t_i)$ has the same form as the deterministic optimal control $\mathbf{u}^{\mathrm{DO}}(t_i)$ for the corresponding problem but with all random variables replaced by their conditional mean values [27, 28, 129, 131, 132, 136, 149, 151, 152]. Thus, if the deterministic optimal control for that associated problem can be expressed as $\mathbf{u}^{\mathrm{DO}}[\mathbf{x}(t_i), t_i]$, the stochastic closed-loop optimal law can be expressed as $\mathbf{u}^{\mathrm{DO}}[\hat{\mathbf{x}}(t_i^+), t_i]$, i.e., the *same* control function but with the current state $\mathbf{x}(t_i)$ replaced by the conditional mean $\hat{\mathbf{x}}(t_i^+)$, conditioned on the observed information time history. If the certainty equivalence property pertains, the optimal closed-loop control law "degenerates" into a feedback form: nothing is to be gained by statistically anticipating future measurements. This is exactly what was observed in Chapter 14—under the LQG assumptions, the stochastic optimal control problem *does* have the certainty equivalence property, and is expressible as

$$\mathbf{u}^*(t_i) = -\mathbf{G}_c^*(t_i)\hat{\mathbf{x}}(t_i^+) \qquad (15\text{-}15)$$

where neither $\mathbf{G}_c^*(t_i)$ nor $\hat{\mathbf{x}}(t_i^+)$ is affected by future measurements, and where the associated deterministic optimal full-state regulator is

$$\mathbf{u}(t_i) = -\mathbf{G}_c^*(t_i)\mathbf{x}(t_i) \qquad (15\text{-}16)$$

It is the readily synthesizable nature of this result that makes the LQG stochastic optimal controller so useful. Moreover, this synthesis capability will motivate one of the *suboptimal feedback* control law design methodologies of this chapter, known as *assumed* (or *forced*, or *heuristic*) *certainty equivalence* design, in which a control is formulated as $\mathbf{u}^{DO}[\hat{\mathbf{x}}(t_i^+), t_i]$ even though certainty equivalence is not a property of the stochastic optimal control problem.

Under weaker though still restrictive assumptions, the stochastic optimal control problem may have the *separation* property [27, 28, 85, 116, 129, 131, 132, 149, 151, 152]: The stochastic closed-loop optimal controller of the system state is expressible as some function of only the conditional mean $\hat{\mathbf{x}}(t_i^+)$ and time $t_i$, though *not necessarily* the function $\mathbf{u}^{DO}[\cdot,\cdot]$ obtained by solving the associated deterministic optimal control problem. Thus, control problems with the certainty equivalence property are a subset of control problems with the separation property. The validity of the separation property does not imply that the optimal closed-loop policy necessarily degenerates to feedback form, and problems with the separation property but without the certainty equivalence property have been shown to arise in physically motivated problems [123].

A control can have a *dual effect* [27, 28, 30, 52, 53, 136] on a system, affecting not only the trajectory of the system *state* but also possibly the future state *uncertainty*. Even though this was not observed for problems adequately described by the LQG assumptions, it may well be an *important* characteristic of controls for more general problems. To define dual effect better, let $\mu_l^k(t_j)$ be the $k$th central conditional moment of $x_l(t_j)$ conditioned on the current measurement history, viewed as a random variable:

$$\mu_l^k(t_j) = E\{[x_l(t_j;\mathbf{u}) - E\{x_l(t_j)|\mathbf{I}(t_j)\}]^k|\mathbf{I}(t_j) = \mathbf{I}(t_j)\} \tag{15-17}$$

Similarly define joint conditional central moments of order $k$ among different components of $\mathbf{x}(t_j)$. Then control $\mathbf{u}[\cdot,\cdot]$ is said to have *no dual effect of order k* ($k \geq 2$) if all such $k$th central conditional moments of the state satisfy

$$E\{\mu^k(t_j)|\mathbf{I}(t_i) = \mathbf{I}(t_i)\} = E\{\mu^k(t_j)|\mathbf{I}(t_i) = \mathbf{I}(t_i; \mathbf{u}[\cdot,\cdot] \equiv \mathbf{0})\} \tag{15-18}$$

for all $t_i \leq t_j$, with probability one. In other words, if we sit at $t_i$ and assess the future uncertainty expressed by that $k$th central moment at $t_j \geq t_i$, that future uncertainty does not depend on whether the control function $\mathbf{u}[\cdot,\cdot]$ was applied or not; note that (15-18) implicitly involves integrating out the dependence on $\mathbf{z}(t_{i+1})$, $\mathbf{z}(t_{i+2}), \ldots, \mathbf{z}(t_j)$ using the appropriate probability densities. If (15-18) is not true for all $k \geq 2$ and all $t_j \geq t_i$, then tne control can affect at least one such central moment and it does have *dual effect*. Whether or not the dual effect is present is a property of a *system model*; system models for which the control has no dual effect are termed *neutral* [28, 52, 129].

Neutrality, or the absence of dual effect, can be related to certainty equivalence in a more general context than the LQG assumptions. For instance, consider a linear discrete-time dynamics system model driven by white (*not* necessarily Gaussian) noise, with measurements modeled as a *nonlinear* function of states and time with white (*not* necessarily Gaussian) corruptive noise, and with quadratic cost criterion to be minimized. Then the optimal stochastic controller has the certainty equivalence property if and only if the control has no dual effect, i.e., if and only if $\mathbf{P}(t_i^+)$ is independent of the previous measurement history (note that generally it does depend on this history with nonlinear measurements and non-Gaussian noises) and so are higher moments [27].

If a system model is not neutral, then a closed-loop control will generally "probe" the system: it will operate in such a manner as to enhance estimation precision, so as to improve the overall performance in the future [28, 29, 52, 70]. In an adaptive control problem, the controller may *purposely* excite certain modes of the system in order to *aid* the identification of uncertain parameters, thereby improving future state estimation (reducing uncertainty) and control performance. For instance, tactical aircraft usually perform "S-turn" maneuvers while aligning the guidance systems of air-to-ground weapons, since such seemingly odd maneuvers (which do not enhance *current* performance of aircraft or weapon) allow greater alignment precision and reduced navigation uncertainty throughout the weapons' time-of-flight, eventually yielding reduced rms miss distance at the target (the performance index, or cost function, to be minimized in this stochastic control problem). This *active learning* is distinct from the *passive* learning of a *feedback* controller: although a feedback controller also learns about the system from real-time measurements, it does not generate commands for the system specifically to assist that learning process—it cannot do so because it is "unaware" of future measurements to exploit the results of such probing action. A system modeled under the LQG assumptions is in fact neutral, so nothing is gained from probing, and the optimal stochastic controller does not exhibit such behavior.

Another characteristic that stochastic controllers can display besides probing is known as *caution*, acting so as to guard against large cost contribution due to excessive *current uncertainties* [28, 29, 70]. Because the controller is uncertain about the consequences of its inputs to the system due to imperfect current knowledge about the system, it exercises more caution in applying those inputs than in the case of knowing the entire state perfectly. Thus, caution is associated with current "operating risk," whereas active learning is associated with a "probing risk" [52].

For instance [28], an open-loop optimal feedback controller will be "more cautious" than a closed-loop optimal controller since the former assumes that no future measurements will be made, whereas the latter fully anticipates taking future observations to generate appropriate corrective actions. This is an issue separate from dual effect and probing. For example, in a perfect measurement problem (thereby removing dual effect), open-loop optimal feedback and optimal closed-loop laws have demonstrated different performance capabilities, due to different amounts of caution [47]. Furthermore, the LQG problem requires no caution of the optimal controller, as well as no probing. In contrast, the linear exponential quadratic problem [123] is neutral, and thus the optimal controller displays no probing, but it does have to exercise caution; as a result, this controller has the separation property but not certainty equivalence.

Having thus characterized some of the properties of stochastic controllers for general applications that do not necessarily meet the LQG assumptions, we now consider *systematic design methodologies* for *efficient* generation of *sub-optimal* control laws. These are to be viewed as potential alternatives to straightforward application of dynamic programming, particularly for problems in which this approach yields formidable (or intractable) solution generation, or results that are too burdensome for real-time implementation.

## 15.3 LINEAR PERTURBATION CONTROL LAWS FOR NONLINEAR SYSTEMS: DIRECT APPLICATION OF LQG SYNTHESIS

To gain some preliminary insights into the control of a system modeled as in (15-1)–(15-3), consider the simple special case of a time-invariant system from which perfect state knowledge is available. The associated noise-free system description is then

$$\dot{\mathbf{x}}(t) = \mathbf{f}[\mathbf{x}(t), \mathbf{u}(t)] \tag{15-19a}$$

$$\mathbf{y}_c(t) = \mathbf{c}[\mathbf{x}(t), \mathbf{u}(t)] \tag{15-19b}$$

Assume also that the dimension of $\mathbf{u}(t)$ equals that of $\mathbf{y}_c(t)$, that $\mathbf{f}[\cdot,\cdot]$ and $\mathbf{c}[\cdot,\cdot]$ are differentiable (to allow Taylor series expansions), and that $\mathbf{u}(t)$ is constrained to be piecewise constant (the output of a sampled-data controller put through a zero-order hold). Now we desire to find an appropriate control law to keep the controlled variables $\mathbf{y}_c(t)$ near a desired setpoint $\mathbf{y}_d$.

Assume that, for each desired setpoint, a unique constant equilibrium solution exists, i.e., that

$$\dot{\mathbf{x}}_0 = \mathbf{0} = \mathbf{f}[\mathbf{x}_0, \mathbf{u}_0] \tag{15-20a}$$

$$\mathbf{y}_{c0} = \mathbf{y}_d = \mathbf{c}[\mathbf{x}_0, \mathbf{u}_0] \tag{15-20b}$$

can be solved so as to define the functions $\mathscr{x}_0(\cdot)$ and $\mathscr{u}_0(\cdot)$ via

$$\mathscr{x}_0(\mathbf{y}_d) = \mathbf{x}_0 \tag{15-21a}$$

$$\mathscr{u}_0(\mathbf{y}_d) = \mathbf{u}_0 \tag{15-21b}$$

Now LQ synthesis is to be used to design a controller to regulate the nonlinear system about the setpoint $\mathbf{y}_d$ (through basic LQ regulator design), or about a setpoint $\mathbf{y}_d(t)$ *near* some nominal $\mathbf{y}_{d0}$ (through PI controllers generated by LQ methods).

For the basic *regulator design* [11, 62, 146], we can conceptually solve for $\mathbf{x}_0$ and $\mathbf{u}_0$ offline (a priori) for each value of $\mathbf{y}_d$. Then the dynamics of the system for any solution $\{\mathbf{x}(t), \mathbf{u}(t), \mathbf{y}_c(t)\}$ for all $t$ that is *near* the equilibrium solution $\{\mathbf{x}_0, \mathbf{u}_0, \mathbf{y}_{c0} = \mathbf{y}_d\}$ can be *approximated* by the *linear time-invariant perturbation system* model

$$\dot{\delta\mathbf{x}}(t) = \mathbf{F}\,\delta\mathbf{x}(t) + \mathbf{B}\,\delta\mathbf{u}(t) \tag{15-22a}$$

$$\delta\mathbf{y}_c(t) = \mathbf{C}\,\delta\mathbf{x}(t) + \mathbf{D}_y\,\delta\mathbf{u}(t) \tag{15-22b}$$

in terms of first-order perturbation variables

$$\delta\mathbf{x}(t) = \mathbf{x}(t) - \mathbf{x}_0; \qquad \mathbf{x}_0 = \mathscr{x}_0(\mathbf{y}_d) \tag{15-23a}$$

$$\delta\mathbf{u}(t) = \mathbf{u}(t) - \mathbf{u}_0; \qquad \mathbf{u}_0 = \mathscr{u}_0(\mathbf{y}_d) \tag{15-23b}$$

$$\delta\mathbf{y}_c(t) = \mathbf{y}_c(t) - \mathbf{y}_{c0}; \qquad \mathbf{y}_{c0} = \mathbf{y}_d \tag{15-23c}$$

with defining matrices as constant precomputable partials of $\mathbf{f}$ and $\mathbf{c}$ evaluated at the equilibrium values:

$$\mathbf{F} = \left.\frac{\partial\mathbf{f}}{\partial\mathbf{x}}\right|_{\mathbf{x}=\mathbf{x}_0,\ \mathbf{u}=\mathbf{u}_0} \qquad \mathbf{B} = \left.\frac{\partial\mathbf{f}}{\partial\mathbf{u}}\right|_{\mathbf{x}=\mathbf{x}_0,\ \mathbf{u}=\mathbf{u}_0}$$
$$\mathbf{C} = \left.\frac{\partial\mathbf{c}}{\partial\mathbf{x}}\right|_{\mathbf{x}=\mathbf{x}_0,\ \mathbf{u}=\mathbf{u}_0} \qquad \mathbf{D}_y = \left.\frac{\partial\mathbf{c}}{\partial\mathbf{u}}\right|_{\mathbf{x}=\mathbf{x}_0,\ \mathbf{u}=\mathbf{u}_0} \tag{15-24}$$

We assume conceptually that $\mathbf{y}_d$ is constant for all time, starting at time $t_0$. In actual application, one can allow infrequent changes in this value, assuming steady state performance to be reached before $\mathbf{y}_d$ changes again: this is treated as a sequence of separate problems with the step changes in $\mathbf{y}_d$ corresponding to new initial conditions at each newly declared "initial time." Such a conceptualization allows for infrequent relinearizations to improve the adequacy of small perturbation assumptions; this will be discussed at greater length later. From (15-22), we can generate the equivalent discrete-time system model (see (14-28))

$$\delta\mathbf{x}(t_{i+1}) = \mathbf{\Phi}\,\delta\mathbf{x}(t_i) + \mathbf{B}_d\,\delta\mathbf{u}(t_i) \tag{15-25a}$$

$$\delta\mathbf{y}_c(t_i) = \mathbf{C}\,\delta\mathbf{x}(t_i) + \mathbf{D}_y\,\delta\mathbf{u}(t_i) \tag{15-25b}$$

where the time-invariant $\mathbf{\Phi}$, $\mathbf{B}_d$, $\mathbf{C}$, and $\mathbf{D}_y$ all generally depend on $\mathbf{x}_0$ and $\mathbf{u}_0$, and thus on $\mathbf{y}_{c0} = \mathbf{y}_d$. Standard LQ techniques are now used to design a discrete-time perturbation full-state regulator in the form of

$$\delta\mathbf{u}^*(t_i) = -\mathbf{G}_c^*(t_i)\,\delta\mathbf{x}(t_i) \tag{15-26}$$

or the steady state constant-gain approximation thereto,

$$\delta\mathbf{u}(t_i) = -\bar{\mathbf{G}}_c^*\,\delta\mathbf{x}(t_i) \tag{15-27}$$

Two possible means of implementing this controller are displayed in Fig. 15.1. In diagram (a), the values of $\mathbf{x}_0$ and $\mathbf{u}_0$ are generated a priori offline (or at least the $\mathscr{x}_0$ and $\mathscr{u}_0$ functions are), and (15-26) or (15-27) is implemented directly once $\delta\mathbf{x}(t_i)$ is produced explicitly by subtraction. In the original coordinates,

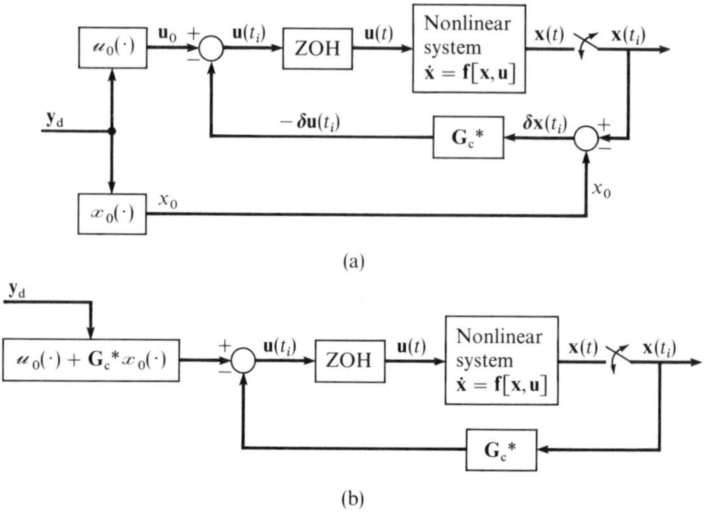


FIG. 15.1 Perturbation regulator.

(15-26) can be expressed as

$$\begin{aligned}\mathbf{u}^*(t_i) &= \mathbf{u}_0 - \mathbf{G}_c^*(t_i)[\mathbf{x}(t_i) - \mathbf{x}_0] \\ &= [\mathbf{u}_0 + \mathbf{G}_c^*(t_i)\mathbf{x}_0] - \mathbf{G}_c^*(t_i)\mathbf{x}(t_i) \\ &= [\mathscr{u}_0(\mathbf{y}_d) + \mathbf{G}_c^*(t_i)\mathscr{x}_0(\mathbf{y}_d)] - \mathbf{G}_c^*(t_i)\mathbf{x}(t_i)\end{aligned} \tag{15-28}$$

This, or the constant-gain approximation to it, can be implemented as in diagram (b). Note the direct similarity of Fig. 15.1 to Fig. 14.9. As in that previous setpoint controller, this regulator requires explicit reevaluation of $\mathbf{u}_0$ and $\mathbf{x}_0$, or $[\mathbf{u}_0 + \mathbf{G}_c^*\mathbf{x}_0]$, whenever $\mathbf{y}_d$ is changed. Unlike the previous result, $\mathbf{G}_c^*(t_i)$ or $\bar{\mathbf{G}}_c^*$ *also* changes when $\mathbf{y}_d$ does, since $\mathbf{y}_d$ affects $\mathbf{\Phi}$, $\mathbf{B}_d$, $\mathbf{C}$, and $\mathbf{D}_y$, which are used to generate the regulator gains. If the *incremental form* of the regulator were used instead of the position form, i.e., for the perturbation variables

$$\delta\mathbf{u}^*(t_i) = \delta\mathbf{u}^*(t_{i-1}) - \mathbf{G}_c^*(t_i)[\delta\mathbf{x}(t_i) - \delta\mathbf{x}(t_{i-1})] \tag{15-29}$$

then conversion to the original coordinates yields

$$\mathbf{u}^*(t_i) = \mathbf{u}^*(t_{i-1}) - \mathbf{G}_c^*(t_i)[\mathbf{x}(t_i) - \mathbf{x}(t_{i-1})] \tag{15-30}$$

Here again the $\mathbf{G}_c^*(t_i)$ or constant $\bar{\mathbf{G}}_c^*$ is a function of $\mathbf{y}_d$, but *no explicit recomputations* of equilibrium $\mathbf{x}_0$ and $\mathbf{u}_0$ are required for real-time usage. The benefit of incremental forms will be accentuated even further in the PI controller forms to follow. As discussed in conjunction with Section 14.7, such controllers are superior to simple regulators for nonzero setpoint control.

To develop a *perturbation PI controller* [34] let $\{\mathbf{x}_0, \mathbf{u}_0, \mathbf{y}_{c0} = \mathbf{y}_{d0}\}$ be an equilibrium solution corresponding to the desired setpoint $\mathbf{y}_{d0}$, and let

$\{\mathbf{x}_0 + \Delta\mathbf{x}_0,\ \mathbf{u}_0 + \Delta\mathbf{u}_0,\ \mathbf{y}_{d0} + \Delta\mathbf{y}_d\}$ be a *nearby* equilibrium solution corresponding to $[\mathbf{y}_{d0} + \Delta\mathbf{y}_d]$, with

$$\mathbf{x}_0 + \Delta\mathbf{x}_0 = \mathscr{x}_0(\mathbf{y}_{d0} + \Delta\mathbf{y}_d) \tag{15-31a}$$

$$\mathbf{u}_0 + \Delta\mathbf{u}_0 = \mathscr{u}_0(\mathbf{y}_{d0} + \Delta\mathbf{y}_d) \tag{15-31b}$$

For any setpoint sufficiently close to $\mathbf{y}_{d0}$, regulating the nonlinear system to $\mathbf{y}_d$ can be reduced to regulating the linear perturbation system to $[\mathbf{y}_d - \mathbf{y}_{d0}] = \Delta\mathbf{y}_d$. Thus, a nonzero setpoint problem arises naturally, for which a PI design is well suited. Conceptually, $\mathbf{y}_{d0}$ is to be regarded as a constant for all time: with only one nominal, only a *single* linearization is required. Changes in setpoint are then reflected as changes in $\Delta\mathbf{y}_d$, i.e., the nonzero setpoint for the single linearized problem. In actual implementation, infrequent changes of the nominal are motivated to maintain adequacy of the linear perturbation approach: this will be discussed further once the forms of the control laws are presented.

Recalling Section 14.9, we can use LQ synthesis to generate PI control laws for the perturbation system model (15-25) defined about the equilibrium corresponding to $\mathbf{y}_{d0}$, simply by replacing $\mathbf{x}$, $\mathbf{u}$, $\mathbf{y}_c$, and $\mathbf{y}_d$ of that section with $\boldsymbol{\delta}\mathbf{x}$, $\boldsymbol{\delta}\mathbf{u}$, $\boldsymbol{\delta}\mathbf{y}_c$, and $\Delta\mathbf{y}_d$, respectively. The preferred incremental form law in terms of these perturbation variables can be written as in (14-209) using regulation error pseudointegral states, (14-228) using control differences, or the generic form (14-237), to yield, respectively,

$$\begin{aligned}\boldsymbol{\delta}\mathbf{u}^*(t_i) = \boldsymbol{\delta}\mathbf{u}^*(t_{i-1}) - \bar{\mathbf{G}}^*_{c1}[\boldsymbol{\delta}\mathbf{x}(t_i) - \boldsymbol{\delta}\mathbf{x}(t_{i-1})]\\ + \bar{\mathbf{G}}^*_{c2}[\Delta\mathbf{y}_d(t_{i-1}) - \boldsymbol{\delta}\mathbf{y}_c(t_{i-1})] + \mathbf{E}[\Delta\mathbf{y}_d(t_i) - \Delta\mathbf{y}_d(t_{i-1})]\end{aligned} \tag{15-32a}$$

$$\boldsymbol{\delta}\mathbf{u}^*(t_i) = \boldsymbol{\delta}\mathbf{u}^*(t_{i-1}) - \mathbf{K}_x[\boldsymbol{\delta}\mathbf{x}(t_i) - \boldsymbol{\delta}\mathbf{x}(t_{i-1})] + \mathbf{K}_\xi[\Delta\mathbf{y}_d(t_i) - \boldsymbol{\delta}\mathbf{y}_c(t_{i-1})] \tag{15-32b}$$

$$\begin{aligned}\boldsymbol{\delta}\mathbf{u}^*(t_i) = \boldsymbol{\delta}\mathbf{u}^*(t_{i-1}) - \mathscr{G}[\boldsymbol{\delta}\mathbf{x}(t_i) - \boldsymbol{\delta}\mathbf{x}(t_{i-1})] + \mathscr{K}_p[\Delta\mathbf{y}_d(t_i) - \boldsymbol{\delta}\mathbf{y}_c(t_i)]\\ + [\mathscr{K}_i - \mathscr{K}_p][\Delta\mathbf{y}_d(t_{i-1}) - \boldsymbol{\delta}\mathbf{y}_c(t_{i-1})]\end{aligned} \tag{15-32c}$$

Then, to express the result in original coordinates, we can make the substitutions

$$\boldsymbol{\delta}\mathbf{x}(t_i) = \mathbf{x}(t_i) - \mathbf{x}_0 = \mathbf{x}(t_i) - \mathscr{x}_0(\mathbf{y}_{d0}) \tag{15-33a}$$

$$\boldsymbol{\delta}\mathbf{u}(t_i) = \mathbf{u}(t_i) - \mathbf{u}_0 = \mathbf{u}(t_i) - \mathscr{u}_0(\mathbf{y}_{d0}) \tag{15-33b}$$

$$\boldsymbol{\delta}\mathbf{y}_c(t_i) = \mathbf{C}\,\boldsymbol{\delta}\mathbf{x}(t_i) + \mathbf{D}_y\,\boldsymbol{\delta}\mathbf{u}(t_i) \tag{15-33c}$$

$$\Delta\mathbf{y}_d(t_i) = \mathbf{y}_d(t_i) - \mathbf{y}_{d0} \tag{15-33d}$$

and also make use of the Taylor series representation

$$\begin{aligned}\mathbf{c}[\mathbf{x}(t_i), \mathbf{u}(t_i)] &= \mathbf{c}[\mathbf{x}_0 + \boldsymbol{\delta}\mathbf{x}(t_i), \mathbf{u}_0 + \boldsymbol{\delta}\mathbf{u}(t_i)]\\ &\cong \mathbf{c}[\mathbf{x}_0, \mathbf{u}_0] + \mathbf{C}\,\boldsymbol{\delta}\mathbf{x}(t_i) + \mathbf{D}_y\,\boldsymbol{\delta}\mathbf{u}(t_i)\\ &= \mathbf{y}_{d0} + \mathbf{C}\,\boldsymbol{\delta}\mathbf{x}(t_i) + \mathbf{D}_y\,\boldsymbol{\delta}\mathbf{u}(t_i)\end{aligned} \tag{15-34}$$

For instance, substitution of (15-33) into (15-32a) yields, after appropriate cancellations,

$$\begin{aligned}\mathbf{u}^*(t_i) = \mathbf{u}^*(t_{i-1}) &- \bar{\mathbf{G}}_{c1}^*[\mathbf{x}(t_i) - \mathbf{x}(t_{i-1})] \\ &+ \bar{\mathbf{G}}_{c2}^*\{[\mathbf{y}_d(t_{i-1}) - \mathbf{y}_{d0}] - [\mathbf{C}\,\delta\mathbf{x}(t_{i-1}) + \mathbf{D}_y\,\delta\mathbf{u}(t_{i-1})]\} \\ &+ \mathbf{E}[\mathbf{y}_d(t_i) - \mathbf{y}_d(t_{i-1})]\end{aligned}$$

Finally, use of (15-34) produces the *regulation error pseudointegral perturbation PI control law*, in *incremental form* using total variables, as

$$\begin{aligned}\mathbf{u}^*(t_i) = \mathbf{u}^*(t_{i-1}) &- \bar{\mathbf{G}}_{c1}^*[\mathbf{x}(t_i) - \mathbf{x}(t_{i-1})] \\ &+ \bar{\mathbf{G}}_{c2}^*\{\mathbf{y}_d(t_{i-1}) - \mathbf{c}[\mathbf{x}(t_{i-1}), \mathbf{u}(t_{i-1})]\} \\ &+ \mathbf{E}[\mathbf{y}_d(t_i) - \mathbf{y}_d(t_{i-1})]\end{aligned} \tag{15-35a}$$

Similarly, the *control difference perturbation PI law* is

$$\begin{aligned}\mathbf{u}^*(t_i) = \mathbf{u}^*(t_{i-1}) &- \mathbf{K}_x[\mathbf{x}(t_i) - \mathbf{x}(t_{i-1})] \\ &+ \mathbf{K}_\xi\{\mathbf{y}_d(t_i) - \mathbf{c}[\mathbf{x}(t_{i-1}), \mathbf{u}(t_{i-1})]\}\end{aligned} \tag{15-35b}$$

and the *generic perturbation PI law* is

$$\begin{aligned}\mathbf{u}^*(t_i) = \mathbf{u}^*(t_{i-1}) &- \mathcal{G}[\mathbf{x}(t_i) - \mathbf{x}(t_{i-1})] \\ &+ \mathcal{K}_p\{\mathbf{y}_d(t_i) - \mathbf{c}[\mathbf{x}(t_i), \mathbf{u}(t_i)]\} \\ &+ [\mathcal{K}_i - \mathcal{K}_p]\{\mathbf{y}_d(t_{i-1}) - \mathbf{c}[\mathbf{x}(t_{i-1}), \mathbf{u}(t_{i-1})]\}\end{aligned} \tag{15-35c}$$

Like the *incremental* form of the *regulator*, these controllers do *not* explicitly involve $\mathbf{y}_{d0}$; as before, the specific gains *are* a function of $\mathbf{x}_0$ and $\mathbf{u}_0$, and thus of $\mathbf{y}_{d0}$. Provided that the system solution remains "close" to the equilibrium associated with $\mathbf{y}_{d0}$, these laws provide adequate performance.

In order to enhance the adequacy of these perturbation PI controllers, we should recall the assumptions used to justify application of LQ synthesis to nonlinear systems, namely, (1) the system solutions $\{\mathbf{x}(t), \mathbf{u}(t), \mathbf{y}_c(t)\}$ for all $t$ stay "near" the equilibrium $\{\mathbf{x}_0, \mathbf{u}_0, \mathbf{y}_{c0}\}$ associated with $\mathbf{y}_{d0}$; (2) the dynamics of the nonlinear system can be adequately approximated by linear time-invariant perturbation equations near $\{\mathbf{x}_0, \mathbf{u}_0, \mathbf{y}_{c0} = \mathbf{y}_{d0}\}$; and (3) the desired setpoint $\mathbf{y}_d(t_i)$ stays "near" the constant $\mathbf{y}_{d0}$. First, we can take measures to ensure that the transients of $\{\mathbf{x}(t), \mathbf{u}(t), \mathbf{y}_c(t)\}$ stay near $\{\mathbf{x}_0, \mathbf{u}_0, \mathbf{y}_{c0} = \mathbf{y}_{d0}\}$: by varying $\mathbf{y}_d(t_i)$ slowly, we thereby keep induced transients small. Second, to make sure $\mathbf{y}_d(t_i)$ is near $\mathbf{y}_{d0}$, we can update $\mathbf{y}_{d0}$ infrequently, thereby changing the linearizations and gain computations, and maintaining the adequacy of the newly generated linear perturbation model. This is *not* accomplished on a frequent basis, since *design validity* is based on $\mathbf{y}_{d0}$ being *essentially constant*, and changes in $\mathbf{y}_{d0}$ are treated as a new declaration of "initial time" and associated initial conditions. Updating $\mathbf{y}_{d0}$ is *simple* for the *incremental* forms of (15-32) and (15-35): only the gains are changed, and the new steady state values of $\mathbf{x}_0$ and $\mathbf{u}_0$ are not needed explicitly for the real-time control generation (the "integration," or summing, process performed on the increments finds these values automatically). On the other hand, position form control laws would require with each $\mathbf{y}_{d0}$ change, not only the gain changes as for the incremental form, but also explicit calculation of new $\mathbf{x}_0$ and $\mathbf{u}_0$, *and* generation of an integrator state correction term. To understand this correction term, assume that steady state performance has been reached for some setpoint $\mathbf{y}_d$ and some nominal value $\mathbf{y}_{d0}$ for linearizations. Then, if $\mathbf{y}_d$ itself is *unchanged* but $\mathbf{y}_{d0}^-$ is changed to $\mathbf{y}_{d0}^+$ at time $t_i$, one would want the control input to the system to be *unaltered*: it should remain at the appropriate steady state $\mathbf{u}_{ss}$ to yield $\mathbf{y}_{css} = \mathbf{y}_d$ and not be affected by internal relinearizations within the control law itself. This desirable property is achieved automatically for the incremental forms: in steady state operation of (15-35a)–(15-35c), all terms premultiplied by gains are zero, so changing the gains due to changed $\mathbf{y}_{d0}$ has no effect on the applied control. However, in position form laws, an integrator state correction term is needed to achieve this desirable characteristic. For instance, consider the position form of (15-35a) in steady state corresponding to $\mathbf{y}_d$, using linearizations $\mathbf{y}_{d0}^-$ or $\mathbf{y}_{d0}^+$:

$$\mathbf{u}_{ss}^- = \mathscr{u}_0(\mathbf{y}_{d0}^-) - \bar{\mathbf{G}}_{c1}^{*-}[\mathbf{x}_{ss} - \mathscr{x}_0(\mathbf{y}_{d0}^-)] - \bar{\mathbf{G}}_{c2}^{*-}\mathbf{q}^- + \mathbf{E}^-[\mathbf{y}_d - \mathbf{y}_{d0}^-]$$
$$\mathbf{u}_{ss}^+ = \mathscr{u}_0(\mathbf{y}_{d0}^+) - \bar{\mathbf{G}}_{c1}^{*+}[\mathbf{x}_{ss} - \mathscr{x}_0(\mathbf{y}_{d0}^+)] - \bar{\mathbf{G}}_{c2}^{*+}\mathbf{q}^+ + \mathbf{E}^+[\mathbf{y}_d - \mathbf{y}_{d0}^+]$$

When $\mathbf{y}_{d0}^-$ is changed to $\mathbf{y}_{d0}^+$, we want $\mathbf{u}_{ss}^+$ to equal $\mathbf{u}_{ss}^-$, yielding the required integrator state correction term for $\mathbf{q}^+$ in terms of $\mathbf{q}^-$. Thus, on the basis of ease of relinearization, the incremental forms are substantially preferable to position form perturbation laws. Note that the enhancements just discussed are based upon $\mathbf{y}_d$ assumed to be varying slowly; rapid command changes may not be adequately handled by these procedures.

EXAMPLE 15.1 Consider the gimbaled platform problem discussed in various Chapter 14 examples, but now let the dynamics include nonlinearities:

$$\dot{\mathsf{x}}(t) = -(1/T)\mathsf{x}(t) + a\mathsf{x}^3(t) + (1/T)u(t) + \mathsf{w}(t)$$

for which the associated noise-free model is

$$\dot{x}(t) = -(1/T)x(t) + ax^3(t) + (1/T)u(t)$$

Let the gimbal angle $x(t)$ be measured clockwise from a vertical reference line, as shown in Fig. 15.2. If $a > 0$, the nonlinear term $[ax^3(t)]$ indicates that positive $\dot{x}(t)$ increases above the simple first order lag type of dynamics when $x(t)$ is a large positive value, and positive $\dot{x}(t)$ decreases when $x(t)$ is a large negative value, as perhaps due to gravity effects on an imbalanced platform.

The equilibrium solutions are found by solving

$$\dot{x}_0 = -(1/T)x_0 + a{x_0}^3 + (1/T)u_0 = 0$$

and, since the desired $y_{c0} = x_0$ is the specified $y_{d0}$, we can define the $\mathscr{x}_0(\cdot)$ and $\mathscr{u}_0(\cdot)$ functions by

$$\mathscr{x}_0(y_{d0}) = x_0 = y_{d0}, \qquad \mathscr{u}_0(y_{d0}) = u_0 = y_{d0} - Tay_{d0}^3$$

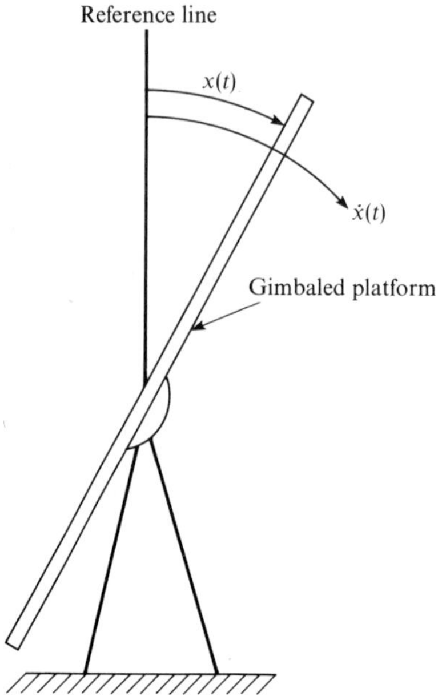


FIG. 15-2 Gimbaled platform and reference line.

where $u_0$ is directly the result of setting $\dot{x}_0 = 0$. This is a unique nominal function definition: only one $\{x_0, u_0, y_{c0}\}$ is generated for a given $y_{d0}$.

Now consider the linearized model. By (15-24), we have

$$F = \left.\frac{\partial f}{\partial x}\right|_{x=x_0,\, u=u_0} = \left[-\frac{1}{T} + 3ax_0^{\,2}\right]$$

$$B = \left.\frac{\partial f}{\partial u}\right|_{x=x_0,\, u=u_0} = +\frac{1}{T}$$

Note that when $a > 0$ and $3ax_0^{\,2} = 1/T$, the effective time constant of the linearized system is infinite: the system linearized about an $x_0$ of large magnitude is more sluggish than one linearized about $x_0 = 0$ or some small magnitude value. If $a < 0$ and $3ax_0^{\,2} = -1/T$, the effective time constant is $T/2$: the linearized system seems less sluggish as $|x_0|$ increases.

At $y_{d0} = 0$, $x_0$ and $u_0$ are both zero, yielding a linear perturbation system model that is identical to the models used in the examples of Chapter 14. Although perturbation regulators of the form (15-30) can be generated, $\bar{G}_c{}^*$ would have to be reevaluated for each new value of setpoint $y_d(t_i)$ and associated $x_0$ and $u_0$. Instead, the incremental form perturbation PI controllers can be implemented as in (15-35a), (15-35b), or (15-35c). For setpoints $y_d(t_i)$ sufficiently close to zero, we can leave $y_{d0} = 0$ and treat $y_d(t_i)$ as a perturbation from zero; i.e., the *gains* in these controllers can be evaluated on the basis of $\{x_0 = 0,\ u_0 = 0,\ y_{c0} = y_{d0} = 0\}$ and do *not* change with $y_d(t_i)$. If $y_d(t_i)$ assumes a value very different from zero, $y_{d0}$ can be changed to a nonzero value to improve the adequacy of the assumed linearized model. This changes $F$ from $[-1/T]$ to $[-(1/T) + 3ay_{d0}^2]$, which alters $\Phi$ and $B_d$ in the equivalent discrete-time model, which changes the controller gains. Now the difference $[y_d(t_i) - y_{d0}]$ is treated as $\Delta y_d(t_i)$ in the perturbation PI controller, but, as seen in (15-35), the *total* $y_d(t_i)$ can be used *directly* in the laws written in *incremental form*. ■

There is further motivation to using the incremental form of PI law instead of position form, namely, more direct compensation of "windup." Windup is an inherent stability problem associated with PI controllers applied to actuators which can saturate. Consider a very large change in desired setpoint: the proportional channel of the PI controller can cause saturation of the system's actuators, and the integrator channel begins to integrate large errors and eventually reaches a commanded control level that would also cause saturation by itself. As the tracking error signal decreases, the proportional channel output also decreases, but the already saturated integrator channel will not "discharge" its commanded control level until *after the error has changed sign*, causing the total PI controller output to stay at or near a saturating value, even though the actual system output $\mathbf{y}_c$ might be very close to the desired value. This can cause very large and undesirable overshoots in observed system behavior. There are two possible antiwindup compensation methods, both of which being considerably easier to implement in incremental forms than in position forms [34]. In all incremental PI controllers, the input to the system can be portrayed as in Fig. 15.3a, where the output of the summing junction is just a direct

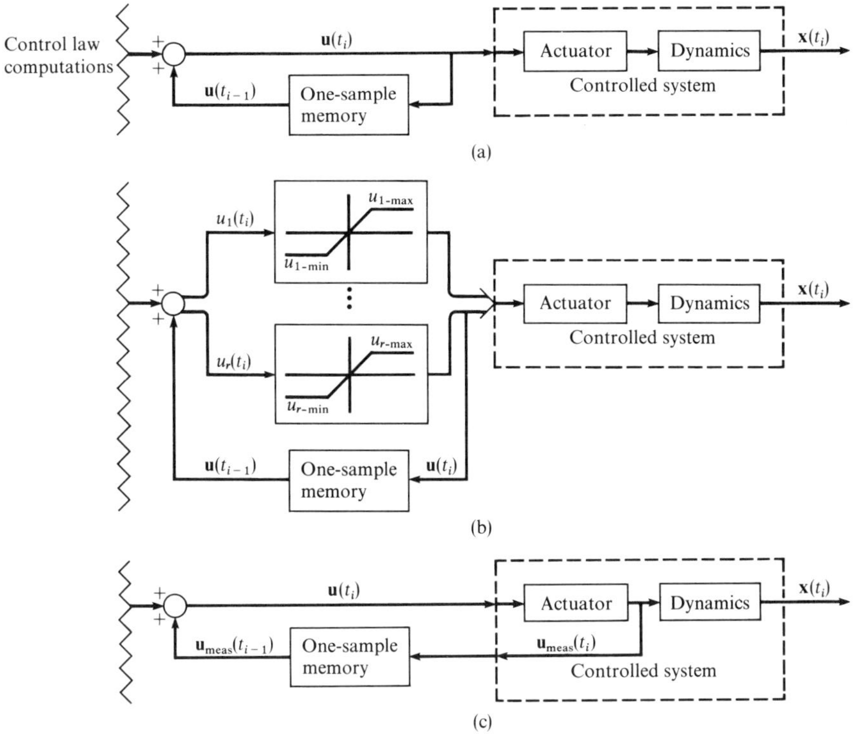


FIG. 15.3 Antiwindup compensation.

implementation of (15-35a), (15-35b), or (15-35c). The first compensation method involves adding a limiter in series with the input to the controlled system, to preclude the control law from commanding values outside the saturation limits of the actuators, as shown in Fig. 15.3b. As an alternative, if the actual actuator outputs can be measured, these are fed back to replace $\mathbf{u}(t_{i-1})$ in the control law, as in Fig. 15.3c; this feeds back saturated actuator outputs rather than preventing inputs from exceeding saturation limits.

Now consider the case in which perfect state knowledge is replaced by noisy measurements of the form (15-2) being available at each sample time. It is then reasonable to generate a linear *perturbation LQG controller* by combining the controllers just developed with the linearized Kalman filter of Section 9.5. If $\mathbf{h}$ is time invariant, then we can define an equilibrium measurement value

$$\mathbf{z}_0 = \mathbf{h}[\mathbf{x}_0] \tag{15-36}$$

which is the measured value one would see if $\mathbf{x}(t_i)$ were at its equilibrium value and if measurements were noise free. This can be subtracted from the actual system measurements to produce measurement deviations from the nominal,

$$\delta\mathbf{z}(t_i) = \mathbf{z}(t_i) - \mathbf{z}_0 \tag{15-37}$$

which can be used as inputs to the perturbation state Kalman filter based on the linearized dynamics model (15-22)–(15-25) and linearized measurement model

$$\delta\mathbf{z}(t_i) = \mathbf{H}\,\delta\mathbf{x}(t_i) + \mathbf{v}(t_i) \tag{15-38a}$$

$$\mathbf{H} = \left.\frac{\partial\mathbf{h}[\mathbf{x}]}{\partial\mathbf{x}}\right|_{\mathbf{x}=\mathbf{x}_0} \tag{15-38b}$$

EXAMPLE 15.2 Consider the problem of Example 15.1, but with discrete-time measurements available as

$$\mathrm{z}(t_i) = \sin[\mathrm{x}(t_i)] + \mathrm{v}(t_i)$$

The nominal measurement is then $z_0 = \sin[x_0]$ and $H = \cos x_0$. If we were to implement a perturbation LQG regulator to drive $\mathrm{y}_c = \mathrm{x}$ to zero ($x_0 = 0$), or a perturbation LQG PI law with a nominal setpoint of $y_{c0} = 0$, we would again replicate the designs of Chapter 14. However, if a different nominal were selected, then not only would the deterministic optimal controller change, but so would the linearized Kalman filter, since it would be based upon altered $F$, $B$, and $H$ values. ■

There is no difficulty in extending the previous discussions from being applicable to time-invariant systems with associated equilibrium solutions to being applicable to *time-varying* systems with *nominal* time-varying state, control, and measurement time histories. Moreover, we can conceive of these "nominal" histories being generated by the solution to the *optimal deterministic* control problem for the corresponding noise-free system model, i.e., for (15-1)

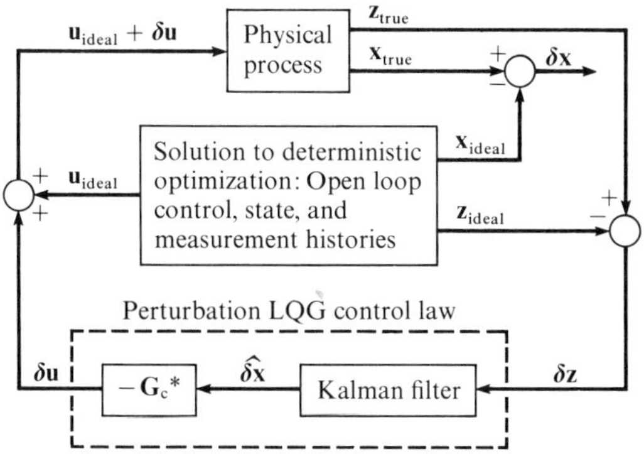


FIG. 15.4 Perturbation LQG regulation about ideal solution.

and (15-2) being replaced by

$$\dot{\mathbf{x}}(t) = \mathbf{f}[\mathbf{x}(t), \mathbf{u}(t), t] \tag{15-39a}$$

$$\mathbf{z}(t_i) = \mathbf{h}[\mathbf{x}(t_i), t_i] \tag{15-39b}$$

By using calculus of variations, Pontryagin's maximum principle, nonlinear programming, or other appropriate means of solving this deterministic optimization problem, one can then specify the *ideal* time histories of states, controls, and measurements in the form of *offline precomputed open-loop law* results [18, 33, 35, 47]. Then a perturbation LQG controller can be developed to regulate deviations from these desired ideal conditions, as shown in Fig. 15.4. The precomputed ideal measurement values are subtracted from measurements drawn from the true physical system to provide inputs, in the form of deviations from desired data, to the perturbation LQG control law. Its outputs, real-time control corrections, are then added to the ideal controls to produce the total control input to the physical system. Thus, there is a *two-level decomposition* of the original problem, into (1) a "strategic level" [10, 11, 15] to produce the *ideal solution* to the associated *nonlinear deterministic* control problem, and (2) a "tactical" or "stabilizing" [10, 11, 15] level to provide real-time *feedback control corrections*, based on linearized models, to keep deviations from ideal conditions small. Although this decomposition is not optimal, it does provide a readily synthesized and implemented result. Note that the "strategic" level in this decomposition can also be interpreted as a generalized command generator in the context of Section 14.10.

EXAMPLE 15.3 Consider the trajectory of a rocket launched from the ground to put a satellite into orbit. On the basis of the nonlinear noise-free models of dynamics, an optimal deterministic control problem solution can yield the ideal pitchover maneuver for the rocket. The ideal controls and associated state values can be precomputed and stored in the guidance system. Then a linear perturbation LQG controller can be generated to keep deviations from the desired flight path as small as possible. This is not the optimal stochastic controller: rather than control the vehicle optimally from wherever the vehicle might be located (as with a closed-loop law provided by dynamic programming based on the full-scale problem description), it instead continually drives the vehicle towards the ideal trajectory selected a priori. ■

Note that in the perturbation LQG controller, $\mathbf{F}(t)$, $\mathbf{B}(t)$, and $\mathbf{H}(t_i)$ are appropriate partials *as evaluated along the precomputed ideal control and state trajectories*, in contrast to the case of extended Kalman filtering in which relinearizations about the current state estimates were accomplished after each measurement update. Such evaluation makes sense here because, at the "strategic" level, you can in fact declare a desirable equilibrium point or ideal time history of values, *toward which you will be controlling at all times*. Thus, unlike the estimation problem, you are taking appropriate action to keep close to the a priori declared values, and thus these provide viable values about which to linearize. It is possible to conceive of a relinearized LQG controller, but this eliminates the possibility of precomputability of gains and the use of steady state gains for all time as an approximation, both of which are important in practice because of their computational advantages; furthermore, you should stay close enough to the ideal values not to require such relinearization. Many practical, operational designs have been in the form of steady state constant-gain perturbation LQG controllers with offline precomputation of both filter gains and deterministic optimal controller gains. For computational reasons, time-invariant models driven by stationary noises, with constant cost weighting matrices, are exploited (treating any time variations quasi-statically), and the steady state result used for real-time application. Gain-scheduling of these constant gains as functions of important parameters (or estimates thereof) that undergo time variations (or that are uncertain) is often accomplished as well; the case of uncertain parameters is addressed in Section 15.6.

Stability and robustness are critical issues for the perturbation LQG controller or any other form of "tactical" controller in this decomposition. First and foremost, this feedback path must provide stability; once that is assured, other performance measures can be considered. The feedback system must be able to tolerate substantial modeling errors, due to such sources as changing parameters, model order reduction and simplification, and failures in sensors or actuators. Even if adaptive designs are to be used for identifying uncertain parameters or failed devices, one often must provide a robust control law to provide at least a stable closed-loop system in the interim period between realizing (by residual monitoring) that *something* happened in the real system that requires adaptation, then isolating exactly what needs adaptation, and finally performing the appropriate change. For instance, a survivable flight control system for military aircraft might be able to reconfigure its control system once it identifies a failed device, but it requires a robust control law to maintain stability of the aircraft, despite major unknown changes in the real system and thus without major control system reconfiguration, until its fault detection logic is able to determine exactly what components have failed. Thus, the stability and robustness analyses of Chapter 14 are especially important for implementation of linear perturbation LQG control laws for nonlinear system applications.

## 15.4 ASSUMED CERTAINTY EQUIVALENCE DESIGN

One common feedback controller synthesis technique is to separate the stochastic controller into the cascade of an estimator and a deterministic optimal control function even when the *optimal* stochastic controller does not have the certainty equivalence property. This is known as the *assumed (or forced, or heuristic) certainty equivalence design* technique [9, 28, 69, 70, 131, 140, 145].

Basically, one sets up and solves the associated deterministic optimal control problem, inoring the uncertainties and assuming perfect access to the entire state, attaining the feedback law $\mathbf{u}^{\mathrm{DO}}[\mathbf{x}(t_i), t_i]$. To achieve a feedback form may be a difficult task in itself: calculus of variations, the maximum principle, and nonlinear programming generally provide solutions in the form of open-loop control and corresponding state time histories [18, 33, 35, 47]. Either these are manipulated into feedback form, or deterministic dynamic programming is used to generate the control law fundamentally as a feedback law. One then solves the state estimation problem to generate $\hat{\mathbf{x}}(t_i^+)$ as the conditional mean of the state conditioned on the current available information, or some approximation to this conditional mean (as provided by the extended Kalman filter of Chapter 9 or any of the nonlinear filters of Chapter 12). Finally, the assumed certainty equivalence control law is generated as

$$\mathbf{u}^{\mathrm{ACE}}(t_i) = \mathbf{u}^{\mathrm{DO}}[\hat{\mathbf{x}}(t_i^+), t_i] \tag{15-40}$$

This is illustrated in Fig. 15.5.

One important special case of this design methodology is the cascading of a *constant-gain extended Kalman filter with a constant-gain linear quadratic state feedback (deterministic optimal) controller* [108]. Consider the special case of

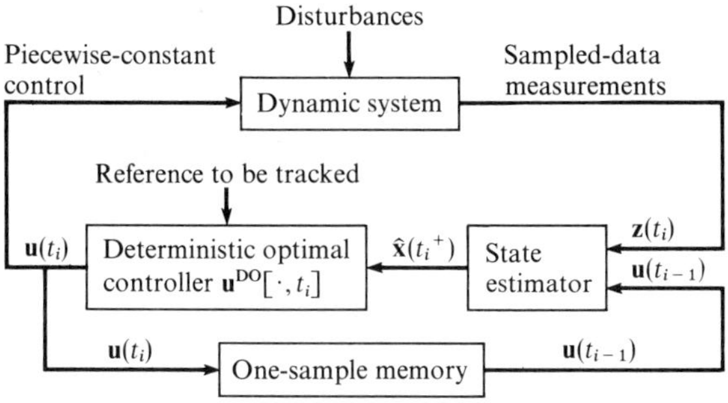


FIG. 15.5 Assumed certainty equivalence design.

(15-1) and (15-2) in which $\mathbf{G}$ is a function only of time, and not the current state; as usual, let $\boldsymbol{\beta}(\cdot,\cdot)$ be of diffusion $\mathbf{Q}(t)$ and $\mathbf{v}(\cdot,\cdot)$ be of covariance $\mathbf{R}(t_i)$ for all times of interest. A nominal linearization is used for design, yielding

$$\dot{\mathbf{x}}(t) = \mathbf{F}\mathbf{x}(t) + \mathbf{B}\mathbf{u}(t) + \mathbf{G}\mathbf{w}(t) \tag{15-41a}$$

$$\mathbf{z}(t_i) = \mathbf{H}\mathbf{x}(t_i) + \mathbf{v}(t_i) \tag{15-41b}$$

As stated previously, we *know* a good location about which to linearize by the basic objective of the control problem. Moreover, (15-41) is of a time-invariant form and $\mathbf{w}(\cdot,\cdot)$ and $\mathbf{v}(\cdot,\cdot)$ are assumed stationary ($\mathbf{Q}$ and $\mathbf{R}$ constant): either $\mathbf{f}$, $\mathbf{h}$, $\mathbf{Q}$, and $\mathbf{R}$ are truly time invariant originally, or time variations are treated quasi-statically in this linearization. Furthermore, the defining matrices in the quadratic cost to be minimized are also assumed to be constant.

First the constant-gain linear quadratic full-state feedback controller is designed for the time-invariant linear perturbation model. To minimize a deterministic cost of the form given in (14-13) over an infinite interval of time ($N \to \infty$ to ignore terminal transients), we would apply the control law given by (15-27)–(15-30), with constant $\bar{\mathbf{G}}_c{}^*$ precomputed offline as

$$\bar{\mathbf{G}}_c{}^* = [\mathbf{U} + \mathbf{B}_d{}^T\bar{\mathbf{K}}_c\mathbf{B}_d]^{-1}[\mathbf{B}_d{}^T\bar{\mathbf{K}}_c\boldsymbol{\Phi} + \mathbf{S}^T] \tag{15-42}$$

where $\bar{\mathbf{K}}_c$ is the solution to the steady state (algebraic) Riccati equation

$$\bar{\mathbf{K}}_c = \mathbf{X} + \boldsymbol{\Phi}^T\bar{\mathbf{K}}_c\boldsymbol{\Phi} - [\mathbf{B}_d{}^T\bar{\mathbf{K}}_c\boldsymbol{\Phi} + \mathbf{S}^T]^T[\mathbf{U} + \mathbf{B}_d{}^T\bar{\mathbf{K}}_c\mathbf{B}_d]^{-1}[\mathbf{B}_d{}^T\bar{\mathbf{K}}_c\boldsymbol{\Phi} + \mathbf{S}^T] \tag{15-43}$$

Whereas the *design system* closed-loop dynamics are

$$\mathbf{x}(t_{i+1}) = \boldsymbol{\Phi}\mathbf{x}(t_i) - \mathbf{B}_d\bar{\mathbf{G}}_c{}^*\mathbf{x}(t_i) + \mathbf{G}_d\mathbf{w}_d(t_i) \tag{15-44}$$

the true system dynamics under such full-state control would be modeled as

$$\mathbf{dx}(t) = \mathbf{f}[\mathbf{x}(t), -\bar{\mathbf{G}}_c{}^*\mathbf{x}(t_i), t]\,dt + \mathbf{G}\,\mathbf{d}\boldsymbol{\beta}(t) \tag{15-45}$$

for all $t \in [t_i, t_{i+1})$. Although both modeling errors and performance index definitions affect robustness, such a linear quadratic *full-state* feedback controller has been shown in Section 14.5 and by practical experience to be inherently very robust [108, 109, 111].

The constant-gain extended Kalman filter has the basic structure of an extended Kalman filter, except that the constant gain is precomputed based on the linearization about the nominal. Again, this makes sense instead of iterative relinearizations because the controller will be attempting to drive the system towards the design nominal. At sample time $t_i$, the filter updates its state estimate according to

$$\hat{\mathbf{x}}(t_i{}^+) = \hat{\mathbf{x}}(t_i{}^-) + \bar{\mathbf{K}}\{\mathbf{z}_i - \mathbf{h}[\hat{\mathbf{x}}(t_i{}^-), t_i]\} \tag{15-46}$$

where $\bar{\mathbf{K}}$ is given by

$$\bar{\mathbf{K}} = \bar{\mathbf{P}}^-\mathbf{H}^T[\mathbf{H}\bar{\mathbf{P}}^-\mathbf{H}^T + \mathbf{R}]^{-1} \tag{15-47}$$

where $\bar{\mathbf{P}}^-$ is the steady state solution to the Riccati recursion

$$\bar{\mathbf{P}}^- = \mathbf{\Phi}\{\bar{\mathbf{P}}^- - \bar{\mathbf{P}}^-\mathbf{H}^T[\mathbf{H}\bar{\mathbf{P}}^-\mathbf{H}^T + \mathbf{R}]^{-1}\mathbf{H}\bar{\mathbf{P}}^-\}\mathbf{\Phi}^T + \mathbf{G}_d\mathbf{Q}_d\mathbf{G}_d^T \quad (15\text{-}48)$$

based on the linearized system. Between sample times $t_i$ and $t_{i+1}$, the propagation equations are

$$\dot{\hat{\mathbf{x}}}(t/t_i) = \mathbf{f}[\hat{\mathbf{x}}(t/t_i), \mathbf{u}(t_i), t] \quad (15\text{-}49)$$

starting from $\hat{\mathbf{x}}(t_i/t_i) = \hat{\mathbf{x}}(t_i^+)$ and yielding $\hat{\mathbf{x}}(t_{i+1}^-) = \hat{\mathbf{x}}(t_{i+1}/t_i)$. Such a constant-gain extended Kalman filter has also been shown theoretically and by practical experience to be rather robust (against divergence) [108, 109, 111]. This approximation to the conditional mean of $\mathbf{x}(t_i)$ is then used to replace $\mathbf{x}(t_i)$ itself in the full-state feedback law.

EXAMPLE 15.4 Recall the problem of Examples 15.1 and 15.2, and consider regulation of $x(t)$ toward zero ($x_0 = y_d = 0$ for linearizations). Then the assumed certainty equivalence design based on the constant-gain extended Kalman filter and full-state controller is

$$u(t_i) = -\bar{G}_c^* \hat{x}(t_i^+)$$

where $\hat{x}(t_i^+)$ is given by

$$\hat{x}(t_i^+) = \hat{x}(t_i^-) + \bar{K}[z_i - \sin \hat{x}(t_i^-)]$$

To propagate to the next sample time, we use

$$\dot{\hat{x}}(t/t_i) = -(1/T)\hat{x}(t/t_i) + a\hat{x}^3(t/t_i) + (1/T)u(t_i)$$

The gains $\bar{G}_c^*$ and $\bar{K}$ are identical to the values of the previous examples, but the propagation relation and residual generation now use $f$ and $h$ explicitly instead of linearizations thereof. To be explicit, the preceding algorithm replaces the linear perturbation result

$$u(t_i) = -\bar{G}_c^* \hat{x}(t_i^+)$$

$$\begin{aligned} \hat{x}(t_i^+) &= \hat{x}(t_i^-) + \bar{K}[z_i - (\sin 0) - (\cos 0)\hat{x}(t_i^-)] \\ &= \hat{x}(t_i^-) + \bar{K}[z_i - \hat{x}(t_i^-)] \end{aligned}$$

$$\begin{aligned} \dot{\hat{x}}(t/t_i) &= [-(1/T) + 3a\,0^2]\hat{x}(t/t_i) + (1/T)u(t_i) \\ &= -(1/T)\hat{x}(t/t_i) + (1/T)u(t_i) \quad \blacksquare \end{aligned}$$

Continuous-measurement versions of such constant-gain controllers are also important practically. As in Section 14.13, a constant-gain linear quadratic full-state feedback controller can be developed for the linearized model (15-41a) in the form of

$$\mathbf{u}(t) = -[\mathbf{W}_{uu}^{-1}\mathbf{B}^T\bar{\mathbf{K}}_c]\mathbf{x}(t) \quad (15\text{-}50)$$

where $\bar{\mathbf{K}}_c$ is the solution to the steady state (algebraic) Riccati equation

$$-\dot{\bar{\mathbf{K}}}_c = \mathbf{0} = \mathbf{F}^T\bar{\mathbf{K}}_c + \bar{\mathbf{K}}_c\mathbf{F} + \mathbf{W}_{xx} - \bar{\mathbf{K}}_c\mathbf{B}\mathbf{W}_{uu}^{-1}\mathbf{B}^T\bar{\mathbf{K}}_c \quad (15\text{-}51)$$

which is evaluated offline. Thus, whereas the *design* closed-loop system dynamics are

$$\dot{\mathbf{x}}(t) = \mathbf{F}\mathbf{x}(t) - \mathbf{B}\mathbf{W}_{uu}^{-1}\mathbf{B}^T\bar{\mathbf{K}}_c\mathbf{x}(t) + \mathbf{G}\mathbf{w}(t) \quad (15\text{-}52)$$

the true system dynamics can be modeled as

$$\mathbf{dx}(t) = \mathbf{f}[\mathbf{x}(t), -\mathbf{W}_{uu}^{-1}\mathbf{B}^{\mathrm{T}}\overline{\mathbf{K}}_{\mathrm{c}}\mathbf{x}(t), t]\,dt + \mathbf{G}\,\mathbf{d}\boldsymbol{\beta}(t) \tag{15-53}$$

Such a full-state control law has *at least* $\pm 60°$ phase margin, infinite gain margin, and 50% (6 dB) gain reduction toleration in each control channel [108, 109], with stronger results for specific applications, as discussed in Section 14.5. This can be cascaded with the constant-gain extended Kalman filter

$$\dot{\hat{\mathbf{x}}}(t) = \mathbf{f}[\hat{\mathbf{x}}(t), -\mathbf{W}_{uu}^{-1}\mathbf{B}^{\mathrm{T}}\overline{\mathbf{K}}_{\mathrm{c}}\hat{\mathbf{x}}(t), t] + \{\overline{\mathbf{P}}\mathbf{H}^{\mathrm{T}}\mathbf{R}_{\mathrm{c}}^{-1}\}\{\mathbf{z}(t) - \mathbf{h}[\hat{\mathbf{x}}(t), t]\} \tag{15-54}$$

where $\overline{\mathbf{P}}$ is the solution to the steady state (algebraic) Riccati equation,

$$\dot{\overline{\mathbf{P}}} = \mathbf{0} = \mathbf{F}\overline{\mathbf{P}} + \overline{\mathbf{P}}\mathbf{F}^{\mathrm{T}} + \mathbf{G}\mathbf{Q}\mathbf{G}^{\mathrm{T}} - \overline{\mathbf{P}}\mathbf{H}^{\mathrm{T}}\mathbf{R}_{\mathrm{c}}^{-1}\mathbf{H}\overline{\mathbf{P}} \tag{15-55}$$

which is also solved offline, with all partials evaluated at the preselected nominal operating point conditions. This yields the final controller,

$$\mathbf{u}(t) = -[\mathbf{W}_{uu}^{-1}\mathbf{B}^{\mathrm{T}}\overline{\mathbf{K}}_{\mathrm{c}}]\hat{\mathbf{x}}(t) \tag{15-56}$$

EXAMPLE 15.5 Consider the same problem as in Example 15.4, but with continuous-time measurements of the same form,

$$\mathsf{z}(t) = \sin \mathsf{x}(t) + \mathsf{v}_{\mathrm{c}}(t)$$

The control law becomes

$$u(t) = -\bar{G}_{\mathrm{c}}^{*}\hat{x}(t) = -[\bar{K}_{\mathrm{c}}/(W_{uu}T)]\hat{x}(t)$$

with

$$\dot{\hat{x}}(t) = -(1/T)\hat{x}(t) + a\hat{x}^{3}(t) - (1/T)\bar{G}_{\mathrm{c}}^{*}\hat{x}(t) + \bar{K}\{z(t) - \sin\hat{x}(t)\}$$

where $\bar{G}_{\mathrm{c}}^{*}$ and $\bar{K}$ are evaluated as in Example 14.25. ■

In either the discrete-time or continuous-time measurement case, when perfect state knowledge is replaced by the state estimate from the filter, the robustness of the feedback controller can be significantly reduced, because of the model mismatch in the filter [45, 111]. To preserve the desired robustness, then, one should devote *substantial design effort* to making $\hat{\mathbf{x}}$ a very good representation of $\mathbf{x}$. The basic robustness of the simple constant-gain linear quadratic full-state law can be exploited, as long as the filter provides a precise estimate of the state under all operating conditions. For this reason, the model upon which the filter is based may well be more complex and of higher dimension (such as by treating uncertain parameters as additional states) than the model used to design the deterministic optimal control law: the additional states and complexity are not meant necessarily to provide additional estimates to the control law, but rather, enhanced precision in estimating the original states. Furthermore, *estimation of uncertain parameters* (see Section 15.6) *and failure detection* [147] are often *crucial* to the *estimation portion* of the stochastic controller structure. When changes occur in the real-world system, the state estimator must be able to *adapt its internal system model* and thus be able to provide an accurate $\hat{\mathbf{x}}$ to the simple and robust optimal controller originally developed as a full-state feedback law for the case of no uncertain parameters. Even for large changes in the real world, it is often adequate to use *constant filter and controller gains*, as long as the system identification and failure detection are incorporated to allow *adaptation* of the filter's *system model*. In this way, the computationally expensive gain calculations can be performed offline. However, if modeling errors and/or actuator or sensor failures are not known to, or compensated by, the filter, then the stability robustness of the cascade combination is totally problem dependent (with no guarantees whatsoever) and often totally inadequate. More will be said about stochastic adaptive control in Section 15.6.

There is no fundamental reason to limit attention to constant-gain designs other than computational burden of the resulting algorithm. By considering finite time durations, filter gain initial transients and deterministic controller gain terminal transients become important and readily computed. Time-varying system, statistics, or cost matrices also generalize the technique with no theoretical difficulty [103, 153, 154].

There are other forms of assumed certainty equivalence designs as well. In Fig. 15.5, the state estimator may be of any of the forms developed in Chapter 10 or 12 (Volume 2) instead of an extended Kalman filter, and the deterministic optimal controller may be totally unrelated to LQ regulators and may be nonlinear in general.

EXAMPLE 15.6 Reconsider Example 15.4, and again use the LQ regulator design, but replace the constant-gain extended Kalman filter with a constant-gain extended Kalman filter with bias correction terms, yielding

$$u(t_i) = -\bar{G}_c{}^* \hat{x}(t_i{}^+)$$

as before, but with

$$\hat{x}(t_i{}^+) = \hat{x}(t_i{}^-) + \bar{K}[z_i - \sin \hat{x}(t_i{}^-) - \tfrac{1}{2}\bar{P} \sin \hat{x}(t_i{}^-)]$$
$$\dot{\hat{x}}(t/t_i) = -(1/T)\hat{x}(t/t_i) + a\hat{x}^3(t/t_i) + (1/T)u(t_i) + 3a\bar{P}\hat{x}(t/t_i)$$

Time-varying gains, adaptive filters, second order filters, or statistically linearized filters could also be used. A time-optimal deterministic control law, in the form of a "bang–bang" nonlinear law, could replace the LQ regulator law used above. ■

## 15.5 CLOSED-LOOP LAW APPROXIMATIONS AND DUAL EFFECT

The previous methods generated control laws in the *feedback* class. It would be desirable to be able to synthesize *closed-loop* control laws, by means other than attempting direct solution of the dynamic programming algorithm, that specifically exhibit the probing and caution characteristics of the full-scale optimal stochastic controller [53, 73, 115, 137].

One such controller with good potential for synthesis and implementation has been denoted as a *dual controller* [28–30, 36, 117, 134, 135, 138, 139]. In this approach, the stochastic dynamic programming relation (see (13-86)–(13-88))

$$\mathscr{C}^*[\mathbf{I}_i, t_i] = \min_{\mathbf{u} \in \mathscr{U}} \left[ E\{L[\mathbf{x}(t_i), \mathbf{u}\{\mathbf{I}(t_i), t_i\}, t_i] + \mathscr{C}^*[\mathbf{I}(t_{i+1}), t_{i+1}] \,|\, \mathbf{I}(t_i) = \mathbf{I}_i\} \right] \tag{15-57}$$

itself is approximated. First of all, the information vector $\mathbf{I}_i$ is replaced by an approximate information description in terms of $\hat{\mathbf{x}}(t_i^+)$ and $\mathbf{P}(t_i^+)$ as produced by some nonlinear estimator, as especially an extended Kalman filter (uncertain parameters can be treated as additional states). Then the control to apply at time $t_i$ is generated by a numerical search as follows. An arbitrary control function, such as an assumed certainty equivalence design, is assumed to be used for times $t_i, t_{i+1}, \ldots, t_N$, thereby generating a *nominal predicted state and control time history* (it is possible to propagate to time $t_{i+M}$ for fixed integer $M$ instead of propagating all the way to the final control application time $t_N$). Stochastic effects are then described via a second order perturbation analysis about this nominal, allowing evaluation of the expected cost associated with the nominal plus a specific perturbation control choice through a set of recursions. An iteration is then conducted to find the control that yields the minimum expected cost of this form. Since the state covariance is precomputed along the nominal trajectory, and this nominal depends on the current control $\mathbf{u}(t_i)$, then the effect of $\mathbf{u}(t_i)$ on the quality of future information $\mathbf{P}(t_j^+)$ for all $t_j > t_i$ is inherently embedded into the determination of the control if the system model is not neutral, as desired.

EXAMPLE 15.7 Let the position of a missile in a plane be denoted by $(r_x, r_y)$, and the associated velocity by $(v_x, v_y)$. Assume it is a point mass moving with constant speed, and that it is controlled by a constrained-magnitude lateral acceleration command $u$, yielding a dynamics model of

$$\begin{bmatrix} \dot{\mathsf{r}}_x(t) \\ \dot{\mathsf{v}}_x(t) \\ \dot{\mathsf{r}}_y(t) \\ \dot{\mathsf{v}}_y(t) \end{bmatrix} = \begin{bmatrix} \mathsf{v}_x(t) \\ -\mathsf{v}_y(t)\mathsf{u}(t) \\ \mathsf{v}_y(t) \\ \mathsf{v}_x(t)\mathsf{u}(t) \end{bmatrix} + \begin{bmatrix} 0 \\ \mathsf{w}_1(t) \\ 0 \\ \mathsf{w}_2(t) \end{bmatrix}$$

At each sampling time, assume only line-of-sight angle measurements are available as

$$\mathsf{z}(t_i) = \tan^{-1}[\mathsf{r}_y(t_i)/\mathsf{r}_x(t_i)] + \mathsf{v}(t_i)$$

It is desired to generate the appropriate control law to drive the missile toward the origin of the coordinate system, based on uncertain initial conditions and real-time measurements.

Assumed certainty equivalence controllers will attempt to drive the missile along a straight line from initial (or current) position toward the origin. However, along such a trajectory, the angle-only measurements do not provide good range information, so that state estimation uncertainty remains high along the line-of-sight direction. In contrast, the dual controller yields a solution in the form of an "S-turn" about that straight-line path, deviating first to one side and then the other, to gain better range information and eventually decrease the rms miss distance by means of enhanced state estimation accuracy [36, 135].

For design purposes, one can generate this dual control law or perhaps seek the full-scale (numerical) solution to the dynamic programming algorithm. Once the structure and characteristics of the resulting closed-loop law are evaluated, an ad hoc offline law computation can be generated to approximate these characteristics, as in this case by performing an offline optimization to find the particular S-turn to minimize terminal rms miss distance for a given scenario or set of scenarios. ■

## 15.6 STOCHASTIC ADAPTIVE CONTROL

Consider the stochastic optimal control of a system in the presence of uncertain parameters, a special case of the general problem formulated in this chapter [1, 6, 22, 25, 31, 40, 49, 66, 68, 72, 73, 75, 83, 84, 86–88, 99, 100, 102, 104, 105, 107, 110, 114, 118, 120, 126, 130, 150]. Thus, in (15-1) and (15-2), $\mathbf{f}$, $\mathbf{G}$, and $\mathbf{h}$ may be functions of a vector of uncertain parameters $\mathbf{a}(t)$, or the second moment matrices $\mathbf{Q}$ and $\mathbf{R}$ describing the dynamics and measurement noises may be dependent on $\mathbf{a}(t)$; the special case of uncertain parameters embedded in a *linear* model [20, 21, 51, 80] was discussed in detail in Chapter 10. Basic models of uncertain parameters include unknown constants, essentially constant values over the most recent $N$ sample periods, and Markov dynamics models, as

$$\dot{\mathbf{a}}(t) = \mathbf{f}_a[\mathbf{a}(t), t] + \mathbf{G}_a[\mathbf{a}(t), t]\mathbf{w}_a(t) \tag{15-58}$$

and especially linear dynamics models driven by white Gaussian pseudonoise:

$$\dot{\mathbf{a}}(t) = \mathbf{F}_a(t)\mathbf{a}(t) + \mathbf{G}_a(t)\mathbf{w}_a(t) \tag{15-59a}$$

$$\mathbf{a}(t_{i+1}) = \mathbf{\Phi}_a(t_{i+1}, t_i)\mathbf{a}(t_i) + \mathbf{w}_{da}(t_i) \tag{15-59b}$$

A particularly useful model of the form of (15-59) is the random constant with pseudonoise added,

$$\dot{\mathbf{a}}(t) = \mathbf{w}_a(t) \tag{15-60a}$$

$$\mathbf{a}(t_{i+1}) = \mathbf{a}(t_i) + \mathbf{w}_{da}(t_i) \tag{15-60b}$$

Typically, the parameter model does *not* depend on state and control values. As a result, there is no effective way to use control to ensure the adequacy of linearizations about an assumed nominal $\mathbf{a}_0$ as we were able to do with nominal $\mathbf{x}_0$ previously. In view of this difficulty, we wish to synthesize implementable stochastic controllers that can provide adequate control despite variations in, or uncertainty in knowledge of, "true" parameter values [72].

In some cases, the natural robustness of a nonadaptive control law admits adequate control in the face of uncertain parameters [24, 108, 109, 111]. Or, as discussed in Section 14.5, a controller may purposely be "robustified" to provide such a characteristic [42, 111]. However, if adequate performance is not achievable in this manner, an adaptive controller must be developed.

One means of generating an adaptive stochastic feedback controller is by *assumed certainty equivalence design* [1, 23, 28, 48, 50, 78, 79, 85, 115, 124, 125, 145]. Consider the special case of a linear system model

$$\mathbf{x}(t_{i+1}) = \boldsymbol{\Phi}(t_{i+1}, t_i; \mathbf{a})\mathbf{x}(t_i) + \mathbf{B}_d(t_i; \mathbf{a})\mathbf{u}(t_i) + \mathbf{w}_d(t_i) \tag{15-61a}$$

$$\mathbf{z}(t_i) = \mathbf{H}(t_i; \mathbf{a})\mathbf{x}(t_i) + \mathbf{v}(t_i) \tag{15-61b}$$

as discussed in Chapter 10. If one treats the parameters as additional states, then (15-59b) can be augmented to (15-61a), with the result being a nonlinear estimation problem because of the product relations in (15-61). As seen in Example 9.9, an extended Kalman filter can be used to generate simultaneous estimates of states and parameters for this problem formulation. Similarly, any of the nonlinear filters of Chapter 12 would be equally applicable. At time $t_i$, one can then generate $\hat{\mathbf{x}}(t_i^+)$ and $\hat{\mathbf{a}}(t_i^+)$. Moreover, one can predict the future parameter values associated with the model (15-59b) via

$$\hat{\mathbf{a}}(t_{j+1}/t_i) = \boldsymbol{\Phi}_a(t_{j+1}, t_j)\hat{\mathbf{a}}(t_j/t_i) \tag{15-62}$$

for all $t_j$ to the end of the control interval of interest, starting from the current value:

$$\hat{\mathbf{a}}(t_i/t_i) = \hat{\mathbf{a}}(t_i^+) \tag{15-63}$$

Using this parameter time history, the deterministic LQ optimal controller gains can be evaluated: the backward Riccati equation can be evaluated by letting $\mathbf{a}(t_j) = \hat{\mathbf{a}}(t_j/t_i)$ in the $\boldsymbol{\Phi}$ and $\mathbf{B}_d$ matrices in this recursion. This yields the time history of controller gains

$$\mathbf{G}_c^*[t_k; \{\hat{\mathbf{a}}(t_j/t_i) \text{ for all } t_j \in [t_k, t_N]\}]$$

for $k = N, N-1, \ldots, i$. *Conceptually*, this entire recursion could be reevaluated online each time a new measurement is taken and processed. Then the control can be generated as

$$\mathbf{u}(t_i) = -\mathbf{G}_c^*[t_i; \{\hat{\mathbf{a}}(t_j/t_i), t_j \in [t_i, t_N]\}]\hat{\mathbf{x}}(t_i^+) \tag{15-64}$$

In the special case of the parameter model (15-60b), the prediction of future parameter values reduces to

$$\hat{\mathbf{a}}(t_j/t_i) = \hat{\mathbf{a}}(t_i^+) \tag{15-65}$$

for all future $t_j$. In other words, until the next measurement updates your estimate, you predict that the current parameter estimate value is the best estimate of parameters for all time into the future. Then the appropriate controller gains to evaluate by backward recursion of the Riccati equation can be expressed simply as

$$\mathbf{G}_c^*[t_k; \hat{\mathbf{a}}(t_i^+)]$$

for $k = N, N-1, \ldots, i$. For computational feasibility, instead of performing this calculation in real time, one can precompute and store $\mathbf{G}_c^*$ (or more simply, steady state $\bar{\mathbf{G}}_c^*$) as a function of $\mathbf{a}$ for various $\mathbf{a}$ values, and evaluate $\mathbf{G}_c^*[t_i; \hat{\mathbf{a}}(t_i^+)]$ by simple interpolation of prestored values. Thus, the basic structure of the adaptive controller is as seen in Fig. 15.6. Note that the parameter estimates are used to evaluate the controller gain $\mathbf{G}_c^*(t_i)$ as well as to enhance the precision in estimating the state $\mathbf{x}(t_i)$. This state estimate is then premultiplied by the controller gain to generate the control to apply to the system. In some applications, it is adequate to use the parameter estimates *only* for state estimation improvement, with $\mathbf{G}_c^*(t_i)$ nonadaptive and evaluated on the basis of a nominal parameter value $\mathbf{a}_{\text{nom}}$, instead of the predictions $\{\hat{\mathbf{a}}(t_j/t_i)\}$ generated in real time. This is computationally advantageous and can often provide reasonable performance due to the inherent robustness of a full-state regulator fed by *precise* state representations, as discussed in Section 15.4.

With parameters treated as additional states, the state and parameter estimator in Fig. 15.6 is naturally an extended Kalman filter or one of the nonlinear filters of Chapter 12. However, for unknown constant or constant-over-$N$-sample-periods models, the maximum likelihood estimators or approximations thereto, as discussed in Sections 10.3–10.6, are equally viable [48]. Or, for discrete or discretized parameters, the multiple model filtering algorithm of Section 10.8 can be employed. If uncertain parameters confined to noise statistics matrices $\mathbf{Q}$ and $\mathbf{R}$ are to be considered, as in Sections 10.7–10.10, the $\mathbf{G}_c^*(t_i)$ value is *not* dependent on $\mathbf{a}$, and estimation of these parameters is performed *only* to enhance the state estimate accuracy.

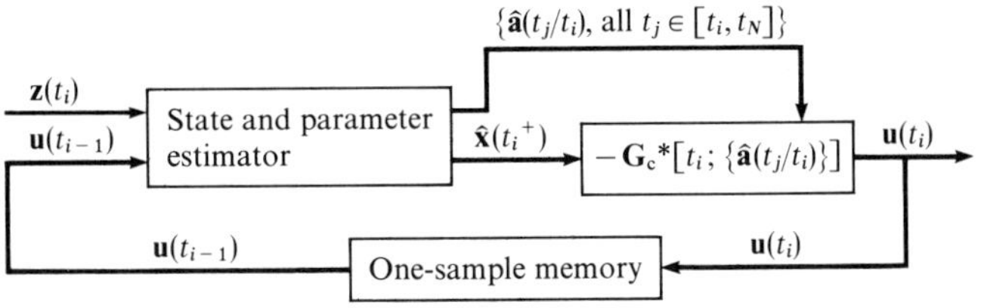


FIG. 15.6 Stochastic adaptive controller generated by assumed certainty equivalence.

EXAMPLE 15.8 Recall the thrust vector control problem of Examples 9.9 and 10.17 [93, 94]. An LQG regulator can be developed in the form of

$$u^*(t_i) = \delta^*_{\text{com}}(t_i) = -\mathbf{G}_c^*(t_i)\hat{\mathbf{x}}(t_i^+)$$

with

$$\hat{\mathbf{x}}(t_i^+) = [\hat{\omega}(t_i^+)\hat{\theta}(t_i^+)\hat{v}_b(t_i^+)\hat{q}(t_i^+)\delta(t_i)]^T$$

for any assumed, known, constant value of the parameter $\omega_b^2$, i.e., the bending mode natural frequency. (Note that $\delta$ is a *deterministic* function of $\delta_{\text{com}}$ by the model in Example 9.9, and thus does not require estimation in a state estimator.) In fact, $\mathbf{G}_c^*[t_i; \omega_b^2]$ was evaluated for sake of demonstration for $\omega_b^2 = 50$, 90, 100, 110, 200, 300, and 400 rad$^2$/sec$^2$, and simple curve-fitted approximations generated to evaluate the five scalar gains $\mathbf{G}_c^*[t_i; \widehat{\omega_b^2}(t_i^+)]$ by interpolation (since $\omega_b^2$ was nominally 100 rad$^2$/sec$^2$, greater interpolation fidelity was provided in the region of this value).

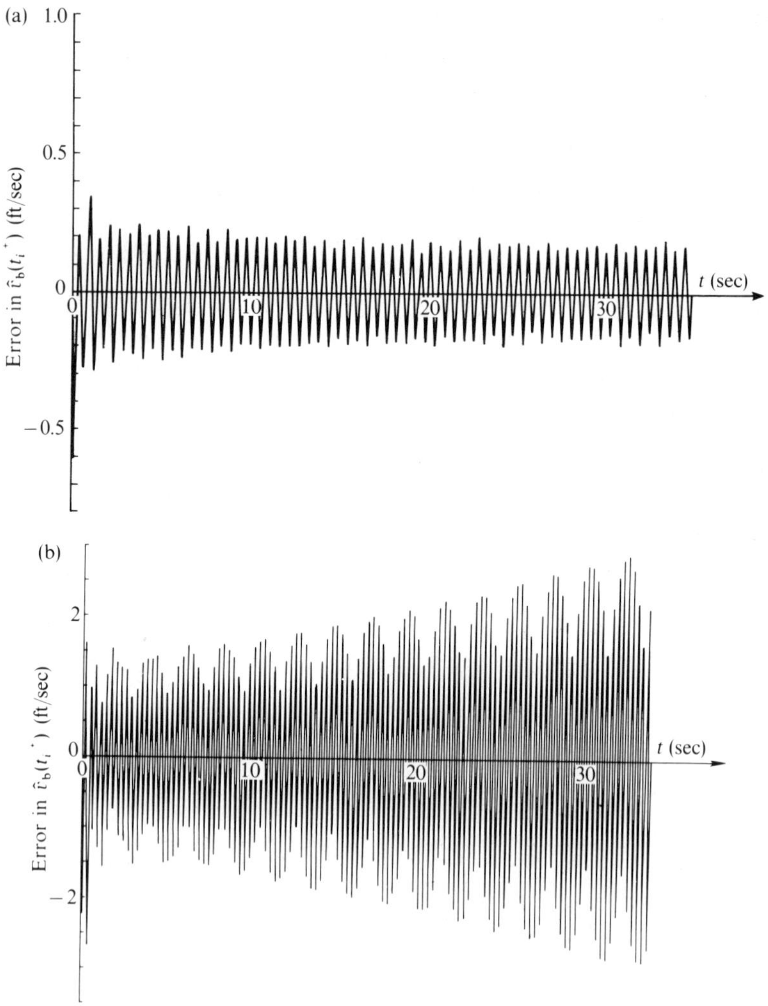


FIG. 15.7 (a) Nonadaptive LQG thrust vector controller based on $\omega_b^2 = 100$ rad$^2$/sec$^2$; (a) Error in $\hat{v}_b(t_i^+)$ for true $\omega_b^2 = 130$ rad$^2$/sec$^2$. (b) Error in $\hat{v}_b(t_i^+)$ for true $\omega_b^2 = 400$ rad$^2$/sec$^2$. From Maybeck [93].

Controllers of various structures were then evaluated by Monte Carlo simulations in which the "truth model" was the same basic system model with $\omega_b^2$ set at various different values (model order reduction was therefore not an issue in controller performance).

A nonadaptive LQG controller that assumed $\omega_b^2 = 100$ rad$^2$/sec$^2$ provided adequate performance as long as the truth model $\omega_b^2$ did not vary too far from the filter assumed value (sensitivity to $\omega_b^2$ is accentuated by the assumption that the uncontrolled system has an undamped, a stable bending mode). Figure 15.7a shows the error in the estimate of the bending velocity coordinate $\hat{v}_b$ in a typical simulation run in which the truth model $\omega_b^2$ was set to 130 rad$^2$/sec$^2$; the corresponding state regulation error was very similar, and approximately three times the magnitude of errors committed when true $\omega_b^2$ matched the assumed 100 rad$^2$/sec$^2$. As the model mismatch increased, performance degraded to that depicted in Fig. 15.7b for the case of true $\omega_b^2 = 400$ rad$^2$/sec$^2$: severe closed-loop instability has been established.

Figure 15.8 presents the corresponding results for a stochastic adaptive controller as in Fig. 15.6, in which the state and parameter estimation was accomplished by a modified Gaussian second order filter as portrayed in Section 12.3, and in which $G_c^*[t_i; \widehat{\omega_b^2}(t_i^+)]$ was evaluated in real time by simple interpolation of the prestored time histories described earlier. In plot (a), the $\hat{v}_b(t_i^+)$ errors (and

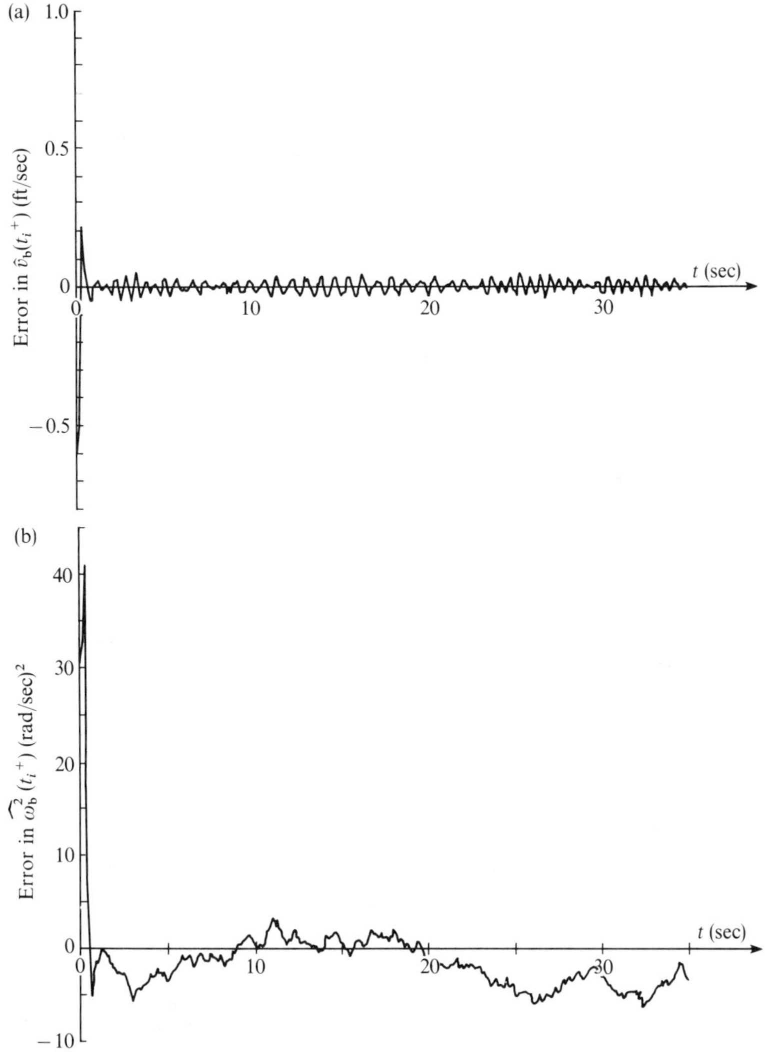


FIG. 15.8 Stochastic adaptive controller: (a) Error in $\hat{v}_b(t_i^+)$ for true $\omega_b^2 = 130$ rad$^2$/sec$^2$. (b) Error in $\omega_b^2(t_i^+)$ for true $\omega_b^2 = 130$ rad$^2$/sec$^2$. From Maybeck [93].

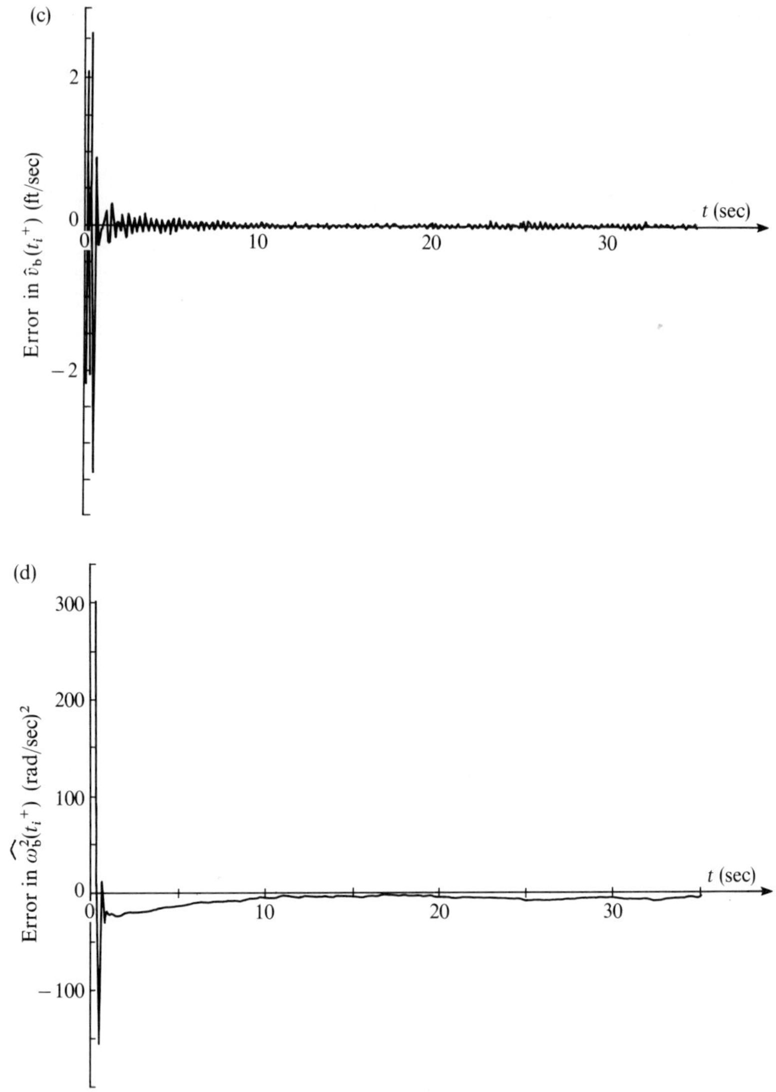


FIG. 15.8 (cont.) Stochastic adaptive controller: (c) Error in $\hat{v}_b(t_i^+)$ for true $\omega_b^2 = 400$ rad$^2$/sec$^2$. (d) Error in $\widehat{\omega_b^2}(t_i^+)$ for true $\omega_b^2 = 400$ rad$^2$/sec$^2$. From Maybeck [93].

corresponding state regulation error) very quickly converge to a level of performance essentially identical to that of the LQG controller based on the correct value of $\omega_b^2$ of 130 rad$^2$/sec$^2$; plot (b) presents the associated error in $\widehat{\omega_b^2}(t_i^+)$, revealing the precision in estimating the uncertain parameter that enabled the adaptive controller to perform so well. The improvement in performance of the adaptive over the nonadaptive controller is more dramatic at true $\omega_b^2 = 400$ rad$^2$/sec$^2$, as seen by comparing Fig 15.8c to Fig 15.7b. Again, the precision in estimating $\omega_b^2$, as depicted in Fig. 15.8d, allowed the adaptive controller to traverse a rapid transient period and attain not only a stable closed-loop system, but tracking performance again very similar to that of an LQG controller based on correct $\omega_b^2$ value.

Very similar results were obtained with state and parameter estimation provided instead by an extended Kalman filter (see Fig. 9.6 for a parameter estimation error time history similar to that of Fig. 15.8b or d) or the full-scale or online approximate implementations of maximum likelihood estimation of Section 10.6 (see Fig. 10.18). A multiple-model filter, as presented in Section 10.8, could also be used readily for this application. Using $\mathbf{G}_c^*(t_i)$ based on a *nominal* value such as 100 rad$^2$/sec$^2$ can also be employed: although the appropriate steady state controller gain on the $\hat{v}_b(t_i^+)$ value more than doubled between the cases of $\omega_b^2 = 100$ and $\omega_b^2 = 400$, the *enhanced state estimates* based on good $\widehat{\omega_b^2}(t_i^+)$ instead of mismodeled nominal $\omega_b^2$ were much more important to improved closed-loop performance than controller gain adjustment. ■

Assumed certainty equivalence is the basis of a class of controllers known as *self-tuning regulators* [8, 9, 30, 87, 100]. First an output regulator (not necessarily a full-state regulator) such as an optimal minimum mean square error controller is designed on the basis of assumed known parameter values. Then the adaptive nature is provided by online estimation of the uncertain parameters, as typically accomplished by simple least squares methods, and the *same* controller structure is used, but with the parameter estimates replacing the "known" values used for design.

Another means of developing an assumed certainty equivalence design adaptive controller employs discrete or discretized parameter values and multiple models. We have already discussed the implementation of a controller as in Fig. 15.6 with $\hat{\mathbf{x}}(t_i^+)$ provided by the multiple model filtering algorithm depicted in Fig. 10.23, and $\hat{\mathbf{a}}(t_j/t_i)$ calculated as in (15-62) and (15-63) with an initial condition $\hat{\mathbf{a}}(t_i^+)$ given by (10-109). A closely related alternative known as the *multiple model adaptive controller* [17, 44, 59, 60] is portrayed in Fig. 15.9. Here a bank of LQG controllers, each based on a particular parameter value $\mathbf{a}_k$, replaces the bank of Kalman filters of Fig. 10.23. Each state estimate $\hat{\mathbf{x}}_k(t_i^+)$ is premultiplied by the appropriate gain $\mathbf{G}_c^*[t_i; \mathbf{a}_k]$ based on the same parameter assumption, to form the optimal control, conditioned on $\mathbf{a}_k$ being the true value of the parameters,

$$\mathbf{u}_k(t_i) = -\mathbf{G}_c^*[t_i; \mathbf{a}_k]\hat{\mathbf{x}}_k(t_i^+) \tag{15-66}$$

Then the adaptive control is generated by adding together the probabilistically weighted $\mathbf{u}_k(t_i)$ values, using (10-104), (10-107), and (10-108).

An *open-loop optimal feedback control law* as described in Section 15.2 can also be developed for adaptive control [21, 24, 26, 80, 133]. It, like the certainty equivalence designs, is in the feedback class, but it is generally somewhat more complex. Consider being at time $t_i$ and having estimates $\hat{\mathbf{x}}(t_i^+)$ and $\hat{\mathbf{a}}(t_i^+)$, and associated conditional covariances $\mathbf{P}_{xx}(t_i^+)$ and $\mathbf{P}_{aa}(t_i^+)$ (and also $\mathbf{P}_{xa}(t_i^+)$ if available) as generated by an estimator operating up to the current time. For an open-loop optimal feedback law, we assume that no more measurements

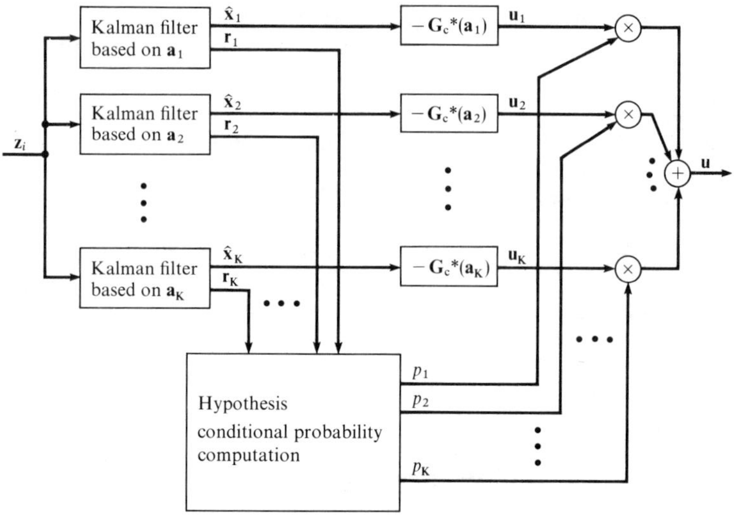


FIG. 15.9 Multiple model adaptive controller.

will be taken, and we obtain the appropriate open-loop optimal control function time history to minimize the conditional expectation of the cost-to-go to final time, conditioned on the measurement history $\mathbf{Z}_i$. That control is applied *only* at the current time $t_i$, and the entire procedure is repeated at the next sample time. In general, the control function depends not only on the $\hat{\mathbf{a}}(t_j/t_i)$ prediction for all $t_j \in [t_i, t_N]$ as for the assumed certainty equivalence design, but also $\mathbf{P}_{aa}(t_j/t_i)$ (and $\mathbf{P}_{xa}(t_j/t_i)$ and higher moments as well if available) as predicted under the assumption of no more measurements being made. This can be seen from the fact that we must be able to evaluate cost function terms such as

$$\int_{-\infty}^{\infty} \int_{-\infty}^{\infty} L[\boldsymbol{\xi}, \boldsymbol{\alpha}, \mathbf{u}, t_j] f_{\mathbf{x}(t_j), \mathbf{a}(t_j)|\mathbf{Z}(t_i)}(\boldsymbol{\xi}, \boldsymbol{\alpha} | \mathbf{Z}_i) \, d\boldsymbol{\xi} \, d\boldsymbol{\alpha}$$

One could approximate $f_{\mathbf{x}(t_j), \mathbf{a}(t_j)|\mathbf{Z}(t_i)}(\boldsymbol{\xi}, \boldsymbol{\alpha} | \mathbf{Z}_i)$ as Gaussian and described by the predicted first and second moments. If a multiple model adaptive filter assuming parameters modeled as unknown *constants* were used, we could write

$$f_{\mathbf{x}(t_j), \mathbf{a}|\mathbf{Z}(t_i)} = f_{\mathbf{x}(t_j)|\mathbf{a}, \mathbf{Z}(t_i)} f_{\mathbf{a}|\mathbf{Z}(t_i)} \tag{15-67}$$

where the first term on the right hand side can be generated via predictions from each filter in the bank of Kalman filters, and

$$f_{\mathbf{a}|\mathbf{Z}(t_i)}(\boldsymbol{\alpha} | \mathbf{Z}_i) = \sum_{k=1}^{K} p_k(t_i) \, \delta(\boldsymbol{\alpha} - \mathbf{a}_k) \tag{15-68}$$

using the notation of Section 10.8. In any event, the optimal open-loop law to apply from time $t_i$ forward may well depend on more than just the predicted *first* moment of $\mathbf{a}(t_j)$. For instance, a controller that assumes $\mathbf{P}_{aa}$ eigenvalues grow in time (as due to slowly varying uncertain parameters) might exhibit more caution at later times than one which assumes $\mathbf{P}_{aa}(t_j/t_i) = \mathbf{P}_{aa}(t_i^+)$ for all $t_j$, while both of these may be considerably more cautious than a controller function based on absolute certainty of parameter values.

In contrast to these feedback laws, a *closed-loop* law will use the knowledge that future measurements are to be taken, in solving the cost minimization problem. In general, it will probe the system, purposely exciting system modes so as to enhance current (and future) estimation of uncertain parameters in order to reduce the total cost of controlling to the final time [54]. The dual controller of Section 15.5 or other approximations to the full-scale dynamic programming algorithm can be used to generate such a law. Another approach that captures this probing characteristic is to perform *optimal input design*. Both open-loop [4, 5, 37, 61, 76, 95–97, 127, 141, 142, 156, 157] and feedback [61, 63, 91, 101, 144] controller designs have been generated that maximize the identifiability of parameters, as by inclusion of an explicit measure of identifiability (trace, weighted trace, or determinant of the parameter estimation error covariance, or inverse covariance or information matrix; volume of error ellipsoids described by such matrices; similar measures associated with sensitivity systems of equations for $\partial\mathbf{x}/\partial a_k$ for all $k$; etc.). For computational feasibility, one may consider looking only $M$ sample periods into the future instead of all the way to the final time, as discussed in Section 15.2.

The issue of model adequacy raised near the end of Section 14.5 is still pertinent even when uncertain parameters are estimated and adaptation is performed [19, 30, 81, 111, 128]. There is typically a frequency range over which a given mathematical model is adequate and the uncertainties are well structured: well represented as uncertain parameters or noises of particular statistical and/or spectral characteristics. However, beyond this range of model adequacy, the uncertainties are unstructured: these are the errors that remain after the best possible adjustment of model parameters is made. The adaptive procedures of this section can only address the structured uncertainties [30, 71]. It is still important to limit the controller bandwidth to the region of structured uncertainties, and to provide any additional rolloff (during or after the controller design procedure) required so as not to induce system response beyond that region. This is due to the fact that the adaptive controller must operate in the presence of the unstructured uncertainities, and if they are substantial, they can confound the adaptation process. Robustness analyses, such as the evaluation of (14-113) or (14-123), provide invaluable insights into frequency regimes in which loop gain must be made particularly small: regions where $\sigma_{\max}[\Delta\mathbf{G}]$ or a bound for $\sigma_{\max}[\Delta\mathbf{G}]$ assumes a large magnitude.

## 15.7 DESIGN PHILOSOPHY

A controller design philosophy evolves from the discussions in preceding sections [13–15, 30]. First of all, *linear perturbation control laws* and *assumed certainty equivalence designs* provide viable synthesis procedures for suboptimal *feedback* control laws. They provide an alternative to *closed-loop* control generated by difficult full-scale solution to stochastic dynamic programming or some approximation thereto such as the dual control of Section 15.5.

For any stochastic controller design, *system modeling* should receive considerable effort. For linear perturbation or assumed certainty equivalence controllers, this effort enhances state estimation accuracy. The more precise the state estimate is, the more robust the stochastic controller will be, since full-state feedback controllers are inherently rather robust [77, 108, 109, 111]. Moreover, the better the model, the smaller the region of unstructured uncertainties; but there is always such a region, and one must explicitly design for small loop gain in those regions.

To ensure good state estimation accuracy, *failure detection* should be incorporated as needed into the estimator (at least, and possibly also into the separate deterministic controller structure) to detect gross changes in the system: first on sensor failures (since this is easier) and then on actuators [39, 43, 74, 147]. Hardware redundancy can be supplemented by functional redundancy and residual monitoring in filters, with event detection performed via hypothesis testing as in Example 12.4 or multiple model adaptation as in Sections 10.8 and 15.6.

Moreover, *adaptation* should be incorporated into the estimator (again, at least) in order to generate a robust stochastic controller. This may entail generating several linearized models about appropriate nominal operating conditions, with each resulting controller allowing gain scheduling as functions of important parameters to be measured or estimated (the functions themselves to be evaluated offline a priori). Real-time adaptation can be accomplished by treating uncertain parameters as additional states (to be estimated via extended Kalman filtering as in Chapter 9 of Volume 2 or other nonlinear filters as in Chapter 12) or by any of the methods discussed in Chapter 10.

The complexity and parallel nature of such adaptations motivate exploiting *parallel architecture of software* and the *distributed computational capability* made possible by microprocessors and minicomputers. System identification and failure detection can be performed in parallel with the basic state estimation and control functions, or multiple model reference adaptation techniques involving parallel banks of individual estimators and/or controllers (perhaps each implemented on a single chip) can be used.

A viable *controller reconfiguration* strategy can also be developed to adapt to major abrupt changes in the system, such as component failures. Through residual monitoring and other means, a failure detection logic can sound an alarm indicating that something (of uncertain origin) has happened in the system. Then a robust controller that can provide adequate control despite major unknown changes in the system can be used to maintain system stability while failure isolation and uncertain parameter estimation is performed. Finally, once the appropriate adaptation is identified, the controller system is reconfigured in order to provide the best possible performance under the newly established conditions. The reconfiguration might entail removal of outputs of failed sensors in the network, or use of alternative components such as standby sensors or multiple-use actuators (for instance, treating aircraft elevators as ailerons if ailerons have failed). Practical aspects that require close attention during design of such a system include (1) the amount of time needed to detect and isolate a failure and then reconfigure the control system, (2) minimizing both false and missed alarms (i.e., removing unfailed components or leaving failed components in the network with resulting degraded performance), (3) ensuring stability in the transition periods and in the degraded modes with failed components removed, (4) computational and subsystem interfacing complexities, and (5) the issue of how "best" to reconfigure the control system. These are still open issues, from both theoretical and practical perspectives, and are currently being researched.

The previous methodologies produced *feedback* controller designs. As such, these controllers do not exhibit probing and other characteristics of a closed-loop control law. To incorporate such features, one should investigate the full-scale (numerical) solution to the stochastic dynamic programming algorithm, or approximations to this solution such as dual control in Section 15.5. If the performance benefit of the closed-loop laws over feedback laws warrants their use, then online implementations can often be developed that exhibit the specific closed-loop law behavior that yields enhanced system properties, but without the online computational burden of the actual optimal closed-loop law.

## 15.8 SUMMARY AND PERSPECTIVE

This chapter considered the design of stochastic controllers for problems with nonnegligible nonlinearities. Section 15.2 formulated the problem and considered the difference between open-loop, feedback, and closed-loop control laws, and it further characterized the caution and dual effect of general stochastic controllers, and the associated probing for active learning that closed-loop controllers exhibit. Generally the solution to the stochastic dynamic programming algorithm developed in Chapter 13 will be a closed-loop law, but this solution is often difficult, if not infeasible, to attain and implement.

Sections 15.3 and 15.4 considered alternative, practical means of synthesizing stochastic *feedback* controllers. First Section 15.3 developed linear perturbation control laws as a means of exploiting LQG synthesis directly. Since control is continually applied to drive back toward design conditions, a single equilibrium or nominal may suffice for linearization. LQG synthesis was used to generate both linear perturbation regulators (15-26)–(15-30) and PI controllers (15-32) and (15-35) to be implemented in conjunction with perturbation Kalman filters based on (15-36)–(15-38). Constant-gain approximations were especially motivated for many applications. A two-level decomposition of the original control problem was also described, with linear perturbation LQG control applied about a nominal generated as the result of applying a deterministic optimal controller to the corresponding noise-free problem, as depicted in Fig. 15.4. Subsequently, Section 15.4 developed assumed certainty equivalence design, in which the associated deterministic optimal controller function $\mathbf{u}^{DO}[\mathbf{x}(t_i), t_i]$ is evaluated, and then a *separate* estimator is established to provide the approximate conditional mean or similar estimate of the state, $\hat{\mathbf{x}}(t_i^+)$, to be incorporated into the final control law for implementation, $\mathbf{u}^{DO}[\hat{\mathbf{x}}(t_i^+), t_i]$. The basic structure was portrayed in Fig. 15.5, and again constant-gain approximations were motivated.

Closed-loop controller design was considered in Section 15.5. One specific "dual control" approximation to the full-scale dynamic programming solution was studied as an alternative to numerical solution of this optimization algorithm.

Section 15.6 concentrated attention on the important special case of stochastic control of systems with models in which uncertain parameters are embedded. The various design methodologies were applied to this problem context, and the resulting controllers were characterized.

A design philosophy resulted in Section 15.7 that emphasized the importance of precise state estimation for robust stochastic control achieved through linear perturbation or certainty equivalence designs. Adequate state modeling in the filter portion is therefore critical, and the limitations of even the best of models must be acknowledged by proper gain rolloff outside the region of model adequacy. Failure detection and adaptive estimation can be viewed as integral parts of providing such precise state estimates, and these enable a reconfigurable controller design as discussed in that section. The magnitude and parallel nature of the algorithms strongly suggested consideration of parallel, distributed computation.

Currently, considerable research and development effort is being extended to develop both the theoretical foundations and engineering practice of such controller designs [13–15, 30]. Existing concepts for adaptive estimation and control are being consolidated and modified, in an effort to achieve a proven, systematic design technique with both outstanding performance capability and on-line implementability. This is true also in the field of failure detection and system reconfiguration. *Optimal* use of microprocessors, minicomputers, and their interconnections for digital compensation is being studied, as an eventual replacement for practically motivated but ad hoc architectures of current systems; asynchronous sampling of sensors and generating of controls, complexity of computations and interfacings, scheduling of resources, and the lack of a single location to which all measurements flow and from which all decisions are made (called "nonclassical information pattern") are but some of the important issues in this optimization process [148]. Closely linked to this is the important developing area of large-scale system design and integration [12, 16, 64, 113, 143]. For instance, we wish to view aircraft aerodynamics, structural characteristics, navigation, guidance (and weapon delivery for military aircraft), and control not as separate entities, but as subfunctions of an overall large-scale integrated system. By interfacing subsystems that have previously been isolated from each other, we allow access to each other's information and capabilities, with a substantial potential for performance enhancement. The impact of microprocessors and distributed computation on eventual implementation is great: for instance, data processing can be performed locally before sending it from a sensor onto the data bus, thereby reducing the overwhelming capacity requirements of transmitting all raw data at sufficiently high bandwidth to provide centralized processing. However, the inherent problem of decentralized control [3, 12, 16, 38, 112, 113, 122] arises: each separate controller (decision-maker, processor, etc.) does *not* have instantaneous access to all measurement information or all other decisions. As a result, performance degradation and instabilities can result that would not be experienced in a conceptually simple but practically infeasible centralized controller.

The conception of the control problem has changed from the simple special cases introduced in Chapter 13. We no longer assume a necessarily centralized controller that accepts noisy sensor outputs and generates appropriate actuator commands to make a physical system behave "properly" despite disturbances and model uncertainties. Instead, we can conceive of a large-scale integrated system with multiple sensor–controller–actuator loops, each designed with a full accounting for uncertainties through stochastic models, with the loops themselves and an overall coordinator of efforts all designed so as to exploit the full computational capabilities and flexibility of digital computers, employing filters, adaptation, redundancy and dispersal, failure detection, and reconfiguration. This changing conception of the estimation and control problem and the maturation of estimation and stochastic control technologies are being dictated by the need for effective, efficient, integrated, and coordinated systems for many applications.

This chapter, and in fact this entire text, is necessarily incomplete. Many unresolved issues remain: many doors are partially open and the paths remain to be traversed, and many more doors and paths have not even been considered [13–15, 32, 55–58, 65, 67, 89, 90, 92, 98, 106, 119, 121, 155]. The purpose of this text has been to lay firm foundations in the fundamentals and to indicate current design practice and future directions. Therefore, it is appropriate to end, not with a false illusion of completeness, but with a deepening and exhilarating awareness that each answer yields many new questions and "paths," that a disciplined and systematic approach equips us with the means for successful conveyance, but that the journey has just begun.

## REFERENCES


1. Advanced Adaptive Optics Control Techniques, Tech. Rep. TR-996-1. The Analytic Sciences Corp., Reading, Massachusetts (January 1978).
2. Aoki, M., "Optimization of Stochastic Systems." Academic Press, New York, 1967.
3. Aoki, M., On decentralized linear stochastic control problems with quadratic cost, *IEEE Trans. Automat. Control* **AC-18** (3), 243–250 (1973).
4. Aoki, M., and Staley, R. M., On input signal synthesis in parameter identification, *Automatica* **6**, 431–440 (1970).
5. Arimoto, S., and Kimura, H., Optimum input test signals for system identification—An information–theoretical approach, *Internat. J. Syst. Sci.* **1**, 279–290 (1971).
6. Asher, R. B., Andrisani, D., II, and Dorato, P., Bibliography on adaptive control systems, *Proc. IEEE* **64** (8), 1226–1240 (1976).
7. Åström, K. J., "Introduction to Stochastic Control Theory." Academic Press, New York, 1970.
8. Åström, K. J., Borisson, U., Ljung, L., and Wittenmark, B., Theory and applications of self-tuning regulators, *Automatica* **13**, 457–476 (1977).
9. Åström, K. J., and Wittenmark, B., On self-tuning regulators, *Automatica* **9**, 185–199 (1973).
10. Athans, M., The role and use of the Linear-Quadratic-Gaussian problem in control system design, *IEEE Trans. Automat. Control* **AC-16** (6), 529–552 (1971).
11. Athans, M., The discrete time Linear-Quadratic-Gaussian stochastic control problem, *Ann. Econ. Social Measurement* **1** (4), 449–491 (1972).
12. Athans, M., Survey of decentralized control methods, *Ann. Econ. Social Measurement* **4** (2), 345–355 (1975).
13. Athans, M., Perspectives in Modern Control Theory, Tech. Rep. ESL-P-676. MIT Electronic Systems Lab., Cambridge, Massachusetts (August 1976).
14. Athans, M., Trends in modern system theory, *Chem. Process Control, AIChE Symp. Ser.* **72** (159), 4–11 (1976).
15. Athans, M., Advances and Open Problems in Multivariable Control System Design, presentation and unpublished notes, MIT, Cambridge, Massachusetts (March 1978).
16. Athans, M. (ed.), Special issue on large-scale systems and decentralized control, *IEEE Trans. Automat. Control* **AC-23** (2) (1978).
17. Athans, M., Castañon, D., Dunn, K., Greene, C. S., Lee, W. H., Sandell, N. R., Jr., and Willsky, A. S., The stochastic control of the F-8C aircraft using a multiple model adaptive control (MMAC) method—Part 1: Equilibrium flight, *IEEE Trans. Automat. Control* **AC-22** (5), 768–780 (1977).
18. Athans, M., and Falb, P. L., "Optimal Control: An Introduction to Its Theory and Applications." McGraw-Hill, New York, 1966.
19. Athans, M., Ku, R., and Gershwin, S. B., The uncertainty threshold principle: Some fundamental limitations of optimal decision making under dynamic uncertainty, *IEEE Trans. Automat. Control* **AC-22** (3), 491–495 (1977).
20. Athans, M., and Tse, E., Adaptive stochastic control for a class of linear systems, *IEEE Trans. Automat. Control* **AC-17** (1), 38 (1972).
21. Athans, M., and Tse, E., On the adaptive control of linear systems using the OLFO approach, *IEEE Trans. Automat. Control* **AC-18** (5), 489–493 (1973).
22. Athans, M., and Varaiya, P. P., A survey of adaptive stochastic control methods, *Proc. Eng. Foundation Conf. Syst. Eng., New England College, Henniker, New Hampshire* ERDA Rep. CONF-750867 (August 1975).

23. Bar-Shalom, Y., Performance adaptive resource allocation in a time-varying environment, *Proc. Conf. Decision and Control, Clearwater Beach, Florida* (December 1976).
24. Bar-Shalom, Y., Effect of uncertainties on the control performance of linear systems with unknown parameters and trajectory confidence tubes, *Proc. IEEE Conf. Decision and Control, New Orleans, Louisiana* pp. 898–902 (December 1977).
25. Bar-Shalom, Y., and Gershwin, S. B., Applicability of adaptive control to real problems—Trends and opinions, *Automatica* **14**, 407–408 (1978).
26. Bar-Shalom, Y., and Sivan, R., The optimal control of discrete time systems with random parameters, *IEEE Trans. Automat. Control* **AC-14** (1), 3–8 (1969).
27. Bar-Shalom, Y., and Tse, E., Dual effect, certainty equivalence and separation in stochastic control, *IEEE Trans. Automat. Control* **AC-19** (5), 494–500 (1974).
28. Bar-Shalom, Y., and Tse, E., Concepts and methods in stochastic control, *in* "Control and Dynamic Systems: Advances in Theory and Applications" (C. T. Leondes, ed.). Academic Press, New York, 1976.
29. Bar-Shalom, Y., and Tse, E., Caution, probing and the value of information in the control of uncertain systems, *Ann. Econ. Soc. Measurement* **5** (3), 323–337 (1976).
30. Bar-Shalom, Y., Wittenmark, B., and Stein, G., Stochastic adaptive control, notes for a tutorial workshop at the *IEEE Conf. on Decision and Control, Albuquerque, New Mexico* (December 1980).
31. Bellman, R., "Adaptive Control Processes: A Guided Tour." Princeton University Press, Princeton, New Jersey, 1961.
32. Birdwell, J. D., Castañon, C. A., and Athans, M., On reliable control system designs with and without feedback reconfigurations, *Proc. 1978 IEEE Conf. Decision and Control, San Diego, California* pp. 419–426 (January 1979).
33. Bliss, G. A., "Lectures on the Calculus of Variations," University of Chicago Press, Chicago, Illinois, 1946.
34. Broussard, J. R., and Safonov, M. G., Design of Generalized Discrete Proportional-Integral Controllers by Linear-Optimal Control Theory, Tech. Rep. TIM-804-1. The Analytic Sciences Corp., Reading, Massachusetts (October 1976).
35. Bryson, A. E., Jr., and Ho, Y., "Applied Optimal Control." Ginn (Blaisdell), Waltham, Massachusetts, 1969.
36. Casler, R. J., Jr., Dual control guidance for homing interceptors with angle-only measurements, *AIAA J. Guidance and Control* **1**, 63–70 (1978).
37. Chen, R. T. N., Input design for parameter identification—Part 1: A new formulation and practical solution, *Proc. Joint Automat. Control Conf., Austin, Texas* (1974).
38. Chong, C. Y., and Athans, M., On the stochastic control of linear systems with different information sets, *IEEE Trans. Automat. Control* **AC-16** (5), 423–430 (1971).
39. Chow, E. Y., and Willsky, A. S., Issues in the development of a general design algorithm for reliable failure detection, *Proc. IEEE Conf. Decision and Control, Albuquerque, New Mexico* pp. 1006–1012 (December 1980).
40. Cruz, J. B., Jr., Workshop on Adaptive Control, Final Report, Rep. R79-1 for AFOSR Contract F49620-79-C-0056. Dynamic Systems, Urbana, Illinois (July 1979).
41. Curry, R. E., A new algorithm for suboptimal stochastic control, *IEEE Trans. Automat. Control* **AC-14** (5), 533–536 (1969).
42. D'Appolito, J. A., and Hutchinson, C. E., Low sensitivity filters for state estimation in the presence of large parameter uncertainties, *IEEE Trans. Automat. Control* **AC-14** (3), 310–312 (1969).
43. Deckert, J. C., Desai, M. N., Deyst, J. J., and Willsky, A. S., F-8 DFBW sensor failure identification using analytic redundancy, *IEEE Trans. Automat. Control* **AC-22** (5), 795–803 (1977).
44. Deshpande, J. G., Upadhyay, T. N., and Lainiotis, D. G., Adaptive control of linear stochastic systems, *Automatica* **9**, 107–115 (1973).

45. Doyle, J. C., Guaranteed margins for LQG regulators, *IEEE Trans. Automat. Control* **AC-23** (4), 756–757 (1978).
46. Dreyfus, S. E., Some types of optimal control of stochastic systems, *SIAM J. Control* **2**, 120–134 (1964).
47. Dreyfus, S. E., "Dynamic Programming and the Calculus of Variations." Academic Press, New York, 1965.
48. Dunn, H. J., and Montgomery, R. C., A moving window parameter adaptive control system for the F8-DFBW aircraft, *IEEE Trans. Automat. Control* **AC-22** (5), 788–795 (1977).
49. Egardt, B., Stochastic convergence analysis of model reference adaptive controllers, *Proc. IEEE Conf. Decision and Control, Albuquerque, New Mexico* pp. 1128–1131 (December 1980).
50. Elliott, H., and Wolovich, W. A., Parameter adaptive identification and control, *IEEE Trans. Automat. Control* **AC-24** (4), 592–599 (1979).
51. Farison, J. B., Graham, R. E., and Shelton, R. C., Identification and control of linear discrete systems, *IEEE Trans. Automat. Control* **AC-12** (4), 438–442 (1967).
52. Fel'dbaum, A. A., "Optimal Control Systems." Academic Press, New York, 1965.
53. Fel'dbaum, A. A., Dual control theory, Parts I–IV, "Optimal and Self-Optimizing Control" (R. Oldenburger, ed.), pp. 458–496. MIT Press, Cambridge, Massachusetts, 1966.
54. Florentin, J. J., Optimal probing, adaptive control of a simple Bayesian system, *J. Electron. Control* **13** (2), 165–177 (1962).
55. Gelb, A., and VanderVelde, W. E., "Multiple-Input Describing Functions." McGraw-Hill, New York, 1968.
56. Gelb, A., and Warren, R. S., Direct statistical analysis of nonlinear systems—CADET, *AIAA J.* **11**, 689–694 (1973).
57. Gilman, A. S., and Rhodes, I. B., Cone-bounded nonlinearities and mean-square bounds, *IEEE Trans. Automat. Control* **AC-18** (3), 260–265 (1973).
58. Gilman, A. S., and Rhodes, I. B., Cone-bounded nonlinearities and mean-square bounds—Quadratic regulation bounds, *IEEE Trans. Automat. Control* **AC-21** (4), 472–483 (1976).
59. Greene, C. S., An Analysis of the Multiple Model Adaptive Control Algorithm, Ph.D. dissertation, ESL-TH-843, Elec. Systems Lab., MIT, Cambridge, Massachusetts (August 1978).
60. Greene, C. S., and Willsky, A. S., An analysis of the multiple model adaptive control algorithm, *Proc. IEEE Conf. Decision and Control, Albuquerque, New Mexico* pp. 1142–1145 (December 1980).
61. Gupta, N. K., and Hall, W. E., Jr., Input Design for Identification of Aircraft Stability and Control Derivatives, Tech. Rep. NASA-CR-2493 (February 1975).
62. Gustafson, D. E., and Speyer, J. L., Design of linear regulators for nonlinear stochastic systems, *J. Spacecr. Rockets* **12**, 351–358 (1975).
63. Gustavsson, I., Ljung, L., and Soderström, T., Identification of process in closed loop—Identifiability and accuracy aspects, survey paper, *Automatica* **13**, 59–75 (1977).
64. Ho, Y. C., and Mitter, S. K., "Directions in Large-Scale Systems." Plenum Press, New York, 1976.
65. Holtzman, J. M., "Nonlinear System Theory: A Functional Analysis Approach." Prentice-Hall, Englewood Cliffs, New Jersey, 1970.
66. Hsia, T. C., Comparisons of adaptive sampling control laws, *IEEE Trans. Automat. Control* **AC-17** (6), 830–831 (1972).
67. Hsu, K., and Marcus, S. I., A general Martingale approach to discrete-time stochastic control and estimation, *IEEE Trans. Automat. Control* **AC-24** (4), 580–583 (1979).
68. Iliff, K. W., Identification and stochastic control of an aircraft flying in turbulance, *AIAA J. Guidance and Control* **1** (2), 101–108 (1978).
69. Jacobs, O. L. R., and Langdon, S. M., An optimal extremal control system, *Automatica* **6**, 297–301 (1970).

70. Jacobs, O. L. R., and Patchell, J. W., Caution and probing in stochastic control, *Internat. J. Control* **16**, 189–199 (1972).
71. Johnson, C. R., Jr., Adaptive modal control of large flexible spacecraft, *AIAA J. Guidance and Control* **3** (4), 369–375 (1980).
72. Johnson, C. R., Jr., The common parameter estimation basis of adaptive filtering, identification, and control, *Proc. IEEE Conf. Decision and Control, Albuquerque, New Mexico* pp. 447–452 (December 1980).
73. Johnson, C. R., Jr., and Tse, E., Adaptive implementation of a one-step-ahead optimal control via input matching, *IEEE Trans. Automat. Control* **AC-23** (5), 865–872 (1978).
74. Jones, H. L., Failure Detection in Linear Systems. Ph.D. dissertation, MIT, Cambridge, Massachusetts (September 1973).
75. Kailath, T. (ed.), Special issue on system identification and time series analysis, *IEEE Trans. Automat. Control* **AC-19** (6), (1974).
76. Kleinman, D. L., and Rao, P. K., An information matrix approach for aircraft parameter-insensitive control, *Proc. IEEE Conf. Decision and Control, New Orleans, Louisiana* (December 1977).
77. Kozin, F., A survey of stability of stochastic systems, *Proc. Joint Automat. Control Conf., Stochastic Problems in Control, Ann Arbor, Michigan*, pp. 39–86 (1968); also *Automatica* **5**, 95–112 (1969).
78. Kreisselmeier, G., Algebraic separation in realizing a linear state feedback control law by means of an adaptive observer, *IEEE Trans. Automat. Control* **AC-25** (2), 238–243 (1980).
79. Kraft, L. G., III, A control structure using an adaptive observer, *IEEE Trans. Automat. Control* **AC-24** (5), 804–806 (1979).
80. Ku, R., and Athans, M., On the adaptive control of linear systems using the open-loop-feedback-optimal approach, *IEEE Trans. Automat. Control* **AC-18** (5), 489–493 (1973).
81. Ku, R., and Athans, M., Further results on the uncertainty threshold principle, *IEEE Trans. Automat. Control* **AC-22** (5), 866–868 (1977).
82. Kushner, H., "Introduction to Stochastic Control." Holt, New York, 1971.
83. Lainiotis, D. G., Partitioning: A unifying framework for adaptive systems, I: Estimation, *Proc. IEEE* **64,** 1126–1142 (1976).
84. Lainiotis, D. G., Partitioning: A unifying framework for adaptive systems, II: Control, *Proc. IEEE* **64** 1182–1197 (1976).
85. Lainiotis, D. G., Deshpande, J. G., and Upadhyay, T. N., Optimal adaptive control: A nonlinear separation theorem, *Internat. J. Control* **15** (5), 877–888 (1972).
86. Landau, I. D., A survey of model reference adaptive techniques—theory and applications, *Automatica* **10**, 353–379 (1974).
87. Landau, I. D., "Adaptive Control—The Model Reference Approach." Dekker, New York, 1979.
88. Landau, I. D., and Silveira, H. M., A stability theorem with applications to adaptive control, *IEEE Trans. Automat. Control* **AC-24** (2), 305–312 (1979).
89. Lauer, G. S., Stochastic Optimization for Discrete-Time Systems, Ph.D. dissertation and Rep. LIDS-TH-949. MIT Lab. for Information and Decision Systems, Cambridge, Massachusetts (October 1979).
90. Lindberg, E. K., An Adaptive Controller Which Displays Human Operator Limitations for a Fighter Type Aircraft. Ph.D. dissertation, Air Force Institute of Technology, Wright-Patterson AFB, Ohio (November 1978).
91. Lopez-Toledo, A. A., Optimal Inputs for Identification of Stochastic Systems. Ph.D. dissertation, MIT, Cambridge, Massachusetts (1974).
92. Luenberger, D. G., "Optimization by Vector Space Methods." Wiley, New York, 1969.
93. Maybeck, P. S., Improved Controller Performance Through Estimation of Uncertain Parameters, Spacecraft Autopilot Development Memo. 17-68. MIT Instrumentation Lab. (now C. S. Draper Lab.), Cambridge, Massachusetts (October 1968).

94. Maybeck, P. S., Combined Estimation of States and Parameters for On-Line Applications. Ph.D. dissertation, MIT, Cambridge, Massachusetts (February 1972).
95. Mehra, R. K., Optimal inputs for linear system identification, *IEEE Trans. Automat. Control* **AC-19** (3), 192–200 (1974).
96. Mehra, R. K., Optimal input signals for parameter estimation in dynamic systems—Survey and new results, *IEEE Trans. Automat. Control* **AC-19** (6), 753–768 (1974).
97. Mehra, R. K., Synthesis of optimal inputs for multiinput-multioutput systems with process noise, *in* "System Identification: Advances and Case Studies" (R. K. Mehra and D. G. Lainiotis, eds.). Academic Press, New York, 1976.
98. Mehra, R. K., and Carroll, J. V., Application of bifurcation analysis and catastrophe theory methodology (BACTM) to aircraft stability problems at high angles-of-attack, *Proc. 1978 IEEE Conf. Decision and Control, San Diego, California* pp. 186–192 (Jan. 1979).
99. Narendra, K. S., and Kudva, P., Stable adaptive schemes for system identification and control—Part II, *IEEE Trans. Syst. Man Cybernet.* **SCM-4** (6), 552–560 (1974).
100. Narendra, K. S., and Monopoli, R. V., "Applications of Adaptive Control." Academic Press, New York, 1980.
101. Olmstead, D. N., Optimal Feedback Controls for Parameter Identification. Ph.D. dissertation, Air Force Institute of Technology, Wright-Patterson AFB, Ohio (March 1979).
102. Pearson, A. E., Adaptive optimal steady state control of nonlinear systems, "Control and Dynamic Systems: Advances in Theory and Applications" (C. T. Leondes, ed.), Vol. 5, pp. 1–50. Academic Press, New York, 1967.
103. Porter, D. W., and Michel, A. N., Input-output stability of time-varying nonlinear multiloop feedback systems, *IEEE Trans. Automat. Control* **AC-19** (4), 422–427 (1974).
104. Price, C. F., and Koeningsburg, W. D., Adaptive Control and Guidance for Tactical Missiles, Tech. Rep. TR-170-1. The Analytic Sciences Corp., Reading, Massachusetts (June 1970).
105. Price, C. F., and Warren, R. S., Performance Evaluation of Homing Guidance Laws for Tactical Missiles, Tech. Rep. TR-170-4. The Analytic Sciences Corp., Reading, Massachusetts (January 1973).
106. Rhodes, I. B., and Snyder, D. L., Estimation and control for space-time point-process observations, *IEEE Trans. Automat. Control* **AC-22** (3), 338–346 (1977).
107. Rohrs, C., Valavani, L., and Athans, M., Convergence studies of adaptive control algorithms; Part I: Analysis, *Proc. IEEE Conf. Decision and Control, Albuquerque, New Mexico* pp. 1138–1141 (December 1980).
108. Safonov, M. G., Robustness and Stability Aspects of Stochastic Multivariable Feedback System Design. Ph.D. dissertation, Rep. ESL-R-763, MIT Electronic Systems Lab., Cambridge, Massachusetts (September 1977).
109. Safonov, M. G., and Athans, M., Robustness and computational aspects of nonlinear stochastic estimators and regulators, *IEEE Trans. Automat. Control* **AC-23** (4), 717–725 (1978).
110. Sage, A. P., and Ellis, T. W., Sequential suboptimal adaptive control of nonlinear systems, *Proc. Nat. Electron. Conf.* **22**, *Chicago, Illinois*, pp. 692–697 (October 1966).
111. Sandell, N. R., Jr. (ed.), Recent Developments in the Robustness Theory of Multivariable Systems, Rep. ONR-CR215-271-1F of the Office of Naval Research, Arlington, Virginia; also Rep. LIDS-R-954 of the MIT Lab. for Information and Decision Systems, Cambridge, Massachusetts (August 1979).
112. Sandell, N. R., Jr., and Athans, M., Solution of some non-classical LQG stochastic decision problems, *IEEE Trans. Automat. Control* **AC-19** (2), 108–116 (1974).
113. Sandell, N. R., Jr., Varaiya, P., Athans, M., and Safonov, M. G., Survey of decentralized control methods for large-scale systems, *IEEE Trans. Automat. Control* **AC-23** (2), 108–128 (1978).
114. Saridis, G. N., "Self-Organizing Control Stochastic Systems." Dekker, New York, 1976.

115. Saridis, G. N., and Lobbia, R. N., Parameter identification and control of linear discrete-time systems, *IEEE Trans. Automat. Control* **AC-17** (1), 52–60 (1972).
116. Schmotzen, R. E., and Blankenship, G. L., A simple proof of the separation theorem for linear stochastic systems with time delays, *IEEE Trans. Automat. Control* **AC-23** (4), 734–735 (1978).
117. Sebald, A. V., Toward a computationally efficient optimal solution to the LQG discrete-time dual control problem, *IEEE Trans. Automat. Control* **AC-24** (4), 535–540 (1979).
118. Siferd, R. E., Observability and Identifiability of Nonlinear Dynamical Systems with an Application to the Optimal Control Model for the Human Operator. Ph.D. dissertation, Air Force Institute of Technology, Wright-Patterson AFB, Ohio (June 1977).
119. Smith, H. W., "Approximate Analysis of Randomly Excited Nonlinear Controls." MIT Press, Cambridge, Massachusetts, 1966.
120. Sorenson, H. W., An overview of filtering and stochastic control in dynamic systems, *in* "Control and Dynamic Systems: Advances in Theory and Applications" (C. T. Leondes, ed.), Vol. 12. Academic Press, New York, 1976.
121. Sorenson, H. W., and Alspach, D. L., Recursive Bayesian estimation using Gaussian sums, *Automatica* **7**, 465–480 (1971).
122. Speyer, J. L., Computation and transmission requirements for a decentralized Linear-Quadratic-Gaussian control problem, *IEEE Trans. Automat. Control* **AC-24** (2), 266–269 (1979).
123. Speyer, J., Deyst, J. J., and Jacobson, D., Optimization of stochastic linear systems with additive measurement and process noise using exponential performance criteria, *IEEE Trans. Automat. Control* **AC-19** (4), 358–366 (1974).
124. Stein, G., Hartmann, G. L., and Hendrick, R. C., Adaptive control laws for F-8 flight test, *IEEE Trans. Automat. Control* **AC-22** (5), 758–767 (1977). Mini-Issue on NASA's Advanced Control Law Program for the F-8 DFBW Aircraft.
125. Stein, G., and Saridis, G. N., A parameter adaptive control technique, *Automatica* **5**, 731–739 (1969).
126. Stengel, R. F., Broussard, J. R., and Berry, P. W., The Design of Digital-Adaptive Controllers for VTOL Aircraft, NASA Rep. CR-144912. The Analytic Sciences Corp., Reading, Massachusetts (March 1976).
127. Stepner, D. E., and Mehra, R. K., Maximum Likelihood Identification and Optimal Input Design for Identifying Aircraft Stability and Control Derivatives, NASA Rep. CR-2200. Systems Control Inc., Palo Alto, Cal. (March 1973).
128. Sternby, J., Performance limits in adaptive control, *IEEE Trans. Automat. Control* **AC-24** (4), 645–647 (1979).
129. Striebel, J., Sufficient statistics in the optimum control of stochastic systems, *J. Math. Anal. Appl.* **12**, 576–592 (1965).
130. Sworder, D., "Optimal Adaptive Control Systems." Academic Press, New York, 1966.
131. Sworder, S. C., and Sworder, D. D., Feedback estimation systems and the separation principle of stochastic control, *IEEE Trans. Automat. Control* **AC-16** (4), 350–354 (1971).
132. Tarn, T. J., Extended separation theorem and exact analytical solution of stochastic control, *Automatica* **7**, 343–350 (1971).
133. Tse, E., and Athans, M., Adaptive stochastic control for a class of linear systems, *IEEE Trans. Automat. Control* **AC-17** (1), 38–52 (1972).
134. Tse, E., and Bar-Shalom, Y., An actively adaptive control for discrete-time systems with random parameters, *IEEE Trans. Automat. Control* **AC-18** (2), 109–117 (1973).
135. Tse, E., and Bar-Shalom, Y., Adaptive dual control for stochastic nonlinear systems with free end-time, *IEEE Trans. Automat. Control* **AC-20** (5), 670–675 (1975).
136. Tse, E., and Bar-Shalom, Y., Generalized certainty equivalence and dual effect in stochastic control, *IEEE Trans. Automat. Control* **AC-20** (6), 817–819 (1975).

137. Tse, E., and Bar-Shalom, Y., Actively adaptive control for nonlinear stochastic systems, *Proc. IEEE* **64**, 1172–1181 (1976).
138. Tse, E., and Bar-Shalom, Y., and Meier, L., Wide-sense adaptive dual control of stochastic nonlinear systems, *IEEE Trans. Automat. Control* **AC-18** (2), 98–108 (1973).
139. Tse, E., Larson, R. E., and Bar-Shalom, Y., Adaptive Estimation and Dual Control with Applications to Defense Communications Systems Controls, Final Rep. on AFOSR Contract F44620-74-C-0026. Systems Control, Inc., Palo Alto, California (February 1975).
140. Tsypkin, Y. Z., "Foundation of the Theory of Learning Systems." Academic Press, New York, 1973.
141. Upadhyaya, B. R., Characterization of optimal inputs by generating functions for linear system identification, *IEEE Trans. Automat. Control* **AC-25** (5), 1002–1005 (1980).
142. Upadhyaya, B. R., and Sorenson, H. W., Synthesis of linear stochastic signals in identification problems, *Automatica* **13**, 615–622 (1977).
143. Varaiya, P. P., Trends in the theory of decision making in large systems, *Ann. Econ. Social Measurement* **1** (4), 493–500 (1972).
144. Wellstead, P. E., Reference signals for closed-loop identification, *Internat. J. Control* **26**, 945–962 (1977).
145. Wieslander, J., and Wittenmark, B., An approach to adaptive control using real-time identification, *Automatica* **7**, 211–218 (1971).
146. Willman, W. W., Some formal effects of nonlinearities in optimal perturbation control, *AIAA J. Guidance Control* **2** (2), 99–100 (1979).
147. Willsky, A. S., A survey of design methods for failure detection in dynamic systems, *Automatica* **12**, 601–611 (1976).
148. Witsenhausen, H. S., A counterexample in stochastic control, *SIAM J. Control* **6** (1), 131–147 (1968).
149. Witsenhausen, H. S., Separation of estimation and control for discrete time systems, *Proc. IEEE* **59** (9), 1557–1566 (1971).
150. Wittenmark, B., Stochastic adaptive control methods: A survey, *Internat. J. Control* **21** (5), 705–730 (1975).
151. Wonham, W. M., On the separation theorem of stochastic control, *SIAM J. Control* **6** (2), 312–326 (1968).
152. Wonham, W. M., Optimal stochastic control, *Proc. Joint Automat. Control Conf., Stochastic Problems in Control*, Ann Arbor, Michigan, pp. 107–120 (June 1968).
153. Zames, G., On the input-output stability of time-varying nonlinear feedback systems: Part I: Conditions using concepts of loop gain, conicity, and positivity, *IEEE Trans. Automat. Control* **AC-11** (2), 228–238 (1966).
154. Zames, G., On the input-output stability of time-varying nonlinear feedback systems—Part II: Conditions involving circles in the frequency plane and sector nonlinearities, *IEEE Trans. Automat. Control* **AC-11** (3), 465–476 (1966).
155. Zarchan, P., Complete statistical analysis of nonlinear missile guidance systems—SLAM (Statistical Linearization Adjoint Method), *AIAA J. Guidance and Control* **2** (1), 71–78 (1979).
156. Zarrop, M. B., A Chebyshev system approach to optimal input design, *IEEE Trans. Automat. Control* **AC-24** (5), 687–698 (1979).
157. Zarrop, M. B., and Goodwin, G. C., Comments on optimal inputs for system identification, *IEEE Trans. Automat. Control* **AC-20** (2), 299–300 (1975).


## PROBLEMS

**15.1** Consider the application discussed in Examples 15.1, 15.2, 15.4, and 15.6. For this application, develop in explicit detail:

(a) the equivalent discrete-time perturbation model, as linearized about the equilibria corresponding to $y_{d0} = k\pi/8$ for $k$ = positive and negative integers.

(b) the perturbation LQG controller based on full-state knowledge of $x(t_i)$ at each sample time, and the constant-gain approximation thereto (consider (15-26)–(15-30)).

(c) the perturbation LQG controller based on measurements

$$\mathrm{z}(t_i) = \mathrm{x}(t_i) + \mathrm{v}(t_i)$$

with $\mathrm{v}(\cdot,\cdot)$ zero-mean white noise of variance $R$ (consider (15-36)–(15-38)). Again obtain the constant-gain approximation. Demonstrate how the controller changes when $y_{\mathrm{d}0}$ changes from 0 to $\pi/4$.

(d) the same as part (c), but based on the measurement model of Example 15.2.

(e) the perturbation PI controller based on perfect knowledge of $x(t_i)$ at each $t_i$ (consider (15-32)–(15-35)). Demonstrate how a change in $y_{\mathrm{d}0}$ from 0 to $\pi/4$ can be accommodated without relinearizations in the controller. Now let $y_{\mathrm{d}0}$ remain at $\pi/4$ and show how internal relinearizations about $\pi/4$ instead of 0 can be accomplished with no change in commanded control. Express the result in terms of the three different types of PI controllers. Add antiwindup compensation to these controllers.

(f) the perturbation PI controllers based on noisy measurements as in part (c); repeat part (e) for these controllers.

(g) the perturbation PI controllers based on noisy measurements as in part (d); repeat part (e) for these controllers.

(h) the controllers as in parts (c), (d), (f), and (g), except that $\hat{x}(t_i^-)$ is used instead of $\hat{x}(t_i^+)$ to replace perfect knowledge of $x(t_i)$. Why might such a controller be used?

(i) a two-level decomposition control law designed to regulate the system about the ideal state solution associated with the ideal open-loop control

$$u_{\mathrm{ideal}}(t_i) = \begin{cases} i/10, & i \in [\ 0, 10) \\ 1, & i \in [10, 30) \\ -1, & i \in [30, 50) \\ 0, & i \in [50, 80) \end{cases}$$

Consider controllers of the forms given in (b)–(h).

**15.2** For the application discussed in Examples 15.1, 15.2, 15.4, and 15.6, develop the following assumed certainty equivalence designs in detail:

(a) precomputed-gain extended Kalman filter with linear quadratic full-state feedback optimal regulator.

(b) constant-gain extended Kalman filter with constant-gain linear quadratic full-state feedback optimal regulator.

(c) controllers of the form given in parts (a) and (b), but in which the extended Kalman filter is replaced by (see Chapter 12, Volume 2):

(1) extended Kalman filter with bias correction terms,
(2) modified truncated second order filter,
(3) modified Gaussian second order filter,
(4) second order assumed density filter based on Gaussian assumption,
(5) third order assumed density filter based on cumulants truncation,
(6) statistically linearized filter.

(d) controllers of the form given in (a) and (b), but in which the LQ full-state feedback regulator is replaced by

(1) PI controller based on LQ synthesis
(2) $u(t_i) = -\hat{x}(t_i^+)$,
(3) $u(t_i) = -\hat{x}(t_{i+1}^-)$,

(4) $$u(t_i) = \begin{cases} -u_{\max} & \text{if} \quad x(t_i) \geq 0 \\ +u_{\max} & \text{if} \quad x(t_i) < 0 \end{cases}$$

(a "bang–bang" controller, typical of some time-optimal controller results),

(5) $$u(t_i) = \begin{cases} -u_{\max} & \text{if} \quad x(t_i) \geq \pi/8 \\ 0 & \text{if} \quad x(t_i) \in (-\pi/8, \pi/8) \\ +u_{\max} & \text{if} \quad x(t_i) \leq -\pi/8 \end{cases}$$

(a "bang–bang" controller with "dead zone," typical of some fuel-optimal controller results).

(e) controllers of the form developed in previous parts, but in which $\hat{x}(t_i^-)$ replaces $\hat{x}(t_i^+)$.

**15.3** Repeat Problem 15.2, but with the measurement model replaced by the linear model

$$\mathsf{z}(t_i) = \mathsf{x}(t_i) + \mathsf{v}(t_i)$$

with $\mathsf{v}(\cdot,\cdot)$ being zero-mean white noise of variance $R$.

**15.4** Repeat Problem 15.2, but with the dynamics model replaced by the linear model

$$\dot{\mathsf{x}}(t) = -[1/T]\mathsf{x}(t) + [1/T]\mathsf{u}(t) + \mathsf{w}(t)$$

**15.5** Consider the problem of the previous examples but in the case of linear dynamics and linear measurement models, as discussed in Example 14.3. However, now let the lag coefficient $[1/T]$ be considered an uncertain parameter.

(a) Develop the stochastic adaptive controller generated by assumed certainty equivalence, as in Fig. 15.6, showing cases in which either $-\mathbf{G}_c{}^*[t_i; \hat{a}(t_i)]$ or $-\mathbf{G}_c{}^*[t_i; a_{\text{nom}}]$ are used for the controller gain block. Also consider $-\bar{\mathbf{G}}_c{}^*[\hat{a}(t_i)]$ and $-\bar{\mathbf{G}}_c{}^*[a_{\text{nom}}]$. Discuss the details of generating these gains. For the estimator, use an extended Kalman filter, treating $[1/T]$ as an additional state. Discuss pseudonoise addition for filter tuning, and the case of constant-gain extended Kalman filtering.

(b) Repeat part (a), adding bias correction terms to the extended Kalman filter (see Chapter 12, Volume 2).

(c) Repeat part (a), replacing the extended Kalman filter with a modified Gaussian second order filter, as in Chapter 12 (Volume 2).

(d) Repeat part (a), replacing the extended Kalman filter with the maximum likelihood estimator of states and parameters of Chapter 10 (Volume 2). Consider online implementations and approximations to enhance computational feasibility.

(e) Repeat part (a), using a multiple model filtering algorithm for the state and parameter estimator (see Chapter 10, Volume 2).

(f) Develop the multiple model adaptive controller for this application. Compare this to the results of part (e) especially.

**15.6** Repeat Problem 15.5, but let the linear dynamics and measurement models be replaced with the nonlinear models of Examples 15.1 and 15.2. Compare these results to those of Problem 15.5. Use linearized models, linearized about a nominal parameter and state value (or one of a discrete number of values, or interpolations thereof) for $\mathbf{G}_c{}^*$ evaluations and parameter estimation in part (d). In parts (e) and (f), consider extended instead of linear Kalman filters in the bank of filters or controllers.

**15.7** Reconsider the tracking problem of Examples 14.5 and 14.6, but now let $Q_{\text{dr}}$ describing the current mean square target maneuver amplitude be considered an uncertain parameter.

(a) Develop the stochastic adaptive controller generated by assumed certainty equivalence, as in Fig. 15.6, using the maximum likelihood estimator of states and parameters of Chapter 10 and $\mathbf{G}_c{}^*(t_i)$ or $\bar{\mathbf{G}}_c{}^*$ of the full-scale feedback LQ optimal controller. Consider online implementations and approximations to enhance computational feasibility.

(b) Repeat part (a), using the correlation method of Section 10.9 or covariance matching techniques of Section 10.10 for adaptive filtering.

(c) Repeat part (a), using a multiple model filtering algorithm for the state and parameter estimation (see Section 10.8, Volume 2).

(d) Develop the multiple model adaptive controller for this application. Compare this to the results of part (c).

**15.8** Consider Example 15.7, in which the control entered the dynamics in a multiplicative manner and in which measurements were very nonlinear (assume a sampling period of 0.1 sec).

(a) Develop the perturbation regulator for this problem, using a nominal of $v_x(t) = -\hat{r}_x(t_0)/10$, $v_y(t) = -\hat{r}_y(t_0)/10$, over a 10-sec flight, using a cost function of

$$J = E\left\{\frac{1}{2}\mathrm{r}_x^2(t_{100}) + \frac{1}{2}\mathrm{r}_y^2(t_{100}) + \sum_{i=0}^{99}\frac{1}{2}[\mathrm{r}_x^2(t_i) + \mathrm{r}_y^2(t_i) + U\mathrm{u}^2(t_i)]\right\}$$

for $U = 0.1$, 1, and 10. Let $\mathrm{w}_1(\cdot,\cdot)$ and $\mathrm{w}_2(\cdot,\cdot)$ be independent zero-mean white Gaussian noises, each of strength 100; let $\mathrm{v}(\cdot,\cdot)$ be independent discrete-time zero-mean white Gaussian noise of variance $R = 0.1$. Let

$$\hat{\mathbf{x}}(t_0) = [-10000 \text{ ft}, 1000 \text{ ft/sec}, -7500 \text{ ft}, 750 \text{ ft/sec}]^{\mathrm{T}}$$
$$\mathbf{P}_0 = \mathbf{0}$$

(b) Develop the assumed certainty equivalence design of a precomputed-gain extended Kalman filter in cascade with the LQ full-state feedback optimal perturbation regulator.

(c) Repeat part (b), adding bias correction terms to the filter.

(d) Assume that an ad hoc offline law has been produced to approximate the characteristics of the dual controller described in Section 15.5, yielding an "S-turn" trajectory instead of straight-line trajectory as used in parts (a)–(c). Develop a two-level decomposition law to regulate the states about this ideal trajectory. Show how the results of (a)–(c) are modified to accomplish this result.

**15.9** Consider the optimal attitude control of a spinning asymmetrical spacecraft. Let $\omega_1$, $\omega_2$, and $\omega_3$ be the angular velocities about the body-fixed principal axes and let $I_1$, $I_2$, and $I_3$ denote the corresponding moments of inertia. Let $u_1$, $u_2$, and $u_3$ denote the torques applied about the principal axes. Thus, the dynamical equations of motion can be written as:

$$I_1\dot{\omega}_1 = (I_2 - I_3)\omega_2\omega_3 + \mathrm{u}_1 + \mathrm{w}_1$$
$$I_2\dot{\omega}_2 = (I_3 - I_1)\omega_3\omega_1 + \mathrm{u}_2 + \mathrm{w}_2$$
$$I_3\dot{\omega}_3 = (I_1 - I_2)\omega_1\omega_2 + \mathrm{u}_3 + \mathrm{w}_3$$

where $\mathrm{w}_1$, $\mathrm{w}_2$, and $\mathrm{w}_3$ are wideband disturbance torques modeled as independent zero-mean white Gaussian noises, each of strength 0.01 $\text{rad}^2/\text{sec}^3$. Assume three rate gyros are available, providing measurements as

$$\mathrm{z}_k(t_i) = \omega_k(t_i) + \mathrm{v}_k(t_i), \qquad k = 1, 2, 3$$

where the $\mathrm{v}_k(\cdot,\cdot)$ noises are independent zero-mean white Gaussian discrete-time noises of variance 0.01 $\text{rad}^2/\text{sec}^4$. Further assume we wish to regulate the angular rates to zero, using a cost function

$$J = \lim_{N\to\infty}\frac{1}{N}E\left\{\sum_{i=0}^{N}\frac{1}{2}[I_1\omega_1^2(t_i) + I_2\omega_2^2(t_i) + I_3\omega_3^2(t_i) + U\mathrm{u}_1^2(t_i) + U\mathrm{u}_2^2(t_i) + U\mathrm{u}_3^2]\right\}$$

(a) Develop the appropriate equivalent discrete-time perturbation models.

(b) Design the LQG perturbation regulator for this problem, particularly the constant-gain approximation.

(c) Develop the LQG perturbation PI controller for this problem, in incremental form (consider the three types of PI controllers).

(d) Design the assumed certainty equivalence controller composed of a steady state extended Kalman filter in cascade with the constant-gain linear quadratic full-state feedback optimal perturbation regulator.

(e) Repeat part (d), but with the extended Kalman filter replaced by (see Chapter 12, Volume 2):

(1) extended Kalman filter with bias correction terms,

(2) modified Gaussian second order filter,

(3) statistically linearized filter.

**15.10** Consider the continuous-time measurement problem of Example 15.5. Develop in explicit detail:

(a) the perturbation LQG controller based on perfect knowledge of $x(t)$, and the constant-gain approximation to it.

(b) the perturbation LQG controller based on measurements of the form $\mathrm{z}(t) = \mathrm{x}(t) + \mathrm{v}_c(t)$, where $\mathrm{v}_c(\cdot,\cdot)$ is zero-mean white Gaussian noise of strength $R_c$. Obtain the constant-gain approximation.

(c) the same as in part (b), but with the measurement described in Example 15.5.

(d) a two-level decomposition control law designed to regulate the system about the ideal state solution associated with the ideal open-loop control

$$u_{\text{ideal}}(t) = \begin{cases} 1 & t \in [0, 30 \text{ sec}) \\ -1 & t \in [30 \text{ sec}, 60 \text{ sec}) \\ 0 & t \in [60 \text{ sec}, 90 \text{ sec}) \end{cases}$$

(e) assumed certainty equivalence design based on precomputed-gain extended Kalman filter with linear quadratic full-state feedback optimal regulator.

(f) controller as in (e), but with constant-gain filter and controller.

(g) controllers as in (e) and (f), but in which the extended Kalman filter is replaced by (see Chapter 12, Volume 2):

(1) extended Kalman filter with bias correction terms,

(2) second order assumed density filter based on Gaussian assumption.

(h) controllers as in (e) and (f), but in which the LQ full-state feedback regulator is replaced by

(1) $u(t) = -x(t)$,

(2) the continuous-time analog of one of the two "bang–bang" nonlinear control laws described in Problem 15.2d.

# Index

## C

* Numbers in parentheses are equation numbers.

**D**

## E

**F**

## G



## H

## L

## M

## N

**O**

## Q



## R

## S

## T



## U